QUANTUM THEORY

QUANTUM THEORY

Reconsideration of Foundations — 4

Växjö, Sweden *11 – 16 June 2007*

EDITORS

Guillaume Adenier
Andrei Yu. Khrennikov
Växjö University, Sweden

Pekka Lahti
University of Turku, Finland

Vladimir I. Man'ko
Lebedev Physicial Institute of Russian Academy of Science, Moscow

Theo M. Nieuwenhuizen
Institute for Theoretical Physics Amsterdam, The Netherlands

SPONSORING ORGANIZATIONS
Swedish Royal Academy of Science
Swedish Research Council
International Center for Mathematical Modeling of Växjö University
EU-Network Quantum Probability

AMERICAN INSTITUTE OF PHYSICS

Melville, New York, 2007
AIP CONFERENCE PROCEEDINGS ■ VOLUME 962

Editors:

Guillaume Adenier
Andrei Yu. Khrennikov

International Center for
Mathematical Modeling, MSI
Vejdes Plats 7
Växjö University
S-35195 Växjö
Sweden

E-mail: guillaume.adenier@vxu.se
andrei.khrennikov@vxu.se

Vladimir I. Man'ko
P. N. Lebedev Physical Institute
Leninskii Prospect 53
Moscow 119991
Russia

E-mail: manko@sci.lebedev.ru

Pekka Lahti
Department of Physics
University of Turku
FIN-20014 Turku
Finland

E-mail: pekka.lahti@utu.fi

Theo M. Nieuwenhuizen
Institute for Theoretical Physics
Valckenierstraat 65
1018 XE Amsterdam
The Netherlands

E-mail: nieuwenh@science.uva.nl

L.C. Catalog Card No. 2007939937
ISBN 978-0-7354-0479-3
ISSN 0094-243X

Printed in the United States of America

CONTENTS

*Italicized name indicates author who presetned the paper.

CONTRIBUTED PRESENTATIONS

PREFACE

The fact that during the past twenty years experimental quantum physics has advanced towards the limit of single objects---manipulating experimentally single molecules, single atoms, single neutrons, single electrons, and even single photons---makes it even more important to develop an interpretation of quantum mechanics that would deal satisfactorily with *individual quantum objects*. This program, which was also the aim of Bohr and Heisenberg, has been pursued by many over the years. Unfortunately it sill faces serious interpretational problems---like the measurement problem---- and it seems more and more to be heading towards the idea of non-objectivity.

Another reason to reconsider the foundations of quantum mechanics is that classical deterministic models in the microscopic domain are being developed, providing an alternative to the conventional quantum description of experiments. Some phenomena which were considered to be "essentially quantum" and unexplainable can now be modeled in a classical framework. An important point is that our computational abilities have greatly improved, especially compared with that of the 1930's-60's. For example, as was remarked by Daniel Cole last year, many problems of Stochastic Electrodynamics for which a solution was out of reach even in the 1990's can now be solved using computer simulation (e.g., the stability of atoms can be modelized with electrons interacting with the zero point field), and a similar point of view was presented in the talk of Yuri Ozhygov.

Mathematics is at the heart of any understanding of the foundations of quantum mechanics; and mathematical reconsiderations, improvements and generalizations of quantum formalism have thus always played a central role in the Växjö conferences. This year we can mention the work of Accardi, Khrennikov, Goyal, Kupczynski, Garola, Sozzo, and Man'ko (on the relation between classical and quantum probability); Bengtsson, Eriksson, Tkadlek, Khots, Popa, Chaturverdi, Hinz, and Hamhalter (on the algebraic structure of quantum mechanics); Lahti (on operator measures and observables); Gustafson (on Hilbert space trigonometry); Hudson and Pulmanova (on quantum probability), Andreev, V. Man'ko, M. Man'ko, Cassinelli and D´Ariano (on the mathematical foundations of quantum tomography); Cattaneo (on Feynman integrals); Nilsson and Vinh Quang (on spectral properties), and Coecke (continuing a series of talks on his algorithmic-like visualization of quantum mechanics); macroscopic quantum models (Grib, D´Hooghe, Morimae, and Ishio); problems in classical and quantum information theory (Rocchi and Nyman); applications of the mathematical formalism of quantum mechanics in cognitive science and economy (Freudenberg, Haven, and we should mention the work of Aerts, Czachor, Choustova, and Khrennikov in the same field during the previous conferences).

An essential role in the understanding of quantum foundations is played by the various possible interpretations of quantum probability. They have always been debated a lot in the Växjö conferences, and a huge diversity of opinions were defended: from the frequency probabilistic views of Khrennikov, to Fuchs' subjective probability interpretation of quantum mechanics. At the present conference, Ballentine, who is well-known for his work on the ensemble (i.e., statistical) interpretation of quantum mechanics, presented an excellent analysis of the role of objective and subjective probability in quantum theory and especially in quantum information theory.

Vladimir Man'ko pointed out in his talk devoted to quantum tomography that it is possible to describe quantum states by fair probability distributions---as an alternative to the standard description by means of wave function---and that such an approach is completely equivalent to all other known *formulations* (mind you, not interpretations!) as path integral or Moyal ones since the approach is a simple mathematical ansatz of applying the analog of well studied integral Radon transform mapping the wave function onto the probability.

Our understanding of quantum theory can of course be greatly improved through the understanding of other connected fields of physics. For instance, the relationship between quantum mechanics and general relativity is certainly complicated, if not conflicting, but it offers a fertile ground for the reconsideration of quantum theory. This was discussed by Pereira, Nieuwenhuizen, Chuprikov, Mardari, Olkhov, Wharton (and we should mention the work of professor Kim during the previous conferences). Another approach is the idea of using random classical field to describe quantum phenomena. This approach has been at the heart of many important contributions in the previous Växjö conferences (notably with the work of Davidson, Cetto, de la Peña, Cole, and Scully), and was tackled on this year in the talks of Boyer, Nieuwenhuizen , Khrennikov, and Roychoudhuri. Another field of research that has always been a subject of intense discussion at the Växjö conferences (notably between Goldstein, Hiley and Scully) is Bohmian mechanics; and this year was no exception with the talks of Nikolic and Akhmeteli, as well as during the discussion at the round table. Finally, von Waldenfels spoke about the radiative transfer of polarized light; and Beretta about the coupling between thermodynamics and quantum mechanics.

Clearly, the Einstein-Podolski-Rosen (EPR) paradox, and its experimental investigation through Bell's inequality tests, remains one of the main sources of debates on the reconsideration of quantum foundations, and has been central throughout the whole series of the Växjö conferences. Unfortunately, a consensus on this matter seems almost impossible to reach. For a first group, that would be more mathematically oriented (e.g., Accardi, Khrennikov, Pitowsky, Kupczynski, Hess, Philipp), the fundamental condition for a violation of Bell's inequality is non-Kolmogorovness---that is, the impossibility to represent a family of observables by random variables on a single Kolmogorov probability space. In this framework, nonlocality appears only as a sufficient condition for non-Kolmogorovness. But, as has been stressed over the years by this first group, it is not a necessary condition! However, this argument is usually hand-waved by another strong group---which would represent a certain orthodoxy of physicists---who consider together with John Bell that the alternative is really either nonlocality or the ``death of reality". In this conference, Bell's inequality was discussed by Accardi, Kupczynski, Carola, Sozzo, Adenier, De Baere, Kracklauer, Cavalcani, Barukcic, and Sanctuary.

From a more philosophical point of view, fundamental problems raised by the interpretation of the wave function, and in particular empiricists interpretation, were discussed in the talks of De Baere and De Muynck---who may incidentally also be considered as "honorary members of the Växjö Club". Another important problem, which has always been vividly discussed during the whole series of the Växjö conferences, is the principle of complementarity, together with the Heinsenberg uncertainty principle. This time, these questions were discussed in the talks of De Muynck, Lahti, De Baere, Denes, Plotnitsky. The physical and mathematical aspects of the measurement problem were analyzed by Sewell, Sadrzadeh, and Mehmani.

Last but not least, experimental results related to quantum information theory, nanotechnology, as well as the fundamental questions of quantum foundations were discussed in the talks of Bauch, Helssing, Jaeger, Lu, and Roychoudhuri.

Since the conference was devoted to 80th university of the Copenhagen interpretation, a number of papers on quantum foundations having historical interest were discussed. In particular, Prof. De Baere discussed papers of Renninger which are practically forgotten nowadays. These papers attracted a huge attention among the participants of the conference. We are grateful to Prof. De Baere for translating these two papers, which are placed in this volume in a special section on the History of the Development of Quantum Mechanics.

Finally, an essential contribution to this conference that should not be overlooked is that of our secretary, Cecilia Eriksson, without whom none of the Växjö conferences would have been possible.

Guillaume Adenier and Andrei Yu. Khrennikov, Växjö University (Sweden).

Pekka Lahti, University of Turku (Finland).

Vladimir Man'ko, Lebedev Physical Institute, Moscow (Russia).

Theo M. Nieuwenhuizen, Amsterdam University (The Netherlands).

HISTORY OF THE
DEVELOPMENT OF QUANTUM MECHANICS

On Wave–Particle Duality*†

M. Renninger

Institute of Crystallography, University of Marburg/Lahn, Germany

Abstract. By means of a thought–experiment, consisting of an interference experiment with two interfering beams, it is shown that it can be demonstrated experimentally that with *one single* particle a wave can be associated that propagates in space and time as a *physical reality*, i.e. that it should not merely be considered as a distribution of probabilities. The notion "physical reality" should be understood such that, when this physical reality is considered in a particular space at a particular time, it should be experimentally possible to influence this reality in such a way that future results of experiments show unambiguously that this reality has been causally influenced by the experimental act in this space and at that time.

Keywords: Wave–particle duality, quantum theory
PACS: 03.65.$-W$, 03.65.Bz

In modern physics it has become more and more the custom, to discuss the last questions about quantum theory, especially the wave–particle dualism, only by means of purely mathematical considerations, and to consider the visualization, or even the mere desire for it, as rudimentary and naive modesty. It is, however, the purpose of the following discussion, to point to some very precise conclusions, which follow merely from purely experimental physical aspects, without any previous knowledge of the mathematical quantum formalism, which to my knowledge have never been obtained in this way before. At the same time the aim is to warn that in considering such issues, the visualization should not be given up *too early*, but should be kept on as long as possible. Clearly there is a limit in so doing, but this limit should not be artificially set so as to exclude interesting possible developments. It is precisely the purpose of this work to present such a development.

In general one speaks of the two different "pictures" or "aspects", the "wave"– and "particle–aspect". Detailed studies report the availability of a series of experiments evidencing the wave–like nature, and another series evidencing the particle–like nature of light and matter. But it has always been stressed, that there exists no experiment in which *at the same time* both wave–like and particle–like properties are observed, and in which the wave–like behaviour of *one single* photon or electron can be shown. According to P. Jordan [1] this were "logically and mathematically nonsense". "It is impossible, that one single indivisible experimental act shows up as well the one as the other appearance of light". Or as stated by W. Heisenberg [2] "Particle– and wave–picture are two different appearances of one and the same physical reality". "Now it is clear, that matter cannot consist simultaneously of waves and particles …". "The seeming double nature has its origin in the inadequacy of our language". "Atomic processes don't have a visual representation. Fortunately for the mathematical description of these processes such a visualization is not necessary altogether; a mathematical formalism of quantum theory is at our disposal, which accounts for all experiments … "

In contrast with these statements I propose the following 3 propositions, the proof of which will be given thereafter; they are first formulated for light in the visual range, but they are also valid, in appropriately modified form, for radiation of other wavelength and for matter radiation:

1. It is possible to demonstrate experimentally, that the *energy* of each photon moves *across space and time* on a single continuous path, and concentrated in the form of a particle, i.e. at each instant it occupies a connected region in space.[1]

---

* Original paper: M. Renninger, "Zum Wellen–Korpuskel–Dualismus", *Zeitschrift für Physik* **136**, 251-261 (1953), translated by W. De Baere, Laboratory for Theoretical Physics, Unit for Subatomic and Radiation Physics, Proeftuinstraat 86, B–9000 Ghent, Belgium, email: willy.debaere@UGent.be.

† This paper has been accepted for publication, after the correspondence of the author with leading quantum physicists has convinced the Editors [of Z.f. Physik, note by translator] of the fundamental importance of the subject.

[1] No specification is given about the form and extension of this region of space, i.e. the "particle".

CP962, *Quantum Theory, Reconsideration of Foundations – 4*,
edited by G. Adenier, A. Yu. Khrennikov, P. Lahti, V. I. Man'ko, and Th. M. Nieuwenhuizen

2. It can also be demonstrated experimentally, that *with each single photon* there may be associated a *guiding wave* (without energy but "causally" influenceable), which obeys precisely the rules for the propagation of an electromagnetic wave (spatial extension, absorption, diffraction, interference, splitting by reflection, refraction, double refraction etc., except energy), hence has unambiguously a *wave nature* that propagates likewise *in space and time*, i.e. as long as the photon is on its way it occupies at each instant a specific region of space that is *not* necessarily single connected.

3. The connection *in space* of the particle of energy and of the wave field is such, that the former can be found in a region where the intensity of the wave is different from zero, with a probability that is proportional to this intensity. Its propagation speed is equal to the group velocity of the wave within the limits set by Heisenberg's uncertainty relations.

Summarized briefly and drastically: *Each quantum consists of a particle of energy, "carried" or "guided" by an energy free wave.*

This image is not new. It was already formulated by de Broglie in 1927 [3] in terms of the by him introduced notion of "pilot wave", but abandoned again because of "des objections très graves"[4, 5].[2] According to a citation from N. Bohr [7], also Einstein once spoke of "ghost fields, guiding photons".

On the other hand the 3 preceding propositions are, however, *neither intended as an explanation, nor as a mere visualization.* Instead, both the energy particle and the guiding field associated with the single photon are considered *simply as experimental facts.* Attempts for explanation should follow *after* observation of the facts!

These statements will be proved by means of a thought experiment, consisting of an elementary interference set–up. In my opinion the usual thought experiments for the analysis of the wave–particle duality (mostly refraction at a single or double slit) fail to penetrate to the deepest possible level, and are not able to reveal the most basic characteristics. Mostly they concern refraction set–ups with a continuous intensity distribution. The main characteristics of our set–up will be: *two* interfering rays, detection in *two* possible places, *two* discrete intensities (probability distributions) that may be chosen as either 1 or 0. This set–up is shown in Fig. 1. (In an Appendix a more detailed description of the set–up is given in order to prove its essential realizability.)

In the proposal it is assumed that two facts are experimentally proven:

1. The outcome of the interference experiment is the same whether many photons contribute *in a short time* or in a long time: each photon interferes *only with itself* (see e.g. Dirac, Principles of Quantum Mechanics, p. 9).

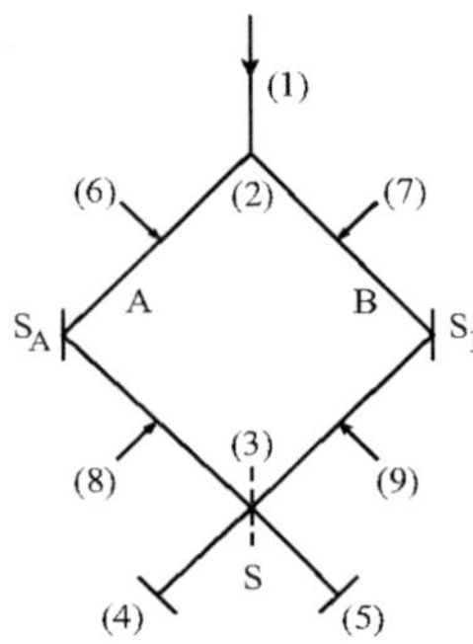

FIGURE 1. Interference set–up, schematic

2. Several coherent sub–rays of a light source *remain* coherent, i.e. are able to interfere with each other after having travelled arbitrary long distances along separated paths. This has been verified experimentally up to distances of 2 km by Michelson and Gale[8].

A parallel ray (1) is split at (2) in 2 rays A and B which move, in separate spaces, through a system of pipes towards (3). There they interfere with each other, e.g. after transmission through and reflection by a beam–splitter S, and arrive in places (4) and (5) where they can be detected.

The path difference between A and B can be set up (e.g. by displacing one of the mirrors S_A or S_B) such that it is either + or $-\lambda/4$, resulting either in brightness in (4) and darkness in (5), or the other way round (the ray that is reflected by the beam splitter S gets a phase shift of $\pi/2$, the phase of the transmitted ray remains unchanged).

The light source should:

1. Allow the transition to such weak intensities, that never (i.e. with a probability tending to zero), more than one photon at the same time is present between (2) and (3).

2. Each time inform the observer by means of a signal the instant – taken each time as $t = 0$ –, when a photon has passed the point (2), within an uncertainty $\pm\Delta t$, which is small compared with the time of passage between (2) and (3).[3]

---

[2] *Remark during the correction.* Only after submission of the present manuscript, the Editor became aware of the fact that de Broglie took up again these former pictures since one year, in connection with relativistic considerations, together with J.-P. Vigier. In an article entitled: "La Physique Quantique restera–t–elle indeterministe?" containing all his publications about this issue so far, he reports in an impressive way about his motivation, which lead him to abandon 25 years ago his former concepts and to take up it again presently. – In the same way the Editor was not aware of the work by D. Bohm [6], in which de Broglie's concept of "pilot–wave" is developed theoretically, and which was the first main reason for de Broglie to develop further his former picture.

[3] With respect to this point, I refer to the Appendix for the essential possibility of such a signal.

The perfection of the optical set–up [parallelism of the ray, reflection and precision of the mirrors, perfection of the detectors in (4) and (5) etc.] should be such that each photon that passes (2) will be observed either in (4) or in (5) (e.g. by means of scintillation or phototube) with a probability very close to 1 (i.e. almost with certainty). Thus if the path difference between A and B is set up such that brightness will be observed in (4), darkness in (5), then each photon entering (1) will be observed in (4).

At different places, such as (6), (7), (8) and (9) it should be possible, to insert in the paths A and B a slide, either an opaque screen with detector, which – as in (4) and (5) – records each photon that is absorbed by the screen, or a $\lambda/2$ plate. The paths A and B should be long enough (say some light–minutes), in order to allow the insertion of such slides on different instants while a photon is moving between (2) and (3).[4]

With this set–up we make then the following observations [Reduced intensity as mentioned above, path lengths adjusted so that brightness is observed in (4), darkness in (5)]:

a) As long as nothing is inserted in the places (6), (7), (8) and (9), each photon passing (1) will be observed in (4), nothing in (5).

b) If in (6) a slide is inserted within the interval between $t_6 - \Delta t$ and $t_6 + \Delta t$, where t_6 is the time the photon needs to move from (2) to (6), $t_6 = l_6/c$ [l_6= distance between (2) and (6)], then for the observation there are two possibilities, each of which is realized with probability 50%:

α) The photon is detected in (6). It is then absorbed and has disappeared, and no further experimental intervention can result in a further observation, e.g. revealing the existence in path B of a further existing physical reality.

β) The photon is *not* detected in (6). Then it can be experimentally verified that with certainty it can be detected e.g. either in (9) at the time $t_9 = l_9/c$, or, when this is not the case, to detect it *either* in (4) *or* in (5), with probability $\frac{1}{2}$. Blocking path A has, when the quantum is not observed in it, yet a definite physical effect, namely that there, where in the case of both open paths no quantum may be observed, now a quarter of all entering quanta will be detected; and it is indifferent, whether path A or path B is closed.

c) Insert now in (6) a fully transparent $\lambda/2$ plate instead of an absorbing slide, again within the time interval $t \pm \Delta t$. The result is: *All photons are now detected in (4) instead of in (5).* The same result is obtained when the plate is inserted in (7) instead of in (6). This means, that it is possible, by this experimental intervention, to send the photon from (4) to (5), *at will from within each of the paths A and B*, i.e. to influence it *causally*. Yes, it is possible, to "cancel" an already decided "bending", by inserting at corresponding later times t_8 resp. t_9 a further – or also the same – $\lambda/2$ plate.

With these results the 3 above–mentioned propositions are proven:

Proposition 1 by experiment b. The photon can always be absorbed, i.e. detected, only in *one* of the paths A and B, never in both together, and further each time at a specific instant, determined by the propagation speed. The fact, that the *certainty*, to "find" the photon in one of both paths – and when it is not found in *one* path, the certainty to find it *then* on a corresponding later instant in the *other* one – moves with speed c, leads to the compelling conclusion, that *the photon itself* (i.e. its total energy) moves inside a confined self contained region on a *continuous path* through space and time. The conclusion of the localization of a *system* in a region, from the *certainty* to *detect* that system in that region, is undoubtedly allowed, and even required by the most elementary logic.

The validity of Proposition 2 follows from observation b, β and especially from c. By the intervention of inserting the phase plate in one of both paths, whatever one, the single photon can be steered at will between both possible detection places. This proves, that with each single photon there is associated a *physical reality that simultaneously moves on both paths*. Additional observations, which will not be discussed further, show, that the evolution of this "physical reality" obeys in all points the laws for the propagation of an *electromagnetic wave*, except the fact, proven by observation b, that it does not contain energy (the total energy moves on *one* of both paths; because the wave moves on *both* paths, it does not contain energy).

Proposition 3, which basically should be considered as a supplementary one, is proven by the combination of all observations, but in particular by the fact, that the experimental interventions of inserting slides or phase plates has the described result, when they happen within the mentioned time interval, which must be the same in each path, independent whether it is an intervention on the energy particle (by means of slides) or an intervention on the wave (by means of phase plates).

Are now, after all these considerations, the wave–"picture" and particle–"picture" really but "two forms of appearance of one and the same physical reality"? Is it not, on the contrary, the case that wave and particle are *two separate*

[4] This does not require large signal speeds, because the paths A and B can be deflected as many times as necessary, so that the positions (2) to (9) are at the same time separated in space and in the immediate neighbourhood of the observer.

physical realities associated with single photons? Did the observation c above not "reveal in one single indivisible experimental intervention both the one and the other picture of the nature of light"? (The result of inserting the phase plate proves the wave nature, the observation in one single point proves the particle nature of one and the same photon).

Of course one is free, to speak of the wave as a pure "probability"–wave. But one should be aware of the fact, that this probability wave propagates in space and time in a continuous way, and in a way that she can be influenced in a finite region of space – and only there! – and also at that time! –, with an unambiguous observable physical effect!

I am not sure, whether the facts and conclusions discussed in this work have some connection with the doubts of A. Einstein about the completeness of the description of elementary processes by quantum mechanics.[5] It is obvious, that precisely the proven reality of the wave associated with the single particle, which quantum mechanics, as it is stated explicitly, is unable to account for, may be considered precisely as an expression of the incompleteness maintained by Einstein.[6] In any case was our definition of "physical reality" inspired by the definition of Einstein, Podolsky and Rosen [10] of the same notion.

In the same way I cannot say, whether there is a connection with the new work by W. Weizel [11], in which it is tried to refute von Neumann's proof of the impossibility of a causal model for quantum theoretical processes.

For sure, the picture of a particle guided by a wave devoid of energy, which is intended to be more than a "picture", namely "reality", nowhere contradicts quantum mechanics, and may moreover be a valuable aid for the visual comprehension of elementary processes and for making exact prognoses about the outcome of experiments.

Heisenberg's uncertainty relation, as far as it has something to do with the motion of a particle as treated here, may be given also a visual significance: it states, that after emission of the particle it is impossible to know where, i.e. *at which point in the wave* the particle, the knot is situated [as well transversally, i.e. *in which direction*, as longitudinally, i.e. *the distance* within the coherence length Δx determined by the uncertainty relation (the expression of the uncertainty relation for light is: $\Delta x \cdot \Delta 1/\lambda \geq 1$)]. Yet, this relation says nothing about our additional statement, which restricts the uncertainty, that within the space made available by the uncertainty relation, the particles follow a continuous path in space and time, and that once a certain direction is taken, it keeps moving in this direction. The "coordinates" of the particle within the guiding wave should be considered as "hidden parameters" in v. Neumann's sense.

It is clear that the concept of a "wave devoid of energy" is disturbing. However, in attempting to avoid this one should realize that the experimental results would allow only much less attractive alternatives: because the existence of the wave can be experimentally verified during the entire motion of the photon, and the existence of the energy only in two points, the emission and the absorption point, the assumption of the existence of a normal electromagnetic wave containing energy would have the unavoidable consequence, that at the moment of absorption the wave would contract with superluminal speed, and moreover through closed walls. Such assumption would be completely unacceptable.

The author is fully aware of the fact, that the uneasiness of the idea of a wave devoid of energy still increases, when the consequences are considered of the fact that the laws of propagation of this wave are in every respect those of an electromagnetic wave. For instance there is refraction (which in the Maxwell–Lorentz theory has its origin in energy exchanges with dipole oscillations of the electrons), double refraction, polarization, scattering etc.; this are all manifestations that come about through interactions with matter. Could this uneasiness be relaxed by the consideration, that also matter must have guiding waves, and that these are the "spooky" entities, to use Einstein's expression, similar with "matter waves"? That the guiding waves "*devoid of energy*" associated with *energy radiation* should correspond with guiding waves *devoid of matter* associated with *matter radiation*? And that the interactions between these spooky entities should co–determine the events resultant from the *energy* interactions between the lumps of energy, resp. matter? A directly unobservable world behind the observable one, guided by the former? But, to stress it again, a world the reality of which can in time and (three dimensional) space be experimentally followed up, and which does not exist only in the form of abstract probabilities.[7]

What happens to the wave devoid of energy of a photon after its absorption? When it is absorbed for example in (6), and when in addition the detectors in (4) and (5) are removed, what happens then with the wave in B? Does she move further towards infinity, or does she disappear at the moment of absorption? Of course this question cannot be answered principally. The former assumption appears to be the more natural one, because it avoids the conclusion to

---

[5] Summarized in Ref. [7].

[6] *Remark during the correction.* As the articles which the Editor became aware of since the submission of the manuscript (see footnote 2 p. 2) show, and in accordance with a kind personal announcement by Mr. Einstein this is indeed the case. In this respect also the work by L. Jánossy [9] should be mentioned, which attempts to give an alternative deterministic account of quantum mechanical processes, although there is a violation of the assumption of the coherence of separated rays over arbitrary distances, the validity of which we accepted in our work (see page 2).

[7] See footnote 8, page 6.

the existence of influences that propagate with infinite speed also through closed walls, a conclusion which within the physical world is inconceivable. In any case were such influences not associated with transport of energy.

It is still possible to consider the following issue: what happens to the particle when the wave is split by reflection at a refracting surface, or at a partially transparent mirror, or by double refraction, i.e. each time a wave is split in more component waves with different propagation vector and mostly also different polarization state? All these cases can be understood without contradiction with the help of the plausible assumption, that each time the particle follows one of the component waves, with a probability that is proportional to the intensity of this component wave, in the same way as a particle in a splitting fluid current. It is not necessary to associate with the particle properties such as polarization. These should be associated only with the guiding wave.

Appendix.

In order to anticipate objections against the *principal* (not the *practical*) feasibility of the described experiments, more precise technical details will be given that support their realizability. The set–up should have the characteristics as shown in Fig. 2:

An almost pointlike monochromatic light source L is situated at the center of a small half spherical convex mirror S_1, which itself is large as compared to L. Furthermore L is in the focus of a parabolic mirror S_2 which again is large with respect to S_1. In this way the spherical wave originating in L is transformed in a parallel bundle (1), which contains the photons.

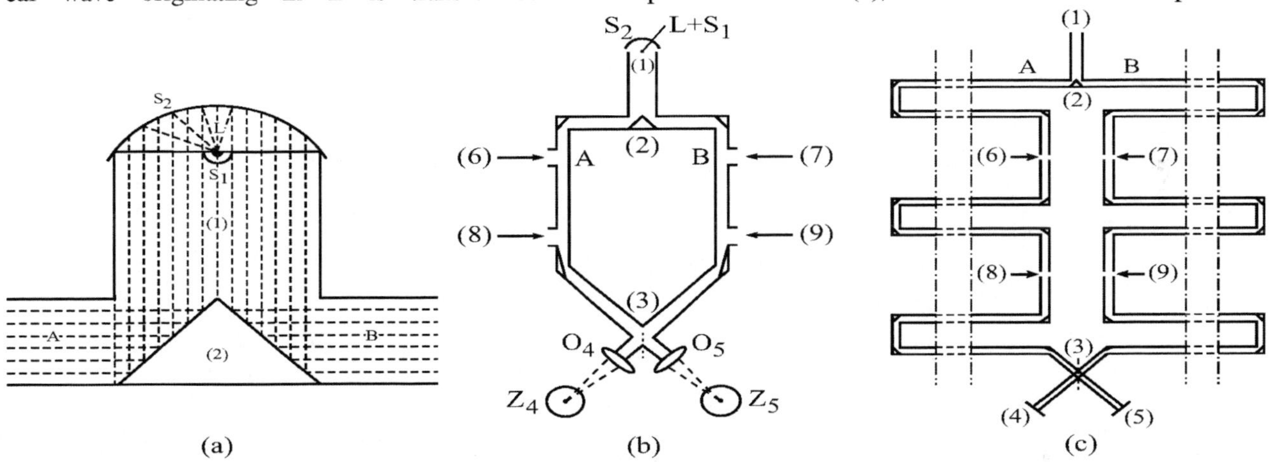

FIGURE 2. Technical details of the set–up. (a) Light source with reflecting mirrors; (b) Complete set–up; (c) Bringing together paths A and B through multiple bending.

This ray (1) is split at (2) in 2 separated rays A and B by means of 2 mirrors which are inclined by 45°. In order to remove the negative effects of the angular divergence of the parallel rays (due to the finite dimension of L) the parabolic mirror may have a large section, i.e. the focus and diameter can be chosen arbitrary large. The following choices may be made: diameter of the light source: 10^{-6} (100 Å), S_1: 1 cm, S_2: 10^6 cm (10 km), total light distance between (2) and (3): 10^{12} cm (30 light–seconds). As already mentioned, for performing experiments b and c it is required that the places (2) to (9) are close to each other. Therefore, the separated paths must come repeatedly in the neighbourhood of the observer, which is illustrated in Fig. 2c. In (6), (7), (8) and (9) the section of the rays may be decreased by optical arrangements, in order to facilitate the insertion of slides and phase plates. – Both lenses O_4 and O_5 focus the outgoing rays (4) and (5) on the ingoing gates of two photomultipliers. The detectors to be inserted in (6), (7), (8) and (9) equally consist of such aggregates (lenses and counters).

An indication is still needed how the observer may get information of the time of emission of a photon within an uncertainty interval that is small compared to the time needed by the photon for moving between (2) and (3). This could be resolved as follows: The light source L consists of a small hollow sphere, which contains a small amount of gas the atoms of which have for simplicity only one excitation level above the ground level. A beam of electrons with known velocity is sent through this sphere containing the gas, and with such an intensity that only one atom is excited in a time interval that is large compared with the time needed for a photon to move between (2) and (3), in our example of the order of hours. A supplementary arrangement measures the energy of the transmitted electrons and informs the observer as soon as an electron is detected that has lost an energy corresponding to the excitation energy of the atoms.

This signal warns the observer, that one of the atoms is excited and a photon will be emitted in the apparatus within the time needed to return to the ground state (of the order of 10^{-8} sec).

The first compositions of the above ideas date already from 20 years ago. The final explanation of the present formulation benefited in a substantial way from a series of very exciting discussions with Dr. S.N. Bagghi and R. Hosemann[8], Fritz Haber–Institute, Berlin–Dahlem, in the course of the last half year.

REFERENCES

1. P. Jordan, *Anschauliche Quantentheorie*, Berlin, 1936, p. 115.
2. W. Heisenberg, *Die physikalischen Prinzipien der Quantentheorie*, Leipzig, 1941, pp. 7, 107, 2nd Ed.
3. L. de Broglie, *Rapport au V. Congres Solvay 1927*, p. 105.
4. L. de Broglie, *Ondes et Corpuscules*, Paris, 1930, p. 34.
5. L. de Broglie, *Introductions à l'Etude de la Mécanique Ondulation*, Paris, 1930, p. 132.
6. D. Bohm, *Phys. Rev.* **85**, 166, 180(1952).
7. P.A. Schilpp, *Albert Einstein, Philosopher–Scientist*, Evanston, Illinois, 1949, p. 206.
8. A. Michelson and Gale, *Nature, Lond.* **115**, 556 (1925).
9. L. Jánossy, *Ann. Phys.* **11**, 323 (1953).
10. A. Einstein, B. Podolsky and N. Rosen, *Phys. Rev.* **47**, 777(1935).
11. W. Weizel, *Z.f. Physik* **134**, 264 (1953).
12. S.N. Bagghi and R. Hosemann, *Z.f. Physik* **135**, 50 (1953).

[8] *Remark during the correction.* In a work in preparation [12] these authors investigate the present issue in terms of their newly introduced "Algebra physikalisch beobachtbarer Funktionen mittels Faltungsoperationen", Part I, and come, following a kind personal announcement, to a confirmation of the guiding wave picture, in which the amplitude of the guiding wave has the dimension of energy. In fact, the average *in time and in space* of the energy density of such a wave should be zero at any time and in each point. How far the related necessary assumption of "negative energy densities" is compatible with the necessity to attribute a kind of "level of energy density" to empty space, will not be discussed here.

Interaction–free Measurements*

M. Renninger

Institute of Crystallography, University of Marburg/Lahn, Germany

Abstract. By means of thought–experiments it is shown that, contrary to common opinion, there exist measurement processes that do not affect at all the measured system. These "negative" measurements consist in the ascertainment of the absence of measurement outcomes that are predicted to occur with a specific probability, and that allow – in analogy with "genuine" measurements – to make new statements about the measured system, and lead, hence, to "reduction of the wave function" in the same way as this is the case with normal, "positive", observations that do disturb the measured system. The compelling consequence is, that the usual justification for the uncertainty relation as an unavoidable retroaction of any measurement process on the measured system breaks down, and that the real justification consists in the permanent influence on the particle of all matter in its immediate or more distant neighbourhood, independent of whether this matter is part of a measurement apparatus or not.

Keywords: Measurement, quantum theory
PACS: 03.65. – W, 03.65.Bz

According to common opinion, Heisenberg's uncertainty relation is considered as expressing the fact – or at least is related to this fact – that the effect of measurements on object systems cannot be made arbitrarily small.[1]
After this point of view has been repeated recently on several occasions, e.g. by Heisenberg [1][2], Brillouin [2], it seems relevant to the present author, to point to a category of measurements, in which there is no interaction at all with the object, or, in order to use Heisenberg's formulation, there is no question of a "disturbance of the object's evolution that, hence, is totally independent of the measurement process itself". This category consists in the experimental observation of the absence of possible outcomes and will be termed further on as "negative" observations. The precise meaning of this will be explained by means of the following Gedankenexperiment.

A photon is emitted from a point P at time $t = 0$ within a known very limited time interval.[3] The point P is at the center of a semi–spherical screen S_1 with radius R_1, in which there is a hole that subtends a solid angle Ω as seen from P, i.e. the solid angle of the semi–spherical screen is $4\pi - \Omega$. Furthermore, S_1 is surrounded by another full screen S_2 that subtends a solid angle 4π.

Now, the wavefunction of the photon is determined precisely by the initial conditions and constraints defined above. Inside S_1 this is a spherical wave, outside S_1 it takes on a more complicated form, including in particular a diffraction pattern due to the hole in S_1. However, also without knowledge of its precise form, the predictions for the probabilities W_1 and W_2 to observe the photon on S_1 and S_2 follow immediately, namely:

$$W_1 = \frac{4\pi - \Omega}{4\pi}, \quad W_2 = \frac{\Omega}{4\pi}.$$

Of course, it is possible to verify this prediction, by imagining the screens S_1 and S_2 as built up by photomultipliers, and by observing a large number of photons. Then the number of registrations in S_1 and S_2 is proportional to W_1 and W_2. All this is trivial.

Now, a possible registration of a photon in S_1 will occur before a registration in S_2. If in S_1 (at $t_1 = R_1/c$) a scintillation is observed, then this is equivalent with what in wave mechanics is known as "reduction of the wavefunction": the probability for the arrival of the photon at the later time R_2/c – until the observation in S_1 still being equal to $\Omega/4\pi$ – instantaneously becomes zero, together with the entire wavefunction in the space between S_1

---

* Original paper: M. Renninger, "Messungen ohne Störung des Meßobjekts", *Zeitschrift für Physik* **158**, 417–421 (1960), translated by W. De Baere, Laboratory for Theoretical Physics, Unit for Subatomic and Radiation Physics, Proeftuinstraat 86, B–9000 Ghent, Belgium, email: willy.debaere@UGent.be.

[1] see e.g. Bohr [3], p. 35, or Jordan [4] p. 307: "It is unavoidable, that each measurement act *disturbs* in a non negligible way the object system *in a lawful way.*"

[2] Heisenberg's opinion is summarized in the Epilogue to the present work.

[3] The fundamental possibility to realize this, has been indicated already earlier by the present author, see [5].

CP962, *Quantum Theory, Reconsideration of Foundations – 4*,
edited by G. Adenier, A. Yu. Khrennikov, P. Lahti, V. I. Man'ko, and Th. M. Nieuwenhuizen

and S_2. This is the usually discussed case, in which one may speak in a justified way of an intervention in the physical process: either the photon is absorbed by the screen S_1 or it is at least scattered inelastically, so that in neither case it is in the same state as before the observation.

However – and this is the main point I want to bring about – "reduction of the wave packet" not only occurs when the photon is *observed* in S_1, but also when it is *not* observed. Or better, expressed in a positive way, when observation reveals that at the critical time $t = R_1/c$ it is *not* found on S_1. Also then does the probability for the later observation on S_2 change instantaneously, however in this case to the value 1 instead as to the value zero! When the photon is not observed in S_1, then it will move *with certainty* towards S_2. Here the statement concerns a prediction about the object system on account of an *observation that did not intervene in the physical evolution*, i.e. a "negative" observation.[4]

When there exist such interaction free observations, then it is proven, that the "lawful necessity" of the influence of each measurement process on the object system does not exist, and that as a consequence it is unjustified to put forward the possibility of a deeper motivation or even an illustration of the uncertainty relation. Rather, this follows immediately from the formalism of quantum mechanics and applies equally well to measurements that do not intervene in the object's evolution, as to those that do intervene.[5] This is also immediately clear in our example, precisely when we make Ω very small, i.e. that S_1 becomes almost a sphere with only a small hole in it: for a photon coming through this hole this amounts to a transversal determination of its position, which results in an accordingly large, by the uncertainty relation determined, uncertainty on the transversal momentum, i.e. on the direction of movement, and this corresponds with a probability distribution (refraction pattern) for the observation on S_2 which has a width that becomes the larger the dimension of the hole in S_1 becomes smaller. Hence also the possible "negative" observation in S_1, i.e. the fact that on time R_1/c a photon has come through the hole, does not allow us to make a better prediction about the observed *position* in S_2 than allowed by the uncertainty relation.

In private discussions with the author, it was often criticized that already the *existence* of the barrier S_1, i.e. the *possibility* of making an observation is equivalent with exercising an influence on the object system, also for the particles passing S_1. I do not deny this. Indeed, it is this influence that is at the origin of the refraction pattern on S_2. But I claim that this is *not provoked by the measurement process itself,* but as already mentioned at the outset, this must follow from the primary wavefunction. In this way no new information is obtained, this being provided only by an actually carried out observation.

In summary:

1. A measurement process, whether a "positive" or a "negative" one, means *un*equivocal "reduction of the wavefunction"; each genuine observation, each gain of information reduces the wavefunction. And vice versa: each reduction of the wavefunction is the result of a gain of information.

2. *Potentiality* of observation and *actual* observation are different things. Essentially, *potentiality* of observation allows for any evolution that is more than interaction free motion of single particles in vacuum (undisturbed spherical wave), such as entangled systems, or motion of particles in absorbing, scattering or refracting media and is nothing else than a consequence of the fact that the considered particle is not alone in the Universe. It is only when the observation is carried out actually that state reduction occurs.

ACKNOWLEDGMENTS

Prof. Süssman, Hamburg, is acknowledged very much for private clarification of his position.

---

[4] Of course, the discussed example is only *one* of many different possibilities for the splitting of ray beams in coherent component beams. In the case of splitting by e.g. partial reflection, by double refraction or by another means, the basic principles governing the behaviour remain the same as here by transversal screening. However, the main characteristic of the experiment discussed here and of the accompanying argumentation is in the first place not the preparation of separated component beams (which is already discussed frequently), but the fact, that the path of one component ray is *blocked*, by an obstacle that at the same time may have the role of measurement apparatus. The fact that in our example this obstacle coincides with the splitting (i.e. that one component ray is intercepted by S_1) is not essential.

[5] According to a kind private communication by Mr. Süssmann, the same prediction may be found in his fundamental paper [6], although rather hidden and not readily accessible. To what extent W. Finkelnburg's formulations [7] are the same, is not unambiguously clear from private discussions.

EPILOGUE

In a personal correspondence Prof. Heisenberg kindly informed me about his opinion with respect to the investigations explained in the present work, and allowed me kindly to summarize his ideas as follows:

It is incorrect to think, that the Copenhagen interpretation of quantum theory claims that the principal unavoidability of the influence of a measurement on the object system is related to a proper measurement–"process", and that, possibly, becoming aware of it later on, has the effect of reducing the wavefunction "retroactively". However, it is not possible at all to objectify a "measurement process" in the sense of Re in all possible cases. Simply and solely it is the circumstance of becoming aware of the measurement result that is objectifyable and responsible for state reduction, which, hence, is relegated to the "cut" between the object system and the apparatus system. However, what is meant by the unavoidable disturbance of the physical process by the measurement, is already the possibility of a measurement, i.e. the existence of the measurement apparatus. It is precisely this existence, that brings about a partly undetermined interaction between the measurement apparatus and the object system, and that after carrying out the experiment leads to the uncertainty relation. On the other hand, the act of the registration, that leads to the reduction, is in fact not a physical, but in a sense a mathematical process. Of course, with the erratic change of our knowledge there corresponds also an erratic change of the mathematical representation of our knowledge.

If Mr. Heisenberg's described view were the common opinion, then my considerations were indeed unnecessary, because basically they are the same, as follows from the last three sentences. At any case, it seems to me that there exists a proper measurement process, and that becoming aware of the result reduces the state (e.g. a – present or absent – to be developed mark on a recorder), at a time instant that is determined more or less exactly by the measurement proper. However, it is certainly not the case that this is the common opinion, as follows from the dominating formulations in the literature. Usually, they speak explicitly of the unavoidable disturbance by the measurement–*act*, the measurement–*process*, or even sharper, of the impossibility, to consider the measurement act "as merely becoming aware of an *already available* real fact." (see Ref. [4], p. 308). So, precisely *the becoming aware of an already available real fact* is realized in the above discussed thought experiment, so that its announcement would be in any case a clarification. The conclusions of this discussion can be justified to full extent and can be considered as supplementary indication, that each statement, such as those just mentioned, going beyond Mr. Heisenberg's expressed limits as mentioned above is inadmissable, and that, hence, it is not possible to speak of an unavoidable disturbance by the measurement *process*.

REFERENCES

1. W. Heisenberg, *Naturwiss.* **45**, 227(1958).
2. L. Brillouin, *Nature, Lond.* **183**, 501(1959).
3. N. Bohr, *"Atomtheorie und Naturbeschreibung"*, Berlin, 1931.
4. P. Jordan, *"Anschauliche Quantentheorie"*, Berlin, 1936.
5. M. Renninger, *Z. Physik* **136**, 251(1953).
6. G. Süssmann, *"Über den Meßvorgang"*, *Bayer. Akad. Ber., München H.*, **88**, 30,(1958).
7. W. Finkelnburg, *"Einführung in die Atomphysik"*, Berlin, 1956.

INVITED PRESENTATIONS

Universality of the EPR-chameleon model

Luigi Accardi* and Satoshi Uchiyama†

*Centro Vito Volterra,
Università degli Studi di Roma "Tor Vergata",
Via Columbia, Rome 00133, Italy
email: accardi@volterra.uniroma2.it
†Department of Life and Creative Sciences
Hokusei Gakuen University Junior College
Atsubetsu-ku, Sapporo 004-8631, Japan
email: uchiyama@hokusei.ac.jp

Abstract. In terms of EPR-chameleon models and local and causal measures, the Bell's argument is reanalyzed. Contrary to Bell, it is shown that the nontriviality of the joint probability measure does not always imply the nonlocality. It is analyzed that under what conditions correlations of distant particles are obtained which are different from the standard correlations. The protocol for the correlations of distant particles admits nontrivial probability measures respecting the locality.

Keywords: EPR-correlation, locality, chameleon effect
PACS: 04.20.Cv, 03.65.Ta, 01.70.+w, 01.30.Cc

INTRODUCTION

In the first section of this paper we briefly recall the basic physical idea of the chameleon effect (i.e. the theory of adaptive dynamical systems) and, in the second one, the main theorem on the reversible, deterministic model constructed in [1, 2] (hereinafter EPR–chameleon model), which reproduces the EPR–Bohm correlations in full respect of causality and locality. Furthermore, as it is apparent from the statement of Theorem (1) below, there is no artificial post–selection in the assumptions of the theorem (no "cospiracy of the detectors").

In the third section we discuss the recent result (see [3]) that the EPR–chameleon model is the only possible model with this property, in a sense that will be specified in the following.

In the EPR–chameleon model, a 'configuration' of the system changes during the measurement process. This change is done locally: action at a distance in the measurement process is unnecessary.

On the other hand, according to a widespread belief, Bell's argument implies that there is no local hidden variable theory that reproduces the EPR–Bohm correlations.

This belief however is based on some assumptions, which were implicit in Bell's argument and which have remained implicit for a long time. The quantum probability approach, started about 30 years ago, has made these assumptions explicit thus introducing, in the debate on the foundations of quantum theory, two main new ingredients, namely:

(i) the chameleon effect, as an intuitive explanation of the mechanism through which non Kolmogorovian statistics can be produced by classical deterministic systems

(ii) the difference between trivial and non–trivial local measures.

Statement (i), which implies that the meaning of the Bell inequality consists in the proof of the existence of sets of experimentally measurable statistical data, coming from similar but incompatible experiments, which cannot be described by a single Kolmogorov model, is nowadays widely accepted in the literature.

Therefore in the present paper we will concentrate our attention on statement (ii), which has a more subtle mathematical nature. This will be done starting from the fourth section below.

The chameleon effect

The chameleon effect gives a mathematically precise formulation to the widely used (and abused) statement that *a measurement determines the value of an observable*: it distinguishes those measurements which register a previously

CP962, *Quantum Theory, Reconsideration of Foundations — 4*,
edited by G. Adenier, A. Yu. Khrennikov, P. Lahti, V. I. Man'ko, and Th. M. Nieuwenhuizen

existing property, independent of the measurement itself, from the adaptive measurements, which register the reaction to a given interaction.

The measurement of the color of a billiard ball provides a good metaphor of the former situation. The measurement of the color of the skin of a chameleon – of the latter.

We speak of *chameleon effect* whenever the dynamics of a system depends on the observable that one measures. This property can be taken as definition of *adaptive dynamical systems*.

Both situations are compatible with EPR's requirement of *pre–determination*, but in the former case this term should be interpreted as the passive reading of a pre–existing property; in the latter – as the active determination of a pre–established response.

The two interpretations are reflected in two different mathematical models and, in the following, we are going to illustrate these differences.

THE EPR–CHAMELEON DYNAMICAL SYSTEM

In this section we briefly recall the EPR–chameleon theorem, proved in [1] which gives in particular a classical, deterministic, reversible, local dynamical system that reproduces exactly the EPR correlations.

Theorem 1 *(i) For $a,b \in [0,2\pi)$ the maps (***local adaptive dynamics***)*

$$T_{1,a}, T_{2,b} \; : \; [0,2\pi)\times \mathbf{R} \to [\mathbf{0},\mathbf{2\pi})\times \mathbf{R} \tag{1}$$

$$T_{1,a}(\sigma_1,\lambda_1) := \left(\sigma_1, \frac{\sqrt{2\pi}|\cos(\sigma_1 - a)|\lambda_1}{4}\right) \in [0,2\pi)\times \mathbf{R} \tag{2}$$

$$T_{2,b}(\sigma_2,\lambda_2) := (\sigma_2, \sqrt{2\pi}\lambda_2) \in [0,2\pi)\times \mathbf{R} \tag{3}$$

are invertible with inverses:

$$T_{1,a}^{-1}(\sigma_1,\lambda_1) := (\sigma_1, \frac{4\lambda_1}{\sqrt{2\pi}|\cos(\sigma_1 - a)|}) \tag{4}$$

$$T_{2,b}^{-1}(\sigma_2,\lambda_2) := (\sigma_2, \frac{\lambda_2}{\sqrt{2\pi}}) \tag{5}$$

(ii) Denote $\hat{T}_{j,x}^{-1}$ the second component of $T_{j,x}^{-1}$ ($j = 1,2,\ x = a,b$) and m_a, m_b arbitrary real numbers. Then the following positive measure on $[0,2\pi)^2\times \mathbf{R^2}$ (initial distribution) is well defined and is a probability measure:

$$P_{a,b}(d\sigma_1 d\sigma_2 d\lambda_1 d\lambda_2) := p_S(\sigma_1,\sigma_2)p_{1,a}(\sigma_1,\lambda_1)p_{2,b}(\sigma_2,\lambda_2)d\sigma_1 d\sigma_2 d\lambda_1 d\lambda_2 := \tag{6}$$

$$:= \frac{1}{2\pi}\delta(\sigma_1-\sigma_2)d\sigma_1 d\sigma_2 \delta(\hat{T}_{1,a}^{-1}(\sigma_1,\lambda_1) - m_a)d\lambda_1 \delta(\hat{T}_{2,b}^{-1}(\sigma_2,\lambda_2) - m_b)d\lambda_2$$

(iii) The $P_{a,b}$–pair correlations of the ± 1–valued **observables**

$$S_a^{(1)}, S_b^{(2)} \; : \; [0,2\pi)\times \mathbf{R} \to \{\pm\mathbf{1}\} \tag{7}$$

$$S_a^{(1)}(\sigma,\mu) := \mathrm{sgn}(\cos(\sigma - a)) \tag{8}$$

$$S_b^{(2)}(\sigma,\mu) := -\mathrm{sgn}(\cos(\sigma - b)) = -S_b^{(1)}(\sigma,\mu) \tag{9}$$

reproduce the EPR correlations for any value of the parameters m_a, m_b, i.e.

$$\int S_a^{(1)}(\sigma_1,\lambda_1)S_b^{(2)}(\sigma_2,\lambda_2)P_{a,b}(d\sigma_1 d\sigma_2 d\lambda_1 d\lambda_2) = -\cos(a-b) \tag{10}$$

Remark .

The physical meaning of the inverse maps $\hat{T}_{j,x}^{-1}$ in the definition of the initial distribution (6) is explained in the paper [4].

THE UNIQUENENSS THEOREM

In the present section we study the most general family of local causal probability measures which reproduce the EPR–Bohm correlations and we prove that, under natural generic conditions and up to convex combinations, they must have the form described in the EPR-chameleon model constructed in [1, 2].

We will use the following notations. The configuration space of the single particle is identified to the unit circle, i.e.

$$S_1 = S_2 = S^1 := \{(x,y) \in \mathbf{R}^2 : \mathbf{x}^2 + \mathbf{y}^2 = \mathbf{1}\} \equiv [\mathbf{0}, \mathbf{2}\pi) \equiv \mathbf{R}/(2\pi\mathbf{Z})$$

and the observables to periodic functions $f : \mathbf{R} \to \mathbf{R}$ with period 2π. We denote

$$T^2 := S^1 \times S^1$$

and define the intervals

$$I_a := \left[-\frac{\pi}{2} + a,\ a + \frac{\pi}{2}\right),$$

$$J_a := \left[a + \frac{\pi}{2},\ a + \frac{3\pi}{2}\right)$$

The random variables $S_a^{(1)}$ and $S_b^{(2)}$ $(a, b \in [0, 2\pi))$ are defined by

$$S_a^{(1)}(s_1) := \chi_{I_a}(s_1) - \chi_{J_a}(s_1) \quad , s_1 \in S_1$$

$$S_b^{(2)}(s_2) := -\chi_{I_b}(s_2) + \chi_{J_b}(s_2) \quad , s_2 \in S_2$$

If $P_{a,b}$ is a local causal probability measure on $S_1 \times S_2 \times M_1 \times M_2$ we denote $R_{a,b}$ its marginal probability on $T^2 = S^1 \times S^1$. Thus:

$$dR_{a,b}(s_1, s_2) = dP_S(s_1, s_2)\ p_{1,a}(s_1) p_{2,b}(s_2) \tag{11}$$

where $s_1, s_2 \in [0, 2\pi)$ are fixed parameterizations of $S_1 = S^1$ and $S_2 = S^1$ respectively, P_S is a probability measure on T^2 and $p_{1,a}(s_1)$, $p_{2,b}(s_2) \geq 0$.

Definition 1 *A family* $\{R_{a,b}\ :\ a, b \in [0, 2\pi)\}$ *of probability measures is said to* **reproduce the EPR statistics** *if for any* $a, b \in [0, 2\pi)$ *one has:*

$$\begin{aligned}
R_{a,b}(I_a \times I_b) &= \frac{1}{2}\cos^2\left(\frac{b-a}{2}\right) =: P_{a,b}^{+-} \\
R_{a,b}(J_a \times J_b) &= \frac{1}{2}\cos^2\left(\frac{b-a}{2}\right) =: P_{a,b}^{-+} \\
R_{a,b}(I_a \times J_b) &= \frac{1}{2}\sin^2\left(\frac{b-a}{2}\right) =: P_{a,b}^{++} \\
R_{a,b}(J_a \times I_b) &= \frac{1}{2}\sin^2\left(\frac{b-a}{2}\right) =: P_{a,b}^{--}
\end{aligned} \tag{12}$$

Remark (1). Notice the rotation invariance of the experimentally measured probabilities.

Remark (2). The class of all families $\mathscr{R} := \{R_{a,b}\ :\ a, b \in [0, 2\pi)\}$, of probability measures which reproduce the EPR statistics, is a closed convex set if topology and convex combinations are defined as follows.

The sequence $(\mathscr{R}_n)$ is said to converge to $\mathscr{R}$ if, for every $a, b \in [0, 2\pi)$ the sequence of probability mesures $(R_{a,b;n})$ converges weakly to the probability mesure $R_{a,b}$.

The family $\mathscr{R}$ is said to be a convex combination of the two families $\mathscr{P}$, $\mathscr{Q}$ if there exists $t \in [0, 1]$ such that, for every $a, b \in [0, 2\pi)$ one has (in obvious notations)

$$R_{a,b} = tP_{a,b} + (1-t)Q_{a,b}$$

Definition 2 *The family of LC probability measures*

$$dR_{a,b}(s_1, s_2) = dP_S(s_1, s_2)\ p_{1,a}(s_1) p_{2,b}(s_2)$$

is said to satisfy the condition of **statistical pre–determination** *if* $\forall(s_1,s_2) \in T^2 \setminus \Delta$ *there exists* $a \in S^1$ *and a neighborhood* G *of* (s_1,s_2), *contained in* $(I_a \times J_a) \cup (J_a \times I_a)$ *such that*

$$p_{1,a}(s_1')p_{2,a}(s_2') > 0 \quad ; \quad \forall(s_1',s_2') \in G$$

Remark.

Statistical predetermination means that, the fact that a configuration is statistically forbidden for all measurements such that the outcomes are precisely (anti-)correlated cannot depend on the local measurements, but it is defined at the source.

Theorem 2 *Suppose that the family of probability measures*

$$dR_{a,b}(s_1,s_2) = dP_S(s_1,s_2)\ p_{1,a}(s_1)p_{2,b}(s_2)$$

satisfies the conditions:
(i) reproduce the EPR statistics
(ii) statistical pre–determination.
Then the source distribution P_S *must have support in the diagonal of* T^2

$$suppP_S \subseteq \Delta := \{(s_1,s_2) \in T^2 : \ s_1 = s_2 (mod\ 2\pi)\}$$

If in addition
(iii) the restriction of P_S *to* Δ *is absolutely continuous with respect to the Lebesgue measure on* Δ
(iv) the local measures $p_{1,a}$ *and* $p_{2,b}$ *are rotation invariant, i.e.*

$$p_{1,a+\delta}(s_1+\delta) = p_{1,a}(s_1) \quad ; \quad p_{2,b+\delta}(s_2+\delta) = p_{2,b}(s_2) \quad ; \quad \forall\delta \in \mathbf{R} \tag{13}$$

(v) the local measures $p_{1,a}$ *and* $p_{2,b}$ *are twice continuously differentiable,*
Then the probability measure $dR_{a,b}(s_1,s_2)$, *defined by (11), must have either the form*

$$dR_{a,b}(s_1,s_2) = \delta(s_1-s_2)ds_1ds_2\frac{1}{4}|\cos(s_1-a)| \tag{14}$$

or the form

$$dR_{a,b}(s_1,s_2) = \delta(s_1-s_2)ds_1ds_2\frac{1}{4}|\cos(s_2-b)|. \tag{15}$$

Proof. Because of rotation invariance

$$\begin{aligned} p_{1,a}(s_1) &= p_{1,0}(s_1-a) =: p_1(s_1-a) \\ p_{2,b}(s_2) &= p_{2,0}(s_2-b) =: p_2(s_2-b). \end{aligned}$$

Using the concentration of P_S on the diagonal, we have

$$dR_{a,b}(s_1,s_2) = \rho(s_1)p_1(s_1-a)p_2(s_2-b)\delta(s_1-s_2)ds_1ds_2.$$

For a and b satisfying $0 \le b-a \le \pi$, $I_a \cap I_b = [-\pi/2+b, a+\pi/2)$, and therefore

$$R_{a,b}(I_a \times I_b) = \int_{-\pi/2+b}^{a+\pi/2} ds_1\rho(s_1)p_1(s_1-a)p_2(s_1-b)$$

From (12) we deduce that,

$$\frac{1}{4}(1+\cos(b-a)) = R_{a,b}(I_a \times I_b) = \int_{-\pi/2+b}^{a+\pi/2} ds_1\rho(s_1)p_1(s_1-a)p_2(s_1-b)$$

Differentiating this with respect to b, we find

$$-\frac{1}{4}\sin(b-a) = -\rho(b-\pi/2)p_1(b-a-\pi/2)p_2(-\pi/2) + \int_{b-\pi/2}^{a+\pi/2} ds_1\rho(s_1)p_1(s_1-a)p_2'(s_1-b)$$

Putting $b=a+\pi$, we obtain

$$0=\rho(a+\pi/2)p_1(\pi/2)p_2(-\pi/2)$$

Since ρ is a probability density, $\rho(a+\pi/2)$ cannot vanish for all a. Since a is arbitrary it follows that

$$p_1(\pi/2)=0 \text{ or } p_2(-\pi/2)=0$$

Let us assume that $p_1(\pi/2)=0$.

Differentiating again with respect to b and putting $b=a+\pi$, we obtain

$$\frac{1}{4}=-\rho(a+\pi/2)p_1'(\pi/2)p_2(-\pi/2)$$

From this we can see

$$p_1'(\pi/2)\neq 0 \text{ and } p_2(-\pi/2)\neq 0$$

and therefore

$$\rho(a+\pi/2)=1/(4p_1'(\pi/2)p_2(-\pi/2))=(\text{const.})$$

since a is arbitrary we deduce that $\rho(s_1)=c$. Since $I_a\cap J_b=[-\pi/2+a,-\pi/2+b)$, (12) implies that

$$\frac{1}{4}(1-\cos(b-a))=R_{a,b}(I_a\times J_b)=c\int_{-\pi/2+b}^{-\pi/2+a}ds_1p_1(s_1-a)p_2(s_1-b)$$

Differentiating this with respect to b, we have

$$\frac{1}{4}\sin(b-a)=-cp_1(b-a-\pi/2)p_2(-\pi/2)+c\int_{-\pi/2+b}^{-\pi/2+a}ds_1p_1(s_1-a)p_2'(s_1-b)$$

Putting $b=a$, we obtain

$$0=-cp_1(-\pi/2)p_2(-\pi/2)$$

Since $p_2(-\pi/2)\neq 0$ it follows that:

$$p_1(-\pi/2)=0$$

Since $(I_a\cap J_b)\cup(I_a\cap I_b)=[-\pi/2+a,\ a+\pi/2)$, by (12) we have

$$\frac{1}{2}=R_{a,b}(I_a\times J_b\cup I_a\times I_b)=c\int_{-\pi/2+a}^{a+\pi/2}ds_1p_1(s_1-a)p_2(s_1-b)$$

Changing variable with $s=s_1-a$, we obtain

$$\frac{1}{2}=c\int_{-\pi/2}^{\pi/2}dsp_1(s)p_2(s-b+a)$$

In the same way, for $\pi\leq b-a\leq 2\pi$ one has,

$$I_a\cap I_b=[-\pi/2+a,-3\pi/2+b) \quad ; \quad I_a\cap J_b=[-3\pi/2+b,a+\pi/2)$$

Therefore we have

$$\frac{1}{2}=R_{a,b}(I_a\times I_b\cup I_a\times J_b)=c\int_{-\pi/2+a}^{a+\pi/2}ds_1p_1(s_1-a)p_2(s_1-b)=c\int_{-\pi/2}^{\pi/2}dsp_1(s)p_2(s-b+a)$$

Since p_1 is continuous and a and b are arbitrary, we conclude that $p_2(s)=$const.$=:c_2$. Thus by renaming

$$\widetilde{p}_1(s_1):=cp_1(s_1)c_2$$

we find

$$dR_{a,b}(s_1,s_2)=\widetilde{p}_1(s_1-a)\delta(s_1-s_2)ds_1ds_2.$$

Our remaining task is to determine the form of $\widetilde{p}_1$.
For a and b satisfying $0 \leq b-a \leq \pi$ the above calculations lead to

$$-\frac{1}{4}\sin(b-a) = -\widetilde{p}_1(-\pi/2+b-a)$$

By putting $s := b-\pi/2$, we obtain

$$\widetilde{p}_1(s-a) = \frac{1}{4}\cos(s-a) \text{ for } -\pi/2 \leq s-a \leq \pi/2$$

Therefore

$$\widetilde{p}_1(s-a) = \frac{1}{4}|\cos(s-a)|,\ -\pi/2 \leq s-a \leq \pi/2.$$

Since $J_a \cap I_b = [a+\pi/2, b+\pi/2)$,

$$\frac{1}{4}(1-\cos(b-a)) = R_{a,b}(J_a \times I_b) = \int_{a+\pi/2}^{b+\pi/2} ds_1 \widetilde{p}_1(s_1-a).$$

By differentiating this with respect to b we have

$$\frac{1}{4}\sin(b-a) = \widetilde{p}_1(b+\pi/2-a).$$

By putting $s = b+\pi/2$, $\widetilde{p}_1(s-a) = \frac{1}{4}\sin(s-a-\pi/2) = -\frac{1}{4}\cos(s-a)$ for $\pi/2 \leq s-a \leq 3\pi/2$. Therefore

$$\widetilde{p}_1(s-a) = \frac{1}{4}|\cos(s-a)|,\ \pi/2 \leq s-a \leq 3\pi/2.$$

Accordingly,

$$dR_{a,b}(s_1,s_2) = \delta(s_1-s_2)ds_1 ds_2 \frac{1}{4}|\cos(s_1-a)|.$$

If we assume that $p_2(-\pi/2) = 0$ instead of $p_1(\pi/2) = 0$, then in the same way we obtain

$$dR_{a,b}(s_1,s_2) = \delta(s_1-s_2)ds_1 ds_2 \frac{1}{4}|\cos(s_2-b)|$$

Summarizing
The genericity assumptions used in the proof of the above uniqueness theorem are the following:
– (i) The local causal structure of the probability measure
– (i) The condition of statistical pre–determination
– (ii) The rotation invariance of the densities describing the local apparata
– (iii) The twice continuous differentiability of these densities
– (iv) The absolute continuity of the source measure with respect to the Lebesgue measure
While conditions (i) and (ii) have a natural physical interpretation, we don't see a natural physical justification for conditions (iii) and (iv).

For example at the moment we have no reasons to exclude the possibility of reproducing the EPR correlations with a source measure having a fractal support.

Therefore it would be interesting to know if, by dropping some of these assumptions, the uniqueness result continues to be true. This problem will be the object of further investigations.

LOCAL, CAUSAL MEASUREMENTS

von Neumann's theory of measurement did not incorporate the locality requirement, which is essential in the discussion of EPR type experiments.

In the following we will briefly outline the main properties of a theory of local, causal measurements in a classical nonrelativistic context. We refer to the survey paper [5] for the quantum formulation and additional information.

We consider a composite system made up of two subsystems, called 'particles' in the following, and denoted with the symbols 1 and 2 respectively. Their 'configuration' (or 'phase') spaces will be denoted by S_1 and S_2 respectively.

The two systems are spatially separated so that the mutual interactions between them can be neglected. Each system interacts locally with a measurement apparatus, i.e. system 1 with apparatus m_1 and system 2 with apparatus m_2. The configuration spaces of the measurement apparata will be denoted by M_1 and M_2 respectively. We use the indices $a, b, \ldots \in I$ to represent settings of the measurement apparata.

Definition 3 *(Ref. [2], Definition 6.) A probability measure $P_{a,b}$ on $S_1 \times S_2 \times M_1 \times M_2$ is called extremal, local and causal (extremal–LC, shortly) if it has the form*

$$dP_{a,b}(s_1, s_2, \lambda_1, \lambda_2) = dP_S(s_1, s_2) P_{1,a}(d\lambda_1; s_1) P_{2,b}(d\lambda_2; s_2) \tag{16}$$

where P_S is a probability measure on $S_1 \times S_2$; for all $s_1 \in S_1$, $P_{1,a}(\,\cdot\,; s_1)$ is a positive measure on M_1; for all $s_2 \in S_2$, $P_{2,b}(\,\cdot\,; s_2)$ is a positive measure on M_2.

It is called local and causal (LC, shortly) if it is a convex combination of extremal–LC measures, i.e. if it has the form

$$dP_{a,b}(s_1, s_2, \lambda_1, \lambda_2) = dP_S(s_1, s_2) \int_X q(dx) P_{1,a}(d\lambda_1; s_1; x) P_{2,b}(d\lambda_2; s_2; x) \tag{17}$$

where P_S, $P_{1,a}(\,\cdot\,; s_1; x)$, $P_{2,b}(\,\cdot\,; s_2; x)$ are as above and (X, q) is a probability space.

Remark (1) We will see in Theorem (2) that, under generic regularity conditions, there exist only two extremal–LC measures and they attribute an asymmetric role to the two particles 1 and 2. Non extremal–LC measures are useful to build statistical models in which the two particles appear in a symmetric way.

Remark (2)

All experimentally measured statistics depend on:

– a preparation,

– an observable

– a dynamics.

In classical physics the mathematical model of an experimental preparation is given by a probability measure (a quantum state in quantum physics). The extreme case of Dirac δ–measures correspond to exact theories.

In EPR type experiments the experimental preparation is localized in different space–time regions:

– the emission time and the interaction times

– the emission region and the local measurement regions

The time separation involves causality: at the emission time, the type of interation that will be met at the measurement time is unknown. Therefore the statistics at the source cannot depend on this interaction. This causality condition is reflected in the factorization condition $P_{a,b} = P_S \cdot Q_{M_{1,a} \times M_{2,b}}$, where $Q_{M_{1,a} \times M_{2,b}}(\,\cdot\,;\, s_1, s_2)$ is a positive measure on $M_1 \times M_2$.

The space separation involves locality: the statistics of the local measurements should be mutually independent **except for possible conservation laws realized at the emission time (pre–determination).**

Locality is reflected in the fact that $Q_{M_{1,a} \times M_{2,b}}(\,\cdot\,;\, s_1, s_2)$ is factorized as $P_{1,a}(\,\cdot\,;\, s_1) P_{2,b}(\,\cdot\,;\, s_2)$

Remark.

Note that it is not required that $P_{1,a}(\,\cdot\,;\, s_1)$ and $P_{2,b}(\,\cdot\,;\, s_2)$ are conditional probability measures. We will come back to this point in section ().

Now we introduce the phase (or configuration) spaces of our systems: for the moment we introduce them in an abstract way, without specifying their structure (this might include position, momentum, spin, ..., and possibly othe *hidden parameters*). We only assume, to simplify the mathematical description, that they are compact Hausdorff spaces:

- the configuration space S_1 of the particle 1,
- the configuration space S_2 of the particle 2,
- the configuration space M_1 of the measurement apparatus for the particle 1,
- the configuration space M_2 of the measurement apparatus for the particle 2.

In terms of these we define the configuration spaces for the composite systems:

$$S := S_1 \times S_2 \,;\; M := M_1 \times M_2 \,;\; \Omega_1 := S_1 \times M_1 \,;\; \Omega_2 := S_2 \times M_2$$

$$\Omega := \Omega_1 \times \Omega_2 = S_1 \times M_1 \times S_2 \times M_2 = S_1 \times S_2 \times M_1 \times M_2 \tag{18}$$

We will use the following notations: $\text{Meas}(\Omega)$ denotes the set of all regular, signed, finite Borel measures on Ω. $\langle \text{Meas}(\Omega),\ C(\Omega)\rangle$ denotes the duality, between $C(\Omega)$ and $\text{Meas}(\Omega) = C(\Omega)^*$, given by the integral:

$$\langle P, f\rangle := \int_\Omega f(\omega)P(d\omega)$$

$\text{Meas}_+(\Omega)$ and $\text{Prob}(\Omega)$ denote the set of all positive measures and the set of all probability measures in $\text{Meas}(\Omega)$ respectively.

Any LC measure on $S_1 \times S_2 \times M_1 \times M_2$, of the form (16), can be written in the following functional form:

$$P_{a,b} := P_S \circ (\overline{P}_{1,a} \otimes \overline{P}_{2,b}) \in (C(\Omega_1) \otimes C(\Omega_2))^* = C(\Omega_1 \times \Omega_2)^* \tag{19}$$

where, for $j = 1,2$ and $x = a, b$, the linear maps
$\overline{P}_{j,x} : C(\Omega_j) = C(S_j \times M_j) \to C(S_j) \subseteq C(\Omega_j)$ are defined by

$$\overline{P}_{j,x}(f)(s_j) := \int_{M_j} f(s_j, \lambda_j) dP_{j,x}(\lambda_j; s_j) \tag{20}$$

for each $f \in C(S_j \times M_j)$.

TRIVIAL EXTREMAL LC PROBABILITY MEASURES

The notion of *trivial LC probability measure* is crucial for the EPR-chameleon models.

Definition 4 *(Ref. [2], Definition 7.) An extremal LC probability measure on the space $S_1 \times S_2 \times M_1 \times M_2$*

$$dP_{a,b}(s_1, s_2, \lambda_1, \lambda_2) = dP_S(s_1, s_2) dP_{1,a}(\lambda_1; s_1) dP_{2,b}(\lambda_2; s_2)$$

is called trivial if, in the notation (20), $\forall a, b \in I$ the map

$$\overline{P}_{1,a} \otimes \overline{P}_{2,b} \ :\ C(\Omega_1 \times \Omega_2) \to C(S_1 \times S_2)$$

is a P_S–conditional expectation i.e.

$$\overline{P}_{1,a}(1_1)(s_1)\overline{P}_{2,b}(1_2)(s_2) \equiv 1 \quad , \quad P_S\text{-a.e.} \tag{21}$$

Denoting

$$p_{1,a}(s_1) := \overline{P}_{1,a}(1_1)(s_1) = \int_{M_1} dP_{1,a}(\lambda_1; s_1) \tag{22}$$

$$p_{2,b}(s_2) := \overline{P}_{2,b}(1_2)(s_2) = \int_{M_2} dP_{2,b}(\lambda_2; s_2) \tag{23}$$

condition (21) becomes equivalent to:

$$p_{1,a}(s_1) p_{2,b}(s_2) = 1 \quad , \quad P_S\text{-a.e.} \tag{24}$$

If a LC measure is trivial, then there exists a positive real number c such that

$$p_{1,a}(s_1) = c,\ p_{2,b}(s_2) = \frac{1}{c} \quad , P_S\text{-a.e.}$$

By redefining $P'_{1,a} := (1/c)P_{1,a}$, $P'_{2,b} := cP_{2,b}$, we can assume without loss of generality that

$$p_{1,a}(s_1) = 1,\ p_{2,b}(s_2) = 1 \quad , P_S\text{-a.e.}$$

The following proposition shows that triviality implies Bell type inequalities.

Proposition 1 *Let I be any index set and let $P_{a,b}$ $(a,b \in I)$ be a family of trivial LC probability measures on the space Ω defined by (18) and let $S_a^{(1)}, S_b^{(2)} : \Omega \to [-1,1]$ $(a,b \in I)$ be a family of random variables satisfying the locality condition*

$$S_a^{(1)}(\omega_1,\omega_2) = S_a^{(1)}(\omega_1)\ ;\ S_b^{(2)}(\omega_1,\omega_2) = S_b^{(2)}(\omega_2)\ ;\ (\omega_1,\omega_2) \in \Omega = \Omega_1 \times \Omega_2\ ,\ a,b \in I \tag{25}$$

For any ordered pair (a,b), with $a,b \in I$, define the $P_{a,b}$–pair correlation of the pair $(S_a^{(1)}, S_b^{(2)})$ in the usual way, i.e.

$$C(a,b) := \int S_a^{(1)} S_b^{(2)} dP_{a,b}$$

(Notice that the probability measure depends on the pair (a,b)). Then the family of pair correlations $\{C(a,b) : a,b \in I\}$ satisfies the Bell inequality in the CHSH form (but not necessarily in the original Bell form).

Proof. See Ref. [1, 3].

Remark (1). The above theorem provides an example of a context dependent family of pair correlations which must satisfy Bell's inequality. This proves that the contextuality condition alone is not sufficient to guarantee the violation of this inequality.

Remark (2). The locality condition (25) is necessary only because of the assumption the probability measures in the family $\{P_{a,b} : a,b \in I\}$) depend on the pair (a,b) (contextuality)). As shown in [6], Theorem (3), in the standard context considered in the literature (a single probability measure and no chameleon dynamics) this assumption is not necessary.

Remark (3). To be a trivial LC measure is a sufficient, but not necessary condition to satisfy Bell's inequality: there are examples of nontrivial LC measures which do not violate Bell's inequality.

Recall that, if Ω, S are topological spaces, a linear map $\mathscr{T}^* : C(\Omega) \to C(S)$ is called a Markov operator if it is positivity preserving ($f \geq 0 \Rightarrow \mathscr{T}^*(f) \geq 0$, $f \in \mathscr{C}(\Omega)$) and identity preserving:

$$\mathscr{T}^*(1_\Omega) = 1_S$$

If on S there is a probability measure P_S and $\mathscr{T}^*$ satisfies the weaker conditions

$$f \geq 0 \Rightarrow \mathscr{T}^*(f) \geq 0\ ;\quad P_S\text{-a.e. } f \in \mathscr{C}(\Omega)$$

$$\mathscr{T}^*(1_\Omega) = 1_S\ ,\quad P_S\text{-a.e.}$$

we call it a *P_S–Markov operator*. Now let

$$\Omega = \Omega_1 \times \Omega_2\ ;\quad S = S_1 \times S_2$$

Lemma 1 *For $j = 1,2$, let $\mathscr{T}_j^* : C(\Omega_j) \to C(\Omega_j)$ be a positivity preserving linear operator The following conditions are equivalent:*

$$\overline{P}_{1,a}(\mathscr{T}_1^*(1))\overline{P}_{1,b}(\mathscr{T}_2^*(1)) = 1\ ;\ P_S - a.e. \tag{26}$$

there exists a constant $c > 0$ such that

$$\overline{P}_{1,a}(c\mathscr{T}_1^*(1)) = \overline{P}_{1,b}(\mathscr{T}_2^*(1)/c) = 1\ ;\ P_S\text{-a.e.} \tag{27}$$

Proof. It is clear that (27) $\Rightarrow$ (26). Let us prove the converse implication. If (26) holds, then

$$P_S \circ ([\overline{P}_{1,a} \circ \mathscr{T}_1^*] \otimes [\overline{P}_{2,b} \circ \mathscr{T}_2^*])$$

is a trivial measure. Therefore, by the remark below Definition (4) there exists a constant $c > 0$ such that

$$c\overline{P}_{1,a}(\mathscr{T}_1^*(1))(s_1) = \frac{1}{c}\overline{P}_{2,b}(\mathscr{T}_2^*(1))(s_2) = 1 \qquad ;\quad P_S - \forall (s_1,s_2) \in S_1 \times S_2$$

and this is (27). □

Definition 5 *A linear positive operator $\mathscr{T}_1^* \otimes \mathscr{T}_2^* : C(\Omega_1 \times \Omega_2) \to C(\Omega_1 \times \Omega_2)$ (or equivalently its dual $\mathscr{T}_1 \otimes \mathscr{T}_2$, acting on measures), which satisfies the conditions of Lemma (1), will be called a $P_{a,b}$–Markovian operator. In such a case, by absorbing the constants c, $1/c$ in the definition of $\mathscr{T}_1^*$ and $\mathscr{T}_2^*$, one can always assume that they are equal to 1.*

Notice that any Markov operator from $C(\Omega_1 \times \Omega_2)$ to $C(\Omega_1 \times \Omega_2)$ is $P_{a,b}$–Markovian for any $P_{a,b}$.

Theorem 3 *Let, for* $j = 1, 2$, $\mathscr{T}_j$ *be a linear mapping of* $\mathrm{Meas}_+(\Omega_j)$ *into* $\mathrm{Meas}_+(\Omega_j)$ *such that* $\mathscr{T}_j^* : C(\Omega_j) \to C(\Omega_j)$ *and let*

$$P_{a,b} = P_S \circ (\overline{P}_{1,a} \otimes \overline{P}_{2,b}) \in \mathrm{Prob}(\Omega_1 \times \Omega_2)$$

be any trivial LC measure. Then if $\mathscr{T}_{1,a} \otimes \mathscr{T}_{2,b}$ *is a* $P_{a,b}$*–Markovian operator,* $(\mathscr{T}_{1,a} \otimes \mathscr{T}_{2,b})(P_{a,b})$ *is a trivial LC measure.*

In particular, if $\mathscr{T}_{1,a} \otimes \mathscr{T}_{2,b}$ *is a Markov operator, it maps trivial LC measures into trivial LC measures.*

Proof. The functional form of $(\mathscr{T}_{1,a} \otimes \mathscr{T}_{2,b})(P_{a,b})$ is:

$$(\mathscr{T}_{1,a} \otimes \mathscr{T}_{2,b})(P_{a,b}) = P_S \circ (\overline{P}_{1,a} \circ \mathscr{T}_{1,a}^* \otimes \overline{P}_{2,b} \circ \mathscr{T}_{2,b}^*). \tag{28}$$

Condition (27) (with $c = 1$) is equivalent to

$$\overline{P}_{1,a}(\mathscr{T}_1^*(1)) = \overline{P}_{2,b}(\mathscr{T}_2^*(1)) = 1 \; ; \; P_S\text{-a.e.}$$

which is equivalent to the triviality of $(\mathscr{T}_{1,a} \otimes \mathscr{T}_{2,b})(P_{a,b})$. ∎

Corollary 1 *Any local reversible dynamics induces a mapping which maps a nontrivial (resp. trivial) LC measure into a nontrivial (resp. trivial) LC measure.*

Proof. The statement about trivial LC measures follows from Theorem (3).

Let μ be a nontrivial LC measure and T be a reversible measurable transformation of $S_1 \times M_1 \times S_2 \times M_2$ into itself. Suppose by contradiction that $\nu := \mu \circ T$ is trivial. The linear mapping $\mathscr{T}$ induced by T is a Markov operator satisfying $\mu = \mathscr{T}(\nu) := \nu \circ T^{-1}$. Its inverse is also a Markov operator satisfying $\nu = \mathscr{T}^{-1}(\mu) := \mu \circ T$.

But if T is local i.e. of the form $T = T_1 \times T_2$ for some $T_1 : S_1 \times M_1 \to S_1 \times M_1$ and $T_2 : S_2 \times M_2 \to S_2 \times M_2$, then $\mathscr{T} = \mathscr{T}_1 \otimes \mathscr{T}_2$ where $\mathscr{T}_1$ and $\mathscr{T}_2$ are Markov operators. By the remark after Definition (5) this contradicts Theorem (3). ∎

BELL IMPLICITLY ASSUMES TRIVIALITY

In "Bertlmann's sock and the nature of reality" [7] Bell requires the condition

$$P(A, B|a, b, \lambda) = P_1(A|a, \lambda) P_2(B|b, \lambda).$$

and argues that it is reasonable to expect that P_1 and P_2 are conditional probability distributions.

Given the following correspondence between the notations in Bell's argument and those of Definition (4):

$$\lambda \leftrightarrow (s_1, s_2) \; , \; P_1(A|a, \lambda) \leftrightarrow \overline{P}_{1,a} \circ \mathscr{T}_{1,a}^* \; , \; P_2(B|b, \lambda) \leftrightarrow \overline{P}_{2,b} \circ \mathscr{T}_{2,b}^*$$

we see that Bell implicitly assumes triviality of the LC measure.

This assumption is equivalent to postulate that the stochastic processes given by the observables of the two systems at the final (measurement) time are conditionally independent given the source.

However such an assumption is physically and probabilistically unwarranted whenever there are constraints (e.g. conservation laws) which are determined both at the source and at the local measurement sites. For example, if $S_a^{(1)}, S_a^{(2)}$ are random variables that depend both on the source and the apparatus variables and if we know that, at the time of measurement, they must satisfy the constraint:

$$S_a^{(1)}(s_1, \lambda_1) + S_a^{(2)}(s_2, \lambda_2) = 0$$

then surely they are not conditionally independent given the source variables (s_1, s_2): here the difference between a pre–existent property and a property pre–determined as a response to a local interaction (the *if* ... *then* ... scheme) is essential.

Since this is precisely the situation in the EPR type experiments, we see that Bell's implicit assumption is not justified in this case.

The class of LC measures is the disjoint union of its two subclasses, of the trivial and the nontrivial LC measures. Corollary (1) implies that a local reversible dynamics induces a map of each of these classes into itself.

Since a local measurement process is described by a local Markov operator, it follows that the initial probability measure,in the EPR-chameleon model, must be nontrivial otherwise one could not have violation of Bell's inequality.

The above discussion proves that the nontriviality of the initial LC measure is precisely what is required by the physical context of the EPR type experiments where locality and causality have to be combined with the existence of constraints (conservation laws). Conversely, Bell's implicit assumption of the triviality of the initial measure, being equivalent to conditional independence given the source, puts the locality and causality requirement in contradiction with the conservation laws.

The dependence of the initial state on the pair (a,b) is not a non–local requirement because, for adaptive dynamical systems, the ensamble to which the individual systems belong is not defined at the emission time but at the time in which the interaction with the local instruments begins to take place.

In other words, since for such systems the measurement locally affects the dynamics, the natural notion of *classical statistical state* is not, as for passive systems, a single probability measure (corresponding to the fact that the state will be independent of the measurement), but a family of probability measures – one for each possible measurement (corresponding to the fact that the state will be determined by the response to a measurement which is not known at the initial time).

The EPR–chameleon model shows that all this is perfectly compatible with a local, deterministic, reversible dynamics.

EMPIRICAL CORRELATIONS OF PAIRS OF DISTANT PARTICLES

The same term 'pair correlation', when referred to pairs of distant particles is often used to describe two completely different experimental procedures. Below we discuss these experimental differences [3].

Standard correlations

The term *standard correlation* is used when the following physical conditions are verified:

1) The total number N of emitted pairs is exactly known.
2) The trajectory of each pair can be followed without disturbance so that, at each time t, the experimenters know exactly to which of the N pairs their measurement is referred. This property will be called *distinguishability*.
3) The observable (f_1, f_2) is measured on each pair of the ensemble. The result of the measurement of (f_1, f_2) on the jth pair will be denoted by
$$(f_{1,j}, f_{2,j});$$
the measurement itself will be denoted by M_j.

Under these conditions the following definition makes sense.

Definition 6 *The empirical correlation between the pair of observables (f_1, f_2), relative to the sequence of measurements $M = (M_j)$ on the ensemble $\{(1_j, 2_j) : j = 1, \ldots, N\}$ is*

$$\langle f_1 \cdot f_2 \rangle_M := \frac{1}{N} \sum_{j=1}^{N} f_{1,j} f_{2,j}. \tag{29}$$

We further specify our context of standard correlations as follows.

4) Each measurement M_j is specified by a time
$$t'_j := t_j + T,$$
where T is independent of j, t_j is the emission time for the pair $(1_j, 2_j)$.
5) The result of the jth measurement does not depend on the interval $[t_j, t_j + T]$ but only on T (time homogeneity).

Under these conditions the correlations (29) are interpreted as the correlations of (f_1, f_2) at time T and T is interpreted as the final time of the single measurement.

Correlations of distant pairs

Suppose that the measurement protocol is the following.

(DP1) It is known that each pair is emitted simultaneously, but the experimenters do not know precisely when the pair is emitted.

(DP2) The experimenters cannot follow the trajectory of each particle, but only register the result of a measurement at time t (*indistinguishability*).

(DP3) The experimenters have synchronized clocks, so the time t is the same for both.

(DP4) The experimenters do not know the total number of emitted particles.

(DP5) The experimenters cannot postulate that, if a particle of a pair reaches one of them, then the other particle reaches the other experimenters.

Conditions (4) and (5) of the previous section are still meaningful because they are referred to single particles. However condition (3) is meaningless because of indistinguishability. Moreover the N, in formula (29) is unknown. In a situation described by the above conditions we speak of *correlations of distant particles*.

In conclusion: under the above described physical conditions, the definition of standard correlations is meaningless and a new one is needed.

Definition 7 *The protocol to define correlations of distant particles is the following:*

(CDP1) The experimenter X, $X \in \{1,2\}$ performs measurements on M_X particles and records
- *the time $t'_{X,j}$ of the jth measurement*
- *the value $f_{X,j}$ of the measured observable f_X*

for $\forall\, j \in \{1,\ldots,M_X\}$.

(CDP2) The two experimenters exchange the sequences

$$\left((t'_{1,j},f_{1,j}) :\ j=1,\ldots,M_1\right) \text{ and } \left((t'_{2,j},f_{2,j}) :\ j=1,\ldots,M_2\right).$$

(CDP3) Each experimenter extracts the sequences

$$\left(f'_{1,h} :\ h=1,\ldots,M_{f_1 f_2}\right) \text{ and } \left(f'_{2,h} :\ h=1,\ldots,M_{f_1 f_2}\right),$$

where

$$\left\{s_h :\ h \in \{1,\ldots,M_{f_1 f_2}\}\right\} := \left\{t'_{1,j} : j \in \{1,\ldots,M_1\}\right\} \cap \left\{t'_{2,j} : j \in \{1,\ldots,M_2\}\right\}$$

and

$$f'_{X,h} := f_{X,j},\ \text{if } s_h = t'_{X,j} \qquad (X=1,2).$$

(CDP4) The empirical correlations of distant pairs are defined by

$$\langle f_1 f_2 \rangle_{DP} := \frac{1}{M_{f_1,f_2}} \sum_{h=1}^{M_{f_1,f_2}} f'_{1,h} f'_{2,h}.$$

In other words: by definition, correlation of distant pairs means conditioned correlations on coincidences.

Practically the totality of the EPR type experiments follows the protocol described in Definition (7).

As far as experiments are performed under this protocol (Definition (7)), it is possible to take the Bell's fifth position named by Gill: "a decisive experiment cannot be done" [8].

REFERENCES

1. L. Accardi, K. Imafuku, and M. Regoli, *Infinite Dimensional Analysis, Quantum Probability and Related Topics*, **5**, 1–20 (2002) quant-ph/0112067, Volterra Preprint N. 494 December (2001) N. 494.
2. L. Accardi, K. Imafuku, and M. Regoli, "Adaptive dynamical systems and the EPR–chameleon experiment", in *Foundations of Probability and Physics – 2*, edited by A. Khrennikov, Mathematical Modeling in Physics, Engineering and Cognitive Science 5, Växjö University Press, Växjö, Sweden-2002, 2003, pp. 11–36.

3. L. Accardi, and S. Uchiyama, *arXiv:quant-ph 0706.1813* (2007). submitted to: IDA–QP (Infinite Dimensional Analysis, Quantum Probability and Related Topics) 14 June (2007)
4. Luigi Accardi, Andrei Khrennikov: *Chameleon effect, the range of values Hypothesis and reproducing the EPR–Bohm correlations*, Foundations of probability and physics–4 AIP Proceedings 889 (2006) 21–29
5. L. Accardi, "Could we now convince Einstein?", in: *Proceedings of the conference: Quantum theory: reconsideration of foundations-3*, Växjö, Sweden 6–11 june (2005), Andrei Khrennikov (ed.), American Institute of Physics, AIP Conference Proceedings 810 (2006) 2–18, Melville, New York, ISBN 0-7354-0301-5.
6. L. Accardi, and M. Regoli, *Volterra Preprint N. 427; arXiv: quant-ph/0007005* (2000).
7. J. S. Bell, *Speakable and Unspeakable in Quantum Mechanics*, Cambridge University Press, Cambridge, 1987.
8. R. D. Gill, "Time, Finite Statistics, and Bell's Fifth Position," in *Foundations of Probability and Physics – 2*, edited by A. Khrennikov, Mathematical Modeling in Physics, Engineering and Cognitive Science 5, Växjö University Press, Växjö, Sweden-2002, 2003, pp. 179–206.

Objective and Subjective Probabilities in Quantum Mechanics

Leslie Ballentine

Simon Fraser University, Burnaby, B.C. V5A1S6, Canada

Abstract. The concept of probability was prominent in the original foundations of quantum mechanics. and continues to be so today. Indeed, the controversies regarding objective and subjective interpretations of probability have again become active. I argue that, although both objective and subjective probabilities have domains of relevance in QM, their roles are quite distinct. Even where both are legitimate, the objective and subjective probabilities differ, both conceptually and numerically. There are quantum probabilities that have no useful subjective interpretations, and there are subjective probabilities that cannot be realized as quantum probabilities.

Keywords: quantum probability, objective, subjective
PACS: 03.65.Ta, 03.67.-a, 02.50.Cw

1. INTRODUCTION

Probability has played an important role in the Foundations of Quantum Mechanics (FQM) from the beginning, and continues to play an important role today. So it is appropriate, at this conference devoted to a reconsideration of FQM, to examine the role, or roles, of probability in QM.

Here are some of the ways in which Probability and FQM are inter-related:

- QM is formulated as an intrinsically probabilistic theory, rather than as a deterministic theory in which probability would play a secondary role.
- The choice of an interpretation of probability affects the interpretation of QM.
- Recent developments in Quantum Information Theory have led to new ways to look at FQM, including a greater emphasis on possible roles for Subjective Probability in QM. [1]

These ideas motivate my paper, although I shall not treat all of them in detail.

2. INTERPRETATIONS OF PROBABILITY

At an earlier conference here in Växjö (25 Nov. - 1 Dec. 2000) [2], I presented, in the spirit of Linnaeus, a taxonomic classification of the many interpretations of probability [3]. These were divided into four major groups:

1. **The Logic of Inductive Inference**. Probability is assigned to *propositions*; $P(A|C)$ is the probability that A is true, given the information C. Inductive logic reduces to deductive logic in the case that the probabilities take on only the values 0 and 1. This point of view was expressed most completely by E. T. Jaynes in his posthumous book *Probability Theory: The Logic of Science* [4].
2. **Ensemble and Frequency interpretations**. Probability is identified with the *measure* on a set, or with the *limit frequency* in an ordered sequence.
3. **Subjective Probability**. A measure of the *degree of reasonable belief* under the condition of incomplete knowledge.
4. **Propensity**. This is a form of causality that is weaker than determinism, and is associated with *events* in an (apparently) indeterministic system. $P(A|C)$ is the propensity for event A to occur under the condition C.

CP962, *Quantum Theory, Reconsideration of Foundations – 4*,
edited by G. Adenier, A. Yu. Khrennikov, P. Lahti, V. I. Man'ko, and Th. M. Nieuwenhuizen

In my 2000 talk, I further suggested that the first of these interpretations (Inductive Logic) might be the master interpretation of probability, from which the others could be derived as special cases.

> "If we specialize to propositions about repeated experiments we obtain the Ensemble–Frequency theory. If we specialize to propositions about personal belief we obtain Subjective probability. If we specialize to propositions about indeterministic or unpredictable events we obtain the Propensity theory." (Ref. [3], p. 72)

I stressed the tentative nature of my classification, and anticipated that some future refinements might be needed.

2.1. The place of Subjective Probability

At a later conference in Växjö (June 1–6, 2003), Rudiger Schack criticized the treatment of subjective probability in my classification.[1] He said that, contrary to my statement quoted above, what changes when we pass to the subjective interpretation is not the nature of the propositions, but what you say about them. The significance of this brief remark can best be illustrated by an example.

Consider a particular quantum state of a spin one–half particle,

$$|\psi\rangle = \frac{4}{5}|+\rangle + \frac{3}{5}|-\rangle \tag{1}$$

and the proposition

A: On the next measurement the spin will be "+".

From the formalism of quantum mechanics, we obtain the uninterpreted probability statement,

$$P(A|\psi) = 0.64 \tag{2}$$

Each interpretation of probability will give a different meaning to this statement.

Propensity –

The probability is 0.64 that the next measurement will yield "spin +".

Frequency –

In a long run of similar measurements on this state, the fraction of "spin +" will be (close to) 64%.

Subjective –

My degree of belief (or confidence) that the next measurement will yield "spin +" is 64%.

Note, however, the subjective interpretation does NOT say

There is a 64% probability that I believe the spin will be "+".

The proposition A is about the spin, not about my belief! Schack was correct in pointing out that the subjective interpretation does not deal with a restricted class of propositions (those which concern personal beliefs), but rather it affects what we say about all propositions, namely, our degree of belief in them.

It is, therefore, necessary to revise the classification of interpretations of probability [3] as it applies to the Subjective interpretation. Rather than being subordinate to the Inductive Logic theory and derived from it by restricting the class of propositions, the Subjective interpretation is parallel to it and can deal with the same class of propositions. The distinction arises from what is said about those propositions; an objective (but possibly imprecise) statement about the truth or falsity of the proposition in the first case, and a subjective statement about someones degree of belief in the second case. Both interpretations are based on the same formalism of Bayesian probability theory, and may reasonably be called the *objective* and *subjective* Bayesian interpretations. Indeed, Jaynes [4] switches back and forth between them effortlessly, and perhaps even unconsciously.

---

[1] Unfortunately, his critical remarks were not included in the conference proceedings.

3. QUANTUM PROBABILITY

Which of the various interpretation of probability are relevant to quantum mechanics? In my earlier paper [3], I showed how both the Frequency and Propensity interpretations are relevant. Regarding the Subjective interpretation, I said, "I have *never* seen a coherent exposition of QM based on a *subjective* interpretation of quantum probabilities as representing *knowledge*." This claim is likely to be controversial because the interpretation of probabilities as knowledge seems to be a tenet of the Copenhagen interpretation, to which at least lip service is still paid. In the rest of this paper, I shall expand upon that statement, with examples to illustrate the point.

First I must define what is meant by the term *quantum probability*. For the purposes of this paper, I shall use it to mean those probabilities that are calculated from the familiar QM formulas, such as:

- $|\psi(x)|^2$ The probability density for the particle to be found at position x.
- $|\langle a_n|\psi\rangle|^2$ The probability that an observable takes on the eigenvalue a_n.
- $Tr(\rho E_\omega)$ The probability that an observable lies within the subset ω of its spectrum. Here ρ is the state operator and E_ω is a projector onto the relevant subset of the spectrum.

Other formulas are also possible, but all assume that a certain state (represented by ψ or ρ) has been prepared, and the probability for an observable to take on some value is calculated. Within the Frequency or Propensity interpretations these quantum probabilities are *objective* in the sense that they do not depend on the observer or his [2] knowledge, but only on the physical situation described by the prepared state. I shall later consider whether they can also be given a *subjective* interpretation as representing someones knowledge.

4. OBJECTIVE AND SUBJECTIVE PROBABILITIES IN QM

Subjective probability is a natural tool to use in cases where different persons have different amounts of information, and hence are likely to make different predictions. Classical and quantum cryptography, whose goal is to share information with some people while denying it to others, are good examples. But the subjective probabilities that occur in quantum cryptography supervene at a high level of application, far above the fundamental quantum probabilities of the previous section. Quantum cryptography uses subjective probabilities because it is cryptography, not because it is quantum.

4.1. Objective and Subjective Density Matrices

To better see the roles of objective and subjective probabilities, we turn to a simple example. A single photon [3] is prepared in the vertical polarization state,

$$|\psi\rangle = |\updownarrow\rangle. \tag{3}$$

This state vector and the probabilities associated with it are *objective*; they describe the result of the physical operation of state preparation, and not anyone's knowledge of it.

We then tell Alice, who is unaware of the state preparation, that the polarization state is either vertical $|\updownarrow\rangle$ or horizontal $|\leftrightarrow\rangle$. We tell Bob that the polarization was prepared in one of four states: vertical $|\updownarrow\rangle$, horizontal $|\leftrightarrow\rangle$, $+45$ degrees $|\nearrow\rangle$, or -45 degrees $|\searrow\rangle$.[4]

Except for a possible phase factor, the vector $|\nearrow\rangle$ is the normalized sum of the vertical and horizontal states, and the vector $|\searrow\rangle$ is their normalized difference. Using the vertical and horizontal state vectors as a basis, Bob's four

---

[2] The word *observer* is a technical term in QM. It denotes an abstract entity having only a few attributes, including a frame of reference and the ability to measure an observable. It should not be thought of as a human person with attributes like consciousness or gender. An observer is traditionally referred to as "he" for the same reason that a ship is referred to as "she" — mere tradition. The hermaphroditic forms "he or she" and "he/she" should be deprecated.

[3] There is nothing special about a photon. Any object with a two dimensional state space would do just as well.

[4] The arrows for $+45$ and -45 should also be double-ended, but unfortunately Latex has no such symbols.

possible states can be represented by the following density matrices:

$$\rho_{\updownarrow} = \begin{pmatrix} 1 & 0 \\ 0 & 0 \end{pmatrix}, \ \rho_{\leftrightarrow} = \begin{pmatrix} 0 & 0 \\ 0 & 1 \end{pmatrix}, \ \rho_{\nearrow} = \begin{pmatrix} 1/2 & 1/2 \\ 1/2 & 1/2 \end{pmatrix}, \ \rho_{\searrow} = \begin{pmatrix} 1/2 & -1/2 \\ -1/2 & 1/2 \end{pmatrix}. \tag{4}$$

Although the actual state of the photon is $\rho_{\updownarrow}$, Alice and Bob do not know this, and so they must use their incomplete information to make a subjective estimate of the density matrix. Alice knows that the polarization may be either vertical or horizontal, so she forms a mixed state from those two components, obtaining the *subjective* density matrix

$$\rho_S = \begin{pmatrix} 1/2 & 0 \\ 0 & 1/2 \end{pmatrix}. \tag{5}$$

Bob knows only that the polarization is in one of the four states (4), so he forms a mixed state from four components, and obtains the same subjective density matrix ρ_S. Notice that, although Alice and Bob have different information, they both arrive at the same subjective density matrix. Nevertheless, their future predictions will not necessarily be the same.

To show this, we perform a measurement that distinguishes between vertical and horizontal polarizations. The result, of course, confirms that the polarization is vertical. We, the readers, knew this in advance, because we knew that the prepared state was $\rho_{\updownarrow}$. But to Alice and Bob, the result of the measurement is new information, which they can use to obtain a better estimate of the photon state. Alice will now correctly infer that the initial state must have been $\rho_{\updownarrow}$, since the other possibility ($\rho_{\leftrightarrow}$) is ruled out by the new data. But Bob, who previously knew only that the initial state was one of four, can now only reduce the number of possibilities to three. However, he can use Bayes theorem to improve his subjective estimate.

4.2. Bayesian Inference of an Unknown State

Since the use of Bayes theorem in quantum theory may be unfamiliar to many readers, I shall give Bob's calculation in full detail. For our purposes, Bayes theorem [5] may be written as

$$P(X_j|Y_i\&C) = P(Y_i|X_j\&C)\frac{P(X_j|C)}{P(Y_i|C)} \tag{6}$$

where "&" denotes the logical "and". Here X_j is one of the four possible initial states (4), and Y_i is one of the possible results of the polarization measurement (either $\updownarrow$ or $\leftrightarrow$). C denotes the background information that Bob possesses, including the rules of QM and the knowledge that the initial state was one of the four in (4), but excluding the result of the polarization measurement. Lacking any information that would prefer any of the four possible states, Bob assigns the *prior probability*[5] $P(X_j|C) = 1/4$. Similarly, before the measurement has been performed, the prior probability for the two possible results is $P(Y_i|C) = 1/2$.

The remaining term on the righthand side of eq.(6), $P(Y_i|X_j\&C)$, is just the probability of the measurement yielding a particular result for a given quantum state, and it can be calculated from the familiar formulas for quantum probability. The probability for obtaining the observed polarization ($\updownarrow$) in each of the possible initial states is, in a slightly modified, but obvious notation,

$$P(\updownarrow||\updownarrow\rangle) = 1, \ P(\updownarrow||\leftrightarrow\rangle) = 0, \ P(\updownarrow||\nearrow\rangle) = 1/2, \ P(\updownarrow||\searrow\rangle) = 1/2. \tag{7}$$

Combining the three factors on the righthand side of (6), we obtain Bob's *posterior probability* for each of the four possible initial states, after the result of the polarization measurement has been taken into account. In an abbreviated notation, Bob's probabilities are

$$P(|\updownarrow\rangle) = 1/2, \ P(|\leftrightarrow\rangle) = 0, \ P(|\nearrow\rangle) = 1/4, \ P(|\searrow\rangle) = 1/4. \tag{8}$$

This calculation confirms the earlier obvious conclusion that Bob will not be able to uniquely infer the initial prepared state of the photon, but it also quantifies his knowledge of what that initial state might have been.

---

[5] The correct term is "prior", meaning "prior to the result of the measurement", and not "a priori", which has a different, metaphysical meaning that is not relevant here.

4.3. Bob's new Subjective Density Matrix

Bob can use his posterior probabilities (8) to form a new mixture as his subjective density matrix. It will be

$$\rho_B = \frac{1}{2}\rho_{\updownarrow} + \frac{1}{4}\rho_{\nearrow} + \frac{1}{4}\rho_{\searrow} = \begin{pmatrix} 3/4 & 0 \\ 0 & 1/4 \end{pmatrix}. \tag{9}$$

Several remarks are in order about the significance of this density matrix ρ_B.

- It is not an estimate of the final state of the photon after the measurement. Indeed, if the measurement was performed as a projective measurement (von Neumann type) then the final state will be $\rho_{\updownarrow}$, and this would be known to Bob and Alice.
- It does not represent Bob's state of belief about the unknown (to him) initial state of the photon. Bob does not believe the initial state was ρ_B. Rather, he believes its was either $\rho_{\updownarrow}$, $\rho_{\nearrow}$ or $\rho_{\searrow}$, with probabilities given by eq.(8).
- However, it is Bob's "safest estimate" of the unknown initial state, in the sense that predictions from it will have the lowest probable error. Even though he now knows that $\rho_{\updownarrow}$ is the most probable value for the initial state, he should hedge his bets in case it turns out to be $\rho_{\nearrow}$ or $\rho_{\searrow}$. The use of ρ_B for prediction accomplishes this "hedging".
- It is not appropriate to interpret ρ_B as "a mixture of 75% vertical polarization and 25% horizontal polarization", because Bob knows that the horizontal polarization state has zero probability. Evidently, the subjective density matrix does not adequately represent Bob's state of knowledge.

5. CONCLUSIONS

Although only a simple example has been treated in this paper, it is sufficient to establish some broad conclusions. The first is that both *objective* and *subjective* states/probabilities may legitimately appear in the same problem. However, they are both conceptually and numerically distinct. The objective state operator (or density matrix) describes the result of the physical operation of state preparation. It is objective in the sense that it does not depend on any observer or on his knowledge.[6] The subjective states and probabilities describe the states of belief of various observers, whose knowledge of the objective situation is incomplete. The objective state of a physical system and the subjective state of someones knowledge are quite different things; neither can replace the other.

Quantum probabilities, as defined in Sec. 3, are naturally to be interpreted as objective if they are calculated from the *objective* state operator. But what if the same formulas are used with *subjective* density matrices? Could we then say the quantum probabilities can also have a subjective interpretation? There are serious difficulties with such a program. Firstly, we have shown that observers who have *different* states of knowledge may, nevertheless, assign the same subjective density matrix. Therefore, the density matrix is inadequate to describe a state of knowledge. Secondly, we have seen that two observers who share the same subjective density matrix may, nevertheless, make different predictions. So, it cannot be correct to assert that quantum states represent states of knowledge.

Another difference between objective quantum states and states of knowledge is illustrated in the mathematical fact that a mixed state operator can be represented as a mixture of pure states,

$$\rho = \sum_i w_i |\psi_i\rangle\langle\psi_i|, \tag{10}$$

in infinitely many ways [6, 7]. Now if ρ represents an objective state (the result of a physical operation of state preparation), then the state operator ρ is sufficient to make all possible predictions about future measurements. In this (and only this) sense, it can be said that the quantum state description is *complete.*[7] However, a subjective density matrix is not adequate to describe a state of knowledge. In the example studied, we saw that Bob's state of belief could be described by a particular mixture representation of his subjective density matrix ρ_S, but other mathematically valid representations of the form (10) did not correspond to his state of belief. In particular, the diagonal representation,

---

[6] To describe the state as objective in this sense is not the same as claiming that the state vector or operator is has ontic status as an "element of reality", a claim that is well known to be problematic.

[7] This does not conflict with the argument of Einstein, Podolsky and Rosen [8] that the quantum description of reality is *incomplete*, because they used a different criterion for *completeness*.

which has some privilege mathematically, does not describe his state of belief. Therefore, if we attempt to use the formalism of quantum states and quantum probability to describe states of knowledge or belief, we find that the subjective quantum state operator is *not* a complete state description. This shows the unnaturalness of forcing a subjective interpretation onto quantum states.

Most of my discussion has dealt with mixed states because the mixture coefficients, w_i, in (10) may sometimes be given a meaningful subjective interpretation. But what about the pure states, such as $\psi(x)$? Can pure state probabilities be given a meaningful interpretation as knowledge? What state of knowledge would lead someone to describe the probability density for the electron in the ground state of hydrogen by its particular analytic form? What state of knowledge would lead to the probability density for an electron in a covalent bond (a function that has no analytic form, but which can be accurately computed numerically)? These probabilities have a natural objective interpretation, being experimentally testable in reproducibly prepared states. But they have no natural interpretation as someones knowledge, except for the contrived case of a person who knows and believes everything QM tells him.

ACKNOWLEDGMENTS

I have enjoyed conversations with Rudiger Schack and Chris Fuchs on this subject. We may still disagree, but at least we now better understand why.

REFERENCES

1. A. Khrennikov, editor, Proceedings of the Conference, *Quantum Theory: Reconsideration of the Foundations*, 17–21 June 2001, Växjö University Press, Växjö Sweden, (2002). This volume contains several papers relevant to the informational approach to QM. Particularly relevant are those by Fuchs and by Mermin.
2. A. Khrennikov, editor, Proceedings of the Conference, *Foundations of Probability and Physics*, 25 Nov. – 1 Dec. 2000, World Scientific, Singapore (2001).
3. L. E. Ballentine, ref. 2, pp. 71–84.
4. E. T. Jaynes, *Probability Theory: The Logic of Science*, Cambridge University Press, Cambridge, (2003).
5. L. E. Ballentine, *Quantum Mechanics: a Modern Development*, World Scientific, Singapore, (1998), sec. 1.5.
6. L. P. Hughston, R. Joza,, and W. K. Wooters, *Phys. Let. A* **183**, 14–18 (1993).
7. K. A. Kirkpatrick, *Found. Phys.* **19**, 95–102 (2006).
8. A. Einstein, B. Podolsky, and N. Rosen, *Phys. Rev.* **47**, 777–780 (1935).

A Curious Geometrical Fact about Entanglement

Ingemar Bengtsson

Fysikum, Stockholm University, 106 91 Stockholm, Sweden

Abstract. I sketch how the set of pure quantum states forms a phase space, and then point out a curiousity concerning maximally entangled pure states: they form a minimal Lagrangian submanifold of the set of all pure states. I suggest that this curiousity should have an interesting physical interpretation.

Keywords: Entanglement
PACS: 02.40.Ky, 02.50.Cw, 03.65.Ca

FROM CLASSICAL TO QUANTUM STATES

I will state a curious fact concerning maximally entangled pure states of a bipartite quantum system [1]. I do not know any physical interpretation of it, but I do expect that there is one. Before I can state my fact, I will have to give a bird's eye view of some aspects of quantum mechanics—aspects that may be worth recalling anyway, in a reconsideration of the foundations of quantum theory.

We begin at the beginning. A classical state—in classical probability theory—is a probability distribution, that is a set of N numbers p_i such that

$$p_i \geq 0 , \qquad \sum_{i=1}^{N} p_i = 1 . \tag{1}$$

The set of classical states forms a simplex; a triangle if $N = 3$, which is a good case to choose as an illustrative example. If you want to introduce dynamics on a probability simplex, the options are rather limited. You can use stochastic maps, tending to shrink the simplex towards some fixed point, and then combine such maps into Markov chains. If we also require the stochastic maps to take pure states into pure states, we only have permutations of the corners to play with.

The set of pure states is isomorphic to the sample space. To define a symplectic structure on sample space, it needs to be a continuous space—the discrete space forming the corners of a finite dimensional simplex simply will not do. To get classical mechanics, with Hamiltonians and symplectic forms, it is necessary to go to the infinite dimensional case. (I am aware that this statement must be qualified [2].)

To generalize the classical case, we first rewrite the definitions in terms of diagonal matrices. Thus a state P is a matrix

$$P = \begin{pmatrix} p_1 & 0 & 0 \\ 0 & p_2 & 0 \\ 0 & 0 & p_3 \end{pmatrix} , \qquad P \geq 0 , \quad \mathrm{Tr}P = 1 . \tag{2}$$

A random variable A is an otherwise unrestricted diagonal matrix,

$$A = \begin{pmatrix} a_1 & 0 & 0 \\ 0 & a_2 & 0 \\ 0 & 0 & a_3 \end{pmatrix} . \tag{3}$$

The expectation value of a random variable, given a state, is then

$$\langle A \rangle = \mathrm{Tr}PA . \tag{4}$$

There is a fairly obvious generalization of all this. We replace the diagonal matrices with "diagonalizable", that is Hermitian, matrices. This still leaves it open whether they should be real or complex matrices, but I will soon argue

CP962, *Quantum Theory, Reconsideration of Foundations – 4*,
edited by G. Adenier, A. Yu. Khrennikov, P. Lahti, V. I. Man'ko, and Th. M. Nieuwenhuizen

that complex numbers are the preferred choice, so let us make it right away. A state is now a positive (and therefore Hermitian) matrix,

$$\rho = \rho^\dagger \geq 0 , \qquad \mathrm{Tr}\rho = 1 . \tag{5}$$

Such a state is known as a density matrix. A random variable is an otherwise unrestricted Hermitian matrix,

$$A = A^\dagger . \tag{6}$$

For expectation values, we keep the classical formula (4).

The quantum generalization is a significant one. The random variables, and the states, now belong to a non-commutative algebra. The $N-1$ dimensional classical state space is turned into an $N^2-1 = (N+1)(N-1)$ dimensional one. This set forms an interesting convex body, which can be understood by observing that a general density matrix can be obtained from a diagonal one by means of a unitary transformation

$$\rho \to U\rho U^\dagger . \tag{7}$$

It is only the $SU(N)$ subgroup that acts effectively on the states. This is a subgroup of $SO(N^2-1)$, so the convex body of density matrix is obtained by performing quite special rotations of the classical simplex. This gives the convex body a subtle structure, but our concern here is with its pure states.

Pure quantum states are density matrices of rank one,

$$\rho = \frac{\psi^\alpha \psi_\beta}{\psi \cdot \bar{\psi}} , \qquad \psi \cdot \bar{\psi} \equiv \psi^\alpha \bar{\psi}_\alpha = \langle \psi | \psi \rangle , \qquad 1 \leq \alpha, \beta \leq N . \tag{8}$$

Equivalently, they are vectors in an N dimensional Hilbert space up to normalization and phase, that is to say they can be regarded as points in complex projective space $\mathbf{CP}^{N-1}$. The real dimension of this space is $2(N-1)$. Hence the pure states in quantum mechanics form a continuous manifold already in the finite dimensional case.

QUANTUM MECHANICS

Because we decided to work with complex numbers, we can now justify the name "quantum mechanics". This happens because, as noted by Strocchi [3] (and later by others [4, 5, 6, 7]), the Hilbert space scalar product conceals a symplectic form.

To see this, split the complex vectors into real and imaginary parts, and reassemble them into real $2N$ dimensional vectors:

$$\psi^\alpha = x^\alpha + iy^\alpha , \qquad X^I = \begin{pmatrix} x^\alpha \\ y^\alpha \end{pmatrix} , \qquad 1 \leq I, J \leq 2N . \tag{9}$$

Similarly, Hermitian operators may be split into real symmetric and imaginary anti-symmetric parts. The scalar product becomes

$$X \cdot Y = X^I g_{IJ} Y^J + iX^I \Omega_{IJ} Y^J , \tag{10}$$

where

$$g_{IJ} = \begin{pmatrix} \mathbf{1} & 0 \\ 0 & \mathbf{1} \end{pmatrix} \qquad \Omega_{IJ} = \begin{pmatrix} 0 & \mathbf{1} \\ -\mathbf{1} & 0 \end{pmatrix} . \tag{11}$$

Therefore g_{IJ} will serve as a metric on state space, while Ω_{IJ} is an anti-symmetric symplectic form.

Recall that a symplectic form on a vector space is simply a non-degenerate anti-symmetric matrix. If Ω^{IJ} denotes its inverse, one can use it to define a Poisson bracket between any two functions on phase space, as

$$\{f(X), g(X)\} = \Omega^{IJ} \partial_I f(X) \partial_J g(X) . \tag{12}$$

If the symplectic form itself depends on X^I, an extra condition—that the form be closed—must be imposed in order to ensure that the Jacobi identities hold.

The point we are driving at is that, once these definitions are in place, the Schrödinger equation takes precisely the form that time evolution always takes in classical Hamiltonian mechanics:

$$i\partial_t \psi^\alpha = H^\alpha_\beta \psi^\beta \quad \Leftrightarrow \quad \dot{X}^I = \Omega^{IJ}\partial_J \langle H\rangle = \{X^I, \langle H\rangle\} \ . \tag{13}$$

It is just that the Hamiltonian function takes a quite special form:

$$\langle H\rangle = \langle \psi|H|\psi\rangle \ . \tag{14}$$

Hence the name quantum mechanics is indeed justified: the space of its pure states is rich enough to support time evolution of Hamiltonian form. We have not identified classical and quantum mechanics, since we admit Hamiltonians of a very special form only. On inspection, one finds that quantum mechanics admits precisely those Hamiltonian flows that preserve the metric which is also present on phase space.

In classical theory, one needs a continuous sample space to support Hamiltonian mechanics. In quantum theory Hamiltonian mechanics is always there, because one uses a continuous set of discrete (or continuous) sample spaces.

We note in passing that nothing like this would work over the real numbers. Real projective space is not a symplectic manifold.

For the record, we will need the symplectic form also on complex projective space. On $\mathbf{CP}^{N-1}$ we have the Fubini-Study metric

$$ds^2 = \frac{\psi\cdot\bar{\psi}d\psi\cdot d\bar{\psi} - d\psi\cdot\bar{\psi}\psi\cdot d\bar{\psi}}{\psi\cdot\bar{\psi}\psi\cdot\bar{\psi}} \ , \tag{15}$$

and its relative, the symplectic form

$$\Omega = i\frac{\psi\cdot\bar{\psi}d\psi\cdot\wedge d\bar{\psi} - d\psi\cdot\bar{\psi}\wedge\psi\cdot d\bar{\psi}}{\psi\cdot\bar{\psi}\psi\cdot\bar{\psi}} \ . \tag{16}$$

Basically this is what we had before, rewritten to be invariant under the transformation $\psi^\alpha \rightarrow z\psi^\alpha$, where z is an arbitrary non-zero complex number.

COMPOSITE SYSTEMS

We are now half way to our fact. Before we get there, we must consider composite systems, described by means of the Hilbert space $\mathscr{H}^{N^2} = \mathscr{H}^N \otimes \mathscr{H}^N$. For this purpose, it is helpful to introduce a notation that regards vectors as square arrays of numbers. Thus we write

$$|\Psi\rangle = \psi^\alpha|e_\alpha\rangle = \frac{1}{\sqrt{N}}\Gamma^{ij}|e_i\rangle|e_j\rangle \ , \qquad 1 \le \alpha \le N^2 \ , \quad 1 \le i,j \le N \ . \tag{17}$$

Although Γ^{ij} is primarily a square array, not a matrix, the terminology of matrix theory applies to it. If the matrix is of rank one, the state is a product state:

$$|\Psi\rangle = \phi^i\lambda^j|e_i\rangle|e_j\rangle = |\phi\rangle|\lambda\rangle \ . \tag{18}$$

Such states are also known as separable. Geometrically, the set of separable states are embedded in $\mathbf{CP}^{N^2-1}$ through the Segre embedding

$$\mathbf{CP}^N \times \mathbf{CP}^N \rightarrow \mathbf{CP}^{N^2-1} \ . \tag{19}$$

They form an orbit under the group of local unitaries, $SU(N)\times SU(N)$, which is a subgroup of $SU(N^2)$.

Once vectors get two indices, density matrices get four. The rank one density matrix describing a pure state is

$$\rho^{ij}{}_{kl} = \psi^\alpha\bar{\psi}_\beta = \frac{1}{N}\Gamma^{ij}\Gamma^*_{kl} \ . \tag{20}$$

A basic point is this: if we control operators of the form $A\otimes\mathbf{1}$ only, then the relevant state is given by the partial trace

$$\mathrm{Tr}_2\rho = \frac{1}{N}\sum_{j=1}^{N}\Gamma^{ij}\Gamma^*_{kj} = \frac{1}{N}(\Gamma\Gamma^\dagger)^i{}_k\ . \tag{21}$$

If the original state is separable, this is still a pure state on $\mathcal{H}^N$. On the other hand, if

$$\Gamma\Gamma^\dagger = \mathbf{1}\ , \tag{22}$$

then $\Gamma \in U(N)$, the reduced state is maximally mixed, and the original state is said to be maximally entangled—all the information in the original state is in (non-classical) correlations.

Projectively the phase is irrelevant, and the set of maximally entangled states in $\mathcal{H}^N \otimes \mathcal{H}^N$ is isomorphic to

$$SU(N)/Z_N \in \mathbf{CP}^{N^2-1}\ . \tag{23}$$

This is a group manifold, and as such it is again an orbit under $SU(N)\times SU(N)$. But precisely what kind of submanifold is it?

THE CURIOUS FACT

Observe that

$$\dim\left[SU(N)/Z_N\right] = N^2-1 = \frac{1}{2}\dim\left[\mathbf{CP}^{N^2-1}\right]\ . \tag{24}$$

This should make you suspicious. If we insert a state vector given by a unitary matrix in the N^2 dimensional version of eq. (16), we find [1] that

$$\Omega_{|SU(N)/Z_N} = 0\ . \tag{25}$$

The dimension of the set of maximally entangled state is one half of the dimension of the set of all pure states, and the symplectic form vanishes when restricted to this submanifold. A submanifold having these two properties is called a Lagrangian submanifold. To a physicist, Lagrangian submanifolds are known as configuration spaces. In a phase space spanned by qs and ps, the subspace spanned by the qs is Lagrangian.

Considered as a fact about differential geometry, our example is well known to mathematicians [8]. They also know that this particular example is a minimal submanifold—if you wiggle it, its volume grows. In a technical sense, its two properties make our submanifold into a special submanifold of a Kähler space.

We have arrived at our curious fact: *The set of maximally entangled pure states is a Lagrangian and minimal submanifold of the set of all pure states.* From a physical point of view, why? I do not know, but I imagine that there should be a reason, perhaps one that can be phrased in control theoretic terms. I did offer a free dinner to any participant who would tell me, but there were no takers. The offer still stands.

ACKNOWLEDGMENTS

I thank Johan Brännlund and Karol Życzkowski for collaboration, Robert Bryant and Vladimir Manko for supplying references, Ansar Fayyazuddin for discussions, and Andrei Khrennikov for being my host in Småland.

REFERENCES

1. I. Bengtsson, J. Brännlund and K. Życzkowski, $\mathbf{CP}^n$, or entanglement illustrated, Int. J. Mod. Phys. **A17** (2002) 4675.
2. S. Chaturvedi, these Proceedings.
3. F. Strocchi, Complex coordinates and quantum mechanics, Rev. Mod. Phys. **38** (1966) 36.
4. T. W. B. Kibble, Geometrization of quantum mechanics, Commun. Math. Phys. **65** (1979) 189.
5. G. W. Gibbons, Typical states and density matrices, J. Geom. Phys. **8** (1992) 147.
6. L. P. Hughston, Geometric aspects of quantum mechanics, in S. Huggett (ed.): *Twistor theory*, Marcel Dekker 1995.

7. A. Ashtekar and T. A. Schilling, Geometrical formulation of quantum mechanics, in A. Harvey (ed.): *On Einstein's path*, Springer 1998.
8. S. H. Wang, On the lifts of minimal Lagrangian submanifolds, arXiv:math.DG/0109214.

Unresolved Classical Electromagnetic Aspects of the Aharonov-Bohm Phase Shift

Timothy H. Boyer

Department of Physics, City College of the City University of New York, Convent Avenue at 138 Street, New York, NY 10031-9198 USA

Abstract. The long-standing controversy regarding the Aharonov-Bohm phase shift is reviewed. The shifts of both optical and particle interference patterns are summarized. It is pointed out that a line of electric dipoles and a line of magnetic dipoles (a long solenoid) both produce experimentally observed phase shifts similar to that produced by introducing a rectangular block of glass behind one slit of a double-slit interference pattern; the double-slit pattern is shifted while the single-slit envelope remains undisplaced. The quantum explanation for the magnetic interference pattern shift introduced by Aharonov and Bohm in 1959 involves completely different ideas from those suggested by a semiclassical analysis. Experiments planned by Caprez, Barwick, and Batelaan should clarify the connections between classical and quantum theories in connection with the Aharonov-Bohm phase shift.

Keywords: Aharonov-Bohm effect, classical electromagnetism
PACS: 03.50.De, 03.65.Sq

Introduction

The foundations of quantum theory are intimately connected with our understanding of the connections between classical and quantum physics. One of the departures between classical and quantum theories occurred in 1959 when Aharonov and Bohm proposed a new phase shift occurring when charged particles pass around opposite sides of a long solenoid [1]. The experimentally observed Aharonov-Bohm phase shift is said to occur in the absence of classical electromagnetic forces and is said to represent a quantum topological effect which has no classical analogue.

Although the Aharonov-Bohm phase shift is well-verified experimentally [2, 3] it has been suggested repeatedly during the past thirty-five years that the phase shift may arise not due to "quantum topology" but rather may be due to classical electromagnetic interactions between the passing charges and the solenoid [4–12]. It is this view which has been sharpened steadily and which very recently has come under experimental investigation [13]. Here we wish to discuss the shifts in particle interference patterns which can arise from electric and magnetic fields.

Interference Pattern Shifts

Particle inference patterns formed when monoenergetic particles pass through two slits to a distant screen are analogous to those obtained when monochromatic light passes through two slits in a wall to a distant screen. In both cases we find a single-slit envelope (associated with the interference of waves passing through a single slit), and then inside the single-slit envelope there is a double-slit interference pattern (associated with the interference of waves passing through two different slits). The interference patterns can be shifted in two distinct arrangements: i) deflection of the entire interference pattern, both the single-slit envelope and the double-slit pattern, or ii) shift of the double-slit pattern leaving the single-slit envelope undisplaced.

i)If a single large prism is place directly behind the two slits so that all light through the slits also passes through the prism, then the interference pattern as a whole is deflected, both the single-slit envelope and the double-slit pattern. An analogous deflection of the particle interference pattern arising from charged particles can be obtained by placing a uniform electric field parallel to both the wall with the slits and to the plane formed by the particle paths through the two slits, or else by placing a uniform magnetic field perpendicular to the plane formed by the the charged particle paths through the slits.

ii)If a rectangular piece of glass is placed behind only one of the slits, then the optical interference pattern will suffer a shift in the double-slit interference pattern leaving the single-slit envelope undisplaced. Indeed, if the piece of glass

CP962, *Quantum Theory, Reconsideration of Foundations – 4*,
edited by G. Adenier, A. Yu. Khrennikov, P. Lahti, V. I. Man'ko, and Th. M. Nieuwenhuizen

is made sufficiently thick or has a sufficiently large index of refraction so that the waves passing through the glass suffer a spatial delay larger than the coherence length of the light, then the double slit interference pattern will break down while the single slit envelope will remain unchanged. Just such a shift (of the double-slit interference pattern leaving the single-slit envelope undisplaced) can be obtained by placing a line of electric dipoles or a line of magnetic dipoles (a long solenoid) between the charged particle beams passing through the two slits, the line of dipoles oriented perpendicular to the plane formed by the two particle beams [12].

Shift Due to a Line of Electric Dipoles

In the electric case, the dipoles are oriented parallel to the wall containing the slits and parallel to the plane formed by the particle beams. The line of dipoles can be obtained by using two line charges λ separated by a small distance 2ε giving a dipole moment per unit length $2\varepsilon\lambda$ [5]. Positive charged particles which pass through one slit are closer to the positive line charge and so are first repelled while approaching and then attracted back on receding, while the charged particles which pass through the other slit are closer to the negative line charge and so are first attracted while approaching and then repelled while receding. Accordingly, the particles through the different slits have different velocities while passing the line of electric dipoles and so have a relative displacement when the particles have passed far beyond the line of dipoles. The electrostatic situation is analogous to the optical situation where (by use of a piece of glass behind one slit) a relative lag is introduced between the waves which pass through the two slits. The electric energy of interaction of a charged particle q with a line of electric dipoles is given by

$$U = qV(\mathbf{r}_q) \tag{1}$$

where $V(r_q)$ is the electrostatic potential of the line of electric dipoles evaluated at the position $\mathbf{r}_q$ of the passing charged particle. The relative lag ΔY between the charges passing on opposite sides of the line of electric dipoles can be calculated either from the electric fields of the line of electric dipoles acting on the trajectories of the charged particles, or by balancing the electrostatic interaction energies against the kinetic energy of the moving charges $mv_q\Delta v_q = -U$. The relative spatial lag is given by [5]

$$\Delta Y = \frac{4\pi q(2\varepsilon\lambda)}{mv_q^2} \tag{2}$$

where $(2\varepsilon\lambda)$ represents the electric dipole moment per unit length. Associated with this spatial lag is also a relative time lag

$$\Delta t = \frac{\Delta Y}{v_q} = \frac{4\pi q(2\varepsilon\lambda)}{mv_q^3} \tag{3}$$

The relative phase shift follows semiclassically as

$$\Delta\phi = \frac{mv_q\Delta Y}{\hbar} = \frac{4\pi q(2\varepsilon\lambda)}{\hbar v_q} \tag{4}$$

Exactly this same phase shift can be obtained by inserting the electrostatic potential energy expression $qV(\mathbf{r}_q)$ for a line of electric dipoles into the Schroedinger equation and solving by the WKB approximation assuming that the electrostatic interaction energy is small compared to the kinetic energy of the particles [10]. This phase shift was observed experimentally by Matteucci and Pozzi [14].

Shift Due to a Line of Magnetic Dipoles (a Solenoid)

In the magnetic case, a long solenoid is placed between the charged particle beams which pass through the slits so that the axis of the solenoid (and so too the magnetic dipoles) are perpendicular to the plane formed by the particle beams. The magnetic energy of interaction of a charged particle q passing a solenoid with constant currents is given by [15]

$$U = \frac{q}{c}\mathbf{v}_q \cdot \mathbf{A}(\mathbf{r}_q) \tag{5}$$

where $\mathbf{v}_q$ is the velocity of the passing charge and $\mathbf{A}(\mathbf{r}_q)$ is the vector potential in the Coulomb gauge of the solenoid evaluated at the position of the passing charge. We can calculate a possible relative lag between the charged particles which pass on opposite sides of the solenoid by using the net Lorentz forces which the passing charges place on the solenoid and then assuming Newton's third law for the forces on the particles, or else by assuming that energy conservation for the magnetic energies occurs by changes in the kinetic energies of the passing particles $mv_q\Delta v_q = -U$. The possilbe relative spatial lag is given by [5]

$$\Delta Y = \frac{q\Phi}{mv_q c} \tag{6}$$

for particles passing on opposite sides of the long solenoid of flux Φ and magnetic moment per unit length $\Phi/(4\pi)$. Associated with this spatial lag is also a relative time lag

$$\Delta t = \frac{\Delta Y}{v_q} = \frac{q\Phi}{mv_q^2 c} \tag{7}$$

The relative phase shift follows semiclassically as

$$\Delta\phi = \frac{mv_q\Delta Y}{\hbar} = \frac{q\Phi}{c\hbar} \tag{8}$$

Alternatively, one can obtain this same phase shift by introducing the vector potential $\mathbf{A}(\mathbf{r}_q)$ of the solenoid into the Schroedinger equation and solving using the WKB approximation assuming that the kinetic energy of the particles is large compared to the magnetic interaction energy $(q/c)\mathbf{v}_q \cdot \mathbf{A}(\mathbf{r}_q)$ [10]. This phase shift in Eq. (8) was predicted by Aharonov and Bohm [1] and first observed experimentally by Chambers [2].

Comparison of the Electric and Magnetic Phase Shifts

We have emphasized that the phase shifts due to a line of electric dipoles and to a line of magnetic dipoles (a solenoid) have both been observed experimentally, and both take the same form corresponding to a shift in the double-slit interference pattern while leaving the single-slit envelope undisplaced. Thus these shifts look just like the optical shift obtained by placing a rectangular piece of glass behind only one slit of the two-slit interference arrangement. Furthermore, semiclassical arguments involving energy conservation lead to exactly the observed phase shifts. This suggests that both phase shifts may involve classical lag effects arising from classical electromagnetic forces and velocity changes for the passing charged particles.

However, this is not at all the accepted explanation in the physics literature. The electrostatic situation (in those rare instances when it is acknowledge in the literature) is indeed accepted as involving classical electric forces and lags analogous to the optical case. However, it is the magnetic case (the Aharonov-Bohm phase shift) which receives virtually all the attention in the literature [16]. It receives this attention precisely because it is supposed to represent a departure from familiar connections with classical physics. The magnetic situation is said to involve a quantum topological effect having no classical analogue. The magnetic phase shift is said to occur in the absence of classical electromagnetic forces on the passing charged particles, and therefore is said to occur without any velocity changes or relative lags for the passing charges. The novelty of the accepted point of view is emphasized by Silverman who regards the Aharonov-Bohm phase shift as one of the new "mysteries" of quantum theory [17]. Indeed, the quantum topological point of view is so deeply entrenched that the lag point of view [4] was published only with difficulty in the Physical Review in 1973; in more recent years, the Physical Review has rejected all manuscripts giving further information about the magnetic phase shift from a classical perspective.

Experimental Implications

The classical and quantum analyses for the magnetic phase shift involve totally different perspectives and lead to different experimental implications. The velocity changes, longitudinal displacements, and time lags implied by the classical analysis are all different from the quantum topological point of view which involves no forces or velocity changes. Thus all these aspects become available for experimental discrimination between the two analyses. There is also a further distinction between the classical and quantum analyses. According to presently accepted quantum theory, the Aharonov-Bohm phase shift arises from an enclosed magnetic flux with no need to discuss any interaction between the passing charges and the sources of the magnetic flux. In contrast, the classical lag analysis depends crucially upon the details of the interaction between the passing charged particles and the current-carriers of the magnetic flux [11]. The quantum point of view leads to the expectation that the interaction of a passing charge and a solenoid should remain the same no matter what the size of the solenoid and no matter how the solenoid currents are produced. Indeed, this perspective encourages experimentalists to search for a classical lag effect for charged particles passing a macroscopic solenoid. Such an experiment was performed recently by Caprez, Barwick, and Batelaan, and no classical lag was seen [18]. However, from the classical perspective involving the detailed interaction of charges, large and small solenoids may not behave the same way; resistance in the wire of the solenoid can make a difference as to whether or not one should expect to observe an energy-conserving force back on the passing charges.

One may ask how two such totally different points of view can lead to the same experimental prediction for the Aharonov-Bohm phase shift. The relative success of the contrasting points of view may reflect a situation analogous to that when a charged particle passes a conductor. As long as the conductor is a good conductor, the forces on a passing charged particle due to image charges will involve only the physical shape of the conductor and not the detailed composition of the conductor. However, if the conductivity is imperfect or if the induced charges correspond to dielectric behavior, then more information regarding the interaction of the passing charge and the material will be required to account for the behavior of a passing charged particle. In a similar vein for the Aharonov-Bohm phase shift, it may well be that when the resistive energy loss of a solenoid is negligible, then all solenoids behave in the same way regarding energy conservation for a passing charged particle. However, when the resistive energy loss is large, then the energy-conserving interaction becomes small, and the interaction of the particle and solenoid becomes quite different. If the frictional forces (solenoid resistance) are large, then we do not expect to find the time lag because the accelerations of the solenoid charges will be small and hence the back forces at the passing charge will be small.

In the experiments of Moellenstedt and Bayh [3], where the Aharonov-Bohm phase shift is clearly present, the passage times of electrons past the solenoids is of the order of 10^{-13} sec (for a 40 keV electron passing 20 microns from the center of the solenoid). This time is not much longer than the collision time 10^{-14} sec in the Drude model for conductivity of a metal. Indeed, Jackson [19] gives γ as of the order of 10^{13} inverse seconds where $-m\gamma v$ is the resistive damping of a particle of mass m and speed v. Thus it is possible that the conservation of energy involving magnetic fields holds in the short-time regime where Moellenstedt and Bayh's experiments were performed, yet would not hold for the much longer passage times for the slower electrons passing the much larger solenoid in the experiment of Caprez, Barwick, and Batelaan.

Crucial Experimental Tests

The crucial test involving time delay corresponds (in a single experiment) to observing the Aharonov-Bohm phase shift and yet not observing the time delay of Eq. (7). Such an observation would rule out the classical lag interpretation of the Aharonov-Bohm phase shift. For the regimes where the Aharonov-Bohm phase shift has been observed, the time delays would be extraordinarily small. Thus for the experiments of Moellenstedt and Bayh, the time delay in Eq. (7) is of the order of 10^{-21} sec.

It is the shifts of particle interference patterns which allow the measurement of extraordinarily small quantities. Thus shift of the interference pattern by a large flux seems the best way to realistically test for the absence of a velocity change. According to the currently accepted quantum point of view, the Aharonov-Bohm phase shift is a purely topological shift, and there is no velocity change and hence no spatial lag for charged particles passing on opposite sides of a solenoid. Accordingly, one can increase the flux in a solenoid by an arbitrary amount and never break down the particle interference pattern. On the other hand, in the classical lag point of view, a sufficiently large solenoid flux will cause a relative lag which is larger than the coherence length associated with the charged particles, and at this point the observed particle interference pattern should break down, just as it does in the optical case. Batelaan's group is hoping to achieve appropriate conditions to observe this possible breakdown [13].

In the classical analysis, the observed Aharonov-Bohm phase shift due to a solenoid (a line of *magnetic* dipoles) is analogous to the observed Matteucci-Pozzi phase shift [14] due to a line of *electric* dipoles [10]. This is not at all the view of currently accepted quantum theory, which accepts the idea that the electrostatic phase shift is due to a classical lag but claims that the magnetic phase shift is due to a new quantum mechanical effect having no classical analogue. Thus ideally one would like to see both these phase shifts (electric and magnetic) tested at the same time with the same coherence length for the electron beam. From both classical and quantum analyses of the electrostatic situation, we believe that we know exactly what is going on involving electrostatic forces leading to a relative lag effect producing the Matteucci-Pozzi electrostatic phase shift. If the interference pattern breaks down for large *electric* dipole magnitude but not for large *magnetic* dipole magnitude, then (in contradiction to the classical analysis) the mechanisms for these two interference pattern shifts are quite different, as is indeed claimed by currently accepted quantum theory. However, if the interference pattern breaks down for both the electric and the magnetic phase shifts, the currently accepted quantum view is in error.

REFERENCES

1. Y. Aharonov and D. Bohm, "Significance of electromagnetic potentials in quantum theory," Phys. Rev. **115**, 485-491 (1959).
2. R. G. Chambers, "A shift of an electron interference pattern by enclosed magnetic flux," Phys. Rev. Lett. **5**, 3-5 (1960).
3. G. Moellenstedt and W. Bayh, "Messung der kontinuierlichen phasenschiebung von Elecktronenwellen in kraftfeldfreien Raum durch das magnetische Vektorpotential einer Luftspule," Naturwissenschaften **49**, 81-82 (1962).
4. T. H. Boyer, "Classical electromagnetic deflections and lag effects associated with quantum interference pattern shifts: considerations related to the Aharonov-Bohm effect," Phys. Rev. D **8**, 1679-1693 (1973).
5. T. H. Boyer, "The Aharonov-Bohm effect as a classical electromagnetic lag effect: an electrostatic analogue and possible experimental test," Il Nuovo Cimento **100**, 685-701 (1987).
6. T. H. Boyer, "Proposed Aharonov-Casher effect: Another example of an Aharonov-Bohm effect arising from a classical lag," Phys. Rev. A **36**, 5083-5086 (1987).
7. T. H. Boyer, "Does the Aharonov-Bohm effect exist?" Found. Phys. **30**, 893-905 (2000).
8. T. H. Boyer, "Classical electromagnetism and the Aharonov-Bohm phase shift," Found. Phys. **30**, 907-932 (2000).
9. T. H. Boyer, "Classical electromagnetic interaction of a point charge and a magnetic moment: Considerations related to the Aharonov-Bohm phase shift," Found. Phys. **32**, 1-36 (2002).
10. T. H. Boyer, "Semiclassical explanation of the Matteucci-Pozzi and Aharonov-Bohm phase shifts," Found. Phys. **32**, 41-49 (2002).
11. T. H. Boyer, "Darwin-Lagrangian analysis for the interaction of a point charge and a magnet: Considerations related to the controversy regarding the Aharonov-Bohm and Aharonov-Casher phase shifts," J. Phys. A:Math. Gen. **39**, 3455-3477 (2006)
12. T. H. Boyer, "Proposed experimental test for the paradoxical forces associated with the Aharonov-Bohm phase shift," Found. Phys. Lett. **19**, 491-498 (2006).
13. G. Gronniger, Z. Simmons, S. Gilbert, A. Caprez, and H. Batelaan, "The Aharonov-Bohm effect, phase or force," (preprint).
14. G. Matteucci and G. Pozzi, "New diffraction experiment on the electrostatic Aharonov-Bohm effect," Phys. Rev. Lett. **54**, 2469-2470 (1985).
15. T. H. Boyer, "Classical electromagnetic interaction of a charged particle with a constant-current solenoid," Phys. Rev. D **8**, 1667-1697 (1973).
16. See for example, J. J. Sakurai, *Modern Quantum Mechanics Revised Edition* (Addison-Wesley, Reading, MA 1994).
17. M. P. Silverman, *More Than One Mystery: Explorations in Quantum Interference* (Springer, New York 1995).
18. A. Caprez, B. Barwick, and H. Batelaan, "A macroscopic test of the Aharonov-Bohm effect," (preprint).
19. J. D. Jackson, *Classical Electrodynamics 3rd ed* (Wiley, New York 1999), p. 312.

Operational Axioms for C*-algebra Representation of Transformations

Giacomo Mauro D'Ariano

QUIT *Group, Dipartimento di Fisica "A. Volta", via Bassi 6, I-27100 Pavia, Italy*
http://www.qubit.it

Abstract. It is shown how a C*-algebra representation of the transformations of a physical system can be derived from two operational postulates: 1) the existence of *dynamically independent systems*; 2) the existence of *symmetric faithful states*. Both notions are crucial for the possibility of performing experiments on the system, in preventing remote instantaneous influences and in allowing calibration of apparatuses. The case of Quantum Mechanics is thoroughly analyzed. The possibility that other no-signaling theories admit a C*-algebra formulation is discussed.

Keywords: Foundations, Axiomatics, Measurement Theory
PACS: 03.65.-w

1. INTRODUCTION

In a set of recent papers [1, 2, 3] I showed how it is possible to derive the mathematical formulation of Quantum Mechanics in terms of complex Hilbert spaces and C*-algebras, starting from a small set of purely operational Postulates concerning experimental accessibility. In the present manuscript I will focus on C*-algebra, showing how a C*-algebra representation of the transformations of a physical system can be derived from two operational postulates only, concerning the existence of: 1) dynamically independent systems; 2) symmetric faithful joint states of two identical systems. Both postulates are crucial for the possibility of performing experiments, the former preventing uncontrollable remote instantaneous influences, the latter allowing calibration of experimental apparatuses.

The C*-algebra representation of the transformations is derived from the postulates via a Gelfand-Naimark-Segal (GNS) construction [4] based on the Jordan decomposition of the symmetric faithful state. The whole construction holds for finite dimensions, but it is valid also for infinite dimensions with the proviso that the Jordan decomposition exists on the Banach space of effects. The notion of *adjoint* of a transformation stems from that of faithful state, and generally depends on it, thus leading to different C*-algebra representations. On the other hand, the two postulates together imply that the linear space of "effects" of two identical independent systems is the tensor product structure of the spaces of the component systems.

A thorough analysis will show that for the case of Quantum Mechanics the adjoint is actually independent on the faithful state. However, as it will discussed in the conclusions, the C*-algebra representation of transformations is not sufficient to derive Quantum+Classical Mechanics, as for the customary operator algebras over Hilbert spaces, and in order to select this case additional postulates are needed. Possible candidates for such postulates, along with the possibility for other no-signaling theories to admit a C*-algebra representation of transformations, are discussed at the end of the paper.

2. THE POSTULATES

The general premise of the present axiomatization is the fact that one performs experiments to gather information on the *state* of an *object* physical system, and the knowledge of such state will then enable to predict the results of forthcoming experiments. Moreover, since we necessarily work with only partial *a priori* knowledge of both system and experimental apparatus, the rules for the experiment must be given in a probabilistic setting. Then, an *experiment* on an object system consists in making it interact with an apparatus, producing one of *a set of possible transformations* of the object, each one occurring with some probability. Information on the state of the object at the beginning of the experiment is gained from the knowledge of which transformation occurred, which is the "outcome" of the experiment

CP962, *Quantum Theory, Reconsideration of Foundations – 4*,
edited by G. Adenier, A. Yu. Khrennikov, P. Lahti, V. I. Man'ko, and Th. M. Nieuwenhuizen

signaled by the apparatus. For the above reasons we can logically *identify the experiment with a set of probabilistic transformations.*

We can now introduce the two postulates.

Postulate 1 (Independent systems) *There exist independent physical systems.*

Postulate 2 (Symmetric faithful state) *For every composite system made of two identical physical systems there exist a symmetric joint state that is both dynamically and preparationally faithful.*

3. THE STATISTICAL AND DYNAMICAL STRUCTURE

The starting point of the axiomatization is the identification **experiment** $\equiv$*set of transformations* that can occur on the object. The apparatus signals which transformation $\mathscr{A}_j$ of the set $\mathbb{A} := \{\mathscr{A}_j\}$ actually occurs. Now, since the knowledge of the state of a physical system allows us to predict the results of forthcoming experiments on the object, then it will allow us to evaluate the probability of any possible transformation in any conceivable experiment. Therefore, by definition, a **state** ω of a system is a rule providing probabilities of transformation, and $\omega(\mathscr{A})$ is the probability that the transformation $\mathscr{A}$ occurs. We clearly have the completeness $\sum_{\mathscr{A}_j \in \mathbb{A}} \omega(\mathscr{A}_j) = 1$, and assume $\omega(\mathscr{I}) = 1$ for the identical transformation $\mathscr{I}$, corresponding to adopting $\mathscr{I}$ as the free evolution (this is the *Dirac picture*, i. e. a suitable choice of the lab reference frame). In the following for a given physical system we will denote by $\mathfrak{S}$ the set of all possible states and by $\mathfrak{T}$ the set of all possible transformations.

When composing two transformations $\mathscr{A}$ and $\mathscr{B}$, the probability $p(\mathscr{B}|\mathscr{A})$ that $\mathscr{B}$ occurs conditional on the previous occurrence of $\mathscr{A}$ is given by the rule for conditional probabilities $p(\mathscr{B}|\mathscr{A}) = \omega(\mathscr{B} \circ \mathscr{A})/\omega(\mathscr{A})$. This sets a new probability rule corresponding to the notion of **conditional state** $\omega_{\mathscr{A}}$ which gives the probability that a transformation $\mathscr{B}$ occurs knowing that the transformation $\mathscr{A}$ has occurred on the object in the state ω, namely $\omega_{\mathscr{A}} \doteq \omega(\cdot \circ \mathscr{A})/\omega(\mathscr{A})$ [1] (in the following the central dot "$\cdot$" will always denote the pertinent variable). We can see that the notion of "state" itself logically implies the identification *evolution$\equiv$state-conditioning*, entailing a *linear action of transformations over states* (apart from normalization) $\mathscr{A}\omega := \omega(\cdot \circ \mathscr{A})$: this is the same concept of **operation** that we have in Quantum Mechanics, which gives the conditioning $\omega_{\mathscr{A}} = \mathscr{A}\omega/\mathscr{A}\omega(\mathscr{I})$. In other words, this is the analogous of the Schrödinger picture evolution of states in Quantum Mechanics (clearly such identification of evolution as state-conditioning also includes the deterministic case $\mathscr{U}\omega = \omega(\cdot \circ \mathscr{U})$ of transformations $\mathscr{U}$ with $\omega(\mathscr{U}) = 1 \forall \omega \in \mathfrak{S}$—the analogous quantum channels, including unitary evolutions.

From the state-conditioning rule it follows that we can define two complementary types of equivalences for transformations: *dynamical* and *informational*. The transformations $\mathscr{A}_1$ and $\mathscr{A}_2$ are **dynamically equivalent** when $\omega_{\mathscr{A}_1} = \omega_{\mathscr{A}_2}$ $\forall \omega \in \mathfrak{S}$, whereas they are **informationally equivalent** when $\omega(\mathscr{A}_1) = \omega(\mathscr{A}_2)$ $\forall \omega \in \mathfrak{S}$. The two transformations are then completely equivalent (write $\mathscr{A}_1 = \mathscr{A}_2$) when they are both dynamically and informationally equivalent, corresponding to the identity $\omega(\mathscr{B} \circ \mathscr{A}_1) = \omega(\mathscr{B} \circ \mathscr{A}_2)$, $\forall \omega \in \mathfrak{S}$, $\forall \mathscr{B} \in \mathfrak{T}$. We call **effect** the informational equivalence class of transformations[2]. In the following we will denote effects with the underlined symbols $\underline{\mathscr{A}}$, $\underline{\mathscr{B}}$, etc., or as $[\mathscr{A}]_{\text{eff}}$, and we will write $\mathscr{A}_0 \in \underline{\mathscr{A}}$ meaning that "the transformation $\mathscr{A}$ belongs to the equivalence class $\underline{\mathscr{A}}$", or "$\mathscr{A}_0$ has effect $\underline{\mathscr{A}}$", or "$\mathscr{A}_0$ is informationally equivalent to $\mathscr{A}$". Since, by definition one has $\omega(\mathscr{A}) \equiv \omega(\underline{\mathscr{A}})$, we will legitimately write $\omega(\underline{\mathscr{A}})$ instead of $\omega(\mathscr{A})$. Similarly, one has $\omega_{\mathscr{A}}(\mathscr{B}) \equiv \omega_{\mathscr{A}}(\underline{\mathscr{B}})$, which implies that $\omega(\mathscr{B} \circ \mathscr{A}) = \omega(\underline{\mathscr{B}} \circ \mathscr{A})$, leading to the chaining rule $\underline{\mathscr{B}} \circ \mathscr{A} \in \underline{\mathscr{B} \circ \mathscr{A}}$ corresponding to the "Heisenberg picture" evolution of transformations acting on effects (notice how transformations act on effects from the right). Now, by definitions effects are linear functionals over states with range $[0,1]$, and, by duality, we have a convex structure over effects, and we will

---

[1] M. Ozawa noticed that the definition of conditional state needs to assume that

$$\sum_{\mathscr{B}_j \in \mathbb{B}} \omega(\mathscr{B}_j \circ \mathscr{A}) = \omega(\mathscr{A}), \qquad \forall \mathbb{B}, \forall \mathscr{A}.$$

Such assumption which seems not implicit in the present axiomatization, would correspond to a kind of *"no-signaling from the future"*. It is presently under consideration if this must be considered as an additional postulate. Notice that such assumption seems to be needed whenever a notion of conditional state is considered which involves transformations of the system. In the present context the notion of conditional state is intimately related to that of "effect" and to the action of transformations over effects.

[2] This is the same notion of "effect" introduced by Ludwig [5]

denote their convex set as $\mathfrak{P}$. An **observable** is just a complete set of effects $\mathbb{L} = \{l_i\}$ of an experiment $\mathbb{A} = \{\mathscr{A}_j\}$, namely one has $l_i = \underline{\mathscr{A}_j}\ \forall j$ (clearly, one has the completeness relation $\sum_i l_i = 1$[3]). We will call the observable $\mathbb{L} = \{l_i\}$ **informationally complete** when each effect l can be written as a real linear combination $l = \sum_i c_i(l) l_i$ of elements of $\mathbb{L}$, and when these are linearly independent we will call the informationally complete observable *minimal.* [4]

The fact that we necessarily work in the presence of partial knowledge about both object and apparatus corresponds to the possibility of incomplete specification of both states and transformations, entailing: a) the convex structure on states; b) the addition rule for **coexistent transformations**, i. e. for transformations $\mathscr{A}_1$ and $\mathscr{A}_2$ for which $\omega(\mathscr{A}_1) + \omega(\mathscr{A}_2) \leqslant 1, \forall \omega \in \mathfrak{S}$ (i. e. transformations that can in principle occur in the same experiment). The addition of the two coexistent transformations is the transformation $\mathscr{S} = \mathscr{A}_1 + \mathscr{A}_2$ corresponding to the event $e = \{1, 2\}$ in which the apparatus signals that either $\mathscr{A}_1$ or $\mathscr{A}_2$ occurred, but does not specify which one. Such transformation is uniquely determined by the informational and dynamical classes as $\forall \omega \in \mathfrak{S}$: $\omega(\mathscr{A}_1 + \mathscr{A}_2) = \omega(\mathscr{A}_1) + \omega(\mathscr{A}_2)$, $(\mathscr{A}_1 + \mathscr{A}_2)\omega = \mathscr{A}_1\omega + \mathscr{A}_2\omega$. The composition "$\circ$" of transformations is distributive with respect to the addition "$+$". We can also define the multiplication $\lambda\mathscr{A}$ of a transformation $\mathscr{A}$ by a scalar $0 \leqslant \lambda \leqslant 1$ as the transformation dynamically equivalent to $\mathscr{A}$, but occurring with rescaled probability $\omega(\lambda\mathscr{A}) = \lambda\omega(\mathscr{A})$. Now, since for every couple of transformations $\mathscr{A}$ and $\mathscr{B}$ the transformations $\lambda\mathscr{A}$ and $(1-\lambda)\mathscr{B}$ are coexistent for $0 \leqslant \lambda \leqslant 1$, the set of transformations also becomes a convex set. Moreover, the transformations make a *monoid* (i. e. a semigroup with identity), since the composition $\mathscr{A} \circ \mathscr{B}$ of two transformations $\mathscr{A}$ and $\mathscr{B}$ is itself a transformation, and there exists the identical transformation $\mathscr{I}$ satisfying $\mathscr{I} \circ \mathscr{A} = \mathscr{A} \circ \mathscr{I} = \mathscr{A}$ for every transformation $\mathscr{A}$. Therefore, the set of physical transformations $\mathfrak{T}$ is a convex monoid.

It is obvious that we can extend the notions of coexistence, sum and multiplication by a scalar from transformations to effects via equivalence classes. In this way also effects make a convex set. As an additional step we can extend the convex monoid of physical transformations $\mathfrak{T}$ to a real algebra $\mathfrak{T}_{\mathbb{R}}$ by taking differences of physical transformations, and multiply them by scalars $\lambda > 1$. We will call the elements of $\mathfrak{T}_{\mathbb{R}}/\mathfrak{T}$ **generalized transformations**. Likewise, we can introduce **generalized effects**, and denote their linear space as $\mathfrak{P}_{\mathbb{R}}$. On generalized effects we can introduce the norm $\|\underline{\mathscr{A}}\| := \sup_{\omega\in\mathfrak{S}} |\omega(\underline{\mathscr{A}})|$, which allows us to introduce also a norm for transformations as $\|\mathscr{A}\| := \sup_{\mathfrak{P}_{\mathbb{R}} \ni \|\underline{\mathscr{B}}\| \leqslant 1} \|\underline{\mathscr{B}} \circ \mathscr{A}\| = \sup_{\mathfrak{P}_{\mathbb{R}} \ni \|\underline{\mathscr{B}}\| \leqslant 1} \sup_{\omega\in\mathfrak{S}} \omega(\underline{\mathscr{B}} \circ \mathscr{A})$. Closure in the respective norm topologies make $\mathfrak{P}_{\mathbb{R}}$ a real Banach space and $\mathfrak{T}_{\mathbb{R}}$ a real Banach algebra.[5]

A purely dynamical notion of **independent systems** coincides with the possibility of performing local experiments. More precisely, we say that two physical systems are *independent* if on the two systems 1 and 2 we can perform **local experiments** $\mathbb{A}^{(1)}$ and $\mathbb{A}^{(2)}$, i. e. whose transformations commute each other (i. e. $\mathscr{A}^{(1)} \circ \mathscr{B}^{(2)} = \mathscr{B}^{(2)} \circ \mathscr{A}^{(1)}$, $\forall \mathscr{A}^{(1)} \in \mathbb{A}^{(1)}$, $\forall \mathscr{B}^{(2)} \in \mathbb{B}^{(2)}$). Notice that the above definition of independent systems is purely dynamical, in the sense that it does not contain any statistical requirement, such as the existence of factorized states. The present notion of dynamical independence is so minimal that it can be satisfied not only by the quantum tensor product, but also by the quantum direct sum [6]. Nevertheless, in Sect. 5 a dimensionality analysis will show that, in conjunction with the existence of

---

[3] With a little notational abuse sometimes we identify $\underline{\mathscr{I}} \equiv 1$, i. e. the identity effect with the constant functional equal to 1.

[4] In previous literature the existence of informationally complete observable has been taken as a postulate. However, in the present context it is easy to show that it is always possible to construct a minimal informationally complete observable starting from a set of available experiments. The proof is by induction, and runs as follows. By definition there must exists a spanning set for $\mathfrak{P}_{\mathbb{R}} = \mathsf{Span}_{\mathbb{R}}(\mathfrak{P})$ that is contained in the convex hull $\mathfrak{P}$ of available effects. The maximal number of elements of this set that are linearly independent will constitute a *basis*, which we suppose has finite cardinality equal to $\dim(\mathfrak{P}_{\mathbb{R}})$. It remains to be shown that it is possible to have a basis with sum of elements equal to 1, and that such basis is obtained operationally starting from the available observables from which we constructed $\mathfrak{P}$.

If all observables are *uninformative* (i. e. with all constant effects $\propto \underline{\mathscr{I}}$), then $\mathfrak{P}_{\mathbb{R}} = \mathsf{Span}_{\mathbb{R}}(\underline{\mathscr{I}})$, $\underline{\mathscr{I}}$ is a minimal infocomplete observable, and the statement of the theorem is proved. Otherwise, there exists at least an observable $\mathbb{E} = \{l\}$ with $n \geqslant 2$ linearly independent effects. If this is the only observable, again the theorem is proved. Otherwise, take a new binary observable $\mathbb{E}_2 = \{x, y\}$ from the set of available ones (you can take different binary observables out of a given observable with more than two outcomes by summing up effects to yes-no observables). If $x \in \mathsf{Span}_{\mathbb{R}}(\mathbb{E})$ discard it. If $x \notin \mathsf{Span}_{\mathbb{R}}(\mathbb{E})$, then necessarily also $y \notin \mathsf{Span}_{\mathbb{R}}(\mathbb{E})$ [since if there exists coefficients λ_i such that $y = \sum_i \lambda_i l_i$, then $x = \sum_i (1-\lambda_i) l_i$]. Now, consider the observable

$$\mathbb{E}' = \left\{\tfrac{1}{2}y, \tfrac{1}{2}(l_1 + x), \tfrac{1}{2}l_2, \ldots, l_n\right\}$$

(which operationally corresponds to the random choice between the observables $\mathbb{E}$ and $\mathbb{E}_2$ with probability $\frac{1}{2}$, and with the events corresponding to x and l_1 made indistinguishable). This new observable has now $|\mathbb{E}'| = n+1$ linearly independent effects (since y is linearly independent on the l_i and one has $y = \sum_{i=1}^n l_i - x = \sum_{i=2}^n l_i + l_1 - x$). By iterating the above procedure we reach $|\mathbb{E}'| = \dim(\mathfrak{P}_{\mathbb{R}})$, and we have so realized an apparatus that measures a minimal informationally complete observable. ■

[5] An algebra of maps over a Banach space can always be made itself a Banach space, also satisfying the bound $\|\mathscr{B} \circ \mathscr{A}\| \leqslant \|\mathscr{B}\|\|\mathscr{A}\|$ defining a Banach algebra. This is true for both the real and the complex cases.

faithful states, dynamical independence agrees only with the quantum tensor product [6]. In Ref. [6] it is shown how the sole dynamical independence implies the impossibility of istantaneous signaling: the no-signaling condition is crucial for experimental control.

In the following, when dealing with more than one independent system, we will denote local transformations as ordered strings of transformations as $\mathscr{A},\mathscr{B},\mathscr{C},\ldots := \mathscr{A}^{(1)}\circ\mathscr{B}^{(2)}\circ\mathscr{C}^{(3)}\circ\ldots$. For effects one has the locality rule $([\mathscr{A}]_{\text{eff}},[\mathscr{B}_{\text{eff}}) \in [(\mathscr{A},\mathscr{B})]_{\text{eff}}$. The notion of independent systems now entails the notion of *local state*—the equivalent of partial trace in Quantum Mechanics. For two independent systems in a joint state Ω, we define the **local state** $\Omega|_1$ (and similarly $\Omega|_2$) as the probability rule $\Omega|_1(\mathscr{A}) \doteq \Omega(\mathscr{A},\mathscr{I})$ of the joint state Ω with a local transformation $\mathscr{A}$ acting only on system 1 and with all other systems untouched.

4. THE C*-ALGEBRA OF TRANSFORMATIONS

Now that we have a real algebra of generalized transformations and a linear space of generalized effects we want to introduce a positive bilinear form over them, by which we will be able to introduce a scalar product via the GNS construction [4]. The role of such bilinear form will be played by a *faithful state*.

We say that a state Φ of a bipartite system is **dynamically faithful** for system 1 when for every transformation $\mathscr{A}$ the map $\mathscr{A}\leftrightarrow(\mathscr{A},\mathscr{I})\Phi$ is one-to-one, namely $\forall\mathscr{A}\in\mathfrak{T}_{\mathbb{R}}$ $(\mathscr{A},\mathscr{I})\Phi=0 \Longleftrightarrow \mathscr{A}=0$. This means that for every bipartite effect $\underline{\mathscr{B}}$ one has $\Phi(\underline{\mathscr{B}}\circ(\mathscr{A},\mathscr{I}))=0 \Longleftrightarrow \mathscr{A}=0$. On the other hand, we will call a state Φ of a bipartite system **preparationally faithful** for system 1 if every joint bipartite state Ψ can be achieved by a suitable local transformation $\mathscr{T}_\Psi$ on system 1 occurring with nonzero probability, i. e. $\Psi=(\mathscr{T}_\Psi,\mathscr{I})\Phi$, with $\mathscr{T}_\Psi\in\mathfrak{T}^+$, $\mathfrak{T}^+$ denoting the positive cone generated by transformations. Clearly a bipartite state Φ that is preparationally faithful is also *locally* preparationally faithful, namely every local state ψ of system 2 can be achieved by a suitable local transformation $\mathscr{T}_\psi$ on system 1.

In Postulate 2 we also use the notion of **symmetric joint state**. This is simply defined as a joint state of two identical systems such that for any couple of effects $\underline{\mathscr{A}}$ and $\underline{\mathscr{B}}$ one has $\Phi(\underline{\mathscr{A}},\underline{\mathscr{B}})=\Phi(\underline{\mathscr{B}},\underline{\mathscr{A}})$. Clearly, for a symmetrical state the notions of dynamical and preparational faithfulness hold for both systems 1 and 2.

For a *faithful* bipartite state Φ, the **transposed transformation** $\tau_\Phi(\mathscr{A})$ of the transformation $\mathscr{A}$ is the generalized transformation which when applied to the second component system gives the same conditioned state and with the same probability as the transformation $\mathscr{A}$ operating on the first system, namely $(\mathscr{A},\mathscr{I})\Phi=(\mathscr{I},\tau_\Phi(\mathscr{A}))\Phi$ or, equivalently $\Phi(\underline{\mathscr{B}}\circ\mathscr{A},\underline{\mathscr{C}})=\Phi(\underline{\mathscr{B}},\underline{\mathscr{C}}\circ\tau_\Phi(\mathscr{A}))$ $\forall\underline{\mathscr{B}},\underline{\mathscr{C}}\in\mathfrak{P}$. Clearly the transposed is unique, due to injectivity of the map $\mathscr{A}\leftrightarrow(\mathscr{A},\mathscr{I})\Phi$, and it is easy to check the axioms of transposition ($\tau_\Phi(\mathscr{A}+\mathscr{B})=\tau_\Phi(\mathscr{A})+\tau_\Phi(\mathscr{B})$, $\tau_\Phi(\tau_\Phi(\mathscr{A}))=\mathscr{A}$, $\tau_\Phi(\mathscr{A}\circ\mathscr{B})=\tau_\Phi(\mathscr{B})\circ\tau_\Phi(\mathscr{A})$) and that $\tau_\Phi(\mathscr{I})=\mathscr{I}$.

The main ingredient of a GNS construction for representing transformations would be a positive form φ over transformations based on a notion of adjoint $\mathscr{A}\to\mathscr{A}^\dagger$ by which one can construct a scalar product as $\langle\mathscr{A}|\mathscr{B}\rangle := \varphi(\mathscr{A}^\dagger\circ\mathscr{B})$ in terms of which we have $\langle\mathscr{A}|\mathscr{C}\circ\mathscr{B}\rangle=\langle\mathscr{C}^\dagger\circ\mathscr{A}|\mathscr{B}\rangle\equiv\varphi(\mathscr{A}^\dagger\circ\mathscr{C}\circ\mathscr{B})=\varphi((\mathscr{C}^\dagger\circ\mathscr{A})^\dagger\circ\mathscr{B})$.[7] We can extract from Φ a positive bilinear form over $\mathfrak{P}_{\mathbb{R}}$ (notice that the bilinear form Φ is actually defined on effects) using its **Jordan decomposition** in terms of its absolute value $|\Phi|:=\Phi_+-\Phi_-$. Indeed, the absolute value can be defined thanks to the fact that Φ is real symmetric, whence it can be diagonalized over $\mathfrak{P}_{\mathbb{R}}$ in the finite dimensional case. Upon denoting by $\mathscr{P}_\pm$ the orthogonal projectors over the linear space corresponding to positive and negative eigenvalues, respectively, [8] one has $|\Phi|(\underline{\mathscr{A}},\underline{\mathscr{B}})=\Phi(\varsigma_\Phi(\underline{\mathscr{A}}),\underline{\mathscr{B}})$, where $\varsigma_\Phi(\underline{\mathscr{A}}):=(\mathscr{P}_+-\mathscr{P}_-)(\underline{\mathscr{A}})$. The map ς_Φ is an involution, namely $\varsigma_\Phi^2=\mathscr{I}$. The fact that the state is also preparationally faithful implies that the bilinear form is *strictly* positive [1] (namely $|\Phi|(\underline{\mathscr{A}},\underline{\mathscr{A}})=0$ implies that $\underline{\mathscr{A}}=0$). The involution ς_Φ over $\mathfrak{P}_{\mathbb{R}}$ corresponds to a generalized transformation $\mathscr{Z}_\Phi\in\mathfrak{T}_{\mathbb{R}}$ defined as $\underline{\mathscr{A}}\circ\mathscr{Z}_\Phi:=\varsigma_\Phi(\underline{\mathscr{A}})$, whence it can be extended to generalized transformations $\mathfrak{T}_{\mathbb{R}}$ via $\underline{\mathscr{B}}\circ\varsigma_\Phi(\mathscr{A})=\varsigma_\Phi(\varsigma_\Phi(\underline{\mathscr{B}})\circ\mathscr{A})$, corresponding to $\varsigma_\Phi(\mathscr{A})=\mathscr{Z}_\Phi\circ\mathscr{A}\circ\mathscr{Z}_\Phi$. Since $\mathscr{Z}_\Phi^2=\mathscr{I}$ the extension of ς_Φ to $\mathfrak{T}_{\mathbb{R}}$ is composition-preserving, i. e. $\varsigma_\Phi(\mathscr{B}\circ\mathscr{A})=\varsigma_\Phi(\mathscr{B})\circ\varsigma_\Phi(\mathscr{A})$.

---

[6] As shown in Refs. [1, 6] the tensor product can be derived from the additional Postulate stating the *local observability principle*.

[7] It is not easy to devise a positive form over generalized transformations $\mathfrak{T}_{\mathbb{R}}$ such that the transposition plays the role of the adjoint on a real Hilbert space. Indeed, if we take φ as the local state of a symmetric faithful state $\varphi=\Phi|_2\equiv\Phi|_1$ we have $\varphi(\tau_\Phi(\mathscr{A})\circ\mathscr{B})=\Phi(\tau_\Phi(\mathscr{A}),\tau_\Phi(\mathscr{B}))$, but the fact that Φ is positive over the convex set $\mathfrak{T}$ of physical transformations doesn't guarantee that its extension to generalized transformations $\mathfrak{T}_{\mathbb{R}}$ is still positive.

[8] The existence of the orthogonal space decomposition corresponding to positive and negative eigenvalues is guaranteed for finite dimensions. For infinite dimensions Φ is just a symmetric form over a real Banach space—the space $\mathfrak{P}_{\mathbb{R}}$ of generalized effects—and the existence of such decomposition needs to be proven.

The explicit form of $\mathscr{Z}_\Phi$ can be obtained in terms of the basis $\{f_j\}$ for $\mathfrak{P}_\mathbb{R}$ reducing the bilinear symmetric form Φ over $\mathfrak{P}_\mathbb{R}$ to the canonical form

$$\Phi(f_i, f_j) = s_i \delta_{ij}, \tag{1}$$

where $s_i = \pm 1$ is the signature of the eigenvector f_i. Then one has

$$\varsigma_\Phi(\underline{\mathscr{A}}) = \underline{\mathscr{A}} \circ \mathscr{Z}_\Phi = \sum_j \Phi(f_j, \underline{\mathscr{A}}) f_j. \tag{2}$$

One can see that $\tau_\Phi \varsigma_\Phi = \varsigma_\Phi \tau_\Phi$. In fact, due to the symmetry of Φ, $\tau_\Phi(\mathscr{Z}_\Phi) = \mathscr{Z}_\Phi$, since for any couple of elements f_k, f_l of the basis

$$\Phi(f_k \circ \tau_\Phi(\mathscr{Z}_\Phi), f_l) = \Phi(f_k, f_l \circ \mathscr{Z}_\Phi) = \Phi(f_l \circ \mathscr{Z}_\Phi, f_k) = \delta_{lk} = \Phi(f_k \circ \mathscr{Z}_\Phi, f_l). \tag{3}$$

whence

$$\begin{aligned} \tau_\Phi(\varsigma_\Phi(\mathscr{A})) &= \tau_\Phi(\mathscr{Z}_\Phi \circ \mathscr{A} \circ \mathscr{Z}_\Phi) = \tau_\Phi(\mathscr{Z}_\Phi) \circ \tau_\Phi(\mathscr{A}) \circ \tau_\Phi(\mathscr{Z}_\Phi) \\ &= \mathscr{Z}_\Phi \circ \tau_\Phi(\mathscr{A}) \circ \mathscr{Z}_\Phi = \varsigma_\Phi(\tau_\Phi(\mathscr{A})). \end{aligned} \tag{4}$$

We now define the **adjoint map** $\mathrm{ad}_\Phi := \varsigma_\Phi \tau_\Phi = \tau_\Phi \varsigma_\Phi$. Here in the following we will also temporarily use the more compact notation $\mathscr{A}^\dagger := \mathrm{ad}_\Phi(\mathscr{A})$, keeping in mind that the definition of the adjoint generally depends on the faithful state Φ with respect to which it is defined. Since ς_Φ is composition preserving whereas τ_Φ is a transposition, one has $(\mathscr{B} \circ \mathscr{A})^\dagger = \mathscr{A}^\dagger \circ \mathscr{B}^\dagger$. Moreover, for $\varphi = \Phi|_1$ we have that $\varphi(\mathscr{A}^\dagger \circ \mathscr{B}) = |\Phi|(\tau_\Phi(\mathscr{A}), \tau_\Phi(\mathscr{B}))$ is a positive bilinear form over transformations (strictly positive over effects, i. e. $|\Phi|(\underline{\mathscr{A}}, \underline{\mathscr{A}}) = 0 \Rightarrow \underline{\mathscr{A}} = 0$), and can be used to define a scalar product over transformations as follows

$$_\Phi\langle \mathscr{A} | \mathscr{B} \rangle_\Phi := \varphi(\mathscr{A}^\dagger \circ \mathscr{B}) = \Phi(\varsigma_\Phi \tau_\Phi(\mathscr{A}), \tau_\Phi(\mathscr{B})). \tag{5}$$

We can then verify that $\mathscr{A}^\dagger := \varsigma_\Phi \tau_\Phi(\mathscr{A})$ works as an adjoint for such scalar product, namely one has $_\Phi\langle \mathscr{C}^\dagger \circ \mathscr{A} | \mathscr{B} \rangle_\Phi = {_\Phi\langle} \mathscr{A} | \mathscr{C} \circ \mathscr{B} \rangle_\Phi$. In this way ς_Φ is identified as the **complex conjugation**, and as usual the adjoint $\mathscr{A}^\dagger := \varsigma_\Phi \tau_\Phi(\mathscr{A}) = \tau_\Phi \varsigma_\Phi(\mathscr{A})$ is the composition of the transposition with the complex conjugation. Now, by taking complex linear combinations of generalized transformations and defining $\varsigma_\Phi(c\mathscr{A}) = c^* \varsigma_\Phi(\mathscr{A})$ for $c \in \mathbb{C}$, we can extend the adjoint to complex linear combinations of generalized transformations, whose linear space will be denoted by $\mathcal{A} \equiv \mathfrak{T}_\mathbb{C}$, which is a complex algebra. On the other hand, we can trivially extend the real linear space of generalized effects $\mathfrak{P}_\mathbb{R}$ to a complex linear space $\mathfrak{P}_\mathbb{C}$ by taking complex linear combinations of generalized effects. The remaining setting up of the C*-algebra representation of $\mathcal{A}$ is just standard GNS construction, starting from the scalar product between transformations in Eq. (5). Symmetry and positivity imply the bounding [1] $_\Phi\langle \mathscr{A} | \mathscr{B} \rangle_\Phi \leqslant \|\mathscr{A}\|_\Phi \|\mathscr{B}\|_\Phi$, where we introduced the norm induced by the scalar product $\|\mathscr{A}\|_\Phi^2 \doteq {_\Phi\langle} \mathscr{A} | \mathscr{A} \rangle_\Phi$. By taking the equivalence classes $\mathcal{A}/\mathfrak{I}$ with respect to the zero-norm elements $\mathfrak{I} \subseteq \mathcal{A}$ we thus obtain a complex pre-Hilbert space equipped with a symmetric scalar product, and, since the scalar product is strictly positive over generalized effects, the elements of $\mathcal{A}/\mathfrak{I}$ are indeed the generalized effects, i. e. $\mathcal{A}/\mathfrak{I} \simeq \mathfrak{P}_\mathbb{C}$ as linear spaces. Being endowed with the scalar product (5) $\mathcal{A}/\mathfrak{I}$ becomes a pre-Hilbert space, whose completion $\mathsf{H}_\Phi := \overline{\mathcal{A}/\mathfrak{I}}$ under the norm induced by the scalar product is then a Hilbert space. In the following we will conveniently denote the equivalence class of transformations containing $\mathscr{A}$ in $\mathcal{A}/\mathfrak{I}$ by the Dirac vector itself $|\mathscr{A}\rangle_\Phi \in \mathsf{H}_\Phi$. From the bounding for the scalar product it follows that the set $\mathfrak{I} \subseteq \mathcal{A}$ of zero norm elements $\mathscr{X} \in \mathcal{A}$ is a left ideal (i. e. $\mathscr{X} \in \mathfrak{I}$, $\mathscr{A} \in \mathcal{A}$ implies $\mathscr{A} \circ \mathscr{X} \in \mathfrak{I}$), whence using our scalar product defined as in Eq. (5) we can represent elements of $\mathcal{A}$ (i. e. generalized complex transformations, since $\mathcal{A} \equiv \mathfrak{T}_\mathbb{C}$) as operators over the pre-Hilbert space of effects $\mathfrak{P}_\mathbb{C}$. The product in $\mathcal{A}$ defines the action of $\mathcal{A}$ on the vectors in $\mathcal{A}/\mathfrak{I}$, by associating to each element $\mathscr{A} \in \mathcal{A}$ the linear operator $\pi_\Phi(\mathscr{A})$ defined on the dense domain $\mathcal{A}/\mathfrak{I} \subseteq \mathsf{H}_\Phi$ as $\pi_\Phi(\mathscr{A})|\mathscr{B}\rangle_\Phi \doteq |\mathscr{A} \circ \mathscr{B}\rangle_\Phi$. The fact that $\mathcal{A}$ is a Banach algebra[9] also implies that the domain of definition of $\pi_\Phi(\mathscr{A})$ can be easily extended to the whole H_Φ by continuity. Being now an operator algebra over a complex Hilbert space, $\mathcal{A}$ becomes a C*-algebra. We just need to introduce the norm on transformations as the respective operator norm over H_Φ, namely $\|\mathscr{A}\|_\Phi := \sup_{v \in \mathsf{H}_\Phi, \|v\|_\Phi \leqslant 1} \|\mathscr{A} v\|_\Phi$, and completion of $\mathcal{A}$ under the norm topology will give a C*-algebra

[9] Indeed norms introduced in Sect. 3 can be extended to the respective complex linear spaces, and the norm completion makes $\mathcal{A}$ also a complex Banach algebra, as explained in the footnote 5.

(i. e. a complex Banach algebra satisfying the identity $\|\mathscr{A}^\dagger \circ \mathscr{A}\| = \|\mathscr{A}\|^2$), as it can be easily proved by standard techniques [1].

I want to emphasize that even though $\mathsf{H}_\Phi \simeq \mathfrak{P}_\mathbb{C}$ as linear spaces, the elements $|\mathscr{A}\rangle_\Phi \in \mathsf{H}_\Phi$ should be regarded as element of the dual space of $\mathfrak{P}_\mathbb{C}$, in the sense that the action of transformations over vectors $|\mathscr{A}\rangle_\Phi \in \mathsf{H}_\Phi$ is from the left—as in the Schrődinger picture—instead of being from the right—as in the Heisenberg picture, e. g. $\pi_\Phi(\mathscr{C})|\mathscr{A}\rangle_\Phi = |\mathscr{C} \circ \mathscr{A}\rangle_\Phi =$, or ${}_\Phi\langle\mathscr{A}|\pi_\Phi(\mathscr{C}) = {}_\Phi\langle\mathscr{C}^\dagger \circ \mathscr{A}|$, as it follows from the identity $\langle\mathscr{B}|\pi_\Phi(\mathscr{C})|\mathscr{A}\rangle_\Phi = \langle\mathscr{B}|\mathscr{C} \circ \mathscr{A}\rangle_\Phi = \langle\mathscr{C}^\dagger \circ \mathscr{B}|\underline{\mathscr{A}}\rangle_\Phi$. The Schrődinger picture is obtained thanks to the transposition in the definition of the scalar product ${}_\Phi\langle\mathscr{B}|\mathscr{A}\rangle_\Phi = |\Phi|(\tau_\Phi(\mathscr{B}), \tau_\Phi(\mathscr{A}))$.

From the definition of the scalar product, and using the fact that the state Φ is also preparationally faithful according to Postulate 2, the Born rule can be written in the GNS representation as $\omega(\underline{\mathscr{A}}) = {}_\Phi\langle\mathscr{A}^\dagger|\rho\rangle_\Phi$, with representation of state $\rho = \tau_\Phi(\underline{\mathscr{T}_\omega})/\Phi(\underline{\mathscr{T}_\omega}, \mathscr{I})$ [1], $\mathscr{T}_\omega$ denoting the transformation on system 2 corresponding to the local state ω on system 1, namely $\omega \propto \Phi(\cdot, \mathscr{T}_\omega)$. Then, the representation of transformations is

$$\omega(\underline{\mathscr{B}} \circ \mathscr{A}) = {}_\Phi\langle\mathscr{B}^\dagger|\mathscr{A}|\rho\rangle_\Phi := {}_\Phi\langle\mathscr{B}^\dagger|\mathscr{A} \circ \rho\rangle_\Phi. \tag{6}$$

4.1. Connecting two faithful states

Suppose that Ω is a symmetric state which is faithful both preparationally and dynamically, and that Φ is another such kind of state. Then, there must exists an invertible generalized transformation $\mathscr{F}$ in the positive cone $\mathfrak{T}^+$ generated by physical transformations, such that

$$\Phi = (\mathscr{F}, \mathscr{I})\Omega. \tag{7}$$

In fact, since Ω is preparationally faithful, there must exists a local physical transformation which transforms Ω into any state with some probability. On the other hand, since Ω is dynamically faithful, in order to have also Φ so, the correspondence between any other joint state and a local map applied to Φ must be one-to-one, which is true iff $\mathscr{F}$ is invertible. If the map $\mathscr{F}^{-1}$ is itself in the positive cone $\mathfrak{T}^+$ generated by physical transformations, then the state is also preparationally faithful, and viceversa. Indeed, any pure joint state Σ must be written as $\Sigma = (\mathscr{S}, \mathscr{I})\Omega$ with $\mathscr{S} \in \mathfrak{T}^+$. Therefore Σ can also be obtained probabilistically from Φ as $(\tilde{\mathscr{S}}, \mathscr{I})\Phi$ using a transformation $\tilde{\mathscr{S}} \propto \mathscr{F}\mathscr{S}\mathscr{F}^{-1} \in \mathfrak{T}^+$ belonging to the convex cone $\mathfrak{T}^+$ generated by physical transformations. Finally, as regards symmetry, the state Φ is symmetric iff $\tau_\Omega(\mathscr{F}) = \mathscr{F}$, since

$$\begin{aligned}\Phi(\mathscr{A}, \mathscr{B}) =& \Omega(\mathscr{A} \circ \mathscr{F}, \mathscr{B}) = \Omega(\mathscr{B}, \mathscr{A} \circ \mathscr{F}) = \Omega(\mathscr{B} \circ \tau_\Omega(\mathscr{F}), \mathscr{A}),\\ \Phi(\mathscr{A}, \mathscr{B}) =& \Phi(\mathscr{B}, \mathscr{A}) = \Omega(\mathscr{B} \circ \mathscr{F}, \mathscr{A}), \quad \forall \mathscr{B}, \mathscr{A} \in \mathfrak{T}\end{aligned} \tag{8}$$

and using preparational faithfulness of Ω one can see that the above identity holds true iff $\tau_\Omega(\mathscr{F}) = \mathscr{F}$ (we remind that two transformations $\mathscr{A}_1$ and $\mathscr{A}_2$ are equal iff $\omega(\mathscr{B} \circ \mathscr{A}_1) = \omega(\mathscr{B} \circ \mathscr{A}_2)$ $\forall \omega \in \mathfrak{S}$ and $\forall \mathscr{B} \in \mathfrak{T}$). Notice now that $\tau_\Omega(\mathscr{F}^{-1}) = \mathscr{F}^{-1}$, since $\mathscr{I} = \tau_\Omega(\mathscr{F}^{-1}) \circ \tau_\Omega(\mathscr{F}) = \tau_\Omega(\mathscr{F}^{-1}) \circ \mathscr{F}$.

The transposed with respect to Φ is obtained as follows

$$(\mathscr{A}, \mathscr{I})(\mathscr{F}, \mathscr{I})\Omega = (\mathscr{I}, \tau_\Phi(\mathscr{A}))(\mathscr{F}, \mathscr{I})\Omega = (\mathscr{I}, \tau_\Phi(\mathscr{A}) \circ \tau_\Omega(\mathscr{F}))\Omega \tag{9}$$

namely $\tau_\Phi(\mathscr{A}) \circ \tau_\Omega(\mathscr{F}) = \tau_\Omega(\mathscr{F}) \circ \tau_\Omega(\mathscr{A})$, which means that

$$\tau_\Phi(\mathscr{A}) = \tau_\Omega(\mathscr{F}^{-1} \circ \mathscr{A} \circ \mathscr{F}) = \mathscr{F} \circ \tau_\Omega(\mathscr{A}) \circ \mathscr{F}^{-1} \tag{10}$$

The canonical basis of eigenvectors $\{f_j\}$ of the bilinear form Φ must satisfy the identities

$$s_j\delta_{ij} = \Phi(f_i, f_j), \quad \delta_{ij} = \Phi(\varsigma_\Phi(f_i), f_j) = |\Phi|(f_i, f_j), \tag{11}$$

and upon multiplying by f_j and summing over the index j one obtains $f_j \circ \mathscr{Z}_\Phi = \sum_j \Phi(f_i, f_j)f_j$, and since $\{f_i\}$ is a basis for $\mathfrak{P}_\mathbb{R}$, one as the identity

$$\underline{\mathscr{A}} \circ \mathscr{Z}_\Phi = \sum_j \Phi(\underline{\mathscr{A}}, f_j)f_j, \quad \forall \underline{\mathscr{A}} \in \mathfrak{P}_\mathbb{R}. \tag{12}$$

For any couple of elements of the complete basis $\{f_j\}$ for $\mathfrak{P}_\mathbb{R}$ one has

$$\delta_{ij} = |\Phi|(f_i, f_j) = \Phi(f_i \circ \mathscr{Z}_\Phi, f_j) = \Omega(f_i \circ \mathscr{Z}_\Phi \circ \mathscr{F}, f_j), \tag{13}$$

and since $\{f_i\}$ is a basis for $\mathfrak{P}_{\mathbb{R}}$, this corresponds to the identity

$$\mathscr{Z}_\Phi \circ \mathscr{F} \circ \mathscr{Z}_{\Omega,\mathbf{f}} = \mathscr{I}, \tag{14}$$

where

$$\underline{\mathscr{A}} \circ \mathscr{Z}_{\Omega,\mathbf{f}} := \sum_j \Omega(\underline{\mathscr{A}}, f_j) f_j, \quad \forall \underline{\mathscr{A}} \in \mathfrak{P}_{\mathbb{R}}. \tag{15}$$

The definition of $\mathscr{Z}_{\Omega,\mathbf{f}}$ generalizes that of $\mathscr{Z}_\Phi$ in specifying the basis $\mathbf{f} := \{f_j\}$ which is generally non canonical for Ω. For $\mathbf{o} := \{o_j\}$ canonical for Ω one has simply $\mathscr{Z}_\Omega \equiv \mathscr{Z}_{\Omega,\mathbf{o}}$. Upon multiplying by f_j and summing over the index j in Eq. (44) we obtain

$$s_i f_i = \sum_j \Phi(f_i, f_j) f_j = \sum_j \Omega(f_i \circ \mathscr{F}, f_j) f_j = f_i \circ \mathscr{F} \circ \mathscr{Z}_{\Omega,\mathbf{f}}. \tag{16}$$

This corresponds to

$$\mathscr{Z}_\Phi = \mathscr{F} \circ \mathscr{Z}_{\Omega,\mathbf{f}}, \tag{17}$$

which, in conjunction with Eq. (14), is a restatement of the involutive nature of $\mathscr{Z}_\Phi$, i. e. $\mathscr{Z}_\Phi \circ \mathscr{Z}_\Phi = \mathscr{I}$, corresponding also to the identities

$$\mathscr{Z}_{\Omega,\mathbf{f}} \circ \mathscr{F} \circ \mathscr{Z}_{\Omega,\mathbf{f}} = \mathscr{F}^{-1}, \quad \mathscr{F} \circ \mathscr{Z}_{\Omega,\mathbf{f}} \circ \mathscr{F} = \mathscr{Z}_{\Omega,\mathbf{f}}^{-1}. \tag{18}$$

Therefore, one also has

$$\mathscr{Z}_\Phi = \mathscr{F} \circ \mathscr{Z}_{\Omega,\mathbf{f}} = \mathscr{Z}_{\Omega,\mathbf{f}}^{-1} \circ \mathscr{F}^{-1}. \tag{19}$$

The complex conjugation obeys the symmetry $\tau_\Phi(\mathscr{Z}_\Phi) = \mathscr{Z}_\Phi$ which is needed for a proper definition of the adjoint. Indeed

$$\tau_\Phi(\mathscr{Z}_\Phi) = \mathscr{F} \circ \tau_\Omega(\mathscr{Z}_\Phi) \circ \mathscr{F}^{-1} = \mathscr{F} \circ \mathscr{Z}_{\Omega,\mathbf{f}} \circ \mathscr{F} \circ \mathscr{F}^{-1} = \mathscr{F} \circ \mathscr{Z}_{\Omega,\mathbf{f}} = \mathscr{Z}_\Phi. \tag{20}$$

One has $\tau_\Omega(\mathscr{Z}_{\Omega,\mathbf{f}}) = \mathscr{Z}_{\Omega,\mathbf{f}}$, since

$$\begin{aligned} \Omega(f_i \circ \mathscr{Z}_{\Omega,\mathbf{f}}, f_j) &= \sum_k \Omega(f_i, f_k)\Omega(f_k, f_j) = \sum_k \Omega(f_j, f_k)\Omega(f_k, f_i) \\ &= \Omega(f_j \circ \mathscr{Z}_{\Omega,\mathbf{f}}, f_i) = \Omega(f_i, f_j \circ \mathscr{Z}_{\Omega,\mathbf{f}}) \end{aligned} \tag{21}$$

We now evaluate the adjoint

$$\mathrm{ad}_\Phi(\mathscr{A}) := \varsigma_\Phi \tau_\Phi(\mathscr{A}) = \mathscr{Z}_\Phi \circ \mathscr{F} \circ \tau_\Omega(\mathscr{A}) \circ \mathscr{F}^{-1} \circ \mathscr{Z}_\Phi = \mathscr{Z}_{\Omega,\mathbf{f}}^{-1} \circ \tau_\Omega(\mathscr{A}) \circ \mathscr{Z}_{\Omega,\mathbf{f}}, \tag{22}$$

and one has

$$\mathrm{ad}_\Phi \equiv \mathrm{ad}_\Omega := (\cdot)^\dagger \quad \Leftrightarrow \quad \mathscr{Z}_{\Omega,\mathbf{f}} \equiv \mathscr{Z}_\Omega, \tag{23}$$

namely if $\mathscr{Z}_\Omega \equiv \mathscr{F}^{-1} \circ \mathscr{Z}_\Phi$. In such case we will also have

$$\mathscr{F}^{-1} = \varsigma_\Omega(\mathscr{F}) = \mathscr{F}^\dagger. \tag{24}$$

5. DYNAMICAL INDEPENDENCE AND TENSOR PRODUCT

As already mentioned, our notion of dynamical independence—i. e. the possibility of performing local experiments—can be satisfied not only by the quantum tensor product, but also by the quantum direct sum. This is shown in detail in Ref. [6]. Here I will show how Postulate 2—the existence of dynamically and preparationally faithful states—in conjunction with dynamical independence, leads to the right dimension for the convex set of states of two independent identical systems according to the tensor product rule.

The state-effect duality leads to the identity $\dim(\mathfrak{P}) = \dim(\mathfrak{S}) + 1$,[10] (we remind that one dimension is blocked by state normalization). Then, the existence of a preparationally and dynamically faithful state guarantees that generalized transformations and generalized joint states are isomorphic as linear spaces, whence $\dim(\mathfrak{T}) = \dim(\mathfrak{S}^{\times 2}) + 1$, $\mathfrak{S}^{\times 2}$ denoting the set of bipartite states of two identical systems, each with set of states $\mathfrak{S}$. Finally, the GNS construction represents generalized transformations as operators over the Hilbert space of generalized effects, whence $\dim(\mathfrak{T}) = \dim(\mathfrak{P})^2$, from which it follows that $\dim(\mathfrak{S}^{\times 2}) + 1 = (\dim(\mathfrak{S}) + 1)^2$. Therefore one has $\dim(\mathfrak{P}^{\times 2}) = (\dim \mathfrak{P})^2$, and $\dim_{\mathbb{C}}(\mathfrak{P}_{\mathbb{C}}^{\times 2}) = (\dim \mathfrak{P}_{\mathbb{C}})^2$ (since $\dim_{\mathbb{C}} \mathfrak{P}_{\mathbb{C}} = \frac{1}{2} \dim \mathfrak{P}_{\mathbb{C}} = \dim \mathfrak{P}_{\mathbb{R}}$), whence $\mathfrak{P}_{\mathbb{R}}^{\times 2} \equiv \mathfrak{P}_{\mathbb{R}}^{\otimes 2}$ and $\mathfrak{P}_{\mathbb{C}}^{\times 2} \equiv \mathfrak{P}_{\mathbb{C}}^{\otimes 2}$. The last identities hold in Quantum Mechanics, as a consequence of the tensor product of complex Hilbert spaces.

---

[10] For convex sets $\mathfrak{C}$, one has $\dim(\mathfrak{C}) := \dim \mathsf{Span}(\mathfrak{C})$, where $\dim \equiv \dim_{\mathbb{R}}$ (if not otherwise stated, the convex sets are always considered real).

6. THE QUANTUM C*-ALGEBRA OF TRANSFORMATIONS

In the following, for given fixed orthonormal basis $\{|j\rangle\}$ for H we will denote by $A^* = \sum_{ij} A^*_{ij}|i\rangle\langle j|$ the operator corresponding to the complex conjugated matrix of $A = \sum_{ij} A_{ij}|i\rangle\langle j|$, and consistently $A^t = (A^*)^\dagger$ will denote the transposed-matrix operator. With the double ket we denote bipartite vectors $|\Psi\rangle\rangle \in \mathcal{H}\otimes\mathcal{H}$, which, keeping the basis $\{|j\rangle\}$ as fixed, are in one-to-one correspondence with matrices as $|\Psi\rangle\rangle = \sum_{ij}\Psi_{ij}|i\rangle\otimes|j\rangle$. We will denote the generalized transformation and the corresponding quantum linear map by the same letter, and we will do so also for state and its corresponding quantum density operator. Moreover, we will write composition of quantum maps as $\mathscr{B}\mathscr{A}$ as usual, instead of using the operational notation $\mathscr{B}\circ\mathscr{A}$. In Quantum Mechanics physical transformations correspond to quantum operations (i. e. trace non-increasing completely positive (CP) maps), effects correspond to positive contractions, generalized transformations $\mathfrak{T}_{\mathbb{R}}$ to differences of CP maps, and generalized effects $\mathfrak{P}_{\mathbb{R}}$ to selfadjoint operators. In the following we will denote by $P_{\mathscr{A}}$ the positive operator describing the effect of the quantum operation $\mathscr{A}$. For example, we will write

$$\rho(\mathscr{A}) = \mathrm{Tr}[\mathscr{A}(\rho)] = \mathrm{Tr}\left[\sum_n A_n\rho A_n^\dagger\right] = \mathrm{Tr}[\rho P_{\mathscr{A}}], \qquad P_{\mathscr{A}} := \sum_n A_n^\dagger A_n. \tag{25}$$

We will also use the notation $\mathscr{A}^\dagger = \sum_n A_n^\dagger \cdot A_n$ for the usual adjoint map of $\mathscr{A} = \sum_n A_n\cdot A_n^\dagger$, and $\mathscr{A}^t = \sum_n A_n{}^t\cdot A_n^*$ for the transposed map.

I will now construct explicitly the C*-algebra $\mathfrak{T}_{\mathbb{C}}$ of c-generalized transformations for a general faithful symmetric quantum state Φ. I first consider the case of the canonical maximally entangled state Ω, and then analyze the general case of faithful symmetric state.

6.1. The maximally entangled state of a qudit

The canonical maximally entangled state of a qudit

$$\Omega = d^{-1}|I\rangle\rangle\langle\langle I|, \tag{26}$$

is faithful, both dynamically and preparationally. The fact that it is dynamically faithful is just the Choi-Jamiolowski representation of CP maps. On the other hand, any pure joint state $d^{-\frac{1}{2}}|S\rangle\rangle$ can be written as $(S\otimes I)d^{-\frac{1}{2}}|I\rangle\rangle$ with $d^{-1}\mathrm{Tr}[S^\dagger S]=1$, $\mathscr{S}\propto S\cdot S^\dagger$ quantum operation (i. e. $\mathscr{S}\in\mathfrak{T}^+$), whence Ω is preparationally faithful. The state Ω is also symmetric, since for any couple of generalized effects one has

$$\Omega(\mathscr{A},\mathscr{B}) = \mathrm{Tr}[\mathscr{A}\otimes\mathscr{B}(\Omega)] = \tfrac{1}{d}\mathrm{Tr}[P_{\mathscr{A}}P^t_{\mathscr{B}}] = \tfrac{1}{d}\mathrm{Tr}[P_{\mathscr{B}}P^t_{\mathscr{A}}] = \Omega(\mathscr{B},\mathscr{A}). \tag{27}$$

The transposition τ_Ω is just the customary transposition $\tau_\Omega \equiv (\cdot)^t$ with respect to any fixed basis $\{|i\rangle\}$ such that Ω has all probability amplitudes equal to $d^{-\frac{1}{2}}$. Indeed, it is easy to check that

$$(\mathscr{A}\otimes\mathscr{I})(|I\rangle\rangle\langle\langle I|) = (\mathscr{I}\otimes\mathscr{A}^t)(|I\rangle\rangle\langle\langle I|). \tag{28}$$

In order to construct an eigenbasis for the Jordan form, consider the following selfadjoint operators

$$X_{kl} = \tfrac{1}{\sqrt{2}}(|k\rangle\langle l| + |l\rangle\langle k|), \quad Y_{kl} = \tfrac{i}{\sqrt{2}}(|k\rangle\langle l| - |l\rangle\langle k|), \quad k<l, \qquad Z_l = |l\rangle\langle l|. \tag{29}$$

One has

$$\mathrm{Tr}[X_{kl}X_{k'l'}] = \delta_{kl'}\delta_{lk'} + \delta_{kk'}\delta_{ll'} \equiv \delta_{kk'}\delta_{ll'}, \tag{30}$$

since for $k=l'>k'$ one has $k'=l>k$. Similarly we have $\mathrm{Tr}[Y_{kl}Y_{k'l'}] = \delta_{kk'}\delta_{ll'}$, and $\mathrm{Tr}[Z_kZ_{k'}] = \delta_{kk'}$, and, moreover

$$\mathrm{Tr}[X_{kl}Y_{k'l'}] = \mathrm{Tr}[Z_lY_{k'l'}] = \mathrm{Tr}[Z_lX_{k'l'}] = 0. \tag{31}$$

Therefore, the following is a canonical basis for the Jordan form of Φ

$$\begin{aligned}[C_j] = \big[&Z_0, Z_1, \ldots, Z_{d-1}, X_{01}, X_{02}, \ldots, X_{0,d-1},\\ &X_{12}, X_{13}, \ldots, X_{1,d-1}, \ldots X_{d-2,d-1}, Y_{01}, \ldots, Y_{d,d-1}\big],\end{aligned} \tag{32}$$

with Jordan form

$$\Omega(C_i,C_j)=\mathrm{Tr}[C_iC_j^*]=\delta_{ij}s_j, \tag{33}$$

Y_{kl} $(0\leqslant k<l\leqslant d-1)$ spanning the eigenspace with negative eigenvalue of the symmetric form Ω. It follows that the transformation ς_Ω corresponds to the complex conjugation $\varsigma_\Omega\equiv(\cdot)^*$ with respect to the same fixed orthonormal basis $\{|i\rangle\}$ used for transposition. We can construct the Kraus form for the corresponding generalized transformation $\mathscr{Z}_\Omega$, passing through the construction of the corresponding Choi-Jamiolowski operator

$$\mathscr{Z}_\Omega\otimes\mathscr{I}(|I\rangle\!\rangle\langle\!\langle I|)=\mathscr{Z}_\Omega\otimes\mathscr{I}\left(\sum_j C_j^*\otimes C_j\right)=\sum_j C_j\otimes C_j=\sum_j C_j^\dagger\otimes C_j=E, \tag{34}$$

which is just the unitary swap operator E, with eigenvectors

$$E|C_j\rangle\!\rangle=|C_j^*\rangle\!\rangle=s_j|C_j\rangle\!\rangle, \tag{35}$$

corresponding to the Kraus form for the generalized transformation $\mathscr{Z}$

$$\mathscr{Z}=\sum_j s_jC_j\cdot C_j. \tag{36}$$

The GNS representation of transformations over effects is provided by the following scalar product

$${}_\Omega\langle\mathscr{A}|\mathscr{B}\rangle_\Omega:=\tfrac{1}{d}\langle\!\langle I|\check{A}^\dagger\check{B}|I\rangle\!\rangle, \tag{37}$$

where corresponding to the map $\mathscr{A}=\sum_n A_n\cdot A_n^\dagger$ we define the operator $\check{A}:=\sum_n A_n\otimes A_n^*$ such that $\check{A}|X\rangle\!\rangle=|\mathscr{A}(X)\rangle\!\rangle$. Indeed, we can check the identities

$${}_\Omega\langle\mathscr{A}|\mathscr{B}\rangle_\Omega:=\Omega(\mathscr{A}^\dagger,\mathscr{B}')=\tfrac{1}{d}\mathrm{Tr}\left[\sum_m A_mA_m^\dagger\sum_n B_nB_n^\dagger\right]=\tfrac{1}{d}\mathrm{Tr}[P_{\mathscr{A}^\dagger}P_{\mathscr{B}^\dagger}], \tag{38}$$

$${}_\Omega\langle\mathscr{A}|\mathscr{C}\circ\mathscr{B}\rangle_\Omega:=\Phi(\mathscr{A}^\dagger,\mathscr{B}'\circ\mathscr{C}')=\tfrac{1}{d}\langle\!\langle P_{\mathscr{A}^\dagger}|\check{C}|P_{\mathscr{B}^\dagger}\rangle\!\rangle=\tfrac{1}{d}\mathrm{Tr}[P_{\mathscr{A}^\dagger}\mathscr{C}(P_{\mathscr{B}^\dagger})] \tag{39}$$

Explicitely, the GNS representation of transformation over effects is

$$|\mathscr{A}\rangle=\check{A}|I\rangle\!\rangle=|\mathscr{A}(I)\rangle\!\rangle=|P_{\mathscr{A}^\dagger}\rangle\!\rangle,\quad \mathscr{B}|\mathscr{A}\rangle=\check{B}|\mathscr{A}(I)\rangle\!\rangle=|\mathscr{B}\mathscr{A}(I)\rangle\!\rangle=|\mathscr{B}(P_{\mathscr{A}^\dagger})\rangle\!\rangle. \tag{40}$$

For qubits the canonical Jordan basis, will be given by the set of four Pauli matrices $\sigma_0\equiv I,\sigma_x,\sigma_y,\sigma_z$ normalized as $C_j=\frac{1}{\sqrt{2}}\sigma_j$, corresponding to the Jordan form

$$\Omega(C_i,C_j)\doteq\tfrac{1}{2}\mathrm{Tr}[\sigma_i\sigma_j^*]=\begin{bmatrix}1&0&0&0\\0&1&0&0\\0&0&-1&0\\0&0&0&1\end{bmatrix}:=\delta_{ij}s_j. \tag{41}$$

Here σ_y spans the eigenspace with negative eigenvalue of Ω.

6.2. General faithful state

According to subsection 4.1 the general form of a joint faithful state of two identical quantum systems with finite dimensional Hilbert space H can be always recast in the following way

$$\Phi=\tfrac{1}{d}\sum_l|F_l\rangle\!\rangle\langle\!\langle F_l|=(\mathscr{F}\otimes\mathscr{I})\Omega \tag{42}$$

with Ω given in Eq. (26), and $\mathscr{F}=\sum_l F_l\cdot F_l^\dagger$ invertible CP map (not necessarily trace-non-increasing), and with $\mathscr{F}^{-1}$ also CP (normalization corresponds to $\mathrm{Tr}[\sum_l F_l^\dagger F_l]=d$, $d=\dim(\mathsf{H})$). Moreover, the state Φ is symmetric iff

$(\mathscr{F}^{-1})^t = \mathscr{F}^{-1}$, corresponding to the operator identity $E\Phi E = \Phi$, E denoting the swap operator. According to Eq. (10) the transposed with respect to Φ is given by

$$\tau_\Phi(\mathscr{A}) = \mathscr{F}\mathscr{A}^t\mathscr{F}^{-1} \tag{43}$$

The canonical basis of eigenvectors $\{C_j\}$ of the bilinear form Φ must satisfy the identity

$$s_j\delta_{ij} = \Phi(C_i,C_j) = \mathrm{Tr}[(C_i \otimes C_j)\sum_l |F_l\rangle\!\rangle\langle\!\langle F_l|] = \mathrm{Tr}[C_j{}^t\mathscr{F}^\dagger(C_i)]. \tag{44}$$

Upon multiplying by C_j and summing over the index j in Eq. (44) we obtain [11]

$$s_iC_i = \sum_j \mathrm{Tr}[C_j{}^t\mathscr{F}^\dagger(C_i)]C_j =: \mathscr{C}\mathscr{F}^\dagger(C_i) \tag{45}$$

where $\mathscr{C} := \sum_j \mathrm{Tr}[C_j{}^t\cdot]C_j$. Identity (44) is equally satisfied by the set $\{C_i^\dagger\}$ with the same eigenvalue. Therefore, it is always possible to choose the operators C_j as selfadjoint, and $\mathscr{C}^\dagger \equiv \mathscr{C}$. It is also easy to check that $\mathscr{C}^t = \mathscr{C}$, since

$$\begin{aligned}\mathscr{C}\otimes\mathscr{I}(|I\rangle\!\rangle\langle\!\langle I|) &= \sum_j C_j \otimes \mathrm{Tr}_1[C_j{}^t \otimes I|I\rangle\!\rangle\langle\!\langle I|] = \sum_j C_j \otimes C_j \\ &= \sum_j \mathrm{Tr}_2[I\otimes C_j{}^t|I\rangle\!\rangle\langle\!\langle I|]\otimes C_j = \mathscr{I}\otimes\mathscr{C}(|I\rangle\!\rangle\langle\!\langle I|).\end{aligned} \tag{46}$$

Using completeness of $\{C_j\}$ and their self-adjointness, it is easy to see that

$$\mathscr{C}(X) = \sum_j \mathrm{Tr}[C_j{}^tX]C_j = \sum_j \mathrm{Tr}[C_jX^t]C_j = \sum_j \mathrm{Tr}[C_j^\dagger X^t]C_j = X^t, \tag{47}$$

namely

$$\mathscr{C} = (\cdot)^t, \tag{48}$$

and using Eq. (46) one can see that $\sum_j C_j \otimes C_j = E$ and $\{C_j\}$ are Hilbert-Schmidt orthonormal. Clearly $\mathscr{C}^2 = \mathscr{I}$, $\mathscr{C}\mathscr{M}\mathscr{C} = \mathscr{M}^*$, i. e. $\mathscr{C} = \mathscr{Z}_\Omega$. According to (23) this will then guarantee that the adjoint will be independent on the faithful state Φ. The map $\mathscr{C}\mathscr{F}^\dagger$ acting on C_i gives their complex conjugated, and since $\{C_i\}$ is a selfadjont basis of the real linear space of selfadjoint operators, $\mathscr{C}\mathscr{F}^\dagger$ is the complex conjugation over all selfadjoint operators, namely [12]

$$\mathscr{Z}_\Phi = \mathscr{C}\mathscr{F}^\dagger = \mathscr{F}\mathscr{C}. \tag{49}$$

The complex conjugation obeys the symmetry $\tau_\Phi(\mathscr{Z}_\Phi) = \mathscr{Z}_\Phi$ which is needed for a proper definition of the adjoint. Indeed

$$\tau_\Phi(\mathscr{Z}_\Phi) = \mathscr{F}\mathscr{Z}_\Phi{}^t\mathscr{F}^{-1} = \mathscr{F}\mathscr{C}\mathscr{F}\mathscr{F}^{-1} = \mathscr{F}\mathscr{C} = \mathscr{Z}_\Phi. \tag{50}$$

Since, by definition, the map $\mathscr{Z}$ is involutive, one has

$$\mathscr{I} = \mathscr{F}\mathscr{C}\mathscr{F}\mathscr{C} = \mathscr{F}\mathscr{F}^* = \mathscr{F}\mathscr{F}^\dagger \tag{51}$$

whence

$$\mathscr{F}^{-1} = \mathscr{F}^* = \mathscr{F}^\dagger. \tag{52}$$

Finally, the adjoint of a map $\mathscr{A}$ is just the usual adjoint, since

$$\varsigma_\Phi\tau_\Phi(\mathscr{A}) = \mathscr{Z}_\Phi\mathscr{F}\mathscr{A}^t\mathscr{F}^{-1}\mathscr{Z}_\Phi = \mathscr{C}\mathscr{A}^t\mathscr{C} = \mathscr{A}^{t*} = \mathscr{A}^\dagger, \tag{53}$$

or, equivalently,

$$\tau_\Phi\varsigma_\Phi(\mathscr{A}) = \mathscr{F}(\mathscr{Z}_\Phi\mathscr{A}\mathscr{Z}_\Phi)^t\mathscr{F}^{-1} = \mathscr{F}\mathscr{C}\mathscr{F}\mathscr{A}^t\mathscr{C}\mathscr{F}\mathscr{F}^{-1} = \mathscr{C}\mathscr{A}^t\mathscr{C} = \mathscr{A}^\dagger. \tag{54}$$

In Table 1 I summarize the most relevant identities and definitions.

---

[11] In the present quantum context the notation $\mathscr{F}^\dagger(X)$ corresponds to the Heisenberg picture $\underline{\mathscr{X}} \circ \mathscr{F}$, with X selfadjoint operator representing the generalized effect $\mathscr{X}$.

[12] On the other hand, for a generic self-adjoint operator it is easy to check that

$$\mathscr{Z}_\Phi(A) = \sum_k \Phi(C_k,A)C_k = \sum_{kl}\mathrm{Tr}[F_l^*AF_l{}^tC_k{}^t]C_k = \sum_{kl}\mathrm{Tr}[F_l^\dagger AF_lC_k{}^t]C_k = \mathscr{C}\mathscr{F}^\dagger(A).$$

TABLE 1. Summary of most relevant identities and definitions

object	definition	identities
Φ	$\sum_l \|F_l\rangle\!\rangle\langle\!\langle F_l\|$	$E\Phi E = \Phi$
$\mathscr{F}$	$\sum_l F_l \cdot F_l^\dagger$	$\mathscr{F} = \mathscr{F}^t,\ \mathscr{F}^{-1} = \mathscr{F}^* = \mathscr{F}^\dagger$
$\mathscr{C}$	$\sum_j \mathrm{Tr}[C_j{}^t \cdot]C_j$	$\mathscr{C} = (\cdot)^t = \mathscr{C}^\dagger = \mathscr{C}^t,\ \mathscr{C}\mathscr{M}\mathscr{C} = \mathscr{M}^*$
$\mathscr{Z}_\Phi$	$\sum_j \Phi(C_j, \cdot)C_j$	$\mathscr{Z}_\Phi = \mathscr{C}\mathscr{F}^\dagger = \mathscr{F}\mathscr{C}$
$\tau_\Phi(\cdot)$		$\tau_\Phi(\mathscr{M}) = \mathscr{F}\mathscr{M}^t\mathscr{F}^{-1}$
$\varsigma_\Phi(\cdot)$		$\varsigma_\Phi(\mathscr{M}) = \mathscr{Z}_\Phi\mathscr{A}\mathscr{Z}_\Phi,\ \tau_\Phi(\mathscr{Z}_\Phi) = \mathscr{Z}_\Phi$
$\mathrm{ad}_\Phi(\cdot)$		$\tau_\Phi(\varsigma_\Phi(\mathscr{A})) = \varsigma_\Phi(\tau_\Phi(\mathscr{A})) = \mathscr{A}^\dagger$
$_\Phi\langle\mathscr{A}\|\mathscr{B}\rangle_\Phi$	$\Phi(\mathscr{A}^\dagger, \tau_\Phi(\mathscr{B}))$	$_\Phi\langle\mathscr{A}\|\mathscr{B}\rangle_\Phi = \frac{1}{d}\sum_l \langle\!\langle F_l\|\check{A}^\dagger\check{B}\|F_l\rangle\!\rangle$

Explicitely, the GNS representation is given by

$$_\Phi\langle\mathscr{A}|\mathscr{B}\rangle_\Phi := \Phi(\mathscr{A}^\dagger, \tau_\Phi(\mathscr{B})) = \tfrac{1}{d}\sum_l \langle\!\langle F_l|\check{A}^\dagger\check{B}|F_l\rangle\!\rangle, \tag{55}$$

where for any CP map $\mathscr{A} = \sum_i A_i \cdot A_i^\dagger$ one has $\check{A} := \sum_i A_i \otimes A_i^*$ (we remind the normalization $\mathrm{Tr}[\sum_l F_l^\dagger F_l] = d$ of state Φ in terms of the Kraus operators of $\mathscr{F}$). For trace-preserving $\mathscr{F}$ one would obtain the same scalar product as in Eq. (38), i. e. $_\Phi\langle\mathscr{A}|\mathscr{B}\rangle_\Phi := \mathrm{Tr}[P_{\mathscr{A}^\dagger}P_{\mathscr{B}^\dagger}]$, however, since $\mathscr{F}^{-1} = \mathscr{F}^*$ is also trace preserving, the only possibility would be $\mathscr{F} = U \cdot U^\dagger$ unitary, and with the additional constraint $U = U^t$ coming from symmetry of Φ.

6.3. The most general quantum scalar product

We start now from the most general scalar product between two quantum transformations and show that it must be of the form (55). The most general form of scalar product between two operators A and B in $\mathsf{B}(\mathsf{H})$ is

$$\varphi(A^\dagger B) = \sum_j \langle v_j|A^\dagger B|v_j\rangle, \quad \sum_j \langle v_j|v_j\rangle = 1 \tag{56}$$

where normalization corresponds to $\varphi(I) = 1$. For quantum transformations the most general scalar product can be constructed upon regarding transformations as operators on $\mathsf{B}(\mathsf{H})$ (in infinite dimensions, more precisely, as operators on the Hilbert space of the Hilbert-Schmidt operators). Therefore, upon considering a complete set of operators $\{E_i\}$, one has

$$(\mathscr{B}, \mathscr{A}) = \sum_i \langle\!\langle \mathscr{B}(E_i)|\mathscr{A}(E_i)\rangle\!\rangle = \sum_i \langle\!\langle E_i|\check{B}^\dagger\check{A}|E_i\rangle\!\rangle, \quad \mathrm{Tr}\left(\sum_i E_i^\dagger E_i\right) = 1, \tag{57}$$

which is exactly of the general form (55). Notice that the general form (55) corresponds to a state Φ that is mixed, being the convex combination $\Phi = \sum_i \mathrm{Tr}[F_i^\dagger F_i]|\tilde{F}_i\rangle\!\rangle\langle\!\langle\tilde{F}_i|$, with $\tilde{F}_i := F_i/\sqrt{\mathrm{Tr}[F_i^\dagger F_i]}$.

7. CONCLUSIONS

In conclusion I want to emphasize that the fact that Postulates 1 and 2 imply a C^*-algebra representation for transformations, and with the correct Born-rule pairing and the correct dimensionality for the tensor-product structure of bipartite systems, is not sufficient to assert that the only possible theory derived from the postulates is Quantum Mechanics. Indeed, as for the general C^*-algebras of operators on Hilbert spaces, Classical Mechanics is also included as special case, corresponding to Abelian $\mathbb{A}$, and, more generally, a combination of both Quantum and Classical in a direct sum of irreducible algebra representations, such as in the presence of constant of motions and/or super-selection rules. Indeed preliminary analysis [7] show that more general theories can satisfy both postulates, such as the non-local no-signaling probabilistic theories generally referred to as *PR boxes* [8]. This is the case, for example, of the model in Ref. [9], which possesses a symmetric faithful state, however with $\dim(\mathfrak{P}_{\mathbb{R}}) = 3$, which cannot be quantum.

As regards additional postulates selecting Quantum Mechanics from the set of theories admitting C^*-algebras representations, one may adopt Postulate 4 in Ref. [1] concerning the possibility of achieving an informationally complete observable by means of a perfectly discriminating observable over system+ancilla. However, such postulate may look quite *ad hoc*, being essentially a restatement of existence of Bell measurements (Bell measurements are locally informationally complete for one system for almost every state-preparation of the other system). Alternative candidates for the quantum-extracting postulate are under study, considering what is specific of the quantum C^*-algebra, e. g. the fact that in the quantum case the C^*-algebra of transformations $\mathcal{A}$ is a kind of *multiplier algebra* [10] of the C^*-algebra $\mathsf{B}(\mathsf{H})$.

I want to stress that the dimensionality identity in Sect. 5 concerning only identical independent systems could be generalized to the case of different systems. This, however, will need to consider transformations between different systems. Thus, also the symmetry of the faithful state must be relaxed, upon considering a suitable transformation that maps the largest to the smallest system. Finally, the faithfulness condition itself may be relaxed, obtaining a generally unfaithful C^*-algebra representation. Thus the C^*-algebra representation of transformations will be just equivalent to the probabilistic framework endowed with the postulated existence of dynamically independent systems. A complete analysis of this direction will be the subject of a forthcoming publication [11].

ACKNOWLEDGMENTS

This research has been supported by the Italian Minister of University and Research (MIUR) under program Prin 2005. I acknowledge very useful discussions with Tomy Short, Dirk Schlingeman, and Masanao Ozawa. Part of the present work has been done in Cambridge UK, during the workshop *Operational probabilistic theories as foils to quantum theory*, July 2-13, 2007.

REFERENCES

1. G. M. D'Ariano, in *Foundations of Probability and Physics -4*, ed. by G. Adenier, C. A. Fuchs, and A. Y. Khrennikov (AIP, Melville, New York, 2007) pag. 79 [also `quant-ph/0611094`].
2. G. M. D'Ariano, *Where the mathematical structure of Quantum Mechanics comes from* Work presented at the conference *Beyond the Quantum* held at the Lorentz Center in Leiden (2006) (quant-ph/0612162)
3. G. M. D'Ariano, *Operational axioms for a C*-algebraic formulation for Quantum Mechanics* in Proceedings of the 8th Int. Conf. on Quantum Communication, Measurement and Computing, ed. by O. Hirota, J. H. Shapiro and M. Sasaki (NICT press, Japan, 2007), p. 345 (also quant-ph/0701219)
4. I. M. Gelfand, and M. A. Neumark, *Mat. Sb.* **12**, 197 (1943).
5. G. Ludwig, *An Axiomatic Basis for Quantum Mechanics I: Derivation of Hilbert Space Structure*, Springer, SPR:adr, 1985.
6. G. M. D'Ariano, J. Phys. A. Math. Theor. **40** 8137 (2007)
7. G. M. D'Ariano and T. Short, unpublished.
8. S. Popescu and D. Rohrlich, Found. Phys. **24** 379 (1994)
9. T. Short, S. Popescu, and N. Gisin, Phys. Rev. A **73** 012101 (2006)
10. G. J. Murphy, *C*-Algebras and Operator Theory*, Academic Press, (London, San Diego, 1990)
11. G. M. D'Ariano, in *Philosophy of Quantum Information and Entanglement*, Alisa Bokulich and Gregg Jaeger (Eds.) (Cambridge University Press, Cambridge UK) forthcoming

Reality and Locality in Quantum Mechanics

W. De Baere

Laboratory for Theoretical Physics
Unit for Subatomic and Radiation Physics
Proeftuinstraat 86, B–9000 Ghent, Belgium
E–mail: willy.debaere@UGent.be

Abstract. It is argued in this work that in order to explain the conflict between Bell's inequality (BI) and quantum mechanics (QM), it is neither necessary to give up the notion of Einstein locality (LOC), nor that of reality, as is done in some recent work. The definite existence of EPR's "elements of reality" is shown by using a result in a former thought experiment by M. Renninger. Einstein locality is saved by sticking to the failure of counterfactual definiteness (CFD) at the level of individual quantum processes.

Keywords: Reality, locality, quantum ontology, quantum interpretation
PACS: 03.65. – *W*, 03.65.*Bz*

INTRODUCTION

In the foundations of QM, "reality" or "realism" is generally defined as something that is independent of observation, and may be represented by physical states in physical theories. Sometimes it is naively assumed that observed values may be considered as pre–existing and possessed properties by object systems, as is the case in classical mechanics. It has been shown explicitly by Kochen and Specker (KS) [1] that this view contradicts basic quantum rules, as has been claimed from the very beginning of QM by Bohr and others. It has also been claimed that the combined assumption of realism and of "Einstein locality" leads to contradictions with QM via Bell's inequality (BI) [2], so that at least one of these assumptions had to be abandoned, usually the assumption of locality. And in recent work, A. Zeilinger et al. [3, 4] even contended that *both* locality *and* realism are in conflict with QM on empirical grounds and, hence, should be abandoned altogether.

In reply to these prevailing common opinions, we have argued in the past that the nonlocality–of–QM claim may be removed immediately by simply giving up, instead, the validity of counterfactual definiteness (CFD) in the domain of QM. Moreover, invoking an old result obtained by M. Renninger [5], we argue in the present work, that neither the notion of "realism" should be questioned, i.e. empirical results show that, on the contrary, "elements of reality" definitely exist *independent of any theory*.

An immediate consequence is the validity of the de Broglie–Einstein–Bohm (dB–E–B) picture of physical reality, which entails that any single quantum system is at the same time *both* a real existing wave *and* a real existing particle embedded in the real wave. However, insisting on the validity of locality, this does not entail the validity of the explicitly nonlocal "Bohmian mechanics" as a kind of HV substitute for standard QM.

REALITY AND LOCALITY IN RECENT WORK

In recent work A. Zeilinger et al. [3, 4] not only criticize Einstein's program to define the goal of physics as consisting in describing "the real factual situation" in the sense of getting an idea about "what is out there" (i.e. with the purpose not only to describe it, but also to understand it), but also the validity of reality and locality. In the style of Bohr, a number of unproven – or even unprovable – statements are made such as:

> "The discovery that individual events are irreducibly random is probably one of the most significant findings of the twentieth century",

CP962, *Quantum Theory, Reconsideration of Foundations – 4*,
edited by G. Adenier, A. Yu. Khrennikov, P. Lahti, V. I. Man'ko, and Th. M. Nieuwenhuizen

"But for the individual event in quantum physics, not only do we not know the cause, there is no cause... There is nothing in the Universe that determines the way an individual event will happen."

And even "non–local realism" is thought to be incompatible with QM:

"Our results suggest that giving up the concept of locality is not sufficient to be consistent with quantum experiments, unless certain intuitive features of realism are abandoned."

Because here it is the notion of reality that is questioned, we will show that the notion of reality is much more rich than what is meant in these statements. Yet, at the end of e.g. [4] the authors seem to start doubting their own conclusion (similar to EPR's self–criticism at the end of their paper):

For example, the concept of ensembles of particles carrying definite polarization could fail. Furthermore, one could consider the breakdown of other assumptions that are implicit in our reasoning leading to the inequality. These include Aristotelian logic, counterfactual definiteness, absence of actions into the past or a world that is not completely deterministic."

It also reminds earlier doubts of H.P. Stapp – who became another advocate of the nonlocality of QM – on the general validity of CFD [6]:

"The key operative concept of the EPR-Bell argument is ... counterfactual definiteness....No satisfactory derivation of nonlocality, or the existence of faster-than-light influences, can be based upon such a CFD assumption: a failure of this assumption is (at least in my opinion) far more likely than the existence of a faster-than-light influence."

The further relationship between the validity of CFD, and the so–called "very deep" questioning by QM of the validity of the notion of reality and the validity of locality will be discussed below.

REALITY AND LOCALITY IN RENNINGER'S THOUGHT EXPERIMENT

Instead of focussing once more on EPR's famous criterion of reality, we shall elaborate below on an alternative definition of reality, in the context of a thought experiment by M. Renninger [5]. Whereas the relevance of EPR's criterion has been discussed strongly, e.g. by Bohr [7], we will see that the criterion introduced by Renninger, leads to the unavoidable conclusion of the real existence of some kind of "external reality" in the sense of Einstein. At the same time it leads, also in an unavoidable way, to the correctness of the de Broglie–Einstein–Bohm picture of the underlying reality.

Renninger's thought experiment

We refer to Ref. [8] for an English translation of Renninger's 1953 paper on wave–particle duality. In this interesting paper, Renninger defines reality or the existence of an "element of reality" as follows:

"The notion 'physical reality' should be understood such that, when this physical reality is considered in a particular space at a particular time, it should be experimentally possible to influence this reality in such a way that future results of experiments show unambiguously that this reality has been causally influenced by the experimental act in this space and at that time."

Renninger's setup (see Fig. 1) consists of a simple Mach–Zehnder interferometer (MZI) through which single photons, or other quantum systems, move one by one. In this setup, * is the source, 1 is the incoming path, A and B are detectors or paths within MZI, 6,7,8 and 9 are points of possible actions, BS_1 and BS_2 are beam splitters, S_A and S_B are mirrors, and C and D are detectors.

As a first action, detectors are installed in the places A and B. Then one of these will fire, i.e. the whole momentum and energy are absorbed by one single detector. According to Renninger, this is evidence for the particle–like behaviour of the reality moving to that detector, after having passed BS_1. Alternatively, one might say that the reality moving along one of the paths in MZI, is able to fire a detector in that path.

In a second stage of the reasoning, detectors A and B are removed, and placed now in C and D in the outgoing paths. If nothing is in the paths, and the path length is equal, then one always finds C=1, D=0. A $\lambda/2$ plate is now put

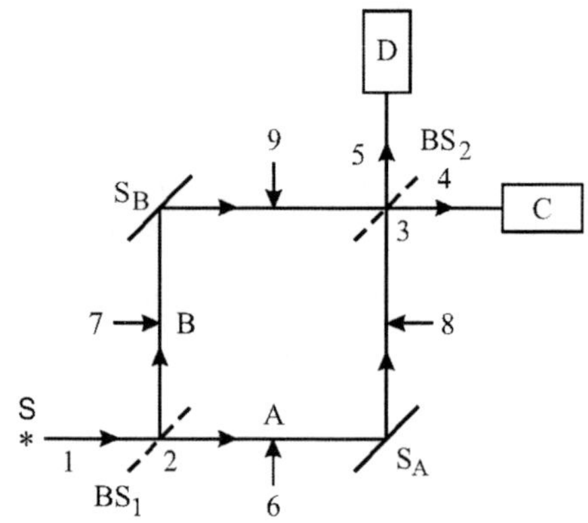

FIGURE 1. Setup in Renninger's thought experiment

at place 7, this produces a phase shift π (or a multiplication with -1 of the wave function), which now always leads to C=0, D=1. The interesting remark here is that the previous change can be undone, by going e.g. to place 6 and put there another – or even the same! – $\lambda/2$ plate in the other path. This will restore the original situation. Of course, this also happens with a $\lambda/2$ plate in both paths.

Because Renninger's reasoning concerns *single* quantum systems, his conclusion is:

> "Are now, after all these considerations, the wave–'picture' and particle–'picture' really but 'two forms of appearance of one and the same physical reality'? Is it not, on the contrary, the case that wave and particle are two separate physical realities associated with single photons? Did the observation c above not 'reveal in one single indivisible experimental intervention both the one and the other picture of the nature of light? (The result of inserting the phase plate proves the wave nature, the observation in one single point proves the particle nature of one and the same photon)."

This is a confirmation of the validity of the dB–E–B picture of reality:

> "This image is not new. It was already formulated by de Broglie in 1927...in terms of the by him introduced notion of 'pilot wave', but abandoned again because of 'des objections très graves' ['very serious objections']"

So, the phenomena happening in both paths, are to be considered as realities that are causally influenceable. These realities may be called a *full wave* (FW) containing the "particle" and an *empty wave* (EW) containing "nothing".

In addition, Renninger adheres to locality in his thought experiment:

> "What happens to the wave devoid of energy of a photon after its absorption? When it is absorbed for example in (6), and when in addition the detectors in (4) and (5) are removed, what happens then with the wave in *B*? Does she move further towards infinity, or does she disappear at the moment of absorption? Of course this question cannot be answered principally. The former assumption appears to be the more natural one, because it avoids the conclusion to the existence of influences that propagate with infinite speed also through closed walls, a conclusion that within the physical world is inconceivable. In any case were such influences not associated with transport of energy."

It also follows that relatively simple quantum processes can now be looked upon, and analyzed, from the point of view of the simultaneous real existence of both a wave *and* a particle, building up the reality of a quantum system.

Renninger's experiment: further expectations

In order to discuss further Renninger's unanswered question above, we modify Renninger's setup in the following way (see Fig. 2). Our argumentation is based on the observations 1) that the realities that correspond with EW and FW interact in BS_2, 2) that a material object, the $\lambda/2$ plate, influences in a causal way both the realities EW and FW, and 3) that also the gravitational field influences both EW and FW, see Colella, Overhauser and Werner [9, 10]. Therefore,

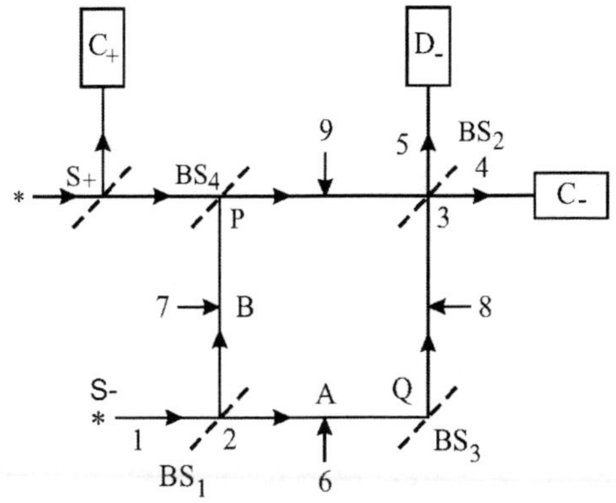

FIGURE 2. Modified setup with two sources

it might be suggested that, maybe, EW and FW are also influenced in an observable way by *another* empty wave EW_1, with observability again behind BS_2. Of course, it is fully agreed that according to standard QM this is not observable, even not behind BS_2. On behalf of this trust in the quantum formalism, this is also Renninger's opinion: "Of course this question cannot be answered principally." Yet, from the point of view of the interacting realities EW and FW, some suspicion might remain.

Consider then a MZI setup with 2 sources S_-, S_+ of quantum systems s^-, s^+. In order to be able to overlap EW_{s+} with EW_{s-} or FW_{s-} at point P and along path 9, we replace the mirror S_B by a beam splitter BS_4. And in order to get the results $C_- = 1$ and $D_- = 0$ every time with certainty in the case of equal path lengths, we have put another beam splitter BS_3 at point Q instead of the mirror S_A. So, with this new setup we have the evolution:

$$|s^-\rangle \longrightarrow -\frac{1}{\sqrt{2}}|C_-\rangle \tag{1}$$

i.e. as before, we have $\text{Prob}(D_-) = 0$ and, with a $\lambda/2$–plate in one of the paths 6 or 7 we again get the evolution

$$|s^-\rangle \longrightarrow -\frac{1}{\sqrt{2}}|D_-\rangle \tag{2}$$

and thus $\text{Prob}(C_-) = 0$.

Add now the source S+, and consider only events where s^+ is observed at C_+. In this case the QM evolution gives rise to a state $-\frac{i}{2}|C_+\rangle|C_-\rangle$. So, the QM conclusion is: EW_{s+} does not spoil the phase relationship between the realities EW_{s-} and FW_{s-} moving resp. in paths 6–8 and 7–9. Yet, if EW_{s+}, EW_{s-} and FW_{s-} are realities (as proven by Rennninger), it might reasonably be expected that somehow they interact with each other, i.e. a possible change of the phase relationship between EW_{s-} and FW_{s-} must be envisaged, with observable effects behind BS_2?

With respect to our proposed modification of Renninger's experiment, we may ask whether there exist as yet any significant result. Maybe, some results may be used from a recent experiment by Irvine et al. [11] that realizes Hardy's thought experiment with photons [12]. We recall that Hardy's setup consists of 2 superposed interferometers (see Fig. 3), and that its purpose was to show the incompatibility of any local HVT with QM. As it is one of our purposes to defend the validity of Einstein locality, we also recall that Hardy's argumentation basically relies on the implicit assumption of the validity of CFD at the level of individual quantum phenomena. However, looking at the setup it is seen that the evolution of the incoming quantum state also implicitly assumes that the phase relationship in each of the 2 interferometers, is only changed by the presence of a particle in the crossing path of the other interferometer, and not if EW is present. This, then, entails the QM predictions $\text{Prob}(d^-v^+) = 0$ and $\text{Prob}(d^+v^-) = 0$. Now, looking at the results of Irvine et al.'s experiment, we note that indeed $\text{Prob}(d^-v^+) = 0$ within statistical errors, but also that there exist a clear deviation from the QM prediction $\text{Prob}(d^+v^-) = 0$, which would mean that the phase relationship between EW_+ and FW_+ has been changed effectively by EW_- moving along path u^-, giving rise to $\text{Prob}(d^+v^-) \neq 0$. Possibly this may be interpreted as an observable effect in my proposal, i.e. detector D_- would now click also sometimes in a real experiment.

Of course, standard QM predicts a negative outcome, i.e. $\text{Prob}(d^+v^-) = 0$, and I can live with it, but then I would like to see an explanation, or at least understand, why the reality EW can influence the reality FW, and not another reality EW'. Otherwise stated, I would like to see a db–E–B model in which the dynamics of EWs and FWs describes their mutual interaction. QM does not afford such an explanation, because it only predicts probabilities for outcomes.

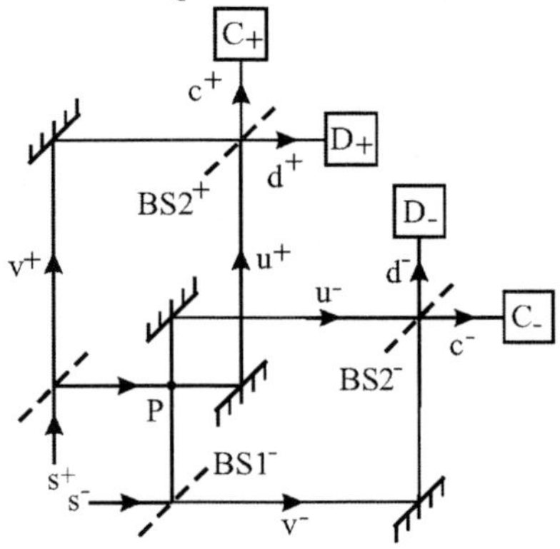

FIGURE 3. Hardy's setup

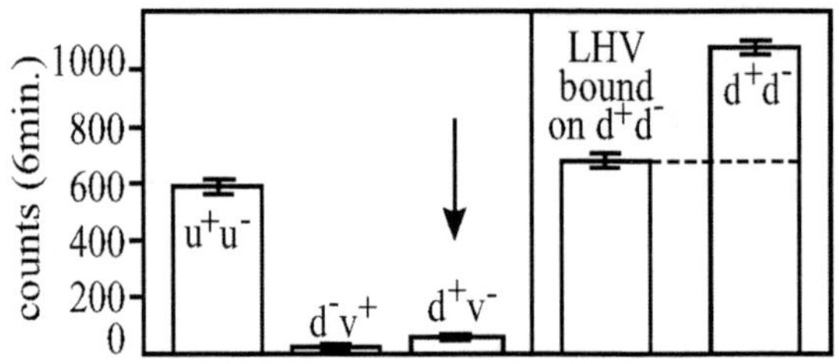

FIGURE 4. Results by Irvine et al. [11]

REPRODUCIBILITY OF INITIAL CONDITIONS AND THE ISSUE OF CFD

For further insight in the significance and importance of CFD, we now mention a former study by Houtappel, Van Dam and Wigner [13]. According to these authors, the initial conditions and the laws of physics play an equally important role

> '...because the laws of nature do not lead to observable consequences unless the initial conditions are given ...".

These initial conditions concern the theory's initial state that, in terms of EPR's or Renninger's terminology, may be considered as representing some "element of physical reality". Conversely, every identified "element of physical reality" (e.g. in one single process) should be described in the physical theory and should correspond with some initial state, as is also stated in EPR's "condition of completeness" [14]:

> "Whatever the meaning assigned to the term complete, the following requirement for a complete theory seems to be a necessary one: every element of the physical theory must have a counterpart in the physical theory".

Here we note that in EPR x and p do not fulfill this requirement for a single process, because they cannot be simultaneously "elements of physical reality". In Renninger's setup, on the contrary, both FW and EW do fulfill this requirement and, hence, should have a representation in a complete theory.

Referring to Houtappel et al.'s study, it may be seen that the assumption of the validity of CFD is unproblematic and used in a justified way in CM and in QM, because in CM and in QM the initial conditions may be reproduced in actual experiments: each time they are represented by identical mathematical symbols or states. Now, bearing in mind that all argumentations of quantum nonlocality deal somehow with *individual* quantum processes, and that they depend crucially on the validity of CFD, it is clear that they depend, in fact, on the implicit assumption that the initial conditions of the underlying HVT (the existence of which has to be assumed from the very beginning) are reproducible in actual practice. I.e. the *reproducibility of localized HV states* representing the physical conditions within a localized region is an essential implicit or explicit assumption in any nonlocality argumentation.

So, although CFD is valid in both CM and QM, it is well thinkable that on a deeper level of description, nature behaves in a lawful local and deterministic way, without initial states being reproducible at that level. It is this failure of CFD, or nonreproducibility of initial states at the HV level, that constitutes our resolution of the apparent nonlocality-of-QM.

ARGUMENTS IN FAVOUR OF QUANTUM LOCALITY

Origins of Quantum Nonlocality

One of many possible definitions of quantum nonlocality is given by A. Peres [15]:

> "...what we call 'the result of a measurement of A' cannot depend only on A and on the state of the system (unless the wave function is an eigenstate of A). It also depends, in a way not yet understood, on the choice of other quantum measurements that may possibly be performed."

In fact, there exist two main origins of the alleged quantum nonlocality. First, there are the *qualitative* grounds: nonlocality argumentations are held within interpretations (such as the Copenhagen interpretation of QM (CIQM)) which claim that chance is an inherent feature of nature, and that a measurement result is "created" at the last moment when the choice of the kind of measurement is made, in a process in which chance, or absence of a cause, plays a crucial role (recall the quotation from Zeilinger at the beginning of this paper). It follows that the emergence of a definite result in an entangled situation where there is a 100% correlation (such as in the spin singlet state) can be explained only by action-at-a-distance, hence by a nonlocal process.

Next there are the *quantitative* grounds: the explicit conflict between QM and Bell–type inequalities (BIs) based on statistics, the Greenberger–Horne–Zeilinger (GHZ) [16] argumentation for which no statistics is needed, and finally the Hardy–type argumentations [17, 18] for non–maximally entangled states. The conclusion of all these argumentations is that the locality assumption is responsible for the conflict. However, a close look at them reveals that all of them are, without exception, based on the assumed unrestricted validity of CFD or, otherwise stated, on the assumption that nature allows the reproducibility of initial conditions in individual quantum cases.

There is also the experiment by A.M. Steinberg et al. [19], that is interpreted as empirical proof that in a cavity photons may travel faster than light. However, as recently shown by H.G. Winful [20, 21], this again rests on an incorrect identification for the cause of the arrival of a photon in a detector behind the cavity, after the quantum process has taken place.

Why Quantum Nonlocality is unacceptible

I summarize here some arguments in favour of the locality property of QM (see also Ref. [22]):

- it is based on questionable interpretations (see above), such as the CIQM for single systems;
- it contradicts observably the relativity principle, which is empirically correct: no one has ever succeeded in transmitting signals or information faster than light;
- it has no general validity and, when retained, can only be considered ad hoc for explaining some results for which there exist alternative explanations in which locality remains valid. Otherwise stated, it would then be the case that in one and the same individual quantum process (such as in correlated measurements) both locality and nonlocality should apply, contradicting Einstein's relativity theories;
- LOC is used *explicitly* in any quantum field theory (QFT) and derivations thereof:

$$[A(x),B(y)] = i\triangle(x-y) = 0 \qquad \text{for}(x-y)^2 < 0\,. \tag{3}$$

  This is the so–called "microscopic causality" of QFT, a basic relation that represents the fundamental belief in the universal validity of causality and locality;
- QM is formally local: the deterministic QM expectation values are independent of far–away actions, as proved e.g. by Ghirardi, Rimini and Weber [23, 24];
- it results from a program that tries to reconstruct QM from a more detailed theory of the classical type, a program that is ill defined because it violates basic QM [25, 26]. This is formally evident from the identity [27]:

$$(\hat{A}\hat{B}+\hat{A}'\hat{B}+\hat{A}\hat{B}'-\hat{A}'\hat{B}')^2 = 4\mathbb{1} - [\hat{A},\hat{A}'][\hat{B},\hat{B}']. \tag{4}$$

  Here the observables $\hat{A}$, $\hat{A}'$, $\hat{B}$, $\hat{B}'$ represent correlated (entangled) measurements along 4 possible combinations of experimental settings, namely incompatible directions $\vec{a},\vec{a}'$ in one region L, and incompatible directions $\vec{b},\vec{b}'$ in another spacelike separated region R.

From Eq. (4) it follows that the expectation value of the left hand side satisfies Cirel'son's inequality [28]:

$$\left|\langle\hat{A}\hat{B}\rangle+\langle\hat{A}'\hat{B}\rangle+\langle\hat{A}\hat{B}'\rangle-\langle\hat{A}'\hat{B}'\rangle\right| \leq 2\sqrt{2}. \tag{5}$$

So, it is seen that if one or both of the commutators vanish, (which implies the existence of a joint probability distribution for the corresponding incompatible observables) the upper bound of the lhs. of (4) reduces to 2, i.e. the usual BI. Moreover, it is the noncommutativity of quantum observables pertaining to a *single* system that plays a crucial role in the violation of the BI by QM. *Nonlocality is not required at all!*

GENERAL STRUCTURE OF LOCAL HIDDEN VARIABLE THEORY

A more complete specification of the individual preparation and measurement may be in terms of HV states $\{\lambda_i(x^\mu), i = 1,2,\ldots,k\}$, with x^μ representing a point in space–time, and i allowing for some complexity so that $\{\lambda_i\}$ may be termed a set of k "field components". Remembering the control one has in some individual quantum processes, such as in Renninger's thought experiment discussed above, these fields obey deterministic field equations, even though initial conditions might are not reproducible. For simplicity, we consider in the following only one field component, $\lambda(x^\mu)$. The general properties of the HV states are then the same as in any other physical theory:

- deterministic space–time evolution of $\lambda(x^\mu)$ (in analogy with the quantum state that obeys the deterministic Schrödinger equation);
- individual deterministic predictions should be given by a kind of "result function" R (to be compared with quantum mechanics' operator and eigenvalue formalism):

$$r(X^\mu) = R[\lambda(x^\mu), x^0 = X^0, \bar{x} \in \Omega_F] \tag{6}$$

X^μ represents a point in space-time at which the result r is measured, and Ω_F is a (set of all points within a) localized region in 3–space of an inertial frame F at time x_0.
In the specific example of correlated spin–component observations along directions d_1, d_2, we can start with the general expressions (in which LOC is not yet imposed):

$$r_1(X^\mu; d_1, d_2) = R[\lambda(x^\mu), x^0 = X^0, \bar{x} \in \Omega_F; d_1, d_2] \tag{7}$$

$$r_2(Y^\mu; d_1, d_2) = R[\lambda(x^\mu), x^0 = Y^0, \bar{x} \in \Omega_F; d_1, d_2] \tag{8}$$

- as to separability, the empirical validity of the finiteness of propagation speed of physical influences (i.e. the validity of both special and general relativity) incites us to propose the following separability or locality hypothesis: there exist regions D_S in 3-space such that the restriction of the HV state $\lambda(x^\mu)$ to D_S propagates in space–time as a relatively autonomous entity (usually called a physical system S or, more commonly, a "particle"). The dynamics is then such that in the course of time and under appropriate conditions this evolution may give rise to one or more observations (as is the case in e.g. parametric down conversion, high energy phenomena, etc.);
- with each observation there corresponds a domain D_M (including the apparatus M) such that the HV state $\lambda(x^\mu)$ in D_M is equivalent with r via the result function R:

$$r(X^\mu) = R[\lambda(x^0 = X^0, \bar{x} \in D_M)]\,. \tag{9}$$

i.e. the physical state outside region D_M is considered to be irrelevant for the measurement result r at X^μ in D_M.
For correlated observations we then have:

$$r_1(X_1^\mu) = R[\lambda_{M_1}(x^\mu; \bar{x} \in D_{M_1})] \tag{10}$$

$$r_2(X_2^\mu) = R[\lambda_{M_2}(x^\mu; \bar{x} \in D_{M_2})] \tag{11}$$

where it is assumed that a common origin has co–determined the localized HV states $\lambda_{M_1}(x^\mu; \bar{x} \in D_{M_1})$ and $\lambda_{M_2}(x^\mu; \bar{x} \in D_{M_2})$.
In Renninger's experiment this would mean that the local HVT predicts either $r_1 = 1$ and $r_2 = 0$, or $r_1 = 0$ and $r_2 = 1$, in contrast with the probabilistic predictions for both results in QM.

Reproducibility and noncontextuality

Consider two identical ensembles of N (identical) individual preparations (corresponding with the same pointer positions at preparation), distinguished by labels (1) and (2):

$$\left\{\lambda(x^{\mu}_{i;(1)}),\, i=1,\ldots,k\right\} \sim |\psi\rangle, \quad \left\{\lambda(x^{\mu}_{i;(2)}),\, i=1,\ldots,k\right\} \sim |\psi'\rangle. \tag{12}$$

Identity is with respect to the QM state vectors, i.e. $|\psi'\rangle \equiv |\psi\rangle$. In contrast, this identity is not with respect to the HV states, so that it is *not* implied that $\lambda(x^{\mu}_{i;(2)}) = \lambda(x^{\mu}_{i;(1)})$, $i = 1,\ldots,k$, i.e. the sets of individual states $\left\{\lambda(x^{\mu}_{i;(1)}),\, i=1,\ldots,k\right\}$ and $\left\{\lambda(x^{\mu}_{i;(2)}),\, i=1,\ldots,k\right\}$ are *not* necessarily identical, even after any possible reorganization of one of the sets.

We have seen that in the case of CM and QM, nature allows the reproducibility of the mathematical representation of physical states. We have argued also that it is not evident that nature allows the same to be true for individual quantum or HV states, i.e. this remains a supplementary assumption that may be invalid at a subquantum level of description. Consider now observables A,B,C, and assume with KS that

$$[A,B]=0, \quad [A,C]=0, \quad [B,C]\neq 0. \tag{13}$$

The KS assumption then reads: A,B,C have values $v(A),v(B),v(C)$ that are revealed on measuring A,B or C, independently of the sequence of their measurement. As is known, the result of the KS argumentation was that incompatibility with QM was demonstrated. However, it is immediately obvious that the classical KS assumption of the irrelevance of the sequence of measurement violates already from the beginning the assumed incompatibility of B and C. In our HV terminology this is explained again in a natural way by the fact that it is necessary that HVs $\lambda\left(x^{\mu}_{AB}\right), \lambda\left(x^{\mu}_{AC}\right)$ are considered at different times, and maybe are unequal and "nonreproducible", to such an extent that their corresponding outcomes $R\left[\lambda\left(x^{\mu}_{AB}\right)\right], R\left[\lambda\left(x^{\mu}_{AC}\right)\right]$ also are not necessarily equal. In this way the KS argumentation does not go through, and the KS contradiction disappears. Once more the issue is the possible invalidity of CFD at the individual quantum level. Here this may be reformulated as the equivalence, in the quantum domain, between what is called "contextuality" and the incompatibility of (some) observables pertaining to the same (quantum) system.

CONCLUSIONS

We conclude:

1. in contrast with recent claims, the concept of reality is not at stake. With EPR this reality should be represented in an appropriate way in an acceptable physical theory, possibly of the HV type;
2. in accordance with empirical reality, this HVT should be local (as the deterministic predictions of both CM and QM already are), yet the property of CFD should be no longer valid, implying that initial conditions are no longer reproducible. This property at automatically removes the known quantum "paradoxes" or "inconsistent quantum conclusions".

REFERENCES

1. S. Kochen and E.P. Specker, *J. Math. Mech.* **17**, 59–87(1967).
2. J.S. Bell, *Physics* **1**, 195–200(1964).
3. A. Zeilinger, *Nature* **438**, 743–743(2005).
4. S. Gröblacher, T. Paterek, Č. Brukner R. Kaltenbaek, M. Żukowski, M. Aspelmeyer, and A. Zeilinger, *Nature* **446**, 871–875(2007).
5. M. Renninger, *Z. Phys.* **136** 251–261(1953).
6. H.P. Stapp, "Quantum nonlocality and the description of nature" in *Philosophical Consequences of Quantum Theory*, page 154, University of Notre Dame Press, 1989, edited by J. Cushing and E. McMullin.
7. N. Bohr, *Phys. Rev.*, **48**, 696–702(1935).
8. W. De Baere, Physics/0504043v1, see also these Proceedings.
9. A.W. Overhauser and R. Colella, *Phys. Rev. Lett.* **33**, 1237–1239(1974).
10. R. Colella, A.W. Overhauser, and S.A. Werner, *Phys. Rev. Lett.* **34**, 1472–1474(1975).

11. W.T.M. Irvine, J.F. Hodelin, C. Simon, and D. Bouwmeester, *Phys. Rev. Lett.* **95**, 030401(2005).
12. L. Hardy, *Phys. Lett. A* **167**, 17–23(1992).
13. R.M.F. Houtappel, H. Van Dam, and E.P. Wigner, *Rev. Mod. Phys.* **37**, 595–632(1965).
14. A. Einstein, B. Podolsky, and N. Rosen, *Phys. Rev.* **47**, 777–780(1935).
15. A. Peres, *Phys. Lett. A* **151**, 107–108(1990).
16. D.M. Greenberger, M. Horne, and A. Zeilinger, "Going beyond Bell's theorem", in *Bell's Theorem, Quantum Theory and Conceptions of the Universe*, pages 69–72, editor M. Kafatos, Kluwer, Dordrecht, 1989.
17. L. Hardy, *Phys. Rev. Lett.* **71**, 1665–1668(1993).
18. L. Hardy, *Phys. Rev. Lett.* **68**, 2981–2984(1992).
19. A.M. Steinberg, P.G. Kwiat, and R.Y. Chiao, *Phys. Rev. Lett.* **71**, 708–711(1993).
20. H.G. Winful, *Phys. Reports* **436**, 1–69(2006).
21. H.G. Winful, quant-ph/0708.3889.
22. W. De Baere, *Found. Phys.* **35**, 33–56(2005).
23. G.C. Ghirardi and T. Weber, *Lett. Nuovo Cim.* **26**, 599–603(1979).
24. G.C. Ghirardi, A. Rimini, and T. Weber, *Lett. Nuovo Cim.* **27**, 293–298(1980).
25. W. De Baere, Ann. Isr. Phys. Soc. **12**, 95–108(1996).
26. W. De Baere, A. Mann, and M. Revzen, *Found. Phys.* **29**, 67–77(1999).
27. L.J. Landau, *Phys. Lett. A* **120**, 54–56(1987).
28. B.S. Cirel'son, *Lett. Math. Phys.* **4**, 93(1980).

Simulating Quantum Computation on a Macroscopic Model

Bart D'Hooghe

Center Leo Apostel, Vrije Universiteit Brussel, Krijgskundestraat 33, 1160 Brussels, Belgium;
bdhooghe@vub.ac.be

Abstract. We present examples of macroscopic systems entailing a quantum mechanical structure. One of our examples has a structure which is isomorphic to the spin structure for a spin 1/2 and another system entails a structure isomorphic to the structure of two spin 1/2 in the entangled singlet state. We elaborate this system by showing that an arbitrary tensor product state representing two entangled qubits can be described in a complete way by a specific internal constraint between the ray or density states of the two qubits, which describes the behavior of the state of one of the spins if measurements are executed on the other spin. Since any n-qubit unitary operation can be decomposed into 2-qubit gates and unary operations, we argue that our representation of 2-qubit entanglement contributes to a better understanding of the role of n-qubit entanglement in quantum computation. We illustrate our approach on two 2-qubit algorithms proposed by Deutsch, respectively Arvind et al.

Keywords: quantum computation, macroscopic simulation of entanglement
PACS: 03.65.Ta, 03.67.-a, 03.67.Ac

INTRODUCTION

Over the past two decades the theory of quantum computation has gained much interest, both for the possible useful applications of quantum systems in computation and cryptography, as well as for the insights it provides in the foundations of quantum mechanics. Feynman remarked that simulating compound quantum systems on a classical computer quickly becomes intractable as the number of particles becomes large, and he suggested how quantum systems themselves could be more suitable to implement such tasks [1, 2]. Deutsch took the next step by defining the Quantum Turing Machine and providing a 2-qubit quantum algorithm, which could solve a problem faster than possible on a classical digital computer [3]. The interest for quantum algorithms got a boost when Shor presented a factoring algorithm for a quantum computer, which has real life application since it would allow to break cryptographic codes faster than by any known means classical possible [4]. After the work of Deutsch, Shor and others it became clear that the possible advantages of computers based on quantum principles could well go far beyond the efficient simulation of quantum systems. Nevertheless, the actual physical implementation of quantum computers encounters many technological challenges such that currently only quantum computers with a very limited number of qubits have been realized.

In this paper we adopt an operational approach to quantum mechanics, in which a physical entity is described by its set of states, its set of properties and a relation of 'actuality' between these two sets which expresses which properties are actual when the system is in a specific state [5]. Following this operational approach, it is a natural step to consider models which exhibit 'quantum behavior' on a structural level, even if they are not microscopic in size. This has lead to a 'quantum-like' model representing the spin properties of a quantum spin 1/2 particle [6]. Putting two such models together in a suitable way by a rod-like connection, one obtains a model for the singlet state of two entangled spin 1/2 [7]. By generalizing this rod-like connection procedure one recovers the full set of states of the compound system of two entangled spin 1/2 [8]. This means that this macroscopic model has the quantum mechanical structure of a 2-qubit quantum computer, such that a quantum computation can be implemented on these models. We illustrate this on two quantum algorithms. The first algorithm is the algorithm of Deutsch [3]. It is well-known that for the 2-qubit case this algorithm only uses product states and does not require entangled states. This is why we consider a second algorithm, proposed by Arvind et al., which does need entangled states to obtain the outcome [9, 10]. In this sense this algorithm is an ideal candidate to show how problems could be solved on our model of a macroscopic 2-qubit quantum computer. One of the advantages of the two qubit case besides its relative simplicity is that it allows for a nice geometrical representation of entanglement, resulting in a more intuitive grasp of what is going on in a two-qubit quantum computation. In general a quantum register contains n qubits. However, it can be shown that any n-qubit interaction can be decomposed into 2-qubit gates and unary operations [11], so in a sense, the 2-qubit quantum computer is already sufficient to demonstrate the most elementary functioning of a general quantum computer.

CP962, *Quantum Theory, Reconsideration of Foundations — 4,*
edited by G. Adenier, A. Yu. Khrennikov, P. Lahti, V. I. Man'ko, and Th. M. Nieuwenhuizen

AN OPERATIONAL APPROACH TO QUANTUM MECHANICS

The SPS-formalism

The basic notions for the description of a physical entity S are as follows [5]. First, we consider that at any moment the entity S is in a (known or unknown to the observer) state $p \in \Sigma$. Also, S has a set of properties $\mathscr{L}$, defined by the set of available experiments which can be performed on S. A property a is either 'actual' or 'potential' for the entity S, meaning that if the property a is actual in the state p, then whenever one would perform the corresponding experiment, one finds the positive outcome with certainty. Between the set of states and (power)set of properties is a relation $\xi : \Sigma \longrightarrow \mathscr{P}(\mathscr{L})$ of actuality that maps each state $p \in \Sigma$ onto the set $\xi(p)$ of those properties that are actual in this state. Dually, one can consider the map $\kappa : \mathscr{L} \longrightarrow \mathscr{P}(\Sigma)$, which maps a property $a \in \mathscr{L}$ onto the set of states $\kappa(a)$ that make this property actual. Depending on the nature of the entity S, one obtains a different structure on the set of states Σ, the set of properties $\mathscr{L}$ and the relation between these two sets. Hence, if we are only concerned with the structural behaviour of an entity, we can focus on the triple $(\Sigma, \mathscr{L}, \xi)$. More abstractly, even without an underlying physical entity S we can consider any two sets Σ and $\mathscr{L}$ and a function $\xi : \Sigma \dashrightarrow \mathscr{L} : p \longrightarrow \xi(p)$ and study the emerging structure. We call $(\Sigma, \mathscr{L}, \xi)$ a State Property Space (**SPS**). If one considers the **SPS** of a quantum entity, one observes that certain 'quantum axioms' hold, i.e., the mathematical structure of the **SPS** needs to obey certain specific rules. Conversely, one could start from an abstract **SPS**, and by imposing a suitable set of axioms one can derive a quantum structure on the set of properties, i.e., it becomes isomorphic with the Hilbert space representation of quantum mechanics, as first demonstrated in the framework of property lattices by Piron [12, 13]. Therefore, in this operational approach a physical system is determined by the structure on its set of states and properties, and the actuality map ξ between these two. This means that it is not necessary (or even meaningful) to make a distinction between quantum and classical systems on basis of their size (classical macroscopic versus microscopic quantum) but rather on the different structure on their **SPS**.

'Quantum' versus 'classical' systems

Clearly, within the operational approach the concept of 'classical' versus 'quantum' has very different meanings depending on the context or framework in which these concepts are used. If one considers the set of states, then quantum states are represented by rays in a complex Hilbert space, while classical states are represented by points in phase space. Quantum properties are represented by closed subspaces in Hilbert space, and give rise to a non-distributive lattice (quantum logic). Classical properties on the other hand can be represented by subsets in phase space and give rise to a Boolean algebra (i.e., distributive lattice). In classical theory, probabilities can only arise due to a lack of knowledge about the state of the system which yields a Kolmogorovian probability calculus [14]. On the other hand, quantum probabilities result in a non-Kolmogorovian probability calculus. Finally, also the description of compound systems is fundamentally different for classical and quantum systems. The classical phase space of a compound classical system is constructed using the Cartesian product, which means that only product states occur. On the other hand, the Hilbert space of a compound quantum system is given by the tensor product of the Hilbert spaces of the subsystems, such that non-product states occur (so-called entangled states).

The possibility of (non)existence of quantum entangled states has been a long debate in physics. Already in their seminal paper of 1935 [15], Einstein, Podolsky and Roosen (EPR) argued that if such entanglement correlation is present in nature then quantum mechanics should be considered an incomplete theory. For almost three decades defenders of local realism and hidden variable theories stood against advocates of the Copenhagen interpretation of quantum mechanics. In 1964 Bell put forward a set of inequalities which can be violated in quantum mechanics, but which need to hold if local hidden variables exist [16]. This shifted the debate from a purely philosophical level to the experimental level. Aspect and his team showed that indeed (generalized) Bell inequalities are violated [17]. On the more mathematical level local hidden variable theories were shown to be incompatible with the structures of quantum mechanics: starting with the first no-go theorems (of von Neumann [18]) till the more elaborated theorems of Kochen–Specker and others [19, 20], which showed that if such hidden variables exist — giving quantum theory an underlying, hidden reality — that then these hidden variables should at least be contextual, i.e., dependent on the measurement context. This fact is taken explicitly into account in the Hidden Measurement Approach following which quantum probabilities can be interpreted as due to an irreducible lack of knowledge in the interaction between system and measurement apparatus [6].

THE QUANTUM SPHERE MODEL

Within the Hidden Measurement Approach various macroscopic models have been proposed with a quantum structure on their set of states and properties, and a quantum probability distribution on the set of outcomes. The sphere model is a generalization of the Bloch sphere representation[1] such that also the measurements are represented [6, 21]. The main differences with the standard use of the Bloch representation are as follows. First, in our approach all points of the Bloch sphere represent states of the spin, such that points on the surface correspond to pure states, while interior points correspond to density states. This is because an arbitrary point $u(r,\theta,\phi)$, $r \in [0,1], \theta \in [0,\pi], \phi \in [0,2\pi)$, of the Bloch sphere can be expressed as a convex linear combination $u(r,\theta,\phi) = ru(1,\theta,\phi) + (1-r)u(0,\theta,\phi)$ from which follows the corresponding density state

$$D(r,\theta,\phi) = \frac{1}{2}\begin{pmatrix} 1 + r\cos\theta & r\sin\theta e^{-i\phi} \\ r\sin\theta e^{i\phi} & 1 - r\cos\theta \end{pmatrix}$$

In this expression $D(1,\theta,\phi) = |\psi\rangle\langle\psi|$ is the usual density state representation of a pure state, while $D(0,\theta,\phi)$ is the density matrix representing the center of the sphere.

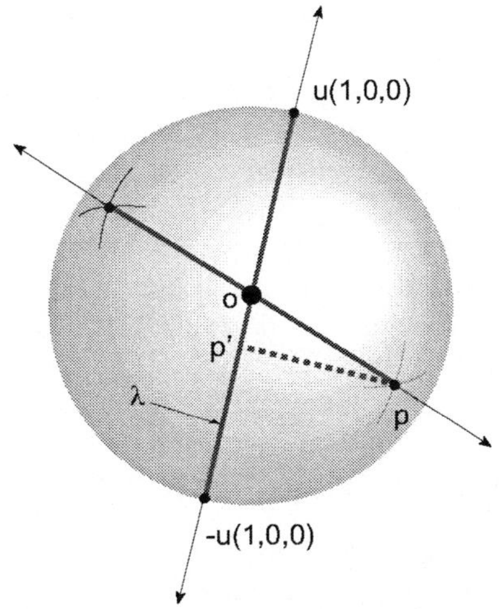

FIGURE 1. The macroscopic spin 1/2 model.

Next to this, the sphere model allows a representation of measurements (see Figure 1). The experiments e_u are defined as follows. We put an elastic of length 2 centered in the origin o of the sphere S^2 between the point u and its antipode $-u$. Let us denote the segment between u and $-u$ with the interval $[-u,u]$. Next, the particle falls from its initial position p orthogonally onto the interval $[-u,u]$ in the point p' where it stays attached to the elastic. Then the elastic breaks randomly in some point $\lambda \in (-u,u)$ such that two possibilities can occur. If the elastic breaks between p' and $-u$, the elastic will pull the point particle towards u where it stays attached and the experiment is said to yield the outcome 'spin up'. If on the other hand the elastic breaks between u and p', then the elastic will pull the particle towards $-u$, where it stays attached, and the measurement is said to yield outcome 'spin down'. To make the description of the experiment e_u complete, one could specify that in the event that the elastic breaks at exactly the point p' where the particle is attached, we assume that the measurement always yields the outcome 'spin up'. Notice however that this event has measure zero to occur such that it is physical irrelevant with respect to the resulting probability distribution over the set of outcomes. Let θ denote the angle between the state p of the system and the direction u of the measurement device. The probability for outcome 'spin up' is given by the length of the elastic between the projection point p' and the point $-u$, normalized by dividing by the total length of the elastic. This yields following probability $P(u \mid p)$ for the 'spin up' outcome and corresponding state transition from initial state p towards

[1] In this representation, a qubit $|\psi\rangle = \cos\frac{\theta}{2}e^{\frac{-i\phi}{2}}|0\rangle + \sin\frac{\theta}{2}e^{\frac{i\phi}{2}}|1\rangle$ is represented by the point $u(1,\theta,\phi) = (\sin\theta\cos\phi, \sin\theta\sin\phi, \cos\theta)$ on the surface of a 3-dimensional sphere, called the Bloch sphere.

final state u, eigenstate of the 'spin up' outcome:

$$P(u \mid p) = \frac{\cos\theta + 1}{2} = \cos^2\frac{\theta}{2}$$

Similarly we can calculate the probability for the outcome 'spin down' as

$$P(-u \mid p) = \frac{1 - \cos\theta}{2} = \sin^2\frac{\theta}{2}$$

These probabilities coincide with the quantum probabilities for a spin experiment on a spin-$\frac{1}{2}$ particle. Note that if one would have knowledge about where the elastic breaks, the measurement procedure happens deterministic. If we call e_u^λ the measurement that consists in performing e_u and such that the elastic breaks in the point λ for some $\lambda \in (-u, u)$, then, each time e_u is performed, it is actually one of the e_u^λ that takes place. We do not control this, in the sense that the e_u^λ are really 'hidden measurements' that we cannot choose to perform. The probability $\mu(e_u, p, o_1)$ that the experiment e_u gives the outcome o_1 if the entity is in state p is a randomization over the different situations where the hidden measurements e_u^λ give the outcome o_1 with the entity in state p. This is exactly the way we have calculated the probability for the quantum sphere model. Remark that if we have a physical entity S and we have a lack of knowledge about the state p of the physical entity S, then a theory describing this situation is necessarily a classical statistical theory with a Kolmogorovian probability model. However, if we have a physical entity S and a lack of knowledge about the measurement e_u to be performed on the physical entity S, and to be changing the state of the entity S, then we cannot describe this situation by a classical statistical theory, because the probability model that arises is non-Kolmogorovian. More generally, one could regard any quantum experiment as a class of 'hidden measurements' such that each hidden measurement by itself is deterministic, but a lack of knowledge about which hidden measurement is actually going on leads to a lack of knowledge on the level of the measurement outcomes, i.e., quantum probability can be explained as due to an uncontrollable (and irreducible) lack of knowledge on the interaction between the measurement device and the system. This interpretation is called the 'Hidden Measurement Approach' of quantum mechanics. Since it locates the lack of knowledge in the measurement interaction, it goes beyond the reach of standard 'no go theorems' for hidden variable theories mentioned in the previous section. Indeed, by its very nature the Hidden Measurement Approach is highly contextual and hence non-local from the very start.

MACROSCOPIC MODEL OF TWO SPIN $1/2$ IN THE SINGLET STATE

Next, let us show how by putting together two sphere models one can construct a macroscopic model for a compound system of two entangled spin 1/2 in the singlet state [7]. The compound system consists of two point particles in the center of two sphere models for a single spin 1/2 which are connected by a rigid but extendable rod, centered in c, i.e., in the middle of the line connecting the two point particles (see Figure 2). The joint experiment $e(a,b)$ is performed as follows. First, the two spheres reach the measurement apparatuses, with measurement direction a respectively b chosen by the experimenter. The spin measurements e_a and e_b are defined similarly as for the single sphere model. When one side is measured, the measurement apparatus draws one of the entities to one of the two possible outcomes with probability $1/2$ (because initially the point particle is in the center of the sphere). Because of the rod the other entity is drawn toward the opposite side of the sphere as compared with the first entity, in which we consider for the sake of illustration the event that e_a has yielded outcome 'spin up'). The connecting rod is then taken away and the second spin measurement is performed. The measurement $e(a,b)$ has 4 possible outcomes $(a,b), (a,-b), (-a,b)$ and $(-a,-b)$. Following Bell [16], we define expectation values $\mathbb{E}(i,j)$:

$$\mathbb{E}(i,j) = +1.P(o_i(up), o_j(up)) + 1.P(o_i(down), o_j(down)) - 1.P(o_i(up), o_j(down)) - 1.P(o_i(down), o_j(up))$$

Giving value $+1$ for outcome (a,b) or $(-a,-b)$, and value -1 for outcome $(a,-b)$ or $(-a,b)$, one finds for the expectation value $\mathbb{E}(a,b) = -a.b$, i.e., this corresponds with the expectation value for the spin measurements along directions a, b on a compound system of two entangled spin 1/2 particles in the singlet state [16]. It should be mentioned that other proposals of macroscopic simulations of quantum-like behaviour have been recently published in [22, 23] (see also references to previous publications by the same authors cited therein).

One could object that the macroscopic violation of Bell-inequalities has a different nature than the quantum violation. Indeed, it is not obvious that the 'faster than light aspect' identified in the quantum experiments is also

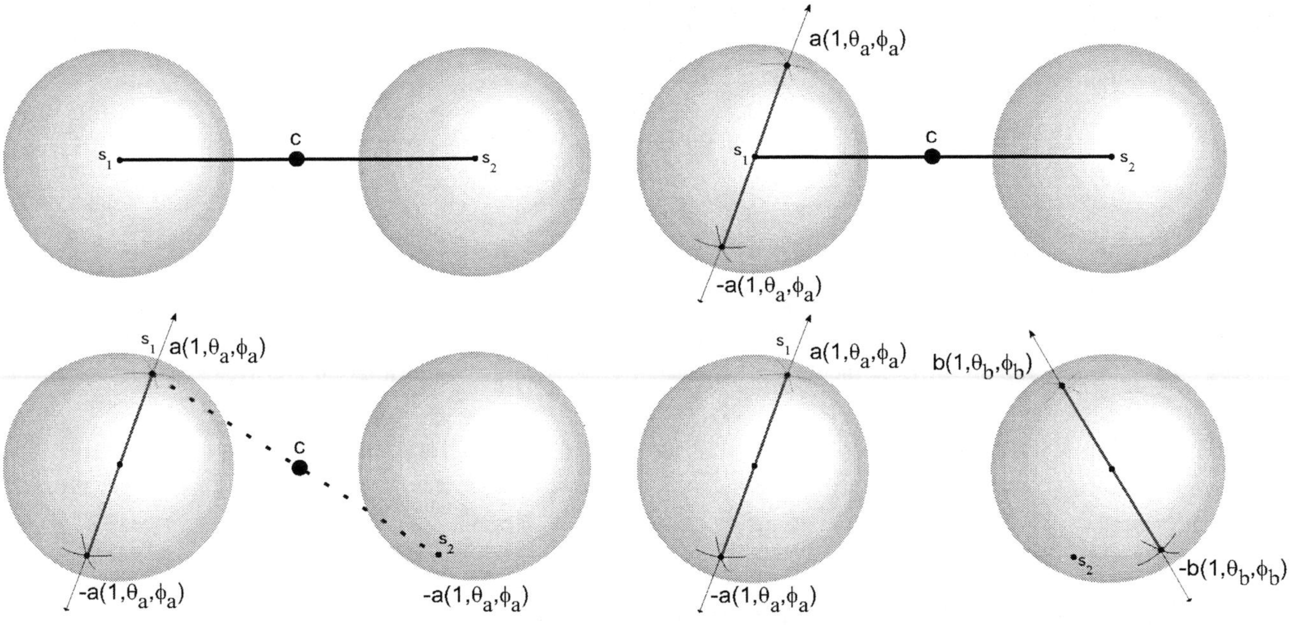

FIGURE 2. (a) The initial configuration of the macroscopic model of two entangled spin $1/2$ in the singlet state. (b) The measurement e_a is performed on the left point particle, drawing it towards the point a. (c) By means of the rigid rod, the right point particle is drawn towards the point $-a$ on its sphere. Next, the connecting rod is taken away. (d) Finally, the measurement e_b is performed on the right point particle.

in the macroscopic singlet model. However, we do not really have to focus on this aspect to make the point we want to make in the realm of the present paper. Indeed, it is not the speed effects present in this entanglement correlation which is vital in a quantum computation, but rather its algebraic structure. In this sense, if one could build a model of 2-qubit entanglement (even with 'slower than light' implementation), it is this possibility to implement entanglement in the compound system which makes it non-classical, and therefore impossible to implement by classical means.

GEOMETRICAL REPRESENTATION OF TWO ENTANGLED SPIN 1/2

Representing entanglement by constraint functions

In the previous section we discussed a compound system consisting of two connected sphere models which represented the singlet state of two entangled spin 1/2. In general, in a 2-qubit quantum algorithm the register will be in other entangled states as well. In [8] we show how this problem is solved for the coupled sphere models by introducing constraint functions. Let us briefly recall how these functions are introduced and some of their properties.

Let $\mathbb{C}_1^2$ and $\mathbb{C}_2^2$ be two copies of $\mathbb{C}^2$, which we label with indices 1 and 2 with the sole purpose of identifying them. A system of two entangled spin $\frac{1}{2}$ is described by means of an arbitrary unit vector $|\psi\rangle \in \mathbb{C}_1^2 \otimes \mathbb{C}_2^2$, which can always be written as $|\psi\rangle = \sum_{ij} \lambda_{ij} |e_1^i\rangle \otimes |e_2^j\rangle$, with $\lambda_{ij} \in \mathbb{C}$, and $\{|e_1^i\rangle\}$ and $\{|e_2^j\rangle\}$ bases of $\mathbb{C}_1^2$ and $\mathbb{C}_2^2$ respectively. Following von Neumann [18], when a measurement is performed on the first spin, it collapses into a spin state described by the unit vector $|x_1\rangle \in \mathbb{C}_1^2$ corresponding with the observed outcome, thus transforming the entangled state $|\psi\rangle$ into $(P_{|x_1\rangle} \otimes I)(|\psi\rangle)$, where $P_{|x_1\rangle}$ is the orthogonal projector on $|x_1\rangle$ in $\mathbb{C}_1^2$, and I is the unit operator in $\mathbb{C}_2^2$. The result is that the entangled spins end up in the product state $|x_1\rangle \otimes \sum_{ij} \lambda_{ij} \langle x_1, e_1^i\rangle |e_2^j\rangle$. This means that as a consequence of the measurement on the first spin, collapsing its state to $|x_1\rangle$, the second spin collapses to the state $\sum_{ij} \lambda_{ij} \langle x_1, e_1^i\rangle |e_2^j\rangle$. In an analogous way we can show that if a measurement is performed on the second spin, resulting in a collapse to the state

$x_2 \in \mathbb{C}_2^2$, the state of the first spin becomes $\sum_{ij} \lambda_{ij} \langle x_2, e_2^j \rangle |e_1^i\rangle$. Because of this, we arrive at the following definition.

Definition 1 (Constraint Functions) *The constraint functions $F_{12}(\psi)$ and $F_{21}(\psi)$ related to ψ are defined in the following way:*

$$F_{12}(\psi) \quad : \quad \mathbb{C}_1^2 \to \mathbb{C}_2^2 : |x_1\rangle \mapsto \sum_{ij} \lambda_{ij} \langle x_1, e_1^i \rangle |e_2^j\rangle$$

$$F_{21}(\psi) \quad : \quad \mathbb{C}_2^2 \to \mathbb{C}_1^2 : |x_2\rangle \mapsto \sum_{ij} \lambda_{ij} \langle x_2, e_2^j \rangle |e_1^i\rangle$$

The constraint functions map the state where one of the spins collapses to by a measurement to the state that the other spin collapses to under influence of the entanglement correlation. A detailed study of the constraint functions gives us a complete picture of how the entanglement correlation works as an internal constraint. Let us list some properties of the constraint functions:

- The constraint functions are canonically defined, i.e., their definition does not depend on the chosen bases.
- $F_{21}(\psi) \circ F_{12}(\psi) = D_1(\psi)$ is the partial trace density matrix over $\mathbb{C}_2^2$ and $F_{12}(\psi) \circ F_{21}(\psi) = D_2(\psi)$ is the partial trace density matrix over $\mathbb{C}_1^2$: $D_1(\psi) \equiv tr_{\mathbb{C}_1^2} |\psi\rangle\langle\psi|$, $D_2(\psi) \equiv tr_{\mathbb{C}_2^2} |\psi\rangle\langle\psi|$.

Constraint functions and the Schmidt diagonal form

To derive a complete view of how entanglement works as an internal constraint for a 2-particle system, we recall the relation between the Schmidt diagonal form [24] and the constraint functions. We begin by choosing the base $\{|x_1^1\rangle, |x_1^2\rangle\} = \left\{(\cos\frac{\theta}{2} e^{-i\frac{\phi}{2}}, \sin\frac{\theta}{2} e^{\frac{i\phi}{2}}), (-i\sin\frac{\theta}{2} e^{-i\frac{\phi}{2}}, i\cos\frac{\theta}{2} e^{\frac{i\phi}{2}})\right\}$ in $\mathbb{C}_1^2$. With respect to this basis, the expression for the density matrix $D(r, \theta, \phi)$ becomes

$$D_1(\psi) = \frac{1}{2} \begin{pmatrix} 1+r & 0 \\ 0 & 1-r \end{pmatrix}$$

Define now the following basis $\{|x_2^1\rangle, |x_2^2\rangle\} = \left\{\frac{\sqrt{2}}{\sqrt{1+r}} F_{12}(\psi)(|x_1^1\rangle), \frac{\sqrt{2}}{\sqrt{1-r}} F_{12}(\psi)(|x_1^2\rangle)\right\}$ in $\mathbb{C}_2^2$. One finds that $\|x_2^1\| = \|x_2^2\| = 1$ and $D_2(\psi)(|x_2^1\rangle) = \frac{1+r}{2}|x_2^1\rangle$, $D_2(\psi)(|x_2^2\rangle) = \frac{1-r}{2}|x_2^2\rangle$. Hence $|x_2^1\rangle$ and $|x_2^2\rangle$ are normalized eigenvectors of $D_2(\psi)$ with eigenvalues $\frac{1+r}{2}$ and $\frac{1-r}{2}$ respectively. Therefore, with respect to the basis $\{|x_2^1\rangle, |x_2^2\rangle\}$, $D_2(\psi)$ is expressed as

$$D_2(\psi) = \frac{1}{2} \begin{pmatrix} 1+r & 0 \\ 0 & 1-r \end{pmatrix}$$

Finally, one can show that the Schmidt diagonal form of $|\psi\rangle$ is given by $|\psi\rangle = \frac{\sqrt{1+r}}{\sqrt{2}}|x_1^1\rangle \otimes |x_2^1\rangle + \frac{\sqrt{1-r}}{\sqrt{2}}|x_1^2\rangle \otimes |x_2^2\rangle$.

Geometrical representation of entanglement

Let us remark that $F_{12}(\psi)$ does not conserve orthogonality (except for the basis vectors $\{|x_1^1\rangle, |x_1^2\rangle\}$ which are mapped onto the orthogonal basis vectors $\{|x_2^1\rangle, |x_2^2\rangle\}$ and in the case that $r = 0$, i.e., ψ is the singlet state). Also, $F_{12}(\psi)$ does not conserve the norm, since $\|F_{12}(\psi)(|x\rangle)\|^2 = \frac{1}{2}(1 + r\cos\theta)$. This means that we have to look at the *normalized* image of a vector $|x\rangle = x(\theta_1, \phi_1)$. This vector is mapped to $|y\rangle = y(\theta_2, \phi_2) = \frac{1}{\|F_{12}(\psi)(|x\rangle)\|} F_{12}(\psi)(|x\rangle)$. One derives the following expression:

$$y(\theta_2, \phi_2) \cdot x_2^1(\theta, \phi) = \frac{r + \cos\theta_1}{1 + r\cos\theta_1}$$

from which one can deduce that straight lines containing the center o_1 on the first sphere are mapped to straight lines through the point $u(r, 0, 0)$ in the second sphere. This gives a geometrical representation of 'stretching' on the second sphere (see Figure 3). Hence the entanglement correlation is represented by a rotation (reflecting the basis vectors in the Schmidt diagonal form) and a stretching depending on the parameter r, i.e., 'amount of entanglement'.

Vice versa, given a general entangled state $|\psi\rangle$ expressed in the canonical basis, we can derive the Schmidt diagonal form using the constraint functions. From the reduced density matrix $D_1(\psi) = F_{21}(\psi) \circ F_{12}(\psi)$

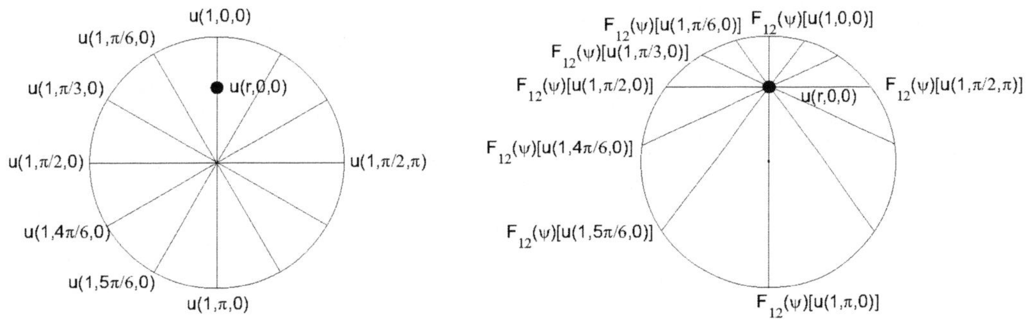

FIGURE 3. Geometrical representation of two entangled spin $1/2$.

one can calculate the basis vectors $\{|x_1^1\rangle,|x_1^2\rangle\}$, and consequently also the second pair of basis vectors by $\{|x_2^1\rangle,|x_2^2\rangle\} = \left\{\frac{\sqrt{2}}{\sqrt{1+r}}F_{12}(\psi)(|x_1^1\rangle), \frac{\sqrt{2}}{\sqrt{1-r}}F_{12}(\psi)(|x_1^2\rangle)\right\}$. By expressing the reduced density matrix in the eigenvectors basis $\{|x_1^1\rangle,|x_1^2\rangle\}$ one obtains the value for the parameter r, i.e., the amount of entanglement.

GEOMETRICAL REPRESENTATION OF 2-QUBIT ALGORITHMS

Deutsch problem

Deutsch problem is as follows [3]: given a $\{0,1\}$-valued function f defined on a two-point domain $\{0,1\}$, determine with a single call of the 'oracle' whether f is constant or balanced. On a classical digital computer, this problem is impossible to solve. However, on a quantum computer this is possible. Consider preparing a quantum register of two qubits in an input state:

$$\psi_0 = (H\otimes H)\,|01\rangle = \frac{|0\rangle+|1\rangle}{\sqrt{2}} \otimes \frac{|0\rangle-|1\rangle}{\sqrt{2}} = \frac{|00\rangle+|10\rangle-|01\rangle-|11\rangle}{2}$$

Calling the oracle corresponds with applying a suitable unitary transformation U_f which maps $|xy\rangle$ onto $U_f\,|xy\rangle = |x(y\oplus f(x))\rangle$, where $x,y\in\{0,1\}$ and $\oplus$ is the sum modulo 2. Denoting $\overline{f(i)} = f(i)\oplus 1$ and noticing that $|f(i)\rangle - \left|\overline{f(i)}\right\rangle = (-1)^{f(i)}(|0\rangle-|1\rangle)$ one obtains that $\psi_1 = U_f(\psi_0) = \frac{|0f(0)\rangle+|1f(1)\rangle-|0\overline{f(0)}\rangle-|1\overline{f(1)}\rangle}{2}$ which can be written as a product state:

$$\psi_1 = \left(\frac{(-1)^{f(0)}\,|0\rangle+(-1)^{f(1)}\,|1\rangle}{\sqrt{2}}\right)\otimes\frac{|0\rangle-|1\rangle}{\sqrt{2}}$$

After applying a Hadamard transformation $H\otimes H$ on both qubits we obtain the following state:

$$\psi_2 = (H\otimes H)U_f(\psi_0) = (-1)^{f(0)}[(1-(f(0)\oplus f(1)))\,|0\rangle + (f(0)\oplus f(1))\,|1\rangle]\otimes|1\rangle$$

- If f is constant, $f(0)=f(1)$ and the state ψ_2 can be rewritten as:

$$\psi_{2,c} = (-1)^{f(0)}\,|0\rangle\otimes|1\rangle$$

i.e., if f is constant, then the value of the first qubit is $|0\rangle$.

- If f is balanced, $f(1) = f(0)\oplus 1$ then the state ψ_2 can be rewritten as:

$$\psi_{2,b} = (-1)^{f(0)}\,|1\rangle\otimes|1\rangle$$

such that if f is balanced, then the first qubit is $|1\rangle$.

This shows that a quantum computer solves Deutsch problem with 1 call of the oracle, by acting on the state of the 2-qubit system as a whole. It is important to notice that Deutsch algorithm uses product states only, which in principle makes it possible to simulate it on a 'classical' computer based on the polarisation states of classical light beams [25].

Deutsch problem on the geometric model

The intermediate states in Deutsch algorithm are product states such that one can represent these states of the compound system in a trivial way on the geometric model by the Cartesian product of the states of the individual qubits. Therefore, in the next section we consider a 2-qubit algorithm exploiting non-product states, such that the geometrical representation by two correlated sphere models becomes less trivial.

Arvind problem

The problem put forward by Arvind and his colleagues is the following [9, 10]: given a $\{0,1\}$-valued function f defined from a 2-qubit domain space to a one-bit range space (i.e., $f(x) : \{0,1\}^2 \to \{0,1\}$), determine with 2 calls of the 'oracle' whether f is even or odd. Since there are 4 possible input values $(00),(01),(10),(11)$ each of which has 2 possible output values $(0,1)$ there are 16 possible functions f. These can be divided into subclasses depending on the number of values 0 respectively 1, i.e., $[0,4],[1,3],[2,2],[3,1],[4,0]$. One has 8 'even' functions $([0,4],[2,2],[4,0])$ and 8 'odd' functions $([1,3],[3,1])$. Clearly, on a classical computer Arvind's problem is impossible to solve, since it would require calling the oracle four times. However, Arvind et al. have shown how to solve this problem on a quantum computer. Again, calling the oracle corresponds with applying a unitary transformation $U_f : |x\rangle_{2-qubit} \to (-1)^{f(x)}|x\rangle_{2-qubit}$

$$U_f = \begin{pmatrix} (-1)^{f(00)} & 0 & 0 & 0 \\ 0 & (-1)^{f(01)} & 0 & 0 \\ 0 & 0 & (-1)^{f(10)} & 0 \\ 0 & 0 & 0 & (-1)^{f(11)} \end{pmatrix}$$

There are 16 possible U_f matrices, of which 8 are separable and 8 are non-separable, i.e., they are entangling, e.g. the matrix with diagonal entries [1,1,1,-1]. One could notice that the subclass of even functions $[0,4],[4,0],[2,2]$ reflects the problem of constant or balanced functions in Deutsch problem. Hence it should come as no surprise that for even f the matrix U_f is separable and no entanglement is used in Arvind's algorithm, just as in Deutsch algorithm. However, for odd f the matrix U_f is non-separable which shows that in that case entanglement is necessary to solve Arvind's problem.

Let the quantum register of two qubits be prepared in the input state $\psi_0 = |00\rangle$, then the consecutive unitary transformations in Arvind's algorithm are as follows:

$$\psi_0 = |00\rangle \longrightarrow (H\otimes H)U_f(1\otimes H)U_f(H\otimes H)|00\rangle$$

such that the consecutive 'intermediate' states in Arvind's algorithm are given by

$$\psi_1 = \frac{|00\rangle+|01\rangle+|10\rangle+|11\rangle}{2}, \quad \psi_2 = \frac{(-1)^{f(00)}|00\rangle+(-1)^{f(01)}|01\rangle+(-1)^{f(10)}|10\rangle+(-1)^{f(11)}|11\rangle}{2}$$

$$\psi_3 = \frac{1}{2\sqrt{2}}\begin{bmatrix} (-1)^{f(00)}+(-1)^{f(01)} \\ (-1)^{f(00)}-(-1)^{f(01)} \\ (-1)^{f(10)}+(-1)^{f(11)} \\ (-1)^{f(10)}-(-1)^{f(11)} \end{bmatrix}, \quad \psi_4 = \frac{1}{2\sqrt{2}}\begin{bmatrix} 1+(-1)^{f(00)+(01)} \\ (-1)^{f(00)+(01)}-1 \\ 1+(-1)^{f(10)+(11)} \\ (-1)^{f(10)+(11)}-1 \end{bmatrix}$$

$$\psi_5 = \frac{1}{2\sqrt{2}}\begin{bmatrix} \left((-1)^{f(00)+f(01)}+(-1)^{f(10)+f(11)}\right)|00\rangle \\ +2|01\rangle \\ +\left((-1)^{f(00)+f(01)}-(-1)^{f(10)+f(11)}\right)|10\rangle \end{bmatrix}$$

- If f is even the final state is given by $\psi_{5,e} = \frac{1}{\sqrt{2}}(\pm|00\rangle+|01\rangle)$
- If f is odd the final state is $\psi_{5,o} = \frac{1}{\sqrt{2}}(\pm|10\rangle+|01\rangle)$

It is important to notice that although $\psi_{5,e} \neq \psi_{5,o}$, these states are *not* orthogonal, which makes it not possible to distinguish them in the standard approach of quantum computation. However, on a NMR based quantum computer these two states *are* distinguishable [10].

Arvind problem on the geometric model

The density matrices for the intermediate states ψ_i in Arvind's algorithm are given as follows.

- If f is even, the reduced density matrices $D_1(\psi_i)$ are given by:

$$
\begin{aligned}
D_1(\psi_1) &= \begin{pmatrix} \frac{1}{2} & \frac{1}{2} \\ \frac{1}{2} & \frac{1}{2} \end{pmatrix}, \; D_1(\psi_2) = D_1(\psi_3) = \begin{pmatrix} \frac{1}{2} & (-)^{f(00)+f(10)}\frac{1}{2} \\ (-)^{f(00)+f(10)}\frac{1}{2} & \frac{1}{2} \end{pmatrix} \\
D_1(\psi_4) &= \begin{pmatrix} \frac{1}{2} & \frac{1}{2} \\ \frac{1}{2} & \frac{1}{2} \end{pmatrix}, \; D_1(\psi_5) = \begin{pmatrix} 1 & 0 \\ 0 & 0 \end{pmatrix}
\end{aligned}
$$

such that the set of eigenstates for each matrix $D_1(\psi_i)$ is given by $\left\{\frac{|0\rangle+|1\rangle}{\sqrt{2}}, \frac{|0\rangle-|1\rangle}{\sqrt{2}}\right\}, i = 1,\ldots,4$; the set of eigenstates for matrix $D_1(\psi_5)$ is $\{|0\rangle, |1\rangle\}$. The consecutive states in Arvind's algorithm are given by

$$
\begin{aligned}
\psi_1 &= \frac{|0\rangle+|1\rangle}{\sqrt{2}} \otimes \frac{|0\rangle+|1\rangle}{\sqrt{2}}, \; \psi_2 = \frac{|0\rangle+(-)^{f(00)+f(10)}|1\rangle}{\sqrt{2}} \otimes (-)^{f(00)} \frac{|0\rangle+(-)^{f(00)+f(01)}|1\rangle}{\sqrt{2}} \\
\psi_3 &= \frac{|0\rangle+(-)^{f(00)+f(10)}|1\rangle}{\sqrt{2}} \otimes (-)^{f(00)} \left[|0\rangle \left(\frac{1+(-)^{f(00)+f(01)}}{2}\right) + \left(\frac{1-(-)^{f(00)+f(01)}}{2}\right)|1\rangle \right] \\
\psi_4 &= \frac{|0\rangle+|1\rangle}{\sqrt{2}} \otimes \left[\left(\frac{(-)^{f(00)+f(01)}+1}{2}\right)|0\rangle + \left(\frac{(-)^{f(00)+f(01)}-1}{2}\right)|1\rangle \right], \; \psi_5 = |0\rangle \otimes \frac{(-)^{f(00)+f(01)}|0\rangle+|1\rangle}{\sqrt{2}}
\end{aligned}
$$

which in fact are all product states, making the representation by two coupled sphere models trivial since 'there is no entanglement correlation'.

- If f is odd the reduced density matrices $D_1(\psi_i)$ are given by:

$$
D_1(\psi_1) = \begin{pmatrix} \frac{1}{2} & \frac{1}{2} \\ \frac{1}{2} & \frac{1}{2} \end{pmatrix}, \; D_1(\psi_i) = \begin{pmatrix} \frac{1}{2} & 0 \\ 0 & \frac{1}{2} \end{pmatrix}, i = 2,\ldots,5
$$

Again, ψ_1 is a product state since applying the Hadamard gate on both qubits in the ground state $|00\rangle$ results in a product state. However, for $i = 2,\ldots,5$ the reduced density matrices $D_1(\psi_i)$ correspond with an entangled state, namely the singlet state. There is no unique Schmidt diagonal form for the singlet state and hence no unique representation of the intermediate states in Arvind's algorithm on the geometric model for odd f. Nevertheless, the fact that these are singlet states also has an unexpected advantage, because these singlet states can be represented by the rigid rod model in a mechanistic way, making it possible to represent Arvind's algorithm 'directly' on the macroscopic singlet model.

CONCLUSIONS

In this paper, we adopt an operational approach to quantum mechanics, making it natural to consider macroscopic 'quantum-like' systems. Although the entanglement correlation present in these models is of a different nature (i.e., slower than light) than quantum entanglement, they reproduce the same probability distributions over the set of outcomes as the standard quantum mechanical systems. This results in a geometric model for entanglement which allows to represent the key intermediate steps in a quantum algorithm. More specifically, we have shown how a compound system of two entangled qubits in a non-product state can be described in a complete way by extracting entanglement into an internal constraint between the two qubits. So-called constraint functions describe the behavior of the state of one of the spins if measurements are executed on the other spin. Therefore the non-product state can be substituted by means of the states of the individual qubits and this internal constraint function, which models the entanglement which is present in a direct way. By making use of the sphere model representation for the spin 1/2, this allows for an easy to grasp visual support for the developed formalism.

We have discussed two 2-qubit quantum algorithms and illustrated them on the geometric model. As our geometric model illustrates nicely, Deutsch algorithm does not require any entanglement between the two qubits. However,

Arvind's algorithm requires entangled states, and we have illustrated the intermediate steps in this algorithm by means of the corresponding entangled states in our geometric model. This shows how in principle modelling entanglement through constraint functions allows for a rigorous method to follow its form and effect in quantum computations. This is very important since at this stage it is not at all clear how entanglement can be exploited in a systematic way. In this sense, our geometric model is a step towards the better understanding of the exploitation of 2-qubit entanglement in a quantum algorithm. Unexpectedly, the considered intermediate entangled states in Arvind's algorithm are all singlet states. On the other hand, this allows us to represent these states with the macroscopic model in which a rigid rod explicitly realizes the entanglement correlation.

ACKNOWLEDGMENTS

This research was supported by Project G.0452.04 of the Fund for Scientific Research–Flanders (Belgium)(F.W.O.–Vlaanderen).

REFERENCES

1. R. Feynman, Simulating physics with computers, *Int. J. Theor. Phys.* **21**, 467–488 (1982).
2. R. Feynman, Quantum mechanical computers, *Found. Phys.* **16**, 507–531 (1986).
3. D. Deutsch, Quantum theory, the Church–Turing principle and the universal quantum computer, *Proc. R. Soc.* **A400**, 97–117 (1985).
4. P.W. Shor, Algorithms for quantum computation, discrete logarithms and factoring, *Proc. 35th Ann. Symp. on Foundations of Computer Science* (Los Alamitos: IEEE Computer Society Press), 124–134 (1994).
5. D. Aerts, E. Colebunders, A. Van der Voorde and B. Van Steirteghem, State property systems and closure spaces: a study of categorical equivalence, *Int. J. Theor. Phys.* **38**, 359–385 (1999).
6. D. Aerts, A possible explanation for the probabilities of quantum mechanics, *J. Math. Phys.* **27**, 202–210 (1986).
7. D. Aerts, A mechanistic classical laboratory situation violating the Bell inequalities with 2sqrt(2), exactly 'in the same way' as its violations by the EPR experiments, *Helv. Phys. Acta* **64**, 1–23 (1991).
8. D. Aerts, E. D'Hondt and B. D'Hooghe, A geometrical representation of entanglement as internal constraint, *Int. J. Theor. Phys.* **44**, 897–907 (2005).
9. Arvind and N. Mukunda, A two-qubit algorithm involving quantum entanglement, *Preprint* quant-ph/0006069 (2000).
10. K. Dorai, Arvind and A. Kumar, Implementation of a Deutsch-like quantum algorithm utilizing entanglement at the two-qubit level on an NMR quantum information processor, *Phys. Rev.* A **63**, 034101/1–4 (2001).
11. V. Vedral, A. Barenco and A. Ekert, Quantum networks for elementary arithmetic operations, *Phys. Rev.* A **54**, 147–153 (1996).
12. C. Piron, Axiomatique quantique, *Helv. Phys. Acta* **37**, 439–468 (1964).
13. C. Piron, *Foundations of quantum physics*, Reading, Massachusetts, 1976.
14. A.N. Kolmogorov, *Grundbegriffe der Wahrscheinlichkeitsrechnung*, Springer, Berlin, 1933.
15. A. Einstein, B. Podolsky and N. Rosen, Can quantum-mechanical description of physical reality be considered complete? *Phys. Rev.* **47**, 777–780 (1935).
16. J.S. Bell, On the Einstein Podolsky Rosen paradox, *Physics* **1**, 195–200 (1964).
17. A. Aspect, J. Dalibard and G. Roger, Experimental test of Bell's inequalities using time-varying analyzers, *Phys. Rev. Lett.* **49**, 1804–1807 (1982).
18. J. von Neumann, *Mathematische Grundlagen der Quantenmechanik,* Springer, Berlin, 1932.
19. S. Kochen and E.P. Specker, The problem of hidden variables in quantum mechanics, *J. Math. and Mech.* **17**, 59–87 (1967).
20. J.M. Jauch and C. Piron, Can hidden variables be excluded in quantum mechanics? *Helv. Phys. Acta* **36**, 827–837 (1963).
21. M. Czachor, On classical models of spin, *Found. Phys. Lett.* **5**, 249–264 (1992).
22. A. Grib, A. Khrennikov, G. Parfionov and K. Starkov, Quantum equilibria for macroscopic systems, *J. Phys. A: Math. Gen.* **39**, 8461–8475 (2006).
23. A. Khrennikov, Quantum-like representation of macroscopic configurations, quant-ph/0705.3898 (2007).
24. M. Nielsen and I. Chuang, *Quantum computation and quantum information*, University Press, Cambridge, 2000.
25. Arvind, Quantum entanglement and quantum computational algorithms, *Proc. of Int. Winter Institute on Found. of Quantum Theory and Quantum Optics (2000 Calcutta) Pramana Jr. of Physics* **56**, 357–365 (*Preprint* quant-ph/0012116) (2001).

From Copenhagen to Neo-Copenhagen Interpretation

Willem M. de Muynck

Theoretical Physics, Eindhoven University of Technology, P.O.B. 513, 5600 MB Eindhoven, the Netherlands

Abstract. Positive and negative features of the Copenhagen interpretation are discussed. As positive features can be mentioned its pragmatism and its awareness of the crucial role of measurement. However, the main part of the contribution is devoted to the negative features, to wit, its pragmatism (once again), its confounding of preparation and measurement, its *classical* account of measurement, its completeness claims, the ambiguity of its notion of correspondence, its confused notion of complementarity. It is demonstrated how confusions and paradoxes stemming from the negative features of the Copenhagen interpretation can be dealt with in an amended interpretation, to be referred to as 'neo-Copenhagen interpretation', in which the role of the measuring instrument is taken seriously by recognizing the quantum mechanical character of its interaction with the microscopic object. The ensuing necessity of extending the notion of a quantum mechanical observable from the Hermitian operator of the standard formalism to the positive operator-valued measure of a generalized formalism yields a sound mathematical basis for a transition from the Copenhagen contextualistic-realist interpretation to the neo-Copenhagen empiricist one. Applications to the uncertainty relations and to the Bell inequalities are briefly discussed.

Keywords: Copenhagen interpretation, generalized formalism
PACS: 01.70.+w, 03.65.Ca, 03.65.Ta, 03.65.Ud

1. INTRODUCTION

Interpretations of physical theories are neither true nor false. Even if they are not completely internally consistent they may be thought to provide useful rules of correspondence between the mathematical formalism and the physical reality a theory purports to describe. It is well-known that the Copenhagen interpretation of quantum mechanics is no exception. Even 80 years after its conception it is impossible to say that there exists a unique and internally consistent interpretation by that name. On the contrary, this interpretation has been characterized (Feyerabend [1]) as "not a single idea but a mixed bag of interesting conjectures, dogmatic declarations, and philosophical absurdities."

In my contribution to a 'conference devoted to 80 years of Copenhagen Interpretation' it therefore is not possible to present a eulogy. On the contrary, my main attention will be directed toward the many imperfections of the Copenhagen interpretation. Among these I do not count the pragmatic attitude the founding fathers of quantum mechanics have displayed while developing the theory so as to be able to come to grips with the "strange" results their experiments confronted them with. On the contrary, at that time a pragmatic approach, not bothering too much about the physical meaning of the applied mathematics, turned out to be advantageous to rapid scientific progress. However, maintaining such a pragmatism for 80 years may have become detrimental both with respect to development of fundamental insights as well as experimental applications. As seen from table 1 the list of negative features is considerably longer than that of the Copenhagen interpretation's positive features. Thus, we have to deal with the Copenhagen

TABLE 1. Positive and negative features of the Copenhagen interpretation

Positive features	Negative features
+1. pragmatism	-1. pragmatism
+2. crucial role of measurement	-2. confusion of preparation and measurement
	-3. classical account of measurement
	-4. completeness claims
	-5. ambiguous notion of correspondence
	-6. confused notion of complementarity

preoccupation with measurement, having as a consequence a confusion of preparation and measurement. Another negative feature is the *classical* account of measurement, overlooking a crucial property of measurements performed on *microscopic* objects by means of *macroscopic* measuring instruments. Moreover, each of the three characteristics of the Copenhagen interpretation, viz. completeness, correspondence, and complementarity, is liable to criticism and will be criticized.

CP962, *Quantum Theory, Reconsideration of Foundations – 4,*
edited by G. Adenier, A. Yu. Khrennikov, P. Lahti, V. I. Man'ko, and Th. M. Nieuwenhuizen

Any of these issues might be sufficient to reject the Copenhagen interpretation as a useful interpretation of quantum mechanics. Yet, there is one important positive feature in the Copenhagen appreciation of quantum mechanics as a description of microscopic reality, viz. its recognition of the *crucial role played by measurement*, which probably outweighs its imperfections. This feature, often considered a weakness of the interpretation, is actually an asset, making quantum mechanics different from classical theory in a truly revolutionary way. It makes quantum mechanics a paradigm of the structuralist methodology (e.g. Suppe [2]) in which the Einsteinian idea of a physical theory as a 'description of an *objective* reality (i.e. being independent of the observer, including his measuring instruments)' has been relinquished.

2. THE COPENHAGEN INTERPRETATION AND MEASUREMENT

2.1. Crucial role of measurement

Classical mechanics is generally presented as representing knowledge about an *objective* reality, i.e., a reality as it is independently of being observed by human observers, and, in particular, not being interfered with by their measuring instruments. The impossibility of an interpretation of quantum mechanics as an *objective* description of reality was a main reason for Einstein to disqualify quantum mechanics as not being an adequate physical theory. On the other hand, Bohr was deeply convinced that the existence of a non-vanishing quantum of action $h \neq 0$ entailed a fundamental impossibility of such objective knowledge: only knowledge could be obtained about 'microscopic reality as it is *in interaction with a measuring instrument*', i.e., *contextual* knowledge (cf. section **4**). The Heisenberg uncertainty relations were considered as evidence of this, the quantized interaction of an object with different measuring instruments (of incompatible observables) being held responsible for the impossibility of obtaining simultaneously sharp knowledge about e.g. position and momentum.

According to Bohr and Heisenberg quantum mechanics did not represent any objective knowledge, and it did not need to do so because such knowledge would not be verifiable. It should nevertheless be stressed here that the discussion between Bohr and Einstein was not over the question of verifiability, but over the question of whether it is possible to obtain knowledge about a microscopic object *without in any way interacting with it* [3], i.e. whether quantum mechanics can yield an *objective* description (Einstein), or whether we should content ourselves with a *contextual* one (Bohr). Only after many years we have gradually become convinced (e.g., by the Kochen-Specker [4] and the Bell [5] theorems) that it is *im*possible to attribute sharp values to quantum mechanical observables as *objective* properties possessed by a microscopic object independently of any measurement. Bohr's contextualism, to the effect that quantum mechanical observables are well-defined only within the context of a measurement serving to measure that very observable, can be seen as an indication that physical theories do not have the absolute property of being either universally true or universally false, but that they are applicable on a certain domain of experimentation, for quantum mechanics its domain of validity being co-determined by the measurement arrangement. The Copenhagen insight that measurement plays an ineradicable role in the interpretation of quantum mechanics is a fundamental epistemological attainment, in a more general form finding application far beyond quantum mechanics and even physics.

2.2. Confusion of preparation and measurement

However, the Copenhagen preoccupation with measurement has its drawback. As a consequence of the emphasis put on 'measurement', there was a tendency to formulate all physical operations in terms of this particular one. Measurement was seen where there is none. Thus, in Heisenberg's version of the Copenhagen interpretation a measurement result actually corresponds to a property of the microscopic object in its *final* state (Heisenberg [6]). Consider, for instance, the Stern-Gerlach experiment. Here, an atom with non-vanishing angular momentum, after traversing an inhomogeneous magnetic field may finally be found in one of a number of different beams (cf. figure 1). Finding the atom in a beam is interpreted as a determination of the value of a component of the atom's angular momentum. It is important to note here that for performing such a measurement it is essential that detectors are set up in the beams. Without these detectors the experimental arrangement does not correspond to a measurement but to a conditional *preparation* (e.g. de Muynck [7], section 3.2.6 and 3.3.4), allowing to perform a measurement of an *arbitrary* observable conditional on the beam (state $|\psi_m\rangle$) the atom happens to be in. Unfortunately, following Heisenberg [6] the Stern-Gerlach arrangement is generally interpreted as a measurement also if no detectors are

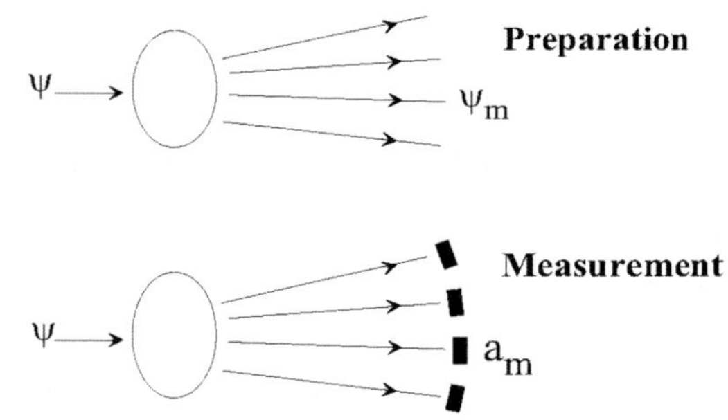

FIGURE 1. Preparation and measurement.

present (sometimes referred to as a 'preparative measurement'), thus allowing the experiment to satisfy (be it only approximately) the von Neumann projection postulate, stating that the conditionally prepared state $|\psi_m\rangle$ equals the eigenstate $|a_m\rangle$ corresponding to the eigenvalue that allegedly would have been found if the measurement had been performed by inserting a detector. It is rather evident that if von Neumann's postulate holds if the detector is absent, its presence will in general change the state appreciably, thus showing the absurdity of the projection postulate as a measurement principle.

As an example of the confusion of preparation and measurement should be mentioned here the EPR setup (figure 2a), presented in [3] as a *measurement* of a property of particle 2 (without in any way interacting with that particle). Unfortunately, Bohr accepted the EPR proposal as such, even though it is not really a measurement of particle 2 but rather a preparation of that particle. The EPR setup should be compared with an EPR-Bell experiment (figure 2b), like the ones performed by Aspect and co-workers [8] to test the Bell inequalities by means of a joint measurement on *both* particles. Unfortunately, experiments of the EPR-Bell type are generally referred to as EPR experiments, thus veiling the fundamental difference of these experiments.

2.3. Bohr's classical account of measurement

The stress laid by Bohr on the macroscopicity of measuring instruments, and the necessity to account for the results of measurement in *classical* terms (cf. section **4**) is well-known and need not be discussed here. It should be emphasized, however, that by thus restricting the attention exclusively to the macroscopic part of the measurement it was neglected that a measurement performed on a microscopic object must also have a *microscopic* part, viz. a part that is sensitive to the *microscopic* information, so as to be able to realize information transfer from the microscopic object to the measuring instrument. For instance, in the Stern-Gerlach experiment the initial phase of the measurement (nowadays generally called the 'pre-measurement') in which the atom interacts with a magnetic field so as to slightly influence the atom's position, is a microscopic process which should be described quantum mechanically rather than classically. In the amplification phase of the measurement the atom's position can be thought to change classically so as to create a sufficient distance between the beams to distinguish one from the other. It therefore seems that the microscopic rather than the macroscopic part of the measurement is the interesting one. Bohr's exclusive attention to

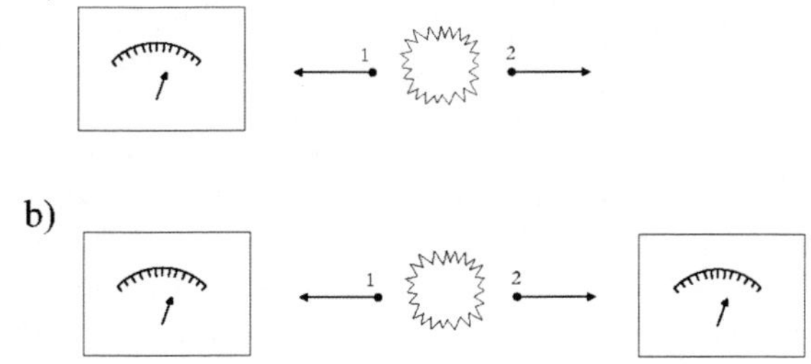

FIGURE 2. a) EPR experiment; b) EPR-Bell experiment.

the macroscopic part has not contributed to achieving this insight, to say the least. In particular, the development of the generalized formalism is a consequence of a rigorous application of quantum mechanics to the interaction of the microscopic object and the measuring instrument (cf. section **3.2**).

It should be noted that von Neumann and Heisenberg did consider a quantum mechanical description of measurement. However, they did not challenge Bohr's classical description; their purpose was only to prove that a quantum mechanical description of measurement is compatible with Bohr's classical one. In particular, reliance on the projection postulate c.q. restriction to measurements approximately satisfying it, was responsible for the fruitlessness of these by itself praiseworthy undertakings.

3. (IN)COMPLETENESS OF QUANTUM MECHANICS

3.1. (In)completeness in a wider sense

Two senses of '(in)completeness of quantum mechanics' should be distinguished, viz. '(in)completeness in a *wider* sense', dealing with the question of 'whether hidden variables are possible', and '(in)completeness in a *restricted* sense', addressing the question of 'whether the standard formalism of quantum mechanics describes all possible measurements within the domain of quantum mechanics'. In the past the main interest has been attracted by the first issue. I will not go into this, but want to restrict myself to demonstrating the relative futility of the discussion by comparing the answer given by Copenhagen physicists A.D. 1935 (viz. "Quantum mechanics is complete; there are no hidden variables") to the answer most practicing physicists would probably give A.D. 2007 (viz. "Quantum mechanics is incomplete; 'no go' theorems of hidden variables theories were found to be defective, possibly at the expense of introducing nonlocality (Bohm, Bell)").

Contrary to the Copenhagen idea that the wave function yields a complete description of an individual object, it seems nowadays to be generally recognized that the wave function does rather describe an *ensemble*. This insight has been gained by careful analysis of interference experiments at low incident particle rates in which impacts of *individual* particles can be registered. From these experiments it can be inferred that what is described by the wave function (i.e. the interference pattern) is not generated by an individual particle but by an ensemble of such particles. The Copenhagen idea that the wave function would be a complete description of an *individual* object does seem to be obsolete by now.

Does this mean that Einstein was right when attempting in the EPR paper [3] to prove the incompleteness of quantum mechanics? Not completely so. Actually, Einstein was bound to fail to do so because he attempted to prove the existence of hidden variables by identifying these with 'values of quantum mechanical observables'. The impossibility of this was rigorously demonstrated, only much later, by Kochen and Specker's proof [4] of the theorem named after them. However, this does not exclude the possibility of subquantum theories involving hidden variables of a different kind. In any case does it seem to be evident that within any theory aspiring at a description of 'reality behind the quantum mechanical phenomena' the contextuality, noticed for the first time by Bohr and rejected by Einstein, will have to be taken into account.

3.2. (In)completeness in a restricted sense

Whereas at this moment the question of '(in)completeness in a wider sense' is not experimentally relevant because we do not have any indication which kind of experiments should be performed so as to transcend the quantum mechanical description in an observationally relevant way, the situation is different as regards '(in)completeness in a restricted sense', at least if we restrict ourselves to the standard (textbook) formalism of that theory (in which measurement probabilities of quantum mechanical observables are represented by expectation values of the projection operators of the spectral representations of Hermitian operators). The standard formalism is easily seen to be 'incomplete in a restricted sense'. For instance, consider the 'which-way polarization measurement of a photon' depicted in figure 3. Here a photon has probability γ to be transmitted by a nonpolarizing semi-transparant mirror toward a polarization measurement setup in direction θ, and probability $1-\gamma$ to be reflected toward a polarization measurement setup in direction θ'. Since the detection probabilities of detectors D and D' are given by $p_D = \gamma\langle E_+^\theta\rangle$ and $p_{D'} = (1-\gamma)\langle E_+^{\theta'}\rangle$, respectively, E_+^θ and $E_+^{\theta'}$ being projection operators, it is evident that the detection probabilities are not described by

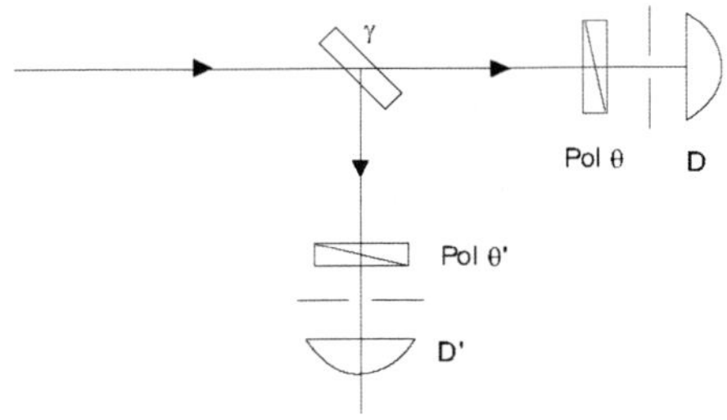

FIGURE 3. Which-way polarization measurement of a photon.

the standard formalism (e.g. $\gamma E_+^\theta \neq (\gamma E_+^\theta)^2$). From the whole range $0 \leq \gamma \leq 1$ of possibilities of the present experiment only the 'set of measure zero $\gamma = (0,1)$' satisfies the standard formalism.

This example can be supplemented by many other ones; actually, there are very few realistic experiments satisfying the standard formalism (see e.g. de Muynck [7]). If it is taken into account that the interaction of object and measuring instrument is a *quantum mechanical* process (cf. section **2.3**), detection probabilities are seen to be expectation values of the operators M_m of a *positive operator-valued measure (POVM)* corresponding to the resolution of the identity $\{M_m\}$, $M_m \geq O$, $\sum_m M_m = I$, rather than a projection-valued measure (PVM) corresponding to the orthogonal decomposition $\{E_m\}$, $E_m^2 = E_m$, $\sum_m E_m = I$ of the standard formalism. Indeed, denoting by ρ_o and ρ_a the initial density operators of microscopic object and measuring instrument, respectively, the final density operator is given by $\rho_{oaf} = U\rho_o\rho_a U^\dagger$, $U = e^{-\frac{i}{\hbar}HT}$, T the interaction time. Then detection probabilities of the pointer positions (represented by a pointer observable having spectral representation $\{E_{am}\}$) are given by

$$p_m = Tr_{oa}\rho_{aof}E_{am} = Tr\rho_o M_m,\ M_m = Tr_a\rho_a U^\dagger E_{am} U.$$

From the physical properties of the distribution $\{p_m\}$ it follows that the operators M_m satisfy all properties of the elements of a POVM defined above. It should be noted that there is no single reason to require idempotency of these operators.

4. THE CORRESPONDENCE PRINCIPLE (MATURE FORM)

Disregarding here Bohr's early use of the correspondence principle (in the sense of requiring a classical limit for defining the notion of a quantum mechanical observable), I want to restrict myself here to his later characterization, according to which the following requirements should be met:

1. A quantum mechanical observable is exclusively defined within the context of the measurement serving to measure that observable.
2. Experimental arrangement and measurement results have to be described in classical terms.

Usually these requirements are considered part of the complementarity principle, while restricting the notion of 'correspondence' to the requirement of a classical limit. It, however, seems more appropriate to stick to Bohr's later use, and restrict 'complementarity' to considerations on *pairs* of 'complementary observables' only (cf. section **5**).

The importance of the first item has been stressed already in section **2.1**, whereas the second item was criticized in section **2.3**. It is my intention in the present section to demonstrate that by not consistently sticking to the first item Bohr introduced an ambiguity in his notion of 'correspondence', thus being responsible for much confusion. The cause of this ambiguity can be traced back to not distinguishing between two different possibilities of what is meant by the word 'phenomenon', viz. either 'a macroscopically observable property of a *microscopic* object', or rather 'a property of a *macroscopic* measuring instrument obtained by letting the instrument interact with the object (e.g. a pointer triggered by the microscopic object to take a certain position on a measurement scale)'. From the way Bohr reacted to the EPR proposal [3] it is evident that he had in mind the first possibility. Indeed, since no measuring instrument is present for particle 2 (cf. figure 2a) Bohr could accept Einstein's proposal to view the EPR measurement setup as a *measurement* on particle 2 only if the measurement result would be taken as a property of that particle.

As a consequence of this Bohr applied his correspondence principle to EPR in an inconsistent way. As a matter of fact, Bohr did not recognize the correlation of two quantum mechanical observables A_1 and A_2, proposed in the

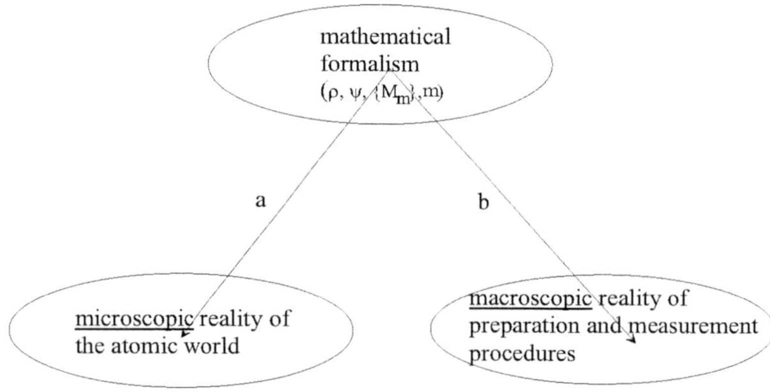

FIGURE 4. *Realist (a) and empiricist (b) interpretations of the mathematical formalism of quantum mechanics.*

EPR paper as a *classical* correlation (A_i either position or momentum of particle i, $i = 1,2$), as an ordinary *quantum mechanical* observable. He granted Einstein well-definedness of correlation observable A_1A_2 although only A_1 is measured in the EPR experiment. But, according to his correspondence principle the correlation observable A_1A_2 is well-defined only in an EPR-Bell experiment. Hence, Bohr's conclusion that Einstein's 'element of physical reality' is ambiguous (due to its dependence on the context of the measurement of either P_1 or Q_1) must be completed by the observation that Bohr's conclusion is itself a consequence of an ambiguity as regards the meaning of a quantum mechanical observable. Actually, like Einstein, also Bohr never considered the possibility of looking upon a quantum mechanical measurement result as a pointer position rather than as a property of the microscopic object. Only by noting the difference between EPR experiments and experiments of the EPR-Bell type (cf. figure 2) the inconsistency of Bohr's reaction to EPR could become obvious.

Unfortunately, negligence of the difference between EPR and EPR-Bell experiments has been perpetuated until the present day. As a consequence it is still widely thought that properties of particle 2 may be nonlocally influenced by the measurement performed on particle 1. It will be seen in section **6** that there is no reason for such a conclusion if quantum mechanical measurement results are assumed to correspond to pointer positions of a measuring instrument.

In order to evade confusions like the one discussed here it is necessary to be as precise as possible with respect to the correspondence between the mathematical formalism of quantum mechanics and physical reality. For this reason two different notions of 'correspondence' should be distinguished here, viz. (restricting ourselves to standard observables)

1. Realist correspondence:
   Quantum mechanical observable A, as well as its values a_m refer to properties of the microscopic object.
2. Empiricist correspondence:
   Quantum mechanical observable A is a label of a measurement procedure, a_m is a label of a pointer position.

More generally, we should distinguish two different interpretations of the mathematical formalism of quantum mechanics, viz. a *realist* interpretation (either objectivistic or contextualistic) interpretation (cf. figure 4(a)) in which quantum mechanics is thought to describe microscopic reality most in the same way classical mechanics is generally thought to describe macroscopic reality, and an *empiricist* interpretation (cf. figure 4(b)) in which state vector and density operator are thought to correspond to preparation procedures, and quantum mechanical observables (either standard or generalized) correspond to measurement procedures and the phenomena induced by a microscopic object in the macroscopically observable pointer of a measuring instrument.

The empiricist interpretation is particularly suited to encompass the generalization from the standard notion of a quantum mechanical observable to the generalized one referred to in section **3.2**. An important application of the empiricist interpretation to the *generalized* formalism is that POVMs can often be interpreted as labeling *nonideal* measurement procedures, not registering 'reality as it objectively is', but taking into account possible disturbing influences introduced by the measurement itself. Thus, of two measurement procedures labeled by POVMs $\{M_m\}$ and $\{N_n\}$, respectively, such that

$$M_m = \sum_n \lambda_{mn} N_n,\ \lambda_{mn} \geq 0, \sum_m \lambda_{mn} = 1, \qquad (1)$$

the first is a nonideal version of the second, the nonideality matrix (λ_{mn}) representing the conditional probability that a measurement of POVM $\{M_m\}$ yields result m if a measurement of POVM $\{N_n\}$ would have given result n. In the following sections applications of this will be discussed. Restricting ourselves to discrete spectra it is possible to quantify the nonideality by means of certain properties of the nonideality matrix. A convenient quantity is the average row entropy

$$J_{(\lambda)} = -\frac{1}{N}\sum_{mn}\lambda_{mn}\ln\frac{\lambda_{mn}}{\sum_{n'}\lambda_{mn'}}. \tag{2}$$

5. THE COMPLEMENTARITY PRINCIPLE

The complementarity principle must be seen as a 'no go' theorem as regards general applicability of 'correspondence as obtains in classical physics': it was soon established that the notion of 'correspondence' as defined in section **4** cannot be applied simultaneously to quantum mechanical position and momentum observables, or, more generally, to *incompatible* standard observables (corresponding to non-commuting Hermitian operators). From discussions of the so-called 'thought experiments' it was seen that measurements of incompatible standard observables had mutually exclusive measurement arrangements, in such a way that measurement of one observable would disturb incompatible ones. The mutual disturbance of measurement results of standard observables, when measured simultaneously, was thought to be described by the Heisenberg uncertainty relation

$$\Delta A \Delta B \geq \frac{1}{2}|\langle\psi|[A,B]_-|\psi\rangle| \tag{3}$$

derived from the standard formalism.

It is important here to remind a criticism by Ballentine [9], to the effect that the Heisenberg uncertainty relation (3) does not refer to 'joint measurement' because it can be tested by means of *separate* ideal measurements of observables A and B. Hence, it is impossible that this inequality is an expression of mutual disturbance as found in the 'thought experiments'; it should be seen as a restriction of our *preparing* abilities rather than of our *measuring* ones. Probably Einstein was closer to the real meaning of the Heisenberg uncertainty relation (viz., as a restriction on the values of standard deviations in an *ensemble*) than were Bohr and Heisenberg, who seem to have been jumping to conclusions by associating the results of their considerations on the 'thought experiments' with equation (3).

Perhaps Bohr and Heisenberg should not be blamed too much for this error because they did not have at their disposal the generalized formalism referred to in section **3.2**, which turns out to be necessary to clarify the situation. It, indeed, is possible by means of the POVMs of the generalized formalism to prove that in trying to jointly measure incompatible standard observables mutual disturbance necessarily occurs (Martens, de Muynck [10]; also de Muynck [7], section 7.10), thus corroborating the results obtained by Bohr and Heisenberg by studying 'thought experiments'.

In order to do so a bivariate POVM $\{R_{mn}\}$ was considered, the expectation values of which describing the joint probability distributions of the two pointers intended to yield the measurement results of the incompatible observables $A = \sum_m a_m E_m$ and $B = \sum_n b_n F_n$ to be measured jointly. Then these latter probability distributions are given by the expectation values of the marginals $\{\sum_n R_{mn}\}$ and $\{\sum_m R_{mn}\}$, respectively. We will call a measurement procedure a joint measurement of A and B if $\sum_n R_{mn} = E_m$ and $\sum_m R_{mn} = F_n$. It is not difficult to prove that this is possible only if A and B are compatible. However, if A and B are incompatible it is possible that the marginals of $\{R_{mn}\}$ correspond to *nonideal* measurements (cf. (1)) of A and B, respectively. Thus, the POVM $\{R_{mn}\}$ describes a joint *nonideal* measurement of standard observables A and B if nonideality matrices $(\lambda_{mm'})$ and $(\mu_{nn'})$ exist such that

$$\begin{array}{l} \sum_n R_{mn} = \sum_{m'} \lambda_{mm'} E_{m'},\ \lambda_{mm'} \geq 0,\ \sum_m \lambda_{mm'} = 1, \\ \sum_m R_{mn} = \sum_{n'} \mu_{nn'} F_{n'},\ \mu_{nn'} \geq 0,\ \sum_n \mu_{nn'} = 1. \end{array} \tag{4}$$

It is easily seen that measurement procedures exist satisfying (4). Thus, the measurement arrangement depicted in figure 3 is a joint nonideal measurement of incompatible standard polarization observables in directions θ and θ', respectively. Indicating a detection event by $+$, and non-detection by $-$, we find

$$(R_{mn}) = \begin{pmatrix} O & \gamma E_+^\theta \\ (1-\gamma)E_+^{\theta'} & I - \gamma E_+^\theta - (1-\gamma)E_+^{\theta'} \end{pmatrix}, \tag{5}$$

event '$++$' having probability zero, and $I - \gamma E_+^\theta - (1-\gamma)E_+^{\theta'}$ corresponding to the possibility that the photon is not detected by either one of the detectors D or D' but is absorbed in one of the analyzers. It is straightforward to calculate

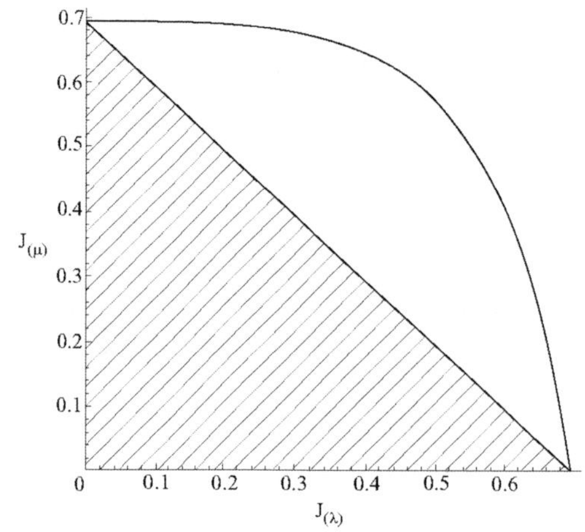

FIGURE 5. Parametric plot of $J_{(\lambda)}$ versus $J_{(\mu)}$ as a function of γ. The shaded area is forbidden by the Martens inequality (7)

the nonideality matrices for this experiment as

$$(\lambda_{mm'}) = \begin{pmatrix} \gamma & 0 \\ 1-\gamma & 1 \end{pmatrix}, \ (\mu_{nn'}) = \begin{pmatrix} 1-\gamma & 0 \\ \gamma & 1 \end{pmatrix}. \tag{6}$$

From (6) it is easily seen how the mutual disturbance of the measurement results of the two standard observables changes if the parameter γ is changed. In particular, complementary behaviour is evident, deviation of one nonideality matrix from the unit matrix being maximal whenever the other's is minimal. This can be expressed by means of the nonideality measure (2). For measurements of the type described by (4) it is possible (Martens, de Muynck [10]) to derive the following general inequality,

$$J_{(\lambda)} + J_{(\mu)} \geq -\ln\{\max_{mn} TrE_mF_n\}, \tag{7}$$

to be referred to as the Martens inequality. It is important to note that, contrary to the Heisenberg inequality (3), the Martens inequality does not depend on the state vector or density operator but in an unambiguous way is a property of the measurement procedure only. From an application of (7) to the joint nonideal measurement of polarization observables depicted in figure 5 it is seen that the Martens inequality yields a perfect illustration of the results obtained by Bohr and Heisenberg when studying the 'thought experiments'. Indeed, their physical observations based on these experiments were completely correct; only their reluctance to exert a full-blown *quantum mechanical* analysis of the measurement process prevented them from a sufficiently complete analysis.

It, incidentally, should be noticed that the mutual disturbance considered here does refer to the final pointer positions of the measuring instruments. This should be distinguished from 'mutual disturbance in a preparative sense' as discussed in the Copenhagen literature, referring to the final state of the microscopic object rather than to the final state of the measuring instrument. Actually, complementarity can manifest itself in two different ways, viz., as a property of 'preparation' described by the Heisenberg inequality, and as a property of 'joint measurement' described by the Martens inequality. To a large extent the Copenhagen confusion with respect to 'complementarity' has its origin in the confusion with respect to preparation and measurement noted in section **2.2**.

6. THE BELL INEQUALITIES AND NONLOCALITY

As is well-known the Bell inequalities are satisfied if all standard observables that are involved are mutually compatible. This entails the conclusion that *in*compatibility is a necessary condition for the Bell inequalities to be violated. But, due to the 'postulate of local commutativity' (stating that observables measured in causally disconnected regions of space-time must be compatible) incompatibility of observables is a local affair. Hence, violation of the Bell inequalities must have a *local* origin. It is the intention of the present section to demonstrate that, indeed, violation of the

FIGURE 6. *Generalized EPR-Bell experiment.*

Bell inequalities may be caused by *local* mutual disturbance in joint nonideal measurements of incompatible standard observables rather than by nonlocal influences.

In order to do so we consider the generalized EPR-Bell experiment depicted in figure 6, the difference with the Aspect experiments [8] being that the mirrors indicated by γ_1 and γ_2 are 'semi-transparant, and stationary' rather than 'completely reflecting, and either stationary or switching'. We consider here two measurements of the type depicted in figure 3, simultaneously performed on particles 1 and 2. The Aspect experiments are special cases of the present experiment, satisfying $(\gamma_1, \gamma_2) = (1,1), (1,0), (0,1)$ and $(0,0)$, respectively. For general (γ_1, γ_2) the (quadrivariate) POVM $\{R_{m_1n_1m_2n_2}\}$ follows directly from the bivariate POVM (5) as

$$R^{(\gamma_1\gamma_2)}_{m_1n_1m_2n_2} = R^{(\gamma_1)}_{m_1n_1} R^{(\gamma_2)}_{m_2n_2}, \tag{8}$$

in which the parameters γ_1 and γ_2 are added to distinguish the two arms of the interferometer. The expectation value of POVM (8) yields the quadrivariate probability distribution of a joint nonideal measurement of four standard polarization observables in directions $\theta_1, \theta_1', \theta_2$, and θ_2', respectively. There is mutual disturbance of the type discussed in section **5** *separately* in each of the arms of the interferometer. As is to be expected from the postulate of local commutativity there is no mutual disturbance between different arms of the interferometer.

Let us finally see where the violation of the Bell inequalities in the Aspect experiments [8] comes from (see de Muynck [7], chapter 9). In order to do so it is important to note that these experiments can be demonstrated to violate the Bell inequalities only by combining results obtained within four *different* arrangements for measuring bivariate probabilities. By contrast, the bivariate probabilities derived from the generalized experiment for fixed (γ_1, γ_2) from the experimentally obtained quadrivariate one do satisfy the Bell inequalities because they are obtained within one single measurement setup. This can be traced back to the fact that each individual particle pair yields a quadruple (m_1, n_1, m_2, n_2) of measurement results (pointer readings); this is a sufficient condition for the statistics to have a Kolmogorovian character. This evidently holds true for each of the Aspect experiments separately, but *not* for measurement results of the four experiments pasted together.

Violation of the Bell inequalities is evidence of the non-Kolmogorovian character of the statistics. Indeed, by changing the measurement arrangement in the experiments performed by Aspect the quadruples are changed in such a way that an *octuple* of measurement results of the four measurements corresponding to $(\gamma_1, \gamma_2) = (1,1), (1,0), (0,1)$, and $(0,0)$, cannot be accommodated into one single quadruple. The cause of this is seen to be a changing of the mutual disturbance in one arm of the interferometer by changing the measurement arrangement in that arm (and analogously in the other arm). These changes are completely accounted for by the the idea of complementarity in a joint measurement of incompatible standard observables as discussed in section **5**. Nonlocal influences between different arms are not necessary to explain the ensuing violation of the Bell inequalities.

It should finally be stressed that the latter conclusion, although derived from (generalized) quantum mechanics, is actually a conclusion about the *physical processes* obtaining in experiments of the kind discussed here. Therefore, the often-heard assertion that quantum mechanical locality might be compatible with subquantum nonlocality (the latter, allegedly, being necessary to cope with derivations of the Bell inequalities from local hidden variables theories), is not applicable: it is utterly unreasonable to suppose that one and the same phenomenon would have completely different *physical* explanations within different theories. If violation of the Bell inequalities has a *local* explanation within a quantum mechanical description of certain phenomena, it should have a *local* explanation within any theory describing the reality behind these phenomena. Rather than acquiescing in an unfortunate marriage of 'observable locality' with 'unobservable nonlocality' by relying on such fancy expressions like 'entanglement', 'inseparability' or 'passion at a distance', it seems more appropriate to perform as precise as possible analyses of measurement processes, and to study the relations between its measurement results and the properties of microscopic objects as described in different theories.

7. FROM COPENHAGEN TO NEO-COPENHAGEN INTERPRETATION (SUMMARY)

Due to Bohr's reference to 'phenomena' the Copenhagen interpretation has an empiricist reputation (e.g. Reichenbach [11]). However, on closer scrutiny Bohr's 'phenomenon' is probably too much inspired by the rather primitive detection methods (like flashes on a scintillation screen) of that time. It does not take into account more modern measurement procedures in which the final observational data are processed so as to be only vaguely reminiscent of their microscopic origin: what is seen by the observer is often just a graph coming out of his printer. Contrary to the Copenhagen interpretation (which, for the main part, is a realist interpretation, be it of a contextualistic kind) the neo-Copenhagen interpretation is an empiricist one, adapted to the generalized formalism of quantum mechanics necessary to also describe more modern experiments (cf. de Muynck [7], chapter 8).

Comparing, in summary, the neo-Copenhagen interpretation with the Copenhagen one with respect to the notions of completeness, correspondence and complementarity, we arrive at the following results:

	Copenhagen interpretation	neo-Copenhagen interpretation
completeness:		
in a wider sense:	yes (claimed)	no
in a restricted sense (standard qm):	yes (claimed)	no
in a restricted sense (generalized qm):		yes
correspondence:		
	contextualistic-realist	empiricist
	partly classical	completely quantum mechanical
complementarity:		
preparation:		Heisenberg inequality
joint measurement:	Heisenberg inequality	Martens inequality

Notwithstanding the considerable differences does it nevertheless seem appropriate to pay a tribute to Copenhagen by referring to that name in the new interpretation, because in the latter the important development away from the classical idea of 'physical theory as an *objective* description of reality', which was started by the Copenhagen interpretation, has been continued, and has been pushed to a logical completion.

ACKNOWLEDGMENTS

The author likes to thank the Faculty of Applied Physics of Eindhoven University of Technology, and in particular professor Thijs Michels, for providing facilities that made this work possible.

REFERENCES

1. P.K. Feyerabend, *Phil. of Science* **35**, 309–331 (1968).
2. F. Suppe, *The structure of scientific theories*, University of Illinois Press, Urbana, Chicago, London, 1977.
3. A. Einstein, B. Podolsky, and N. Rosen, *Phys. Rev.* **47**, 777–780 (1935).
4. S. Kochen and E.P. Specker, *J. Math. and Mech.* **17**, 59–87 (1967).
5. J.S. Bell, *Physics* **1**, 195–200 (1964).
6. W. Heisenberg, *Zeitschr. f. Phys.* **43**, 172–198 (1927).
7. Willem M. de Muynck, *Foundations of quantum mechanics, an empiricist approach*, Fundamental theories of physics, vol. 127, Kluwer Academic Publishers, Dordrecht, Boston, London, 2002.
8. A. Aspect, P. Grangier, and G. Roger, *Phys. Rev. Lett* **47**, 460–463 (1981); A. Aspect, J. Dalibard, and G. Roger, *Phys. Rev. Lett.* **49**, 1804–1807 (1982).
9. L.E. Ballentine, *Rev. Mod. Phys.* **42**, 358–381 (1970).
10. H. Martens and W. de Muynck, *Found. of Phys.* **20**, 255–281, 357–380 (1990).
11. H. Reichenbach, *Philosophical Foundations of Quantum Mechanics*, University of California Press, 1944.

On a Mathematical Model of Brain Activities

K.-H. Fichtner*, L. Fichtner†, W. Freudenberg** and M. Ohya‡

*Friedrich Schiller Unversity Jena, Institute of Applied Mathematics, E.-Abbe-Platz 2, 07743 Jena, Germany
†Friedrich Schiller Unversity Jena, Institute of Psychology, Am Steiger 3, 07743 Jena, Germany
**Brandenb. Techn. University Cottbus, Dep. of Mathematics, PO box 10 13 44, 03013 Cottbus, Germany
‡Tokyo University of Science, Department of Information Science, Noda City, Chiba 278-8510, Japan

Abstract. The procedure of recognition can be described as follows: There is a set of complex signals stored in the memory. Choosing one of these signals may be interpreted as generating a hypothesis concerning an "expexted view of the world". Then the brain compares a signal arising from our senses with the signal chosen from the memory leading to a change of the state of both signals. Furthermore, measurements of that procedure like EEG or MEG are based on the fact that recognition of signals causes a certain loss of excited neurons, i.e. the neurons change their state from "excited" to "nonexcited". For that reason a statistical model of the recognition process should reflect both – the change of the signals and the loss of excited neurons. A first attempt to explain the process of recognition in terms of quantum statistics was given in [1]. In the present note it is not possible to present this approach in detail. In lieu we will sketch roughly a few of the basic ideas and structures of the proposed model of the recognition process (Section). Further, we introduce the basic spaces and justify the choice of spaces used in this approach. A more elaborate presentation including all proofs will be given in a series of some forthcoming papers [2, 3]. In this series also the procedures of creation of signals from the memory, amplification, accumulation and transformation of input signals, and measurements like EEG and MEG will be treated in detail.

Keywords: Fock space, coherent state, quantum statistics, beam splitter
PACS: 03.65Db, 05.30Jp, 87.10Mn

INTRODUCTION

Specialists in modern brain research are convinced that signals in the brain should be coded by populations of excited neurons (cf. [4, 5, 6, 7]). Considering models based on classical probability theory states of signals should be identified with probability distributions of certain random point fields located inside the volume G of the brain. [1] represents an attempt to explain the process of recognition in terms of quantum statistics. Here (pure) states of signals are described by complex functions Ψ of point configurations inside of G where $|\Psi|^2$ is the probability density of a random point field. Thus one obtains again a probability distribution of a random point field called the position distribution corresponding to the quantum state. On the other hand, the probability distribution of each random point field can be identified with the position distribution corresponding to a certain quantum state. In this sense quantum models are more general than classical models, i.e. the use of quantum theory gives rise to a more detailed description of reality.

We discuss a quantum model of signals in the brain based on the so-called bosonic Fock space. The use of that Fock space has several advantages. For instance, as it was pointed out in [6, 7] the space G representing the whole volume of the brain can be divided into disjoint regions $G_1, \ldots, G_n$ responsible for different aspects of the signals. More precisely, corresponding to that decomposition of the space the signal should be decomposed into different parts representing a special type of information contained in the signal. This can be reflected in the fact that the bosonic Fock space corresponding to G can be identified with the tensor product of the bosonic Fock spaces corresponding to the subsets $G_1, \ldots, G_n$. Furthermore, this property of the Fock space enables us to reflect in a clear sense another fundamental experimental experience, namely that the brain is acting parallel in the disjoint regions. As it was pointed out in [7] the procedure of recognition of signals can be described as follows: There is a set of complex signals stored in the memory. Choosing one of these signals may be interpreted as generating a hypothesis concerning an "expexted view of the world". Then the brain compares a signal arising from our senses with the signal chosen from the memory. That changes the state of both of the signals in such a manner that after the procedure the signals coincide in a certain sense. Furthermore, measurements of that procedure like EEG or MEG are based on the fact that recognition of signals causes a certain loss of excited neurons, i. e. the neurons change their state from "excited" to "non-excited" (firing of the neurons). For that reason a statistical model of the recognition process should reflect both, that change of the signals, and the loss of excited neurons. The activity in different parts of the brain is described by so-called beam

CP962, *Quantum Theory, Reconsideration of Foundations – 4*,
edited by G. Adenier, A. Yu. Khrennikov, P. Lahti, V. I. Man'ko, and Th. M. Nieuwenhuizen

splitters well-known in quantum optics. Application of such a beam splitter may be interpreted as an exchange of the support of the input signals and the signal choosen from the memory, a procedure which was mentioned in [7]. Recognition takes place if one of the outputs of the splitting procedure is collapsing. This is described by the projection onto the vaccum state in the considered region of the brain. One can show (cf. [2]) that for sufficiently high intensities of the signals both operators are approximately equal. Let us remark that - in contrary to classical models - we do not need to specify a special region of the brain where the memory is located. Furthermore, due to the fact that quantum theory is non-local activity in one part of the brain causes immediately certain changes of the other parts of the brain. For that reason our quantum model of the brain may give an answer to one of the main questions of modern brain resarch, namely, how to explain that activities in one region of the brain have immediate consequences concerning the other regions (cf. [8, 9, 10, 11, 7]).

We have chosen the bosonic Fock space as our basic space for modelling the activities of the brain mainly because of the above mentioned property that the space corresponding to G can be identified with the tensor product of the spaces corresponding to the subsets $G_1, \ldots, G_n$. However, to avoid elaborate formulae in the present note we consider only one area G, and we confine ourselves to a simplified description of the basic spaces and operators used in this approach. Details of the sketched approach including all proofs are given in [12, 1, 2, 3]. In this series of papers also the memory space, the procedures of creation of signals from the memory, amplification, accumulation and transformation of input signals, and measurements like EEG and MEG will be treated in detail.

THE SPACE OF SIGNALS

Starting point will be a set G representing the space where the process of recognition and processing of the signals takes place. For the mathematical model it is irrelevant what is the concrete structure of G. To start with a general space G has the advantage that it can be used as a model for very different aspects of the recognition processes in the brain. So let G be an arbitrary complete separable metric space and $\mathfrak{G}$ its σ-algebra of Borel sets. Further, let μ be a fixed finite diffuse measure on $[G, \mathfrak{G}]$, i.e. $\mu(G) < \infty$ and $\mu(\{x\}) = 0$ for all singletons $x \in G$. Especially, we are concerned with the case where G is a compact subset of $\mathbb{R}^d$ and μ is the $d-$dimensional Lebesgue measure restricted to G. The elements of the Hilbert space $\mathscr{H} = L^2(G, \mu)$ can be interpreted as *functions of the excited neurons*. From this space of square integrable complex-valued functions on G we pass over to the Hilbert space $\mathscr{M} := L^2(M, F)$ where $M := \{\{x_1, \ldots, x_n\} : \ x_j \in G, \ n \geq 0\}$ is the set of all finite symmetric point configurations (including the void configuration $\mathfrak{o}$). The measure F (defined on a canonical σ-algebra $\mathfrak{M}$ over M) fulfills $F(\{\mathfrak{o}\}) = 1$ and $F(M) = \exp\{\mu(G)\}$, and restricted to n-point configurations it is (up to symmetrisation) just the product measure $\mu^{\otimes n}$. One can show that the space $\mathscr{M}$ is isomorphic to the usual symmetric (bosonic) Fock space $\Gamma(\mathscr{H})$ with $\mathscr{H} = L^2(G, \mu)$. For details we refer for instance to [13]or to [14]. The space $\mathscr{M}$ consists of (complex-valued)*functions of configurations of excited neurons*, and will be interpreted as the *space of signals*. A quantum state on a Hilbert space $\mathscr{H}$ is a positive linear normalized functional ω on the set $\mathscr{L}(\mathscr{H})$ of all bounded linear operators on $\mathscr{H}$. In our special case a general (normal) state ω on $\mathscr{M}$ can be written in the form

$$\omega(A) = \sum_{n=1}^{N} \alpha_n \langle \Psi_n, A \Psi_n \rangle \qquad (A \in \mathscr{L}(\mathscr{M})) \tag{1}$$

where $(\Psi_n)_{n=1}^N$ is a given orthonormal sequence from $\mathscr{H}$ and $(\alpha_n)_{n=1}^N$ is a sequence of non-negative numbers with $\sum_{n=1}^N \alpha_n = 1$, $N \leq \infty$.

To each state ω on $\mathscr{M}$ we can relate its position distribution. The following result was shown (for a more general case) in [13, Theorem 2.15]

Proposition 1 *To each normal state ω on $\mathscr{M}$ there exists a unique probability measure P_ω on $[M, \mathfrak{M}]$ such that*

$$P_\omega(Y) = \omega(O_Y) \qquad (Y \in \mathfrak{M}). \tag{2}$$

P_ω is called the position distribution *of the state ω.*

The position distribution P_ω of a normal state is absolutely continuous with respect to the Fock space measure F. If ω is given in the form (1) we obtain F-a.e.

$$\frac{dP_\omega}{dF} = \sum_{n=1}^{N} \alpha_n |\Psi_n|^2. \tag{3}$$

We see that P_ω is the probability distribution of a random (classical) configuration of points, namely the distribution of the positions of the excited neurons.

An especially important class of functions in $\mathscr{M}$ are the so-called *exponential vectors*. The notion of a signal will be defined with the aid of such exponential vectors. For a measurable function $g : G \longrightarrow \mathbb{C}$ we define $\exp\{g\} : M \longrightarrow \mathbb{C}$ by setting

$$\exp\{g\}(o) = 1 \qquad \exp\{g\}(\varphi) = \prod_{x \in \varphi} g(x) \qquad \text{if } \varphi \neq o. \tag{4}$$

The function $\exp\{g\}$ is called *exponential vector* corresponding to g. The set of all exponential vectors is total in $\mathscr{M}$. This means that each function from $\mathscr{M}$ can be approximated by linear combinations of exponential vectors.

Definition 2 *For $g \in L^2(G)$ we denote by ω_g the pure state given by*

$$\omega_g(A) := e^{-\|g\|^2} \langle \exp\{g\}, A \exp\{g\} \rangle. \tag{5}$$

The state ω_g is called the coherent state *corresponding to g.*

Roughly speaking, coherent states describe states of systems of quantum particles where each particle is in the same one-particle state. Because of

$$\exp\{0\}(o) = 1 \qquad \exp\{0\}(\varphi) = 0 \qquad \text{if } \varphi \neq o. \tag{6}$$

the state ω_0 is called the *vacuum state* in $\mathscr{M}$.

There is a close connection between coherent states and *Poisson point processes*. The position distribution P_{ω_g} of the coherent state ω_g with $g \in L^2(G)$ is the Poisson point process with intensity measure Λ_g. Hereby, Λ_g is the finite measure on $[G, \mathfrak{G}]$ absolutely continuous with respect to μ with Radon-Nikodym derivative $(d\Lambda_g / d\mu)(x) = |g(x)|^2$ ([cf. instance [13]).

Let $f \in L^2(G)$ be normalized and $0 \neq \lambda \in \mathbb{C}$. Denoting by $\mathscr{N}$ the number operator it is easily checked that $\omega_{\lambda f}(\mathscr{N}) = |\lambda|^2$. Thus $|\lambda|^2$ is the expectation of the total number of particles of the quantum system being in the coherent state $\omega_{\lambda f}$. The state f (with $f \in L^2(G)$, $\|f\| = 1$) represents a function of an excited neuron, the state $\omega_{\lambda f}$ on the Fock space represents a *signal* of the type f with intensity $|\lambda|^2$. So for $\lambda_1, \lambda_2 \in \mathbb{C} \setminus \{0\}$ the coherent states $\omega_{\lambda_1 f}$ and $\omega_{\lambda_2 f}$ are states where all *one-particle states* are in the same state f but with different intensities $|\lambda_1|^2$ resp. $|\lambda_2|^2$. Observe that because of $P^{\lambda f}(\{o\}) = e^{-|\lambda|^2}$ the *vacuum part* of the coherent state $\omega_{\lambda f}$ will have a very small probability to occur if the intensity $|\lambda|^2$ is very large.

Signals are states on $\mathscr{M}$. The *vacuum state* ω_0 represents the *empty signal*. A signal is described by a state on $\mathscr{M}$. The Fock space $\mathscr{M}$ is an infinite dimensional Hilbert space. It seems to be natural to restrict all considerations to finite dimensional subspaces. Let us be given an orthonormal system $f_1, \ldots, f_N$, $N \geq 1$ of functions from $L^2(G)$:

$$\|f_k\| = 1, \ \langle f_l, f_k \rangle = 0 \qquad (l, \ k \in \{1, \ldots, N\}, \ l \neq k\}). \tag{7}$$

Observe that there do not exist exponential vectors being orthogonal. To obtain an orthogonal system of "exponential vectors" in $\mathscr{M}$ we have to restrict them to non-void configurations, i. e. we have to take out the vacuum part from $\exp\{f\}$. Given a fixed orthonormal system according to (7) and a positive number λ we set

$$|0\rangle := \exp\{0\}, \qquad |j\rangle := \frac{\exp\{\lambda f_j\} - \exp\{0\}}{\sqrt{e^{\lambda^2} - 1}} \qquad (j \in \{1, \ldots, N\}) \tag{8}$$

Definition 3 *Given an orthonormal system $f_1, \ldots, f_N$, $N \geq 1$ of functions from $L^2(G)$ (satisfying (7)) the elements $|j\rangle \in \mathscr{H}$, $j \in \{1, \ldots, N\}$ are called* elementary signals. *The Hilbert space $\mathscr{H}_{\text{sig}} \subset \mathscr{M}$ (consisting of superpositions of elementary signals and eventually the empty signal $|0\rangle$) is called the* space of regular signals.

THE MEMORY

In the above section we passed from the *one-particle space* $L^2(G)$ to the *configuration space* $\mathscr{M}$ which occured to be again a L^2-space. We passed from single points $x \in G$ to the set M of finite configurations $\varphi = \{x_1, \ldots, x_n\}$ from

G. Signals were defined to be elements from $\mathscr{M}$, i. e. functions on the configuration space. The information stored in the memory is more complex. We have to consider sets of configurations. First we replace single points x from G by configurations φ from $\tilde{M} := M \setminus \{o\}$. These configurations φ will be our starting point and we pass from the single configurations $\varphi_1, \ldots, \varphi_n$, $n \geq 1$ from $\tilde{M}$ to sets $\Phi = \{\varphi_1, \ldots, \varphi_n\}$ of configurations. The entirety of all finite sets of (non-void) configurations we denote by $\mathbb{M}$. Such a construction is a very useful tool in point process theory especially for a characterization of branching processes (cf. [15] or also [16]). In [15] the counting measures $\varphi \in \tilde{M}$ are called *families* and the sets of counting measures Φ are said to describe the *family structure*.

In full analogy to the considerations above we define the L^2-space over $\mathbb{M}$ as the basic space for states in the memory. Analogously to F the corresponding measure $\mathbb{F}$ on $\mathbb{M}$ restricted to families consisting of n configurations will be (up to symmetrisation) the product measure $F^{\otimes n}$, and one has $\mathbb{F}(\mathfrak{O}) = 1$, $\mathbb{F}(\mathbb{M}) = \exp\{F(M \setminus \{o\})\}$ ($\mathfrak{O}$ denotes the empty set of configurations). One can show that $L^2(\mathbb{M}) := L^2(\mathbb{M}, \mathbb{F})$ can be identified with the symmetric Fock space $\Gamma(\mathscr{M})$. So the space $L^2(\mathbb{M})$ can be understood as the Fock space over the Fock space $\mathscr{M}$. We will call it the *compound (symmetric) Fock space over G*. This space is not mysterious at all. The stimulus for the physicist V. A. Fock in his pioneering paper [17] introducing second quantization was the search for a way of passing from quantum states of single objects to states of collections of these objects. In our case this new space is the usual second quantization of a Hilbert space representing the space of signals.

In full analogy to the space $\mathscr{M}$ we introduce for functions $R \in \mathscr{M}$ exponential vectors $\mathrm{Exp}\{R\}$ in the compound Fock space $L^2(\mathbb{M})$. For a (not necessarily normalized) function $R \in L^2(\tilde{M}(G))$ (satisfying $R(o) = 0$) we define a function $\mathrm{Exp}\{R\} \in L^2(\mathbb{M}, \mathbb{F})$ by setting

$$\mathrm{Exp}\{R\}(\mathfrak{O}) := 1, \qquad \mathrm{Exp}\{R\}(\Phi) := \prod_{\varphi \in \Phi} R(\varphi) \qquad \textit{for} \quad \Phi \in \mathscr{M} \setminus \{\mathfrak{O}\}. \tag{9}$$

The compound Fock space introduced above will serve as the basic space for states in the memory. Each state on the Hilbert space $L^2(\mathbb{M})$ one could call a state of the memory. As in the case of the space of signals we will restrict our considerations to a subspace connected with the defined space of signals.

Definition 4 *The space of memory is the Hilbert space given by*

$$\mathscr{H}_{\mathrm{mem}} := \mathrm{Lin}\left\{\mathrm{Exp}\{R\} - \mathrm{Exp}\{0\} : \; R \in \mathscr{H}_{\mathrm{sig}}, \; R(o) = 0\right\}. \tag{10}$$

where $\mathrm{Lin}(A)$ *denotes the closure of the linear span of the elements of A.*

THE PROCESS OF RECOGNITION

Step 1: The recognition process is based on a comparison of signals: one signal will be the input signal obtained by accumulation of elementary signals, the other one is taken arbitrarily from the memory. If both signals coincide (completely or partially) there will be a *collapse* indicating the recognition of the signal. A detailed discussion and interpretation of this mechanism one can find in [1, 2, 3]. First we define an operator describing the selection of a signal from the memory.

Definition 5 *Let* $\mathscr{D} : L^2(\mathbb{M}) \longrightarrow L^2(\tilde{M} \times \mathbb{M})$ *be defined for all* $W \in L^2(\mathbb{M})$ *by*

$$\mathscr{D}W(\varphi, \Phi) := \frac{1}{\sqrt{|\Phi| + 1}} \cdot W(\Phi + \delta_\varphi) \qquad (\varphi \in \tilde{M}(G), \; \Phi \in \mathscr{M}) \tag{11}$$

where $|\Phi|$ denotes the number of configurations in the set Φ. Now, let $W \in \mathscr{H}_{\mathrm{mem}}$ be a (pure) state of the memory.

Proposition 6 *The operator* $\mathscr{D}$ *is an isometry, Moreover,* $W \in \mathscr{H}_{\mathrm{mem}}$ *implies* $\mathscr{D}W \in \mathscr{H}_{\mathrm{sig}} \otimes \mathscr{H}_{\mathrm{mem}}$.

Now, let ρ be a state on $\mathscr{H}_{\mathrm{sig}} \subseteq \mathscr{M}$ describing an input signal obtained by an accumulation of elementary signals. Further, let ν be a state of the memory. Using the operator $\mathscr{D}$ a signal is separated from the memory.

This yields to a state $\rho\otimes\mathcal{D}\nu\mathcal{D}^*$ on

$$\mathcal{H} := \mathcal{H}_{\text{sig}}\otimes\mathcal{H}_{\text{sig}}\otimes\mathcal{H}_{\text{mem}} \subseteq \mathcal{M}\otimes\mathcal{M}\otimes L^2(\mathbb{M}). \tag{12}$$

↑	↑	↖
in-com-ing sig-nal	signal chosen from memory	memory before and after processing

States on $\mathcal{H}$ will be the starting point for the next step.

Step 2: Let us concentrate our considerations to the first two spaces $\mathcal{H}_{\text{sig}}\otimes\mathcal{H}_{\text{sig}}$ of the tensor product space $\mathcal{H}$. Now the two signals are compared with the aid of the following operator:

Definition 7 *The operator $\mathcal{V} : \mathcal{M}\otimes\mathcal{M} \longrightarrow \mathcal{M}\otimes\mathcal{M}$ defined on exponential vectors by*

$$\mathcal{V}(\exp\{f\}\otimes\exp\{g\}) = \exp\{\frac{1}{\sqrt{2}}\cdot(g-f)\}\otimes\exp\{\frac{1}{\sqrt{2}}\cdot(g+f)\} \quad f,\, g\in L^2(G) \tag{13}$$

is called symmetric beam splitter.

Observe that $\mathcal{V}$ is unitary and self-adjoint. Beam splitters are well-known operators in quantum optics. More general beam splitting operators (being unitary but not self-adjoint) were considered e. g. in [18, 19, 20, 21]. Now, the signals will be processed with the aid of $\mathcal{V}$. Let the input signal ρ_1 and the state ρ_2 describing the signal chosen from the memory be coherent states:

$$\rho_1 \;=\; \langle\exp\{\alpha f\},\cdot\exp\{\alpha f\}\rangle\cdot e^{-|\alpha|^2}, \qquad \rho_2 = \langle\exp\{\lambda g\},\cdot\exp\{\lambda g\}\rangle\cdot e^{-|\lambda|^2}$$

with $f,\, g\in L^2(G)$, $\|f\|=1$, $\|g\|=1$, $\alpha,\,\lambda\in\mathbb{C}\setminus\{0\}$. Observe that $|\alpha|^2$ is the intensity of the signal ρ_1, $|\lambda|^2$ the intensity of ρ_2.

As it was pointed out in [7] the signals in the brain permanently exchange their supports being sets of excited neurons. This procedure is described by the above beam splitters. If the two input signals coincide ($\alpha=\beta$ and $f=g$) the first of the output signals will be empty (the vacuum state) though it is a superposition of two non-empty signals. Thus the output will be the vacuum state. Applying twice the operator $\mathcal{V}$ we always return to the original state. But never a state can remain the same after one application of $\mathcal{V}$.

Step 3: The changed signal finally has to be stored in the memory. The last step corresponds to an application of the operator $\mathcal{D}^*$. Summarizing, (one step of) the whole procedure is done with the operator $A_\varepsilon : \mathcal{M}\otimes L^2(\mathbb{M}) \longrightarrow \mathcal{M}\otimes L^2(\mathbb{M})$ given by

$$A_\varepsilon := (I_{\mathcal{M}}\otimes\mathcal{D}^*)(\mathcal{V}\otimes I_{L^2(\mathbb{G})})(I_{\mathcal{M}}\otimes\mathcal{D}) \tag{14}$$

where I denotes the identical operator on the corresponding spaces.

Let us remark that in a more refined model one has to tak into account that it needs certain time until the input signal ρ_1 will be completely located in the the brain. There is a process of accumulation of the signal. Up to a certain fixed time τ in a discrete tact the intensity of the input signal grows up. Step by step the beam splitter $\mathcal{V}$ is applied to this changing input signal and the (unchanged) signal chosen from memory. With a certain probability in the first component of the output after the application of $\mathcal{V}$ will be the vacuum $\exp\{0\}$. In this case the process is finished (recognition took place). If there will be no vacuum in the first term of the output (*no collapse*) applying a second time the beam splitter one returns to the original input, that means nothing has changed. After the next time interval the same procedure will be repeated - now with the amplified input in the first component. If no collapse occurs during this unit of time no recognition took place.

REFERENCES

1. K.-H. Fichtner and L. Fichtner. Bosons and a quantum model of the brain. Jenaer Schriften zur Mathematik und Informatik, Math./Inf./08/05 (2005), Friedrich Schiller University, Jena, 2005, 26 pages.
2. K.-H. Fichtner, L. Fichtner, W. Freudenberg, and M. Ohya. Quantum models of the recognition process - a convergence theorem. Technical report, 2007. 32 pages.
3. K.-H. Fichtner, L. Fichtner, W. Freudenberg, and M. Ohya. Quantum models of the recognition process - mathematical prerequisites. Technical report, 2007. 44 pages.
4. J.W. Philips et al. Imaging neural activity using MEG and EEG. *IEEE Engineering in Medicine and Biology*, pages 34–42, May/June 1997.
5. R. Hari and O.V. Lounasmaa. Neuromagnetism: tracking the dynamics of the brain. *Physical World*, pages 33–38, May 2000.
6. Wolf Singer. Consciousness and the structure of neuronal representations. *Phil. Trans. R. Soc. Lond.*, B 353:1829–1840, 1998.
7. Wolf Singer. *Der Beobachter im Gehirn. Essays zur Hirnforschung*. Suhrkamp Verlag, Frankfurt a.M., 2002.
8. A.K. Engel and W. Singer. Temporal binding and the neural correlates of sensory awarenes. *Trends in Cogn. Sci.*, 5(1):16–25, 2001.
9. Wolf Singer. Neuronal synchrony: a versatile code for the definition of relations? *Neuron*, 24:49–65, 1999.
10. Wolf Singer. Sriving for coherence. news and views. *Nature*, 397:391–393, 1999.
11. P.R. Roelfsema and W. Singer. Detecting connectedness. *Cerebral Cortex*, 8:385–396, 1998.
12. K.-H. Fichtner and L. Fichtner. Über eine mathematische Modellierung von Hirnaktivitäten. Technical report, Friedrich Schiller University, Jena, 2005, 22 pages.
13. K.-H. Fichtner and W. Freudenberg. Point processes and the position distribution of infinite boson systems. *J. Stat. Phys.*, 47:959—978, 1987.
14. H. Maassen. Quantum Markov processes on Fock space described by integral kernels. In L. Accardi and W. von Waldenfels, editors, *Quantum Probability and Applications II*, volume 1136 of *Lecture Notes in Mathematics*, pages 361 – 374, Berlin, Heidelberg, New York, 1985. Springer.
15. A. Liemant, K. Matthes, and A. Wakolbinger. *Equilibrium Distributions of Branching Processes*. Akademie-Verlag, Berlin, 1988.
16. K.-H. Fichtner. On the position distribution of the ideal Bose gas. *Math. Nachr.*, 151:59 – 67, 1991.
17. V. A. Fock. Konfigurationsraum und zweite Quantelung. *Zeitschrift für Physik A*, Bd. 75, N 9-10:622–647, 1932 (Erratum: Zeitschrift für Physik Phys. 1932, Bd. 76, N 11-12, S. 852).
18. K.-H. Fichtner, W. Freudenberg, and V. Liebscher. Time Evolution and Invariance of Boson Systems Given by Beam Splittings. *Infinite Dimensional Analysis, Quantum Probability and Related Topics*, 1(4):511 – 531, 1998.
19. K.-H. Fichtner, W. Freudenberg, and V. Liebscher. Characterization of Classical and Quantum Poisson Systems by Thinnings and Splittings. *Math. Nachr.*, 218:25–47, 2000.
20. K.-H. Fichtner, W. Freudenberg, and V. Liebscher. On exchange mechanisms for bosons. *Random Operators and Stochastic Equations*, 12:no. 4, 331–348, 2004.
21. W. Freudenberg, M. Ohya, and N. Watanabe. On quantum logical gates on a general Fock space. In M. Schürmann and U. Franz, editors, *Quantum Probability and Infinite Dimensional Analysis. From Foundations to Applications*, volume XVIII of *QP–PQ: Quantum Probability and White Noise Analysis*, pages 252 – 268, New Jersey London Singapore, 2005. World Scientific.

Quantum Logic and Macroscopic Quantum Games

A.Grib

Department of Theoretical Physics and Astronomy, A.I.Herzen State Pedagogical University of Russia, 48 Moika Emb., St. Petersburg 191186, Russia

Abstract. Examples of macroscopic situations when stochasticity is described by the quantum probability amplitude instead of the Kolmogorovian probability are given. They are based on the nondistributive quantum logical lattices obtained from some Boolean distributive ones by putting away some of its elements as unobservable. The possibility of the macroscopic quantum game based on these examples is discussed.

Keywords: Quantum logic, nondistributive lattices, Hasse diagram, wave function.
PACS: 03.65.Ta, 03.67.-a, 89.65.Gh

In this paper we continue the investigation [1] of macroscopic situations described by the same mathematical formalism as some simple (spin one half and spin one) quantum systems. In these situations stochasticity is described by some wave function as vector in finite dimensional Hilbert space with noncommuting operators in it as observables. So one has the complementarity property for such systems. These situations can occur in economics [2, 3], biology [4] etc so that chance in these sciences must not necessarily be described by the standard Kolmogorovian probability measure as it is usually supposed to be but by the more general quantum formalism. In [1] it was shown that new type of Nash equilibria can arise in these cases. Differently from the microworld the Planck's constant does not play any role in our examples.

Here we consider some other than in our papers [1, 5, 6] example named "the firefly in a box" [7]. This example will be used by us in order to show the connection between the Boolean lattice with probabilistic description of chance and the non Boolean nondistributive lattice with the quantum probability measure on it.The example is the simplification of the well known more complex Foulis "'firefly in a box"' case [8, 9]. We take it because in our paper [1] we found Nash equilibria for a quantum game based on Hasse diagram for this simplified case.It occurred that in quantum case there are Nash equilibria more profitable than the classical ones. These Nash eqilibria correspond to realizations of the nondistributive lattice in terms of certain noncommutative projectors in Hilbert space of the spin one half system.We shall not discuss other known examples (Wright urn [10])leading to nondistributive lattices because we did not investigate Nash equilibria for these examples.

The rule for the quantum macroscopic game for our example can be formulated as follows.

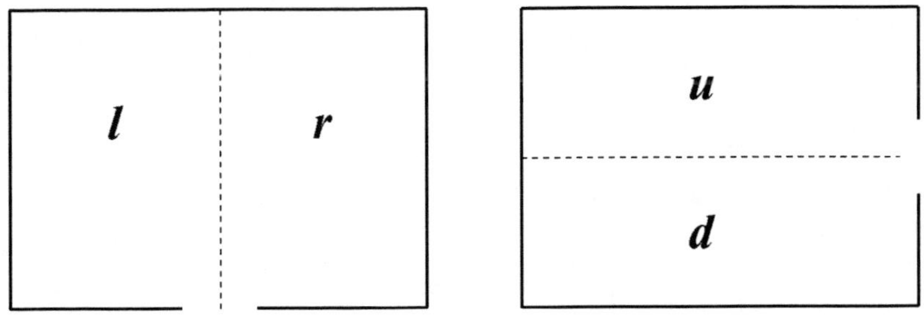

FIGURE 1. The firefly in a box with two windows.

A firefly (surely a macroscopic agent) is roaming around a box and some observer (other macroscopic agent) can see it either through the window at the bottom of the box or through the window at the right side of it. Each window has a thin line perpendicular to it drawn at it's centre so that an observer can see the firefly in one or another halves of the box.

CP962, *Quantum Theory, Reconsideration of Foundations – 4*,
edited by G. Adenier, A. Yu. Khrennikov, P. Lahti, V. I. Man'ko, and Th. M. Nieuwenhuizen

An observer cannot look at the same moment at two windows at once so that one has two incompatible experiments. Let us call possible observable situations as "left", "right", "up" and "down". The outcomes of the experiments can be described by the nondistributive lattice employed by Birkhoff and von Neumann [11] for the spin one half system with two complementary observables - (different spin projections) - being measured. This is an orthocomplemented lattice. All rigorous mathematical definitions can be found in [12]. However for better understanding of the paper we must give some necessary definitions and conjectures here.

The lattice L is a partially ordered set (S, $\leq$)with two operations $\vee, \wedge$ so that each pair $x, y \in L, x \neq y$ has a supremum $x \vee y$ and an infimum $x \wedge y$. There are elements $\oslash \in L, I \in L$ such that $x \vee \oslash = x$, $x \wedge \oslash$, $x \vee I = I$, $x \wedge I = x$.

The lattice is complemented if for $\forall x \in L$ exists at least one complement x' such that $x \wedge x' = \oslash$, $x \vee x' = I$. The elements of the lattice are orthogonal $x \perp y$ if $x \leq y'$. The operations $\wedge, \vee$ can sometimes be understood as logical disjunction and conjunction.Then it is supposed that if x is true than $x \vee y$ is true, if y is true then $x \vee y$ is true. However in 'quantum logic "'if"' is not equal to "'always if"' (iff). So in general it is not correct to think that from $x \vee y$ true follows that either of them is true. Negation in quantum logic is realized through orthogonality. There is some discussion in literature on the problem of logical interpretation of lattices(see [13]) for the general case so that different views arise from different definitions.

One can draw the so called Hasse diagram for the lattice so that lines correspond to partial order,going up one can obtain intersection of lines at $\vee$, going down one can obtain intersection at $\wedge$.

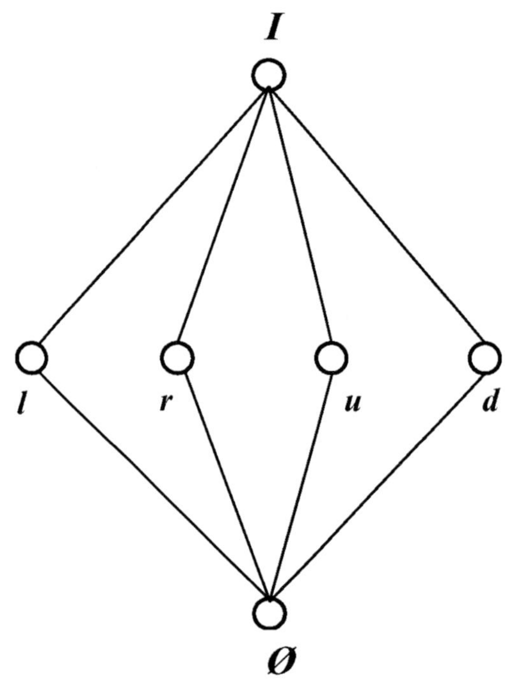

FIGURE 2. Hasse diagram of the nondistributive lattice of the firefly in a box example.

Here "l", "r", "u", "d" are elements (logical atoms) of the lattice describing different experimentally testable propositions for the firefly on Fig.1. Elements "l" and "r" as well as "u" and " d" are orthogonal. One also has for the two lattice operations $\wedge$ ("and"), $\vee$ ("or")

$$l \vee r = u \vee d = l \vee u = r \vee d = l \vee d = r \vee u = I$$

$$l \wedge r = l \wedge u = l \wedge d = r \wedge u = u \wedge d = r \wedge d = \oslash$$

which means that "l" or "r" is always true while "l" and "r" is always false etc. For example "r" and "u" is always false because "experimentally" there is no such observable element at the disposal of the observer due to the impossibility of simultaneous observation of the corresponding situations. The lattice is nondistributive because

$$l \wedge (r \vee d) = l \wedge I = l \neq (l \wedge r) \vee (l \wedge d) = \oslash \vee \oslash = \oslash.$$

If the firefly randomly moves inside the box the observer can describe the outcomes of his observations as some representation of the nondistributive lattice in terms of projectors ("yes - no" questions) in two dimensional real space. He (she) defines the "quantum" probability of the outcomes from some wave function $|\Psi\rangle$ taken as the two dimensional vector by

$$w_\alpha = \langle \Psi | \hat{P_\alpha} | \Psi \rangle, \alpha \in \{l, r, u, d\}, \qquad (1)$$

here one has

$$w_l + w_r = w_u + w_d = 1. \qquad (2)$$

So one has the wave function and noncommuting operators – the projectors $\hat{P_\alpha}$ for the "firefly in a box" example.

For macroscopic situations of this type one can be interested to find answers on the questions: what is the meaning of the "preparation" of the wave function and what is the meaning of measuring different noncommuting observables from the point of view of our macroscopic agents?

To find the answer one must take into account the fact that this lattice can be embedded into some Boolean lattice. The physical meaning of the embeddement is simple: the observer cannot check some situations described by some more general Boolean lattice due to the character of his (her) experiments. These elements of the lattice are "hidden variables" for the observer and the nondistributivity of the lattice is the payment for his (her) "ignorance". A general theory of realizing of quantum logical lattices as "‘concrete logics"’ obtained from Boolean lattices was developed in [14]15))

Surely due to Kochen-Specker"s theorem [15] not all quantum logical lattices can be embedded into Boolean lattices, that is why quantum theory of microparticles is basic and is not a hidden variable theory. The other objection is breaking of Bell's inequalities for relativistic systems if entangled states are considered [16].

However for our aim to find macroscopic situations with macroscopic agents with behaviour described by the quantum rules (i.e. by the Born-Luders rule for calculation of probabilities) the class of such "quasiclassical" lattices [5, 7] embedded in Boolean ones is wide enough.

To construct the Boolean lattice divide the box on four parts.

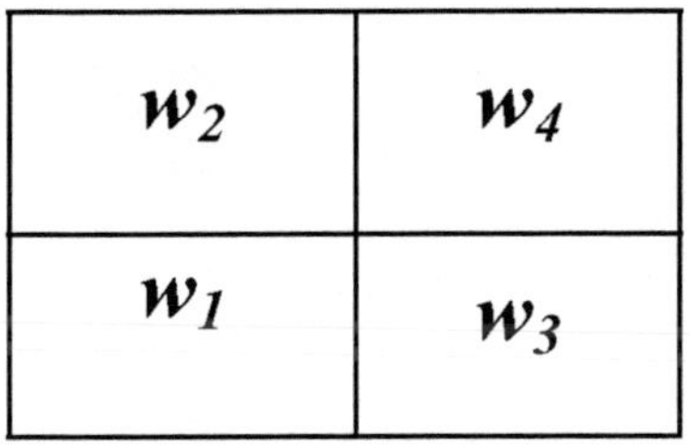

FIGURE 3. Probability of different firefly in a box positions from it's "own" point of view.

The firefly can occur in any of the four parts. Construct the Boolean lattice based on elements 1, 2, 3, 4 as it's atoms.

It is seen from the Fig.4 that one obtains the nondistributive lattice of the Fig.2 by considering only composite elements "l", "r" ",u", "d" of the second row while the atomic elements 1, 2, 3, 4 as well as the third row composed from triples are unobservable.

So the main lesson is that one can expect obtaining of the quantum rules in situations when one has stochastic processes which are secondary to some basic non observable ones. This is typical for situations on the stock market, in some complex biological systems etc. The other important feature is complementarity as impossibility of checking all properties at the same moment of time.

What is the connection of the probabilities on the Boolean lattice and the quantum probability on the nondistributive one?

Denoting $w_a : w_1, w_2, w_3, w_4$ the probabilities for atoms of the Boolean lattice one obtains the equations

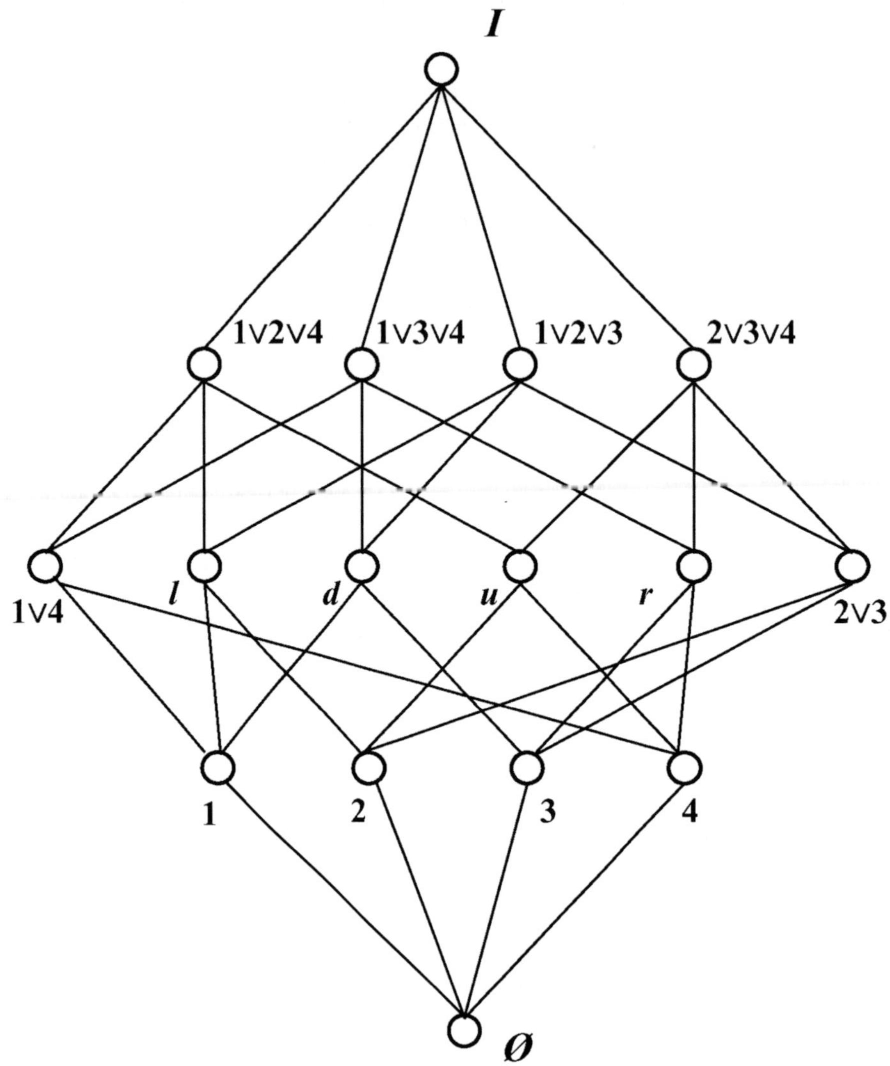

FIGURE 4. Hasse diagram of the firefly in a box example for the Boolean case.

$$\begin{aligned} w_1 + w_2 &= w_l \\ w_3 + w_4 &= 1 - w_l \\ w_1 + w_3 &= w_d \\ w_2 + w_4 &= 1 - w_d \end{aligned} \tag{3}$$

From these equations it is easy to see that the "strange" quantum rule for one and the same object(the firefly)

$$w_l + w_r + w_u + w_d = 2 \tag{4}$$

is not strange at all if it is considered not for the elementary events but for the complex ones!

It seems that any distribution w_1, w_2, w_3, w_4 leads to some fixed wave function and some fixed representation of the lattice in terms of projectors on some fixed directions on the plane. However one can see that distributions leading to appearance of two ones or zeros for complex events of the second floor are prohibited due to the definition of disjunction in the quantum logical lattice while it is possible in the Boolean case. This means that logical atoms of the first floor cannot be definitely determined from the point of the observer for the quantum logical case being totally "hidden" for him(her). One can ask the question: what prohibits the firefly to occur in the corner 1?

Is it possible for the observer to get 1for the outcome "left" and 1 for the outcome "down"? The answer is surely positive. But for the quantum system it is impossible to get eigenfunction for noncommuting spin operators. Is it a contradiction?

The answer is "no"! It is here where something like the wave packet collapse idea comes into the play. Quantum theory does not forbid to get in observations of complementary observables positive results. It only says that if one prepares the wavefunction as an "eigenfunction of the projector on the "left" part and performs many observations of complementary observables then one obtains some probability distribution for the complementary observable "up"-"down". This distribution is not 1, 0 because the firefly has some freedom and the only instruction for him given on the preparation stage is to be "somewhere" in the "left" part. The observer cannot always let him be in the corner1,so in many experiments with the same "left" instruction the results will be sometimes "up", sometimes "down" with frequencies obtained from the wave function and the representation of observables in the form of operators.

If the distributions w_l, w_d are known the distribution w_1, w_2, w_3, w_4 obtained from it is not unique: it is defined up to some arbitrary w_i where w_i is some probability for fixed i.

This is just the manifestation of "indefiniteness" of the quantum situation in comparison to Boolean one as we just said before.

The macroscopic quantum game considered previously in [1] called the macroscopic Stern-Gerlach quantum game can be organized for the "firefly in a box" case as follows.

There are two partners Alice and Bob and two boxes with fireflies there. Alice can try to choose some classical probability distribution for Boolean elementary outcomes 1, 2, 3, 4 with limitations discussed before. This can be done by "training" of the firefly stimulating it to come more often to this or that part of the box. For example any observation by Alice of the part of the box is accompanied by the flash of light with some prolongation in time stimulating the firefly to react. Different times of observation result in different frequencies for the firefly to occur in some part of the box. Supposing that Alice has at her disposal only l, r, u, d outcomes she cannot define exact distributions for Boolean elementary outcomes but only some class of it. However from the quantum point of view definition of frequencies for l, r, u, d means definition of the wave function and the representation of the atoms of the non Boolean lattice in terms of noncommuting projectors, i.e. definition of the angle between different spin projections. This can be called "the preparation stage".

In special cases getting 1,0 for some of the complementary possibilities she can speak about some fixed wave function as an eigenfunction of some spin projection operator. However in general case it is not the case.

Same manipulations are made by Bob. However neither Alice nor Bob have knowledge of the training procedures of the partner.

Then the game begins. Alice and Bob with their trained fireflies look at the results of their observations accompanied by flashes of light. In cases defined by the rules of the game when for example Alice gets some fixed result "$'\alpha$"' while Bob gets "β" Bob must pay money to Alice etc.

There is some payoff matrix defining in what cases Bob pays Alice some money as it was defined in [1]. The profit depends on the frequencies of the outcomes. The average profit is calculated by using the quantum rule for projectors $\hat{P}^a_\alpha$ for Alice, $\hat{P}^b_\alpha$ for Bob

$$\begin{aligned} \overline{H} &= \langle \Psi_A | \langle \Psi_B | \hat{H} | \Psi_B \rangle \Psi_A \rangle; \\ \hat{H} &= c_3 \hat{P}^a_u \otimes \hat{P}^b_d + c_1 \hat{P}^a_d \otimes \hat{P}^b_u + c_4 \hat{P}^a_l \otimes \hat{P}^b_r + c_2 \hat{P}^b_r \otimes \hat{P}^b_l. \end{aligned} \tag{5}$$

which in terms of the "Boolean philosophy" means calculation of

$$\overline{H} = c_3 w_u w^b_d + c_1 w_d w^b_u + c_4 w_l w^b_r + c_2 w_r w^b_l, \tag{6}$$

where w^b_α are probabilities used by Bob.

Here one must make some remarks about some peculiar features of this calculation.

The representation of the nonBoolean lattice in terms of noncommuting projectors is not unique. It is defined up to some angle. This angle can be defined by Alice and Bob by choosing w_α for the pure case. Existence of different representations of the non Boolean lattice parametrised by the angle is the manifestation of the freedom of the observer to choose measurement of any spin observable for the real quantum system. There is also another freedom for the observer manifested in the choice of the wave function.

If after this definition of the angle Alice and Bob cannot change the angles then their freedom is now limited by choosing only some special distributions for the firefly in the complementary experiment. The "quantum logic" leads to arising of a special "quantum "correlation" between complementary observations. This correlation according to [1] can be expressed by the constraint

$$\frac{(w_u + w_l - 1)^2}{cos^2(\theta)} + \frac{(w_l - w_u)^2}{sin^2(\theta)} = 1. \tag{7}$$

Different choice of the angle leads to the different constraint. Different Nash equilibria can be found for different angles.It was shown in[14]1)for the Hasse diagram of the simplified version of the firefly in a box considered in this paper that some of these equilibria are more profitable for partners others are less.So one can put the hypothesis that it is the more profitable equilibria that can play the role of the principle of choice of the representation of the lattice.

Can one understand quantum rules from the classical Boolean point of view? Strange as it is but the answer in general is negative. Why after all the firefly must respect the constraint (7)? Equations (3) do not lead to such a constraint. This means that it occurs only if it is obtained by some training of the firefly by external agents like Alice and Bob. In general case one can only have a set of patterns of probabilities parametrised by different angles without a criterium for the choice of this or that pattern. There is no such ambiguity for quantum particles where particles do not have freedom of choice for this or that observable. This choice is made by the observer.

Absence of isomorphism between non Boolean and Boolean lattices leads to some other differences.

One of the differences is existence of so called Szabo's "counter-intuitive probabilities" [17]. This means that one can have probabilities close to one for complementary observations while the joint probability vanishes. In Boolean lattice embedding the nonBoolean one this is impossible for the complex events of the second row.

The other feature is the impossibility of obtaining the quantum rules from the Boolean lattice using the standard conditional probabilities and the Bayes formula for conditional probabilities. This is due to the fact that nondistributuvity breaks the formula for the full probability and the conditional probability calculus.

An interesting problem arises if one thinks not about a game with restricted rules but about some biological situation when the fireflies "seeing" one another as in our example fixing only complex events due to getting more "profit" in the case of the quantum Nash equilibria (see [1]) spontaneously choose some quantum correlation with a fixed angle.This can lead to explanation of some synchronous effects in the behaviour of real fireflies in nature [18].

ACKNOWLEDGMENTS

The author is indebted to the Ministry of Science and Education of Russia, grant RNP 2.1.1.6826 and to the Royal Academy of Sweden, grant on collaboration with states of the former Soviet Union, for financial support of the work on this paper. The author thanks Prof. A. Khrennikov for useful discussion of main ideas of the work and Dr. N. Vasil'ev for help in preparation of the Latex file of the paper.

REFERENCES

1. A.Grib, A.Khrennikov, G.Parfionov and K.Starkov, Quantum equilibria for macroscopic systems, *Journal of Physics A*, **39**, pp.8461-8475 (2006).
2. G.Soros, *Alchemy of finances*, Wiley, 2004.
3. O.Choustova, Quantum Bohmian model for financial market, *Physica* A, **374**, 016138 (Part 2) (2005).
4. A.Khrennikov, Quantum-like brain. Interference of mind, *Biosystems*, **84**, pp.225-241, (2006).
5. A.Grib and Jr. W.A.Rodrigues , *Nonlocality in Quantum Physics*, (New York: Dodrecht Plenum/Kluwer), 1999.
6. A.Grib, A.Khrennikov, G.Parfionov, K.Starkov, Probability amplitude in quantum like games *Quantum Theory: Reconsideration of Foundations-3* (Vaxjo University Press), 2004, pp.108-114.
7. K.Svozil, *Quantum Logic*, (Singapore: Springer), 1998.
8. D.J.Foulis, A half century of quantum logic, What have we learned? *In Einstein meets Magritte*, (KluwerDordrecht), **7**, 1999, pp.1-36.
9. D.Cohen, *An introduction to Hilbert space and quantum logic*, (New York: Springer), 1989.
10. R. Wright, Generalized urn models, *Found. Physics*, **20**, pp.881-903 (1991).
11. G.Birkhoff and J. von Neumann, The logic of quantum mechanics *Ann. Math.*, **37**, pp.823-842 (1936).
12. G.Birkhoff, *Lattice theory*, (AMS, Providence, Rhode Island), 1993.
13. J.Pykacz, Conjunctions, disjunctions and Bell-type inequalities on orthoalgebras, *International Journal of Theoretical Physics* **35**, pp.2355-2363 (1996).
14. 14) P.G. Ochinnikov, Measures on finite concrete logics, *Proc. Amer. Math. Soc.*, **127**, 1957-1966 (1999).

15. S.Kochen and E.P.Specker, The problem of hidden variables in quantum mechanics, *J. of Rat. Mech. and Analysis* **17**, pp.59-87 (1967)
16. J.S.Bell, *Speakable and unspeakable in quantum mechanics*, (Cambridge: Cambridge University Press), 1988.
17. L.Szabo, Is there anything non-classical? *E-print* quant-ph/9606006 (1986).
18. Y.Zhou, W.Gall, and K. Nabb, Synchronizing fireflies, *College Math. J.*, pp.187-193 (2006).

The Born Rule

Karl Gustafson

Department of Mathematics, University of Colorado, Campus Box 395, Boulder, CO 80309-0395, USA

Abstract. I review the Born probabilistic interpretation of quantum mechanics from its inception to the present, including my recent sharpenings. The paper looks closely at a few issues, and does not even mention many others.

Keywords: Born probabilistic interpretation, Measurement Problem, Operator Domains
PACS: 03.65-w, 03.65.Ta, 03.65.Db, 01.65.+g

INTRODUCTION

Born's 1926 interpretation of the Schrödinger equation wavefunction as a probability density came from his Born approximation of electron scattering experiments. He concluded that one could not know the state of an electron after colliding with an atom, one could know only "how probable is the specific outcome of the collision" and "I myself am inclined to renounce determinism in the world of atoms". A year earlier, in 1925 Born was the first to write down the Heisenberg uncertainty relations in matrix form. Later, Born's probabilistic interpretation of quantum mechanics was formalized operator-theoretically with Von Neumann's Hilbert space formulation of quantum mechanics. More background and history of Born's thinking and writings related to his probabilistic interpretation of quantum mechanics may be found in my recent review [1] of the recent biography of Born [2] and the intimately related book [3] of Born–Einstein letters. For example, on Born's gravestone [2] "is carved $pq-qp = h/2\pi i, \ldots$ first written down by Born in July 1925," and [2, p. 126] Born was proud that he was "the first person to write a physical law in terms of non-commuting symbols." From [3, p. 186], Born tells Einstein that $\psi(t)$ does not really describe a state of a single system. Rather "For what is really meant is, of course, that you take all individuals of 67 and count the percentage who live for a certain length of time [to say his life expectation was 4.3 more years, the example Born was using]. This has always been my own concept of how to interpret $|\psi|^2$. Instead you [Einstein] propose a system of a large number of identical individuals—a statistical total. It seems to me that the difference is not essential, but merely a matter of language."

In this paper it is not my intent to repeat all the extensive historical research about Max Born, e.g. such as that reported on in [1], [2], [3], except insofar as how some related aspects might valuably impinge on my discussions below. Rather, here I want to emphasize a few other thoughts, by others and myself, as to the extent of validity of Born's probabilistic interpretations of quantum mechanics, how well the various embodiments of such interpretations into the various versions of Born's Rule that one finds in the literature, really should be accepted or questioned. Born was heavily influenced by Bohr's view, that one must always take the statistical view of quantum mechanics. Thus Born and Heisenberg along with Bohr were, we could say, the three foundational pillars of the Copenhagen Interpretation. One could add J. Von Neumann, (and even others, especially Pauli), who helped in 1927 culminate the thinking into its final Hilbert space formulation.

My historical re-readings included, among others, some classic sources to quantum thought [4]–[16]. Some further works will be cited later as they pertain to certain issues. Rather than outline the paper here, let us just proceed.

NON-COMMUTING PHYSICS

Born's 1925 writing down of Heisenberg's uncertainty relations into matrix form, referred to above, was not the first noncommuting symbols of physical law. Of course he probably meant in the statement quoted above that the relation $pq-qp = h/2\pi i$ was unavoidable in quantum mechanics. When I first encountered the uncertainty principle many many years ago, I saw that its essence was just the differentiation by parts rule of Newton's calculus:

$$d(xu(x))/dx = xdu(x)/dx + u(x) \tag{1}$$

hence $pq-qp =$ identity operator. Many others perhaps have understood this fact, but it is good to remember.

CP962, *Quantum Theory, Reconsideration of Foundations – 4*,
edited by G. Adenier, A. Yu. Khrennikov, P. Lahti, V. I. Man'ko, and Th. M. Nieuwenhuizen

To further emphasize this basic noncommutativity of physics, be it classical or quantum physics, let us recall two classical examples of great importance, fluid dynamics and electromagnetism, respectively.

Consider for example Green's first identity, so essential to both fluid dynamics and electricity and magnetism. The derivation, (e.g., see [17] is just

$$(v(x)u'(x))' = v'(x)u'(x) + v(x)u''(x) \tag{2}$$

where $'$ denotes partial derivative with respect to x_i. One then uses the Gauss Theorem (fundamental theorem of calculus) on the left side to arrive at the Green Identity

$$\int_{\Omega} \text{grad}\, u \cdot \text{grad}\, v = \int_{\partial\Omega} v\frac{\partial u}{\partial n} - \int_{\Omega} v\Delta u. \tag{3}$$

Here u in the fluids situation is velocity vector. In an $E+M$ situation (e.g., see [18]) such equations relate electric field intensity E and electric displacement D. Noncommuting physics has been with us since the beginning of science, and classically so.

However, Born and Heisenberg were enmeshed in the puzzles of quantum mechanics, from which emerged their quantum mechanical canonical commutation relations. As described by Heisenberg [4], he considers the wave equation of Schrödinger

$$H\left(\frac{h}{2\pi i}\frac{\partial}{\partial q}, q\right)\psi(q) + \frac{h}{2\pi i}\frac{\partial \psi(q)}{\partial t} = 0 \tag{4}$$

and derives "the probability amplitude" ψ. Then [4, p. 158] he wants to regard it as the space-time (i.e., classical) equation for matter waves. For one (corpuscular) electron the total Hamiltonian consists of kinetic energy $E_k = (p_x^2 + p_y^2 + p_z^2)/2\mu$ and potential energy $E_p = -eV$ where e and μ are the charge and mass of the electron and V is the electrostatic potential. In short order (using Green's Identity, plus assuming fall-off to zero at 'infinity') he arrives at the time-dependent Schrödinger equation

$$\frac{h^2}{8\pi^2\mu}\nabla^2\psi + eV\psi - \frac{h}{2\pi i}\frac{\partial \psi}{\partial t} = 0. \tag{5}$$

As he has used differentiation by parts (Green's Identity) to get here, is there any uncertainty built into the derivation? No, it is a classical $E+M$ argument, which, however, does depend very much on noncommutative (classical) physics. Heisenberg goes on to argue that by dropping the potential V in the equation, one gets De Broglie's equations for the interference of matter waves at very low densities. Thus one accounts for the important quantum experiments of Davisson et al. (diffraction of matter waves) and Wilson photographs (α-particles), in a completely classical way.

Heisenberg [4, p. 161] is quick to point out that "one can regard these achievements of classical theories only as proof of the similarity of the classical and quantum theories, in the sense of the correspondence principle". By this he means of course Bohr's 1923 Correspondence Principle, one version of which asserts that quantum mechanics in the limit of large quantum numbers should tend to a classical physics description.

As mentioned in the introduction to this paper, I have no intention of attempting here any complete, or chronological account, of probability in quantum mechanics. Such would be a book. But let me bring to your attention the wonderful argument of Bohr [5] when he countered one of Einstein's most ingeneous attacks. This was in response to the famous EPR paper [19]. Bohr responded [20] on the basis of his Complementarity Principle, one version of which is the assertion that in wave-particle duality, one may have one or the other but not both at the same time. Bohr's principal point as he recounts the argument in [5] is that measurements such as those proposed in the EPR paper demand mutually exclusive experimental arrangements.

To emphasize this point, and this is the point I wish to make here, Bohr [5] goes on to point out "In fact, although any pair q and p of conjugate space and momentum variables obeys the rule of noncommutative multiplication [expressed by $pq - qp = \sqrt{-1}h/2\pi$] and can thus only be fixed with reciprocal latitudes [given by $\Delta q \Delta p \approx h$], the difference $q_1 - q_2$ between two space coordinates referring to the constituents of the system will commute with the sum $p_1 + p_2$ of the corresponding momentum components, as follows directly from the commutativity of q_1 with p_2 and q_2 with p_1. Both $q_1 - q_2$ and $p_1 + p_2$ can, therefore, be accurately fixed in a state of the complex system and, consequently, we can predict the values of either q_1 or p_1 if either q_2 or p_2, respectively, are determined by direct measurements."

Let me write this in a classical, Newtonian calculus, differentiating by parts way. One wonders if

$$(x-y)\left(\frac{\partial}{\partial x} + \frac{\partial}{\partial y}\right)u(x,y) = \left(\frac{\partial}{\partial x} + \frac{\partial}{\partial y}\right)(x-y)u(x,y). \tag{6}$$

It does.

Feynman [6] describes how Schrödinger thought also in such classical terms. The picture was that of a stationary nucleus, electric currents moving around it, charges P and currents J, to generate electromagnetic fields, and these would radiate light. Feynman [6, III.21-6] states that then "It was Born who correctly (as far as we know) interpreted the ψ of the Schrödinger equation in terms of a probability amplitude—that very difficult idea that the square of the amplitude is not the charge density but is only the probability per unit volume of finding an electron there, and that when you do find the electron some place the entire charge is there. That whole idea is due to Born." Feynman goes on (I paraphrase a bit here) to argue that if one has a very large number of particles, and the probability of finding one at a point is $\psi\psi^*$, then $\psi\psi^*$ can be interpreted as a probability density, and if each particle carries the same charge, then indeed we may interpret $\psi\psi^*$ as a charge density, and its flow under a Schrödinger equation as a current of probability, seen classically as a charge current density.

THE BORN PROBABILITY INTERPRETATION

Born's exact interpretation is described many places but I shall refer to Hund [7] and Hendry [8]. Hund [Chapter 10] describes how attempts in 1922–24 to calculate the ground and ionized states of the Helium atom led to the probabilistic interpretation two years later by Born. Of course the fundamental earlier ingredients were Planck's black body radiation quanta [21] and Einstein's light emission quanta [22]. How atomic bound states and photons interacted was still not understood. The Bohr–Kramers–Slater 1924 paper [23] introduced the hypothesis that the light waves corresponding to absorption and emission in quanta are of the nature of probability waves, that they do not represent reality in a classical sense, but rather the "possibility" of such a reality. Here I am paraphrasing Heisenberg's 1955 account in [9], Chapter 2, to augment Hund's account. The waves defined the probability at every point that an atom present there is emitting or absorbing radiation. This hypothesis abandoned the conservation of energy for individual processes, later reinstated. But it was a first step in what has come to be known as the Born probability interpretation. So we may see in it already in 1924 those ideas of quantum probability, and later used terms such as possibility, potentiality, propensity, proclivity, catalogue, other terms I have observed in the quantum literature, to describe what Schrödinger's wave functions might mean. Then Heisenberg in July 1925 argued that one must give up formal rules using the terms position and frequency because they were not observable. He was strongly influenced [9] by the ideas in the Bohr, Kramers, Slater 1924 paper. Also Heisenberg formulated his observable amplitudes as Fourier coefficients. Because he wanted real (probability) amplitudes, he inferred from real classical multiplication $xy = z$ that multiplication of analogous Fourier series should yield something like [7, p. 134]

$$\sum_k x(n,k)y(k,\ell) = z(n,\ell). \tag{7}$$

Since (using Hund's terminology here) it need not be true that $xy = yx$, one must replace the classical product xy by $(xy + yx)/2$. This interests me because this would, in modern terms, allow me to consider the product observable

$$\mathrm{Re}(AB) = \frac{AB + (AB)^*}{2} \tag{8}$$

given two (self-adjoint) observables $A = A^*$ and $B = B^*$. I will return to this point later.

There is much more in Hund's account. In particular, he details [7, Chapter 11] how Schrödinger first was attempting quantum explanations using the classical vibrating string equation, with its second order time derivative $\partial^2\psi/\partial t^2$, which did not quite work. Hence he decided to write a periodic wave in the complex form

$$\psi \sim e^{\mp i\omega t} \tag{9}$$

which then gave him his Schrödinger equation

$$-\frac{h^2}{2m}\Delta\psi + V\psi \mp ih\frac{\partial\psi}{\partial t} = 0 \tag{10}$$

with its first order time derivative. As to its interpretation, Schrödinger thought he was representing particles as a tight bundle of waves, even as a distribution $\psi^*\psi$ in coordinate space, somewhat as I described above.

Born then improved upon this, working on models of the collision of electrons with atoms (now called, scattering theory). Born saw the De Broglie or Schrödinger wave as a kind of conducting field for electrons. The field determined the probabilities for the transition of an atom into other states and for the deflection of an electron in particular directions. He found solutions to a time-independent Schrödinger equation with spherical potential that took one incoming plane wave $\psi_{in} = e^{ikz-i\omega t}$ to one outgoing scattered wave $\psi_{out} = f(k,\theta)e^{ikr-i\omega t}/r$. He then regarded the quantity $|f(k,\theta)|^2 d\Omega$ as the probability that an electron would be deflected through an angle θ into the solid angle $d\Omega$. The $\psi^*\psi$ was the probability that the coordinates would lie in the region $d\tau$.

I must interject here that the above picture is the old Rayleigh, classical scattering, picture. See my book [17, p. 234]. The difference is replacing a small sphere scatterer by a spherically symmetric potential scatterer.

Then Born considered the discrete case, expanding $\psi = \sum c_n u_n(x)e^{-i\omega_n t}$ as the Fourier series solution to the time-dependent Schrödinger equation mentioned above (take the $-$ sign). He interpreted $|c_n|^2$ as the probability that the system was in the stationary state n. This corresponds mathematically to Heisenberg's more intuitive reasoning I gave above. Pauli jumped in and extended the probabilistic interpretation to a probability $|\phi(p)|^2 dp$ for a momentum value in momentum space.

Born also, with the help of Jordan, if you allow me to use the term 'cleaned up', cleaned up Heisenberg's new 'kinematics of radiation', the result being the casting of Heisenberg's relations into matrix form. Such is treated in Hund's book [7] but I refer the reader to a more mathematical account in Hendry [8]. The quantum transition expressions were noncommutative, and Born was well-versed as a mathematician in matrix theory and its noncommutative multiplication. He soon saw [8, p. 70] how to put Heisenberg's virtual oscillators into matrix terminology. Jordan saw how to put them into diagonal form. We note incidentally that this was Pascual Jordan, not the (earlier) mathematician Marie Jordan of the more general diagonal matrix canonical form. In any case, there resulted the fundamental Born-Heisenberg-Jordan 1926 paper [24].

As related better in Hendry [8], many opinions and fixed epistemological beliefs both helped and hindered these great pioneers of quantum mechanics. Heisenberg sometimes [8, p. 93] dismisses Born as just "a mathematical methods man". Pauli is even more dismissive [8, p. 73]. Heisenberg dismisses [8, p. 87] Schrödinger's attempts at visualizing his wave function as "rubbish". Dirac refers [8, p. 87] to Schrödinger's interpretations as "metaphysics". Of course Dirac's use of delta function [8, p. 109] could be regarded, from a mathematician's point of view, as a "metafiction", if I may coin such a word. Von Neumann [25] entered the picture and started to put the whole theory into Hilbert space. This obviated nonrigorous Dirac delta function arguments and also, by showing ℓ_2 and $\mathscr{L}_2$ to be isomorphic, combined the Heisenberg and Schrödinger pictures.

What is really great to read is how they all kept writing and talking and meeting with each other. The public likes to create heros and celebrities but it seems to me the truth is in the other direction of (even informal) teamwork.

THE BORN RULE

Most formulations of what are now called 'The Born Rule' seem to me to be best cast in the Von Neumann language of Hilbert space quantum mechanics, and that is the language I will use here. For a good brief mathematical background of the canonical commutation relations, the Stone-Von Neumann theorem on uniqueness of Schrödinger couples, the essential (up to some underlying group theory differences in the assumptions, I should warn) equivalence of the Schrödinger, Heisenberg, and Weyl representations, I recommend Putnam's little book [26].

The easiest statement of the Born rule would be: a measurement must collapse a wavefunction to one of the Hamiltonian's eigenfunctions. Although such view is widely used, I am as opposed to it as is Penrose [10], Chapter 22. Nonetheless, I want to avoid quibble here and just state some versions of conventionally accepted Born Rule.

Usually one states the Born rule first for a Hamiltonian H with discrete spectrum $\{\lambda_i\}$. Then one asserts (as above) that the measurement will be an eigenvalue λ_i and the probability of a given wave function being in the corresponding eigenstate ϕ_i is:

$$||P_i\psi||^2 = \langle P_i\psi, P_i\psi\rangle = \langle P_i\psi, \psi\rangle = |\langle\psi,\phi_i\rangle|^2. \tag{11}$$

I dislike the Dirac bra/ket notation and will not use it so let me rewrite the above in both functional analytic Hilbert space, and also matrix theorists language. Here $\langle u,v\rangle \equiv v^*u$ signifies an inner product in a Hilbert space. Then the above Born probability can be rewritten:

$$\begin{aligned} P_i &= \phi_i\phi_i^* = \text{outer product} = \text{rank one projection} \\ P_i\psi &= (\phi_i\phi_i^*)\psi = \phi_i(\phi_i^*\psi) = (\phi_i^*\psi)\phi_i \quad \text{associativity} \\ \langle P_i\psi,\psi\rangle &= \psi^*P_i\psi = (\psi^*\phi_i)(\phi_i^*\psi) = |\langle\phi_i,\psi\rangle\langle\psi,\phi_i\rangle|. \end{aligned} \tag{12}$$

Here $P_i = P_i^2 = P_i^*$ is an orthogonal selfadjoint eigenprojector.

Bub [11] gives a precise description of generalizing the Born rule, first from eigenvalues to a Borel subset E of the real line which may contain several eigenvalues λ_i:

$$\text{Prob}_\psi(a \in E) = \sum_{\lambda_i \in E} ||P_i \psi||^2. \tag{13}$$

If ψ is a convex combination of unit vectors ψ_i with convex combination weights w_i, this becomes

$$\text{Prob}_\psi(a \in E) = \sum_i w_i ||P_A(E)\psi_i||^2 \tag{14}$$

where here I use Bub's notation A for the Hamiltonian H representing a quantum observable. As then carefully developed by Bub [11, pp. 254–260], one obtains a generalized Born rule for the probability of values a of an observable A in a Borel set E by a pure or mixed state, expressed in terms of the trace of the density operator W representing that state ψ, as

$$\text{Prob}_W(a \in E) = tr(WP_A(E)). \tag{15}$$

Gleason's theorem shows that the set of all pure and mixed states is complete: all possible probability measures μ definable on the lattice of orthogonal projectors on a Hilbert space are generated by the density operators of the generalized Born rule above. Here the probability measures μ under consideration are the maps from the lattice onto the interval (0,1) satisfying the usual probability measure axioms for each Boolean sublattice of the lattice. Thus we may say that the Born rule uniquely defines the Hilbert space probabilities if we accept the usual propositional calculus of quantum theory and the projective geometry formulation of quantum mechanics, in which quantum states are represented by rays, and convex combinations of rays, in the Hilbert space.

Jauch [12] gives good further explanations of this just described picture. Moreover he describes in detail the propositional calculus of quantum propositions. I am quite interested in the notion of compatibility in quantum mechanics. Most books treat it lightly, that is, they just assume that it must be there. Bub [11] takes the usual view: two observables A_1 and A_2 are compatible if and only if they commute: $A_1 A_2 = A_2 A_1$. If they do, they are both a function of a third observable A_3. Jauch [12] relates this to compatibility of propositions: they must generate a Boolean sublattice of projections. One can equivalently state compatibility in terms of the closed subspaces of the Hilbert space, as they are in 1–1 correspondence with the orthogonal projections onto them.

THE MEASUREMENT PROBLEM

It is generally accepted that since the "completion" of the new quantum mechanics in 1927 (80 years ago), the outstanding remaining difficulty is the measurement problem. Penrose [10] does a good job of describing the possible ways out of the dilemmas, paradoxes, difficulties, that one meets in this problem. Huge amounts have been written on the problem. Here I bring out only a few comments.

Bell [13] gives a great discussion in his famous "Against Measurement" article. This treatment brings out the pros and cons and opinions of many famous physicists. Bell is particularly concerned with the "shifty split" between atomic system and (classical) measuring apparatus. He ends up more-or-less suggesting that we accept either the De Broglie-Bohm (linear) theory, or the Ghirardi-Rimini-Weber (nonlinear) theory. I call the former a 'stream-line' view, with a problematic quantum potential put in by hand. The latter theory [27] postulates that you need to add spontaneous stochastic localizations.

I must point out that the GRW approach has similarities to the views of I. Prigogine, see my articles [28], [29]. Prigogine often discussed with me his view, in which what we mathematicians call bifurcation theory, occurs only spontaneously at certain transition times in physical processes. I have pointed out in [29] that after considerable effort I have concluded that Prigogine's thinking, and that of Einstein and others, came down to their assuming a microscopic detailed balance, even of, probabilities.

Aspect [14] presents beautiful case studies of actual experiments on wave-particle duality. These obtain experimental confirmation of particle-like behavior of light emitted by a one photon sources, contrasted with a non-particle-like behavior of light emitted by a usual source (pulsed light-emitting diode). For the one photon sources, they find interference fringes building up "one photon at a time". Thus single photons do interfere, a wave-like behavior. A conclusion is that Bohr's Complementarity principle is supported. For a given experimental set up, only one behavior will appear.

Primas and Esfeld [15], in an extensive (55 pages) questioning of Wigner's formulations of quantum theory, conclude that the measurement problem is ill-posed. They assert that one should stay with the Expectation-value postulate to avoid all contradictions. They take a C^* algebra ansatz: all physical systems are characterized by an appropriate C^* algebra. There is a one–one correspondence between the properties of the physical system and the commutative C^* subalgebras. Every observable A generates a unique commutative C^* subalgebra. The valuation of every physical property is given (independent of any measurement) by a continuous set function defined on the spectral measures of the observable A. Thus physical properties are not dependent on measurements, and are also not probabilistic in any way. Then the Expectation Postulate gives the probability that a measurement gives a value a in a Borel set E as given by the Kolmogorov probability measure $\mu(E) = tr(WP_A(E))$. Here I took the liberty of changing the [15] notation to conform to the generalized Born rule I gave above. The Von Neumann Projection Postulate is consistent with their formulation up to the time of the measurement, but the Primas-Esfeld ansatz insists that you say nothing about the state after measurement. So the issue of wave function collapse upon measurement is 'finessed', if I might use that term. Again in my own terms, it seems to me that the approach of [15], which goes back originally to Pauli [30], doesn't really address the measurement problem. It avoids it. By assuming intrinsic value distributions, in my picture of it, one is saying there is an a priori population, with a posteriori statistics, as in classical statistics. Also I confess to not liking the open systems formulations [31] that underly the formulations of [15]. On the other hand, I also ascribe to an intuition in which "there really aren't any unitary evolutions". See my arguments in [32]. I have enjoyed discussing physics several times with the great theoretical chemist Hans Primas. But I do not agree that we should view a physical particle as just an irreducible representation of the Poincare group!

Griffiths [16] refers to the wave function ψ as a pre-probability, and entirely nonphysical. However, in his theory of (many) consistent histories, he claims to have enough histories (my interpretation of his treatment, here) so that he can choose from among them to explain away the paradoxes. The famous Gedanken thought experiments of Einstein, Wigner, usually involve counterfactuals, and these may be avoided or treated by choosing the right consistent history. In same manner, wave function collapse is not needed when describing the interaction picture between a quantum system and a measuring device. The Born rule does not apply when conditional probabilities are measured in a way that involves two times. Penrose [10] compares the many-worlds model (everything is a unitary evolution) with the consistent histories model (even more general, one can do measurements, but no problem). But I do agree that we all must now seriously question the Von Neumann projection postulate and its corollary wave function collapse.

OPERATOR DOMAINS AND QUANTUM PROBABILITIES

In this section, I will put forth a view which is my own. However, it utilizes the Hilbert space formulation of Von Neumann as extended further by Kato [33]. The mathematical prerequisite is unbounded operator theory, see [33], [34], [35] among others. Basically, such an approach can be described here as asking you to account for all wave functions at all times when describing any physical dynamics. In particular, for the physics here, I will stress preparation of quantum state, and its consequences. I will end with a sharpening (in my opinion) of Born's rule, or more descriptively, a mathematization of it in terms of "all possible probabilities and how they must behave dynamically".

Because this paper is becoming long, I will just use facts from unbounded operator theory. Also I will refer to my recent papers [36], [37], [38], [39], from which more background, history, mathematics, may be obtained.

Briefly, with B. Misra, in 1974 we embarked upon an attempt to model a quantum counter. I have given an account in [36]. Interesting unbounded operator questions were encountered and not resolved. However, shortly thereafter, Misra and Sudarshan put out their important Zeno Paradox paper [40]. By going to the density operator description, they avoided some issues, but not others. I made some further progress in [36]; but not much.

Then in [37] I returned to the Counter, or if you prefer, the Zeno problem. Consider a central problem, that of when the following "continual observation" strong operator limit exists:

$$s - \lim_{n\to\infty} (PU(t/n)P)^n. \tag{16}$$

When P (representing the measurement) commutes with the Hamiltonian H which generates the unitary evolution $U(t) = e^{itH}$, the situation is trivial. My intuition in 1974, and today, does not permit me to put both the apparatus and the quantum system all under one big Hamiltonian system in which everything commutes. From 1974–75 I made clear in [37] a few mathematical results (i.e., facts). Let P be an arbitrary but orthogonal Hilbert space projection, and consider the evolution $Z_t = PU_tP$ where U_t is unitary as above. Then Z_t is a semigroup evolution (e.g., Markov process) for all $t \geq 0$ iff P projects onto a subspace M which is a 'subspace without regeneration': $PU_sP^{\perp}U_tP = 0$ for all $t, s \geqq 0$. If M is a reducing subspace (re H), then Z_t is a unitary evolution. The idea of subspace without regeneration is related to a model of particle decay which Misra and I developed in 1976 [41]. The decay products are not allowed to feed back into the evolution. Notice the statement is iff: you don't even get a semigroup without that assumption.

It turns out that the relationship of your measuring apparatus (the projector P) with the domain $D(H)$ is quite critical. For example, if you want the projected Hamiltonian PHP to generate any kind of quantum evolution, you must have $D(HP)$ dense. See [37] for some results. In [38] I elaborate this point. For example, a mathematical analysis gives the following results: $D(HP)$ is dense iff PH is closeable. Then $(HP)^*$ is defined, $(HP)^* = \overline{PH} \supset PH$ has domain at least a large as $D(H)$, and $(PHP)^* = \overline{PHP}$. There is further discussion in [38]. In [38] and [39] I criticize almost all Zeno papers as failing in rigor about these matters.

For example, if you look closely at [40], you will see at least four passages in which the "continuous measurement limit" (16) above is just assumed. Going to the density operator formulation somewhat obscures these critical issues. I am saying this less as a criticism and more as an opportunity. In my opinion, if we could better understand these rigorous mathematical difficulties, we would better understand the physics. In similar vein, in [38] and [39] I point out how critical domain issues are glossed over in recent Zeno papers (see [38], [39] for citations) by Exner and Ichinose (2003), Facchi, Pascazio, Tasaki, Tokuse (2004), Schmidt (2002), Spedalieri, Lee, Florescu, Kapale, Yurtsever, Dowling (2005), among others. In some of these papers, one finds a decomposed Hamiltonian such as $H_{signal} + H_{bath} + H_{interaction}$. This of course is very convenient but in my opinion the critical physics, hence the need for rigorous mathematics, is still hidden within the domain of the interaction Hamiltonian.

Recall that it is essential to know or characterize exactly the operator domain $D(T)$ for any densely defined differential operator T occurring in an initial value problem

$$\begin{cases} \frac{du(t)}{dt} & = Tu, \quad t > 0, \\ u(0) & = u_0 \quad \text{given.} \end{cases} \tag{17}$$

In terms of quantum physics, the whole class of initial states $\{u_0\}$ upon which you can solve this differential equation to speak of the solution

$$u(t) = Z_t u_0 = e^{Tt} u_0 \tag{18}$$

are exactly those upon which you can suitably differentiate. In other words [33–35], the acceptable initial states are exactly the domain $D(T)$. Such requirement is implicit in the (unbounded) commutativity

$$Z_t T \subset T Z_t \tag{19}$$

which you must use to prove the solution satisfies the differential equation.

For a quantum, e.g., Schrödinger's equation evolution, that is, the solution $\psi(t) = U(t)\psi_0$ of the initial value problem

$$\begin{cases} \dot{\psi}(t) & = H\psi(t), \quad t > 0 \\ \psi(0) & = \psi_0 \end{cases} \tag{20}$$

where H is the Schrödinger operator, we know, for example for hydrogen, that $D(H) = W^{2,2}(\mathbb{R}^3)$, the Sobolev spaces of wave functions $\psi(x)$ which have all derivatives up to and including order two, square integrable, i.e., in $L^2(\mathbb{R}^3)$. In [38] I stated the following Theorem: Any unitary group U_t necessarily exactly preserves its infinitesimal generator H's $D(H)$. That is, U_t maps $D(H)$ one to one onto itself, for all $-\infty < t < \infty$. To feel the content of this Theorem, contrast to the Heat equation initial value problem

$$\begin{cases} \dot{u}(t) & = \Delta u(t), \quad t > 0 \\ u(0) & = u_0 \end{cases} \tag{21}$$

where Δ is the Laplacian operator. As is well-known (e.g., see [17]), the solution semigroup Z_t immediately smooths any initial heat data u_0 into C^∞ functions at all instants $(t > 0)$.

The proof of this Theorem ([38], Thm. 4.1; [39], Thm. 5.2) is quick, and let me give it here. For every $-\infty < t < \infty$ the linear isometry property $||U_t x|| = ||x||$ and the commutativity property $U_t H \subset H U_t$ guarantee that U maps $D(H)$ one to one into $D(H)$. Suppose it is not onto $D(H)$. Then for some t there is some x in $D(H)$ which is not in $U_t(D(H))$. Apply U_t^{-1} to x. Again using the commutativity, we know $z = U_{-t}x$ is in $D(H)$. But then $x = U_t z$ was in $U_t(D(H))$.

I have not found this Theorem in any of the books on my shelves. Perhaps it is published and I would welcome knowing where. One reason it may have been missed by some experts is that traditionally one develops the general semigroup theories for Z_t, where generally the mapping Z_t takes $D(T)$ strictly into $D(T)$. In fact, Z_t maps $D(T)$ one to one onto $D(T)$ iff Z_t is unitary. See [38, 39] for the technical details. Thus I do not find my theorem in such treatises as [26], [33], [35], [42], [43], [44], for example.

Another reason it is missed is the historical tendency to establish the selfadjointness of quantum mechanical Hamiltonians H by first establishing their essential selfadjointness on a dense subset $D \subset D(H)$. This procedure was key to the success of T. Kato and F. Rellich in establishing approximately 60 years ago the selfadjointness of the atomic Schrodinger Hamiltonians. See Kato [33] for the exact references. One finds the same procedure in current classical treatments. For example, in Reed and Simon [43] you find the following theorem. Let D be a dense domain invariant (note: that means, into) under U_t, on which U_t is strongly differentiable: then i^{-1} times the strong derivative of U_t is essentially selfadjoint on D. Also the following theorem is there. Let A be selfadjoint and D a dense subdomain in $D(A)$. If for all t, e^{itA} maps D into D, then D is a core for A. That means that A was essentially selfadjoint on D. One finds the same result in Amrein [44].

Another taste for how closely my theorem has been missed may be found in the open systems formulations of quantum mechanics. There a favorite setting, see Davies [31], is the interaction picture in quantum fields models, in which

$$U_t(\rho) = e^{-itH} \rho e^{itH} \tag{22}$$

where H is an unbounded selfadjoint operator and ρ is a selfadjoint trace class operator. Let $T = -iad_H$ (the latter is Davies' notation) be the infinitesimal generator of the strongly continuous one parameter group of isometries U_t on the state space V of trace class operators. Then [31, p. 82], the domain $D(T)$ is all selfadjoint trace class operators ρ such that $\rho(D(H)) \subseteq D(H)$ and such that $H_\rho - \rho H$ on $D(H)$ has an extension to a trace class operator σ on the whole Hilbert space. But my theorem makes the situation much more precise, for I know that e^{itH} and e^{-itH} map $D(H)$ one to one exactly onto $D(H)$. I may further develop this theme elsewhere.

As a further example here of the sharpenings of the Born interpretation my theorem can provide, consider the (Weyl form) of the canonical commutation relations

$$U_t V_s = e^{its} V_s U_t \tag{23}$$

where U_t and V_s are the momentum and position unitary evolutions. Then, since $W^{1,2}(\mathbb{R}^n)$ is the exact domain of both the momentum and position (admittedly for the latter, under a Fourier transform unitary equivalence) Hamiltonian infinitesimal generators of those evolutions, both factors on both sides of (23) map $W^{1,2}$ one to one onto, at each instant of time. So the CCR exactly preserve the Born total probability ensemble. One can impose the same sharpening on the time-dependent Dirac equation unitary evolutions, and elsewhere.

I am not sure whether I should call my Theorem [38]: quantum reversibility iff quantum probability regularity preservation, or if I should refer to it as the quantum probability no-smoothing theorem. In fact, I don't really know how all those prepared initial quantum states ψ_0 will mix among themselves as the Schrödinger evolution proceeds. But they must remain in $W^{2,2}$. The situation becomes more complex when you consider Helium and higher atoms. The usual dodge in mathematics is to use the French trick (Laurent Schwartz, among others), of just always working with the dense subspace C_0^∞ of infinitely differentiable functions of compact support, and then speaking of the Hamiltonian H as the closure operator. In other words, one may not know in the literature (it must be there, but I do not find it at this writing) exactly what $D(H)$ is. For Helium, one can see that it will be in $W^{2,2}(\mathbb{R}^6)$ but may be smaller, due to the two $\mathbb{R}^3$ interacting electrons in the Schrödinger operator. I would like to pursue this technical question elsewhere. There is some information in [45], although they also fall back onto the C_0^∞ dodge.

But I do know: whatever be the exact $D(H)$ for those atoms, those are the initial probability states ψ_0 which you must prepare before you can speak of their quantum evolution as a 'Born' ensemble. You are not permitted to lose any one of those probabilities under the evolution U_t as you approach the measuring apparatus. And if you lose any of those probabilities in the act of measuring, not only will you lose unitarity, you might not even have a semigroup evolution Z_t afterward.

There is more. By the Sobolev Lemma, e.g., see [17], ψ in $W^{[n/2]+1+k}(\Omega)$ will be $C^k(\Omega)$. Here $[n/2]$ means: rounded down to an integer, where n is the dimension of the space $\mathbb{R}^n$ in which the physical domain Ω resides. Thus for hydrogen $D(H) = W^{2,2}(\mathbb{R}^3)$ means that what I may call the "candidate Born initial prepared probabilities" ψ_0 are C^0, i.e., continuous in space variable. The Sobolev Lemma asserts that domains $W^{2,2}$ as quantum probabilities, will also lose more and more continuity as the chemical element, like Helium, lives in a higher dimensional configuration space.

RECAPITULATION AND COMMENTARY

In Section 2 I emphasized how much classical calculus and classical physics and classical mechanics was implicit within the thinking and calculations of Born, Heisenberg, Schrödinger, and the others as they developed the new and 'final' quantum mechanics. In Section 3 I emphasized how classical Fourier series coefficients somehow became quantum probability amplitudes. The classical vibrating string led to thinking of electrons as classical systems of oscillators. Born's quantum probabilities were classic forward cone scattering amplitudes.

I also illustrated how noncommutativity is the general situation in physics, be it classical or quantum. But to me, a remaining central issue is our insistence that two quantum observables must commute. As I mentioned in Section 3, Heisenberg wanted *real* probability amplitudes, so he went to $(xy+yx)/2$, even when x and y are operators. Hund [7] does not elaborate this point, so let me do so a bit here.

One will notice in reading quantum physics history that at critical moments, a resort to justifications in terms of classical intensities provides the bridge in thinking. For example, Hund [7, p. 131] describes how the thinking was that of an atom's state as being determined in terms of its radiation intensities. I mention that this is typical "inverse scattering" thinking, such as determining a brain state from nuclear resonance. Then one extended the views of an atom's state as determined also by radiation intensities associated with transitions between internal vibration modes. Then Hund [7, p. 134, top] we have Heisenberg thinking "under certain circumstances" $xy \rightarrow (xy+yx)/2$. What exactly, we may ask, are those circumstances?

Also, as I mentioned in Sections 3 and 4, it seems to me we continue to be cavalier in our attitude of just accepting that we will not speak in quantum mechanics of the composition of two observables AB unless they commute. This is called *compatibility*. Of course this issue has been much discussed in the quantum literature. I already gave examples about joint probabilities, compatible propositional logics, and the discussion in Bub [11] and Jauch [12] are quite good. Bub [11, p. 239] ends up postulating that quantum mechanics is evolution in a "non-Boolean possibility space". Of course this means we really do not know what is going on. Underlying all these deliberations and postulates is our inability to escape from the notions of real-valued measurements and real-valued probabilities. That is, in particular, it seems to me, that we just "like" observables A and B which as selfadjoint operators, commute. Then BA and AB are also candidate (selfadjoint, that is) observables. Likewise, our projections P that are already in the Hamiltonians spectral family. Similarly as described in Section 4 we want a real probability $P_W(a \in E) = tr(WA)$. Then upon measurement of property a we can say the quantum system changed into a new density operator $W_{new} = AWA/tr(WA)$. It is easy to show, by the way, that generally, when AB is not selfadjoint, it is not even a normal operator. If one adds some other condition, e.g. $A > 0$, then AB normal will imply AB selfadjoint. Here I am speaking only of bounded operators A and B. The point is that we have reason to heavily invest in speaking only of commuting observables. But is it really the physical quantum reality?

In Section 5 I mentioned the extremely important experiments discussed in Aspect [14]. In Section 6 I prove, to paraphrase that result here, that one may only speak of evolving states $\psi(t) = U_t\psi_0$ in terms of their having "the same analytical probability properties" as had the prepared initial state ψ_0. What underlies my theorem is a deep physical ansatz of quantum state evolutions ψ_t as generalized recoordinatizations of the initial state ψ_0. So if one takes a Born-like probabilistic view of quantum mechanics, then one is just recoordinatizing the initial prepared probabilities. I therefore conjecture that we should develop a quantum measurement theory based upon the same ansatz. Let me call this conjecture a metatheorem.

Metatheorem 1 *The sharper view of Born probabilities is that of recoordinatizations of the ensemble of prepared initial states. Correct quantum measurement theory must also account for exactly which evolving coordinatizations we are able to measure, and how.*

As a corollary to help support my ansatz and conjecture, let me state

Corollary 1 *The MetaTheorem explains the Aspect [14] experiments. When the source was particle-like, the initial coordinatization was particle-like, and the evolution was particle-like. When the source was wave-like, the initial coordinatization was wave-like, and the evolution was wave-like.*

REFERENCES

1. K. Gustafson, Review of *The End of the Certain World: The Life and Science of Max Born*, by Nancy Thorndike Greenspan, Basic Books (2005). To appear in J. of Scientific Exploration (2007).

2. N. Greenspan, *The End of the Certain World: The Life and Science of Max Born*, Basic Books (2005).
3. G. Born, I. Newton-John, M. Pryce, *The Born–Einstein Letters*, Walker and Company (1971).
4. W. Heisenberg, *The Physical Principles of the Quantum Theory*, U. of Chicago Press (1930).
5. N. Bohr, *Discussion with Einstein on Epistemological Problems in Atomic Physics*, in Albert Einstein: Philosopher-Scientist, (ed. P. A. Schilpp), The Library of Living Philosophers, Evanston (1949), 200–241.
6. R. Feynman, *The Feynman Lectures on Physics*, Addison–Wesley (1965).
7. F. Hund, *The History of Quantum Theory*, Harper and Row (1974).
8. J. Hendry, *The Creation of Quantum Mechanics and the Bohr–Pauli Dialogue*, D. Reidel (1984).
9. W. Pauli (ed.), *Niels Bohr and the Development of Physics*. McGraw–Hill (1955).
10. R. Penrose, *The Road to Reality*, A. Knopf (2005).
11. J. Bub, *Interpreting the Quantum World*, Cambridge (1997).
12. J. Jauch, *Foundations of Quantum Mechanics*, Addison–Wesley (1968).
13. J. Bell, *Against Measurement*, Physics World, August (1990), 33–40.
14. A. Aspect, P. Grangier, *Wave-Particle Duality: A Case Study*, in Sixty Two Years of Uncertainty (ed. A. Miller), Plenum (1990).
15. H. Primas, M. Esfeld, *A Critical Review of Wigner's Works on the Conceptual Foundations of Quantum Theory*, available online (1997).
16. R. Griffiths, *Consistent Quantum Theory*, Cambridge (2002).
17. K. Gustafson, *Partial Differential Equations and Hilbert Space Methods*, Dover (1999).
18. H. Newell, *Vector Analysis*, McGraw–Hill (1955).
19. A. Einstein, B. Podolsky, N. Rosen, *Can Quantum-Mechanical Description of Physical Reality be Considered Complete*?, Phys. Rev. **47** (1935), 777-780.
20. N. Bohr, *Can Quantum-Mechanical Description of Physical Reality be Considered Complete*?, Phys. Rev. **48** (1935), 696–702.
21. M. Planck, *Verhandl. Deutsch. Phys. Ges* **2** (1900) 2.
22. A. Einstein, *Ann. Physik* **17** (1905), 132.
23. N. Bohr, H. Kramers, J. Slater, *Z. Phys.* **24** (1924), 69.
24. M Born, W. Heisenberg, P. Jordan, *Z. Phys.* **35** (1926), 557.
25. D. Hilbert, J. v. Neumann, L. Nordheim, *Math. Ann* **98** (1927), 1.
26. C. R. Putnam, *Commutation Properties of Hilbert Space Operators and Related Topics*, Springer (1967).
27. G. Ghirardi, P. Pearle, A. Rimini, *Markov Processes in Hilbert Space and Continuous Spontaneous Localization of Systems of Identical Particles*, Phys. Rev. A **42** (1990), 78–89.
28. K. Gustafson, *Professor Ilya Prigogine: A Personal and Scientific Remembrance*, Mind and Matter **1** (2003), 9–13.
29. K. Gustafson, *Microscopic Irreversibility*, Discrete Dynamics in Nature and Society **8** (2004), 155–168.
30. W. Pauli, *Uber Gasentartung und Paramagnetismus*, Zeit Phys. **41** (1927), 81–102.
31. E. B. Davies, *Quantum Theory of Open Systems*, Academic Press (1976).
32. K. Gustafson, *Time–Space Dilations and Stochastic–Deterministic Dynamics*, in Between Choice and Chance (eds. H. Atmanspacher, R. Bishop), Imprint Academic (2002), 115–148.
33. T. Kato, *Perturbation Theory for Linear Operators*, Springer (1980).
34. S. Goldberg, *Unbounded Linear Operators*, McGraw–Hill (1966).
35. E. Hille, R. Phillips, , *Functional Analysis and Semigroups*, Amer. Math. Soc. (1957).
36. K. Gustafson, *The Quantum Zeno Paradox and the Counter Problem*, in Foundations of Probability and Physics-2 (ed. A. Khrennikov), Växjö University (2002), 225–236.
37. K. Gustafson, *Bell and Zeno*, Int. J. Theor. Physics **44** (2005), 1931–1940.
38. K. Gustafson, *Reversibility and Regularity*, Int. J. Theor. Phys. (2007), to appear.
39. K. Gustafson, *Noncommutative Trigonometry and Quantum Mechanics*, in Advances in Deterministic and Stochastic Analysis (eds. N. Chuong, P. Lax, D. Mumford, P. Ciarlet, D. Phong), World Scientific (2007).
40. B. Misra, G. Sudarshan, *The Zeno's Paradox in Quantum Theory*, J. Math. Phys. **18** (1977), 756–763.
41. K. Gustafson, B. Misra, *Canonical Commutation Relations of Quantum Mechanics and Stochastic Regularity*, Letters in Math. Physics **1** (1976), 275–280.
42. K. Yosida, *Functional Analysis*, Springer (1969).
43. M. Reed, B. Simon, *Methods of Modern Mathematical Physics*, Academic Press (1980).
44. W. Amrein, *Non-Relativistic Quantum Dynamics*, D. Reidel (1981).
45. K. Jorgens, J. Weidmann, *Spectral Properties of Hamiltonian Operators*, Springer (1973).

Decoherence, Disentanglement and Foundations of Quantum Mechanics

Gregg Jaeger* and Kevin Ann†

*Quantum Imaging Laboratory and College of General Studies
Boston University, Boston MA 02215, U. S. A.
†Department of Physics
Boston University, Boston MA 02215, U. S. A.

Abstract. Decoherence and disentanglement are phenomena central to quantum mechanics. Here, we consider the relative rates of decoherence and disentanglement in two-qubit, three-qubit, and two-qutrit systems when subject to pure dephasing noise alone, and a very recent result for $d \times d$ systems. Of particular interest is the specific counterintuitive effect related to the nonadditivity of such weak noises, known as *Entanglement Sudden Death (ESD)*, in which the entanglement of a composite quantum system goes abruptly to zero in finite time, coherence only exponentially decaying. We discuss these results in the context of the foundations of quantum mechanics.

Keywords: Decoherence, Disentanglement, Dephasing, Damping, Entanglement Sudden Death, Foundations of Quantum Mechanics
PACS: 03.65.Ud, 03.65.Yz, 03.67.Ua

INTRODUCTION

When considering fundamental issues of quantum theory, the phenomena of quantum decoherence and disentanglement are of great significance [1, 2]. Quantum coherence is the property of a quantum system permitting superposition states to persist. However, in natural environments for most physical systems, this property is relatively delicate, being easily degraded by interaction of a system with its environment [3]. The resulting decoherence of the quantum system typically increases state mixedness and causes uniquely quantum behavior to disappear. In particular, the closely related property of quantum entanglement, which is responsible for nonclassical features, is lost. This is referred to as "disentanglement," and "entanglement death" when entanglement is entirely lost [4–8]. To study the behavior of a quantum physical system interacting with its environment, the environment may be modeled as a noise source. In studying these uniquely quantum phenomena, it is important to characterize the relationship between decoherence and disentanglement. Here, we consider recent work examining this relationship in few-qubit open quantum systems with their environments modeled as white Markov noise [4–10].

Intuitively, one might expect that weak noise influences, for example, acting locally on the components of a composite quantum system to be additive, that is, for the combined effect of these weak noises to be the sum of the various local effects as separately manifested. Surprisingly, however, this has been found not to be the case. Historically, this was first found in a very simple continuous system, a pair of gaussian states of position [11], and the simplest possible discrete quantum system, a two-qubit system [5]. Here, we focus on systems described by finite-dimensional Hilbert spaces. It has been shown that when such systems are subjected to two different sorts of multi-local noise, dephasing and damping, each individual source serves to reduce component subsystem coherence as well as joint-system coherence and entanglement, but only asymptotically. However, when these two noise sources act simultaneously on a system appropriately prepared, entanglement abruptly reaches zero, that is, it does so in finite time, whereas the quantum coherence decays only exponentially in time, a qualitatively significant effect termed *Entanglement Sudden Death (ESD)*. Here, we critically comment on both the models used and interpret the results in a broader context. Theoretical results such as those of the Żurek group, *e.g.* [12, 13], as well as experimental results have been produced with the hope that decoherence might hold the key to ultimate foundational questions in quantum mechanics. Here, we suggest rather that finite-time disentanglement, that is ESD, rather than decoherence may be key to some important questions in quantum foundations, such as the measurement problem. The confirmation of ESD in photonic [14] and atomic systems [15], showing that this effect indeed occurs in practice, lends greater importance to the exploration of this possibility. Moreover, we explain that local dephasing noise alone is sufficient to induce ESD.

CP962, *Quantum Theory, Reconsideration of Foundations – 4*,
edited by G. Adenier, A. Yu. Khrennikov, P. Lahti, V. I. Man'ko, and Th. M. Nieuwenhuizen

DISENTANGLEMENT AND DECOHERENCE

The simplest model in which to study quantitatively the relationship between decoherence and disentanglement is a bipartite two-level (qubit) system. Qubits may be represented by a variety of quantum physical systems, such as electron spins, photon polarization states, and atomic energy levels. Yu and Eberly [9] first considered a two-qubit system as an open quantum system subjected to multi-local and collective pure dephasing Markovian noise using the operator-sum decomposition within the general approach we discuss here. We later extended these results by examined a three-qubit system subjected to dephasing noise in three ways, differentiated by scope: multi-locally in which the noise acting on each qubit is independent, collectively in which all qubits experience the same noise, and sub-collective noise (in which some subset of two qubits experiences the same noise and the remaining qubit may experience a different sort of noise) [10]. We then further extended this further to a pair of three-level systems subjected to multi-local dephasing noise [8]. Let us comment on these models and summarize the problems and solutions to each in turn.

Two qubits

The first finite-dimensional case considered was that of the two-qubit system, coupled to an environment giving rise to white Markovian *dephasing noise* alone, both local and collective in nature [9]. The interaction Hamiltonian was taken to be

$$H(t) = -\frac{1}{2}\mu\left[B(t)\left(\sigma_z^{\mathrm{A}} + \sigma_z^{\mathrm{B}}\right) + b_{\mathrm{A}}(t)\,\sigma_z^{\mathrm{A}} + b_{\mathrm{B}}(t)\,\sigma_z^{\mathrm{B}}\right], \tag{1}$$

where μ is, for example, the gyromagnetic ratio of the particle, $\hbar = 1$, σ_z is the well-known Pauli matrix, and $b_{\mathrm{A}}(t)$, $b_{\mathrm{B}}(t)$, and $B(t)$ represent Markovian dephasing noise acting locally on qubit A, qubit B, and collectively on both, respectively. Although this Hamiltonian was constructed for spin qubits, the model is readily generalized to any physical system consisting of distinct component subsystems subjected to multi-local and collective noise. The noise terms satisfy the following relations.

$$\begin{aligned}
\langle B(t)\rangle &= 0\,, && (2)\\
\langle B(t)B(t')\rangle &= \frac{\Gamma_{\mathrm{AB}}}{\mu^2}\delta(t-t')\,, && (3)\\
\langle b_X(t)\rangle &= 0\,, && (4)\\
\langle b_X(t)b_X(t')\rangle &= \frac{\Gamma_X}{\mu^2}\delta(t-t')\,, && (5)
\end{aligned}$$

with $X = \mathrm{A},\mathrm{B}$. Eqs. 2 and 4 impose the condition that the ensemble average $\langle\cdot\rangle$ of the noise fields are zero; the relations Eqs. 3 and 5 require the temporal evolution to be Markovian. For this model and all subsequent similar models involving dephasing noise fields discussed, these two conditions will continue to hold, although the associated relations will not be explicitly exhibited.

Note that there exists two distinct noise terms here, multilocal noises $b_{\mathrm{A}}(t)$ and $b_{\mathrm{B}}(t)$ acting independently at the level of each individual qubit and collective noise $B(t)$ acting on both qubits in the same way. These situations may be physically interpreted as qubits in spatially separated environments or a single global environment, respectively. Markovian evolution assumes there is no back-action of the system to its environment and that there is no memory of interaction after the effective timescale of interaction. Furthermore, the nature of this noise is such that the relaxation timescales of either the qubit or its environment are much shorter than the timescales of the qubit-environment interaction between them, allowing one to disregard the internal dynamics of both the system and environment. In constructing a model for dephasing, the relative timescales and effective dynamics must be specified. We have constructed our model as one of an open quantum system. This involves what corresponds to the tracing out of environmental degrees of freedom under a closed system construction, discarding fine-grained features in favor of coarse-grained ones assumed to be Markovian in character. That is, one can think of any quantum system as the reduced subsystem of a larger system. After purifying this system by a unitarily embedding into a higher-dimensional system, with the support of the Stinespring dilation theorem [16], one can unitary time-evolve the larger system, over which one traces to get the system state as a reduced state. Although the dynamics of the composite system is still unitary, the reduced subsystem dynamics may not be. This naturally lends itself to the use of completely positive trace-preserving (CPTP) maps of statistical operators, described below.

To represent states in the density matrix formalism, one must consider a specific base in which to represent matrices. In the present case, it is convenient to use the basis

$$|1\rangle_{AB} = |++\rangle_{AB}\,,\ |2\rangle_{AB} = |+-\rangle_{AB}\,,\ |3\rangle_{AB} = |-+\rangle_{AB}\,,\ |4\rangle_{AB} = |--\rangle_{AB}\,. \tag{6}$$

It is important also to note that quantum coherence has typically been studied in a basis-dependent fashion in the literature on decoherence, to which the studies considered here belong. Any discussion of the rate of decoherence requires a specification of the basis whenever this is done. Similarly, although entanglement is basis *independent*, there are a number of entanglement measures used for its quantification; different measures may give different numerical values of entanglement, although all good measures must be monotones of one another. Thus, one must take care to specify both the basis and entanglement measure used in any such analysis. In determining the time-evolved density matrix, one must take the ensemble averages over all three noise fields,

$$\rho(t) = \langle\langle\langle\rho_{st}(t)\rangle\rangle\rangle \quad , \text{ where } \ \rho_{st}(t) = U(t)\rho(0)U^{\dagger}(t) \ \text{ and } \ U(t) = \exp\left[-i\int_0^t dt' H(t')\right] \tag{7}$$

are the standard unitary evolution with the unitary operator given by the solution of the Schrodinger equation. We have used the very practical operator-sum decomposition to find the time-evolved state, in which one expresses dynamical equations via completely positive trace preserving (CPTP) maps. In particular, operators satisfying the completeness relation, $\sum_\mu \overline{K}_\mu^{\dagger}\overline{K}_\mu = I$ and the trace-preserving condition, $\sum_\mu \overline{K}_\mu\overline{K}_\mu^{\dagger} = I$, in which in general $\left[\overline{K}_\mu, \overline{K}_\mu^{\dagger}\right] \neq 0$, may be used to find time-evolved the density matrix. The time-evolved density matrix takes the form of $\rho(t) = \mathscr{E}(\rho(0)) = \sum_{\mu=1}^{N}\overline{E}_\mu^{\dagger}(t)\rho(0)\overline{E}_\mu(t)$, where $\mathscr{E}$ is a superoperator representing the CPTP map, which is an operator acting on operators, the $\overline{E}_\mu$ represent the non-unique Kraus operators, and N is not necessarily fixed, because the $\overline{E}_\mu$ are typically nonunique. In the two-qubit case under the conditions specified above, the temporally evolved state and operators are

$$\rho(t) = \sum_{i,j=1}^{2}\sum_{k=1}^{3}\left(D_k^{\dagger}E_j^{B\dagger}E_i^{A\dagger}\right)\rho(0)\left(E_i^{A}E_j^{B}D_k\right), \tag{8}$$

$$E_1^A = \begin{pmatrix} 1 & 0 \\ 0 & \gamma_A(t) \end{pmatrix}\otimes I,\ E_2^A = \begin{pmatrix} 0 & 0 \\ 0 & \omega_A(t) \end{pmatrix}\otimes I,\ E_1^B = I\otimes\begin{pmatrix} 1 & 0 \\ 0 & \gamma_B(t) \end{pmatrix},\ E_2^B = I\otimes\begin{pmatrix} 0 & 0 \\ 0 & \omega_B(t) \end{pmatrix}, \tag{9}$$

$$D_1 = \begin{pmatrix} \gamma(t) & 0 & 0 & 0 \\ 0 & 1 & 0 & 0 \\ 0 & 0 & 1 & 0 \\ 0 & 0 & 0 & \gamma(t) \end{pmatrix},\ D_2 = \begin{pmatrix} \omega_1(t) & 0 & 0 & 0 \\ 0 & 0 & 0 & 0 \\ 0 & 0 & 0 & 0 \\ 0 & 0 & 0 & \omega_2(t) \end{pmatrix},\ D_3 = \begin{pmatrix} 0 & 0 & 0 & 0 \\ 0 & 0 & 0 & 0 \\ 0 & 0 & 0 & 0 \\ 0 & 0 & 0 & \omega_3(t) \end{pmatrix}, \tag{10}$$

and the elements in the matrices are given by

$$\gamma_A(t) = e^{-t/2T_A},\ \gamma_B(t) = e^{-t/2T_B},\ \omega_A(t) = \sqrt{1-\gamma_A^2},\ \omega_B(t) = \sqrt{1-\gamma_B^2}, \tag{11}$$

$$\gamma(t) = e^{-t/2T_{AB}},\ \omega_1(t) = \sqrt{1-\gamma^2},\ \omega_2(t) = -\gamma^2\sqrt{1-\gamma^2},\ \omega_3(t) = \sqrt{(1-\gamma^2)(1-\gamma^4)}, \tag{12}$$

where $T_X = \frac{1}{\Gamma_X}$ $(X = A, B, AB)$. In a similar way that Eqs. 2, 3, 4, and 5 specifying the white noise and Markovian conditions are assumed for subsequent discussions, Eqs. 11 and 12 defining the terms in the matrices will be assumed when defining subsequent dephasing operators. For tractability of notation, time does not explicitly appear as an argument for these quantities from hereon, but rather is implied. These operators act to dephase the system, both at the joint-system and the individual subsystem level, in the sense that the off-diagonal elements in the respective density matrix (at least asymptotically) decay to zero.

The timescales in which density matrix elements decay define the pertinent decoherence timescales, which we compare to disentanglement timescales. For example, these timescales may represent the transverse spin-spin dephasing timescales T_2, differing from the longitudinal spin-lattice damping timescales T_2. In general, $T_1 = 2T_2$. This and subsequent similar analyses comparing decoherence and disentanglement timescales focus only on the T_2 timescales and make no mention of T_1 timescales. The reason for this is to keep the analysis as simple as possible, retaining only the uniquely quantum dephasing aspect of noise and thereby avoiding complications associated with amplitude damping, such as such as energy exchange between system and environment and modification of relative state populations. This simplicity is not merely a matter of convenience; as shown below during the examination of the effects of two different sorts of weak noises, there is not necessarily the expected additivity.

As evidenced by the ever-expanding literature concerning entanglement measures, there are many complications. Nonetheless, the entanglement of a two-qubit system *is* easily quantified by the concurrence measure, as defined by Wootters [17], as an intermediate quantity, from which one can calculate a value for this system the entanglement of formation E_f. The entanglement of formation is the canonical entanglement measure; it corresponds to the minimum entanglement of the weighted average of possible ensembles that may form a given two-qubit mixed state. Since concurrence is a monotonic function of the entanglement of formation, it is a valid entanglement measure in its own right and is typically used as such. The concurrence may be written

$$C(\rho_{AB}) = \max\left[0, \sqrt{\lambda_1} - \sqrt{\lambda_2} - \sqrt{\lambda_3} - \sqrt{\lambda_4}\right], \tag{13}$$

where λ_i are the eigenvalues of the matrix

$$\rho_{AB}\tilde{\rho}_{AB} \equiv \rho_{AB}\left(\sigma_y^{\mathrm{A}} \otimes \sigma_y^{\mathrm{B}}\right)\rho_{AB}^{*}\left(\sigma_y^{\mathrm{A}} \otimes \sigma_y^{\mathrm{B}}\right), \tag{14}$$

ordered from largest to smallest, with ρ^* the component-wise complex-conjugation of the density matrix, and σ_y is the well-known Pauli matrix. Entanglement of formation is thus defined as

$$E_f(\rho) = h\left(\frac{1+\sqrt{1-C^2(\rho)}}{2}\right), \tag{15}$$

where $h(x) = -x\log_2 x - (1-x)\log_2(1-x)$ is the binary Shannon entropy. Note that, formally, this suggests that the concurrence serves as a probability. Under this interpretation, which to our knowledge we introduce for the first time here, the concurrence is the probability that a density matrix under some abstract conjugation operation is indistinguishable from its original unconjugated state [18]: a large overlap implies a value close to unity indicating high indistinguishability and thus high entanglement, whereas low overlap implies a value close to zero indicating distinguishability and thus low entanglement. Although the conjugation used to define concurrence may be seen as a time-reversal operation, it may be taken to be any operation which brings a density matrix defined on some Hilbert space into another subspace whose distinguishability defines the degree of entanglement.

The state of a generic two-qubit system in the computational basis is

$$|\Psi\rangle = a|1\rangle_{\mathrm{AB}} + b|2\rangle_{\mathrm{AB}} + c|3\rangle_{\mathrm{AB}} + d|4\rangle_{\mathrm{AB}}, \tag{16}$$

with the usual normalization condition $a^2+b^2+c^2+d^2=1$. Although this general state may be analyzed explicitly, it is vastly more illuminating and computationally tractable to define two classes of states, *fragile* and *robust*, according to their behavior when subjected to collective dephasing noise. The fragile states, are defined as those which lose coherence asympototically and contain the Bell states $|\Phi^{\pm}\rangle$. The robust states, in contrast, maintain coherence for all time and contain $|\Psi^{\pm}\rangle$.

(i) The *fragile* class has the forms

$$|\phi_1\rangle = a|1\rangle_{\mathrm{AB}} + b|2\rangle_{\mathrm{AB}} + d|4\rangle_{\mathrm{AB}} \quad \text{and} \quad |\phi_2\rangle = a|1\rangle_{\mathrm{AB}} + c|3\rangle_{\mathrm{AB}} + d|4\rangle_{\mathrm{AB}}\,. \tag{17}$$

(ii) The *robust* class has the forms

$$|\psi_1\rangle = a|1\rangle_{\mathrm{AB}} + b|2\rangle_{\mathrm{AB}} + c|3\rangle_{\mathrm{AB}} \quad \text{and} \quad |\psi_2\rangle = b|2\rangle_{\mathrm{AB}} + c|3\rangle_{\mathrm{AB}} + d|4\rangle_{\mathrm{AB}}\,. \tag{18}$$

The main features of this model are as follows: operator-sum decomposition representing Markovian dephasing noise acting at the local and collective level, dephasing and damping timescales, the definition of concurrence used to compute the amount of entanglement, and finally the classes of states used. For multi-local and collective Markov dephasing noise, when disentanglement does occur, as quantified by the decay of the concurrence, it always proceeds at least as fast as decoherence as quantified by the decay of the off-diagonal elements of the state density matrix.

$$t_{\mathrm{dis}} \leq t_{\mathrm{dec}} \tag{19}$$

This is important because it shows that the rate at which the pair of systems loses correlation may be faster than the rate at which each individually loses quantum coherence. It is valuable to relate this result to properties of decoherence-free subspaces (DFS) in order to qualify the phrase "when disentanglement occurs" in the previous paragraph. DFS are protected subspaces in the entire Hilbert space due to certain symmetries of the state. It effectively makes the state immune to the effects of the environment. As a specific example, let us examine the singlet state $|\Psi^-\rangle$ and let the environmental fields be magnetic fields. Since there is no net spin for such a field to act upon, this state does not decohere and thus protected from a magnetic environment. Such situations are a core component of the "decoherence program," because the coupling of environment to the specified degrees of freedom of the system, effectively chooses a basis in which to decohere the system. This is discussed further in the final section.

Three qubits

More complex systems than qubit pairs are those containing a greater number of component subsystems and/or greater dimensionality in each subsystem. In general, there is no simple way to quantify the entanglement of multipartite systems of arbitrary dimensions. One reason is that entanglement is contingent upon the definition of a subsystem, which in general is not uniquely defined; the entanglement between different partitions of different subsystems cannot be described by a single scalar, effectively making it a vector quantity [19, 20]. Interestingly, when the number of dimensions of each subsystem is infinite as in continuous-variable systems, much is known about the entanglement for a special class of states, the Gaussian states. Although an analogous situation doesn't occur when the number of component subsystems is increased, we may consider instead a large number of *copies* of a state, from which asympototic entanglement measures may be defined.

The system immediately more complex than the above simple system of two-qubits involves three qubits. After examining this case, in which we increase the number of subsystems, we will then examine the case in which the dimension of each system increases, that of a two three-level (qutrit) system. The Hamiltonian describing environmental dephasing noise acting on the three-qubit system is given by

$$\begin{aligned} H(t) = -\tfrac{1}{2}\mu\Big[& B_{\mathrm{A}}^{(1)}(t)\,\sigma_z^{\mathrm{A}} + B_{\mathrm{B}}^{(1)}(t)\,\sigma_z^{\mathrm{B}} + B_{\mathrm{C}}^{(1)}(t)\,\sigma_z^{\mathrm{C}} \\ & + B_{\mathrm{AB}}^{(2)}(t)\left(\sigma_z^{\mathrm{A}}+\sigma_z^{\mathrm{B}}\right) + B_{\mathrm{BC}}^{(2)}(t)\left(\sigma_z^{\mathrm{B}}+\sigma_z^{\mathrm{C}}\right) + B_{\mathrm{AC}}^{(2)}(t)\left(\sigma_z^{\mathrm{A}}+\sigma_z^{\mathrm{C}}\right) \\ & + B^{(3)}(t)\left(\sigma_z^{\mathrm{A}}+\sigma_z^{\mathrm{B}}+\sigma_z^{\mathrm{C}}\right)\Big], \end{aligned} \tag{20}$$

where μ is, for example, the gyromagnetic ratio of the particle, $\hbar = 1$, σ_z is the well-known Pauli matrix, and $B^{(1)}(t)$, $B^{(2)}(t)$, and $B^{(3)}(t)$ represent noise fields acting locally on a single qubit, a subset of two qubits, and collectively on all three, respectively, exhausting all subsets of qubits that noise may act on. As stated earlier, all dephasing noise fields satisfy the analogous relations given in Eqs. 2, 3, 4, and 5 to ensure Markov time evolution. The time evolution for this system given in the operator-sum decomposition is

$$\rho(t) = \sum_{i,j,k=1}^{2}\ \sum_{l,m,n,p=1}^{3} \left(F_p^{\mathrm{ABC}\dagger} E_n^{\mathrm{AC}\dagger} E_m^{\mathrm{BC}\dagger} E_l^{\mathrm{AB}\dagger} D_k^{\mathrm{C}\dagger} D_j^{\mathrm{B}\dagger} D_i^{\mathrm{A}\dagger}\right)\rho(0)\left(D_i^{\mathrm{A}} D_j^{\mathrm{B}} D_k^{\mathrm{C}} E_l^{\mathrm{AB}} E_m^{\mathrm{BC}} E_n^{\mathrm{AC}} F_p^{\mathrm{ABC}}\right), \tag{21}$$

where the D_i are multi-local operators acting on the level of a single qubit defined similarly to Eq. 9, E_j acts collectively on a subset of two qubits defined similar to Eq. 10, and F_k acts collectively on the entire qubit system and is given by

$$F_1(t) = \mathrm{diag}(\gamma(t),1,1,1,1,1,1,\gamma(t)), \tag{22}$$

$$F_2(t) = \mathrm{diag}(\omega_1(t),0,0,0,0,0,0,\omega_2(t)), \tag{23}$$

$$F_3(t) = \mathrm{diag}(0,0,0,0,0,0,0,\omega_3(t)) \tag{24}$$

with the elements in the density matrix defined as in Eq. 12. These operators act to decay the off-diagonal elements in the respective density matrices to zero.

The basis used to describe the joint system is again the "computational basis," which is the one naturally selected by the environment, say for electron spins in a magnetic environment.

$$\begin{aligned} &|1\rangle_{\mathrm{ABC}} = |000\rangle_{\mathrm{ABC}}, |2\rangle_{\mathrm{ABC}} = |001\rangle_{\mathrm{ABC}}, |3\rangle_{\mathrm{ABC}} = |010\rangle_{\mathrm{ABC}}, |4\rangle_{\mathrm{ABC}} = |011\rangle_{\mathrm{ABC}}, \\ &|5\rangle_{\mathrm{ABC}} = |100\rangle_{\mathrm{ABC}}, |6\rangle_{\mathrm{ABC}} = |101\rangle_{\mathrm{ABC}}, |7\rangle_{\mathrm{ABC}} = |110\rangle_{\mathrm{ABC}}, |8\rangle_{\mathrm{ABC}} = |111\rangle_{\mathrm{ABC}}, \end{aligned} \tag{25}$$

For a system of three qubits, the analogous robust and fragile states are the W and GHZ states, respectively. The W-state has the property that it has no genuine tripartite entanglement [21] according to the tangle (squared concurrence) measure, yet becomes a maximally entangled two-qubit state under the loss of one qubit. The GHZ-state, in contrast, has maximal tripartite entanglement according to the tangle measure, yet has zero bipartite entanglement at the two-qubit level. Let us note that these simple classifications according to robustness and fragility do not in general exist either in systems of higher numbers of qubits or higher dimensions. For example, it was shown by [22, 23] that there are *nine* inequivalent classes under stochastic local operations with classical communication. There is much evidence supporting the view that the structure of entanglement is of a complex nature. The tripartite classes are given by

$$|W^g\rangle = \bar{a}_1|001\rangle + \bar{a}_2|010\rangle + \bar{a}_4|100\rangle \quad \text{and} \quad |GHZ^g\rangle = \bar{a}_0|000\rangle + \bar{a}_7|111\rangle, \tag{26}$$

Since there is no entanglement measure for mixed three-qubit states, the pairwise entanglement content was examined using the concurrence. In a similar fashion to the two-qubit analysis, the rate of decay of off-diagonal elements of the joint-system and component subsystem density matrices were used to quantify decoherence. In all cases, it was shown that disentanglement, when it occurs between any two qubits of the system subjected to dephasing noise on some subset of qubits, takes place at a faster rate than decoherence, in agreement with results obtained for the two-qubit system [9].

Two qutrits

The next step in this line of investigation is the analysis of a bipartite three-level (qutrit) system which may represent, for example, two three-level atoms in the "V" configuration. This extends the two-qubit system by increasing the dimension of each subsystem instead of increasing the number of component subsystems. Here, the Hamiltonian is given by

$$H(t) = -\frac{\mu}{2}\left[b_{\rm A}^{(1)}(t)\,\sigma_z^{\rm A} + b_{\rm B}^{(1)}(t)\,\sigma_z^{\rm B} + b_{\rm AB}^{(2)}(t)\left(\sigma_z^{\rm A} + \sigma_z^{\rm B}\right)\right], \tag{27}$$

where $\hbar = 1$, μ is the gyromagnetic moment of the atom, and $\sigma_z = \mathrm{diag}(1, e^{\frac{i2\pi}{3}}, e^{\frac{i4\pi}{3}})$ is the dephasing operator for three-level systems with subscripts denoting qutrits A, B, or both. σ_z represents the raising operator for spin-one systems, since we are now dealing with $SU(3)$ operators instead of those contained in $SU(2)$ as in the previous two-level systems. As before, the noise fields $b_X^{(i)}(i = 1,2)$ satisfy conditions for white noise and Markov evolution.

The basis eigenstates for each qutrit are given by $\{|0\rangle, |+1\rangle, |-1\rangle\}$, representing the ground state, first-excited state, and second-excited state of the atom, respectively. Since we are assuming the "V" configuration the states $|+1\rangle, |-1\rangle$ couple to the ground state but not to each other. The basis eigenstates for the two-qutrit system is given by

$$\begin{aligned}&\{|1\rangle,|2\rangle,|3\rangle,|4\rangle,|5\rangle,|6\rangle,|7\rangle,|8\rangle,|9\rangle\}\\ &\quad \doteq \{|00\rangle,|0,+1\rangle,|0,-1\rangle,|+1,0\rangle,|+1,+1\rangle,|+1,-1\rangle,|-1,0\rangle,|-1,+1\rangle,|-1,-1\rangle\}.\end{aligned} \tag{28}$$

The temporal evolution of the density matrix $\rho(t)$ is again given via a completely positive trace preserving (CPTP) linear map,

$$\rho(t) = \sum_{i,j=1}^{2}\sum_{k=1}^{3}\left(D_k^{\rm AB\dagger}E_j^{\rm B\dagger}E_i^{\rm A\dagger}\right)\rho(0)\left(E_i^{\rm A}E_j^{\rm B}D_k^{\rm AB}\right), \tag{29}$$

where the operators of the operator-sum decomposition are

$$E_1^{\rm A} = \mathrm{diag}(1, \gamma_{\rm A}(t), \gamma_{\rm A}(t)) \otimes I_3\,,\ E_2^{\rm A} = \mathrm{diag}(0, \omega_{\rm A}(t)\,\omega_{\rm A}(t)) \otimes I_3\,, \tag{30}$$

$$E_1^{\rm B} = I_3 \otimes \mathrm{diag}(1, \gamma_{\rm B}(t)\ ,\ \gamma_{\rm B}(t))\,,\ E_2^{\rm B} = I_3 \otimes \mathrm{diag}(0, \omega_{\rm B}(t), \omega_{\rm B}(t))\,, \tag{31}$$

$$D_1^{\rm AB} = \mathrm{diag}(\gamma_{\rm AB}(t), 1, 1, 1, \gamma_{\rm AB}(t), 1, 1, 1, \gamma_{\rm AB}(t))\,, \tag{32}$$

$$D_2^{\rm AB} = \mathrm{diag}(\omega_{\rm AB1}(t), 0, 0, 0, \omega_{\rm AB2}(t), 0, 0, 0, \omega_{\rm AB2}(t))\,, \tag{33}$$

$$D_3^{\rm AB} = \mathrm{diag}(0, 0, 0, 0, 0, 0, 0, 0, \omega_{\rm AB3}(t))\,, \tag{34}$$

where the time-dependent parameters are as specified in Eq. 11 and Eq. 12. Although decoherence is quantified as in the previous case, now we must use a different entanglement measure, since the Wootters concurrence is no longer applicable. However, the negativity measure, which is the sum of the negative eigenvalues of the partially transposed density matrix is applicable: $\mathcal{N}(\rho) = \left(\left\|\rho^{T_A}\right\|_1 - 1\right)/2$, where ρ^{T_A} is the partial transpose of the density matrix with respect to qutrit A and $\|\cdot\|_1$ denotes the trace norm [24]. Mathematically, the negativity is the degree to which a density matrix fails to be positive under a positive map but not completely positive map, the positive map in this case being the partial transpose with respect to a particular subsystem.

The general state of a bipartite three-level system is $|\Psi\rangle_{AB} = \bar{a}_1|1\rangle + \bar{a}_2|2\rangle + \bar{a}_3|3\rangle + \bar{a}_4|4\rangle + \bar{a}_5|5\rangle + \bar{a}_6|6\rangle + \bar{a}_7|7\rangle + \bar{a}_8|8\rangle + \bar{a}_9|9\rangle$. In a similar fashion to the previous analyses, we define two classes of states characterized by their asymptotic behavior in the presence of collective dephasing noise.

(i) The *fragile* class $|\phi\rangle = \bar{a}_1|1\rangle + \bar{a}_5|5\rangle + \bar{a}_9|9\rangle$, in which $\bar{a}_1, \bar{a}_5$, and $\bar{a}_9$ may be non-zero and all other terms $\bar{a}_i = 0$, has the forms

$$|\phi_1\rangle = \bar{a}_1|1\rangle + \bar{a}_5|5\rangle \quad , \quad |\phi_2\rangle = \bar{a}_1|1\rangle + \bar{a}_9|9\rangle \quad , \quad \text{and} \quad |\phi_3\rangle = \bar{a}_5|5\rangle + \bar{a}_9|9\rangle \ . \tag{35}$$

(ii) The *robust* class $|\psi\rangle = \bar{a}_2|2\rangle + \bar{a}_3|3\rangle + \bar{a}_4|4\rangle + \bar{a}_6|6\rangle + \bar{a}_7|7\rangle + \bar{a}_8|8\rangle$, in which all $\bar{a}_i$ listed may be non-zero and $\bar{a}_1 = \bar{a}_9 = 0$, has the forms

$$|\psi_1\rangle = \bar{a}_2|2\rangle + \bar{a}_4|4\rangle \quad , \quad |\psi_2\rangle = \bar{a}_3|3\rangle + \bar{a}_7|7\rangle \quad , \quad \text{and} \quad |\psi_3\rangle = \bar{a}_6|6\rangle + \bar{a}_8|8\rangle \ . \tag{36}$$

In this system–environmental-noise model, as in previous models [9, 10], we found that under pure dephasing noise, the timescale of disentanglement is *never* slower than that of decoherence.

In principle, we could extend this method of analysis to an even greater number of constituent subsystems and to higher dimensions, taking multi-local and collective noise to act on subsets of N d−level subsystems from 1 to N. However, this has the drawback that in physical systems of higher dimensions, either due to a greater number of constituent subsystems or due to each subsystem being of high dimension, there exist many different entanglement classes, which are not amenable to the simple fragile and robust classifications considered. Furthermore, there are no widely used mixed-state entanglement measures for multipartite systems and none that are rigorously defined, to our knowledge. Although the simplest two-qubit states are, experimentally, merely difficult to produce, protecting them from degrading to classical states in light of other types of decoherence is extremely difficult. Furthermore, identifying components and reconstructing the entire $d_A d_B \times d_A d_B$ density matrix is an extraordinarily difficult resource-intensive procedure. Any further analysis along this path must be carefully considered in light of conditions that simplify the situation, such as symmetries of the state and the noise.

ENTANGLEMENT SUDDEN DEATH

Let us now consider the phenomenon of the complete loss of entanglement in finite time. Intuitively, the combined effect of weak noises would be expected, on the basis of classical physical experience, to be additive. However, this is not always the case in quantum systems, even subjected to dephasing noise which can affect classical as well as quantum systems. Yu and Eberly [4] first found that in finite-dimensional systems that when a two-qubit system is subjected to two different types of multi-local noises, dephasing and damping, disentanglement occurred in finite time, despite only asymptotically occurring when only individual noise sources were present. They quantified disentanglement by the reduction of concurrence and decoherence by the decay of off-diagonal elements in the two-qubit system joint density matrix. They termed this phenomenon *Entanglement Sudden Death (ESD)*. After its initial discovery in a bipartite qubit system, we demonstrated its existence in a qubit-qutrit system [7], the only other finite-dimensional bipartite quantum system in which the only fully general mixed state entanglement measure, negativity, is applicable. The negativity which is based on the Peres-Horodečki criterion [24–27].

Two-qubit

To understand the evolution of this direction of investigation, we begin by summarizing the two-qubit density matrix ansatz considered by Yu and Eberly [4].

$$\rho_\lambda^{AB} = \frac{1}{9}\begin{pmatrix} 1 & 0 & 0 & 0 \\ 0 & 4 & \lambda & 0 \\ 0 & \lambda & 4 & 0 \\ 0 & 0 & 0 & 0 \end{pmatrix} . \tag{37}$$

The concurrence of an arbitrary two-qubit density matrix is given by

$$C^{AB} = 2\max\{0, \sqrt{\rho_{14}\rho_{41}} - \sqrt{\rho_{22}\rho_{33}}, \sqrt{\rho_{23}\rho_{32}} - \sqrt{\rho_{11}\rho_{44}}\}. \tag{38}$$

For this class, there is an initial concurrence of $C_\lambda(0) = 2\lambda/9$. The relevant decomposition operators for dephasing noise are those of Eq. 9, whereas the amplitude damping noise that serve to modify the population of the so that all states end up in the ground state is described by

$$E_1^{'A} = \begin{pmatrix} 1 & 0 \\ 0 & \gamma_A'(t) \end{pmatrix} \otimes I,\ E_2^{'A} = \begin{pmatrix} 0 & \omega_A'(t) \\ 0 & 0 \end{pmatrix} \otimes I,\ F_1^{'B} = I \otimes \begin{pmatrix} 1 & 0 \\ 0 & \gamma_B'(t) \end{pmatrix},\ F_2^{'B} = I \otimes \begin{pmatrix} 0 & \omega_B'(t) \\ 0 & 0 \end{pmatrix}, \tag{39}$$

with $\gamma_A'(t) = \gamma_B'(t) = \exp[-\Gamma_1 t] \equiv \gamma'$ and $\omega_A'(t) = \omega_B'(t) = \sqrt{1-\gamma'^2(t)} \equiv \omega'$, in contrast to the Γ_2 terms corresponding to the elements in the dephasing matrices.

Let us now review the effects of each of the dephasing, damping, and both combined in turn, as found by Yu and Eberly. When only dephasing exists, the diagonal elements are unaffected $\{a(0),\ b(0),\ c(0),\ d(0)\} = \{a(t),\ b(t),\ c(t),\ d(t)\} = \{1/9,\ 4/9,\ 4/9,\ 0\}]$ and only the off-diagonal elements change according to $z(0) = \frac{\lambda}{9} \rightarrow z(t) = \frac{\lambda}{9}\exp[-\Gamma_2 t]$ when plugging into Eq. 38 yields $C_\lambda^{ph.} = (2\lambda/9)\exp[-\Gamma_2 t]$. Note that this expression approaches zero only in the limit $t \rightarrow \infty$, so there always remains some amount of bipartite entanglement under dephasing noise. When only amplitude damping exists, the elements of the density matrix are affected in the following way for the range $3 \le \lambda \le 4$: $\{a(0),\ d(0),\ z(0)\} = \{\frac{1}{9},\ 0,\ \frac{\lambda}{9}\} \rightarrow \{a(t),\ d(t),\ z(t)\} = \{\frac{1}{9}\exp[-2\Gamma_1 t], \frac{1}{9}\omega'^4 + \frac{8}{9}\omega'^2, \frac{\lambda}{9}\exp[-\Gamma_1 t]\}$, with the values of b and c unaffected. These values yield a concurrence of $C_\lambda^{amp.} = (2/9)[\lambda - \sqrt{\omega'^4 + 8\omega'^2}]\exp[-\Gamma_1 t]$. Note that this also approaches zero only asymptotically in time under this damping noise. When both dephasing and damping are combined, an interesting qualitatively new effect occurs. $C_\lambda^{ph.+amp.} = 2\max\left\{0,\ \lambda e^{-\Gamma_2 t} - \sqrt{\omega'^4 + 8\omega'^2}\right\}$ since the off-diagonal element $z = \lambda/9$ now falls off as a result of both the effects of dephasing and amplitude damping noise. Thus, one sees that one does not have an additivity of noise as expected, because each of the dephasing and damping rates go smoothly to zero separately, whereas they cause the combined expression to go to zero abruptly in finite time.

Qubit-qutrit

After this result, we were interested to know whether ESD phenomenon is specific to a two-qubit system or a generic phenomenon. The two-qubit density matrix was chosen so that the analysis would be as simple as possible, yet still be sufficiently complex to demonstrate the desired effect. However, it is possible that complex effects may occur in higher dimensions that prevent the existence of ESD. Let us remark that our focus in on finite dimensional systems. For infinite-dimensional continuous-variable bipartite systems, ESD was shown to exist [11] for Gaussian systems, though these are qualitatively different from finite-dimensional systems. The original analysis of bipartite ESD was extended to a four-qubit model [5], however, that system was only analyzed according to bipartite dynamics since there are no multipartite entanglement measures for arbitrary mixed four-qubit states. Thus, the first extension to higher finite-dimensional systems was ours for a qubit-qutrit system in [7]. Incidently, this is the only other bipartite finite-dimensional case in which an entanglement measure existed for arbitrary mixed state systems, namely, the negativity. In order to show the existence of ESD clearly in a simple ansatz similar to [4], we derive an ansatz in the same way as [28] which has the feature that it only looks at terms responsible for joint-system coherence by setting those solely responsible for individual subsystem coherence to zero.

The joint qubit-qutrit system is described by the density matix $\rho_{AB} = [\rho_{ij}]$, where $\rho_{ji}^* = \rho_{ij}$, $\Sigma_i \rho_{ii} = 1$, with $i, j = 1, \ldots, 6$. We wanted to focus on the terms that had only density matrix terms containing joint system coherence. Thus, the general density matrix of the qubit, by tracing out the qutrit.

$$\rho_A = \begin{pmatrix} (\rho_{11}+\rho_{22}+\rho_{33}) & (\rho_{14}+\rho_{25}+\rho_{36}) \\ (\rho_{41}+\rho_{52}+\rho_{63}) & (\rho_{44}+\rho_{55}+\rho_{66}) \end{pmatrix}. \tag{40}$$

Now, when there is no coherence at all at the reduced density matrix level, all the off-diagonal elements can be set to zero and thus gives: $\rho_A' = \mathrm{diag}((\rho_{11}+\rho_{22}+\rho_{33}), (\rho_{44}+\rho_{55}+\rho_{66}))$. Similarly, the qutrit reduced density matrix is

$$\rho_B = \begin{pmatrix} (\rho_{11}+\rho_{44}) & (\rho_{12}+\rho_{45}) & (\rho_{13}+\rho_{46}) \\ (\rho_{21}+\rho_{54}) & (\rho_{22}+\rho_{55}) & (\rho_{23}+\rho_{56}) \\ (\rho_{31}+\rho_{64}) & (\rho_{32}+\rho_{65}) & (\rho_{33}+\rho_{66}) \end{pmatrix}, \tag{41}$$

Again, when there is no subsystem coherence, we have $\rho'_{\mathrm{B}} = \mathrm{diag}((\rho_{11}+\rho_{44}),(\rho_{22}+\rho_{55}),(\rho_{33}+\rho_{66}))$. The full joint-system density matrix corresponding to those reduced density matrices are given by the following.

$$\rho'_{\mathrm{AB}} = \begin{pmatrix} \rho_{11} & 0 & 0 & 0 & \rho_{15} & \rho_{16} \\ 0 & \rho_{22} & 0 & \rho_{24} & 0 & \rho_{26} \\ 0 & 0 & \rho_{33} & \rho_{34} & \rho_{35} & 0 \\ 0 & \rho_{42} & \rho_{43} & \rho_{44} & 0 & 0 \\ \rho_{51} & 0 & \rho_{53} & 0 & \rho_{55} & 0 \\ \rho_{61} & \rho_{62} & 0 & 0 & 0 & \rho_{66} \end{pmatrix}, \tag{42}$$

The task now is to show when this generic joint-system density matrix exhibits ESD. In order to achieve this, we make use of the negativity measure consisting of the sum of the negative eigenvalues of the joint system density matrix as defined in the qutrit analysis. To simplify the analysis even further, we choose the following ansatz,

$$\rho'_{AB}(\bar{x}) = \begin{pmatrix} \frac{1}{4} & 0 & 0 & 0 & 0 & \bar{x} \\ 0 & \frac{1}{8} & 0 & 0 & 0 & 0 \\ 0 & 0 & \frac{1}{8} & 0 & 0 & 0 \\ 0 & 0 & 0 & \frac{1}{8} & 0 & 0 \\ 0 & 0 & 0 & 0 & \frac{1}{8} & 0 \\ \bar{x} & 0 & 0 & 0 & 0 & \frac{1}{4} \end{pmatrix}, \tag{43}$$

where $0 \le \bar{x} \le \frac{1}{4}$ in order to satisfy the positivity requirement. The time-evolution is obtained again, much like the previous analyses, according to the operator sum decomposition. $\rho(t) = \mathscr{E}(\rho(0)) = \sum_\mu K_\mu(t)\rho(0)K_\mu^\dagger(t)$. For local and multi-local dephasing environments, the $K_\mu(t)$ are of the form $K_\mu(t) = F_j(t)E_i(t)$, so that

$$\begin{aligned}
\rho_{\mathrm{AB}}(t) &= \mathscr{E}(\rho(0)) = \sum_{i=1}^{2}\sum_{j=1}^{3} F_j(t)E_i(t)\rho_{\mathrm{AB}}(0)E_i^\dagger(t)F_j^\dagger(t), && (44)\\
E_1(t) &= \mathrm{diag}(1,\gamma_{\mathrm{A}})\otimes\mathrm{diag}(1,1,1) = \mathrm{diag}(1,1,1,\gamma_{\mathrm{A}},\gamma_{\mathrm{A}},\gamma_{\mathrm{A}}), && (45)\\
E_2(t) &= \mathrm{diag}(0,\omega_{\mathrm{A}})\otimes\mathrm{diag}(1,1,1) = \mathrm{diag}(0,0,0,\omega_{\mathrm{A}},\omega_{\mathrm{A}},\omega_{\mathrm{A}}), && (46)\\
F_1(t) &= \mathrm{diag}(1,1)\otimes\mathrm{diag}(1,\gamma_{\mathrm{B}},\gamma_{\mathrm{B}}) = \mathrm{diag}(1,\gamma_{\mathrm{B}},\gamma_{\mathrm{B}},1,\gamma_{\mathrm{B}},\gamma_{\mathrm{B}}), && (47)\\
F_2(t) &= \mathrm{diag}(1,1)\otimes\mathrm{diag}(0,\omega_{\mathrm{B}},0) = \mathrm{diag}(0,\omega_{\mathrm{B}},0,0,\omega_{\mathrm{B}},0), && (48)\\
F_3(t) &= \mathrm{diag}(1,1)\otimes\mathrm{diag}(0,0,\omega_{\mathrm{B}}) = \mathrm{diag}(0,0,\omega_{\mathrm{B}},0,0,\omega_{\mathrm{B}}), && (49)
\end{aligned}$$

with the factors in the matrices defined as before and the $E_i(t)$ and $F_j(t)$ operators again satisfy the CPTP relations.

For a specific example, let us analyze the case in which there exists local dephasing noise on both the qubit and the qutrit, with the qualification that dephasing noise on either subsystem individually have the same consequences. The time-evolved matrix in this case then given by

$$\rho'_{\mathrm{AB}}(x,t) = \begin{pmatrix} \frac{1}{4} & 0 & 0 & 0 & 0 & x\gamma_{\mathrm{A}}\gamma_{\mathrm{B}} \\ 0 & \frac{1}{8} & 0 & 0 & 0 & 0 \\ 0 & 0 & \frac{1}{8} & 0 & 0 & 0 \\ 0 & 0 & 0 & \frac{1}{8} & 0 & 0 \\ 0 & 0 & 0 & 0 & \frac{1}{8} & 0 \\ x\gamma_{\mathrm{A}}\gamma_{\mathrm{B}} & 0 & 0 & 0 & 0 & \frac{1}{4} \end{pmatrix}, \tag{50}$$

for which $\bar{x} = x\gamma_{\mathrm{A}}\gamma_{\mathrm{B}}$. The eigenvalues of this matrix are given by $\{\lambda_k(x,t)\} = \{\frac{1}{8},\frac{1}{8},\frac{1}{8},\frac{1}{8},\frac{1}{4}(1+4x\gamma_{\mathrm{A}}\gamma_{\mathrm{B}}),\frac{1}{4}(1-4x\gamma_{\mathrm{A}}\gamma_{\mathrm{A}})\}$, of which only the last one can possibly be negative. According to the positivity condition, this is a valid density matrix for values of x such that all eigenvalues are positive, $0 \le x \le \frac{1}{4}$, since $\gamma_{\mathrm{A}},\gamma_{\mathrm{B}} \le 1$. The eigenvalues $\left\{\lambda_k^{T_{\mathrm{A}}}(x,t)\right\}$ of the partial transpose with respect to qubit A are $\{\frac{1}{4},\frac{1}{4},\frac{1}{8},\frac{1}{8},\frac{1}{8}(1+8x\gamma_{\mathrm{A}}\gamma_{\mathrm{B}}),\frac{1}{8}(1-8x\gamma_{\mathrm{A}}\gamma_{\mathrm{B}})\}$. As before, the only eigenvalue of interest is the last one, which takes on a negative eigenvalue when $\frac{1}{8} < x \le \frac{1}{4}$. Thus, we find

$$\mathscr{N}(\rho'_{\mathrm{AB}}(x,t)) = \max\{0, x\gamma_{\mathrm{A}}(t)\gamma_{\mathrm{B}}(t) - 1/8\}, \tag{51}$$

may reach zero in finite time as the first term decays according to the γ_A and γ_B factors. Thus, we showed that ESD can take place in the next largest system finite-dimensional bipartite system, supporting the conjecture that it is not specific to the simplest quantum system and may be a generic effect. Most recently, we have extended this result to pairs of d-dimensional systems for all d in the case of initially isotropic entangled systems, exploiting their symmetry [29].

RELATION TO FOUNDATIONS OF QUANTUM MECHANICS

An attempt to solve the quantum measurement problem was made 25 years ago by Żurek, invoking a process termed environmentally induced superselection or *einselection* [30]. The idea is that the environment of the system under measurement naturally chooses the basis in which state is measured by the measuring apparatus, for the class of interactions that include measurements. Although this approach has encountered a number of difficulties in solving the measurement problem, such as the failure of *coherence* to be entirely lost in finite times under local noise, the Żurek group has continued to pursue the study of decoherence and to produce increasingly subtle treatments of system–apparatus–environment interactions to better understand its effects. We have been following a related direction of investigation that focuses not on decoherence, but rather on *disentanglement* which, in comparison with decoherence, can relatively easily occur and can even do so in finite time under local noise, as we have discussed here. Because the measurement process requires the entanglement between measured system and measurement apparatus and occurs in a larger noisy environment, we are suggesting that if a sufficient formal description of physical conditions for a successful completed measurement process is characterized primarily by disentanglement then environmental noise may play the key role in measurement and its study may allow for the solution the quantum measurement problem, through its ability to induced disentanglement between measuring apparatus and measured system.

REFERENCES

1. M. A. Nielsen and I. L. Chuang, *Quantum Computation and Quantum Information* (Cambridge University Press; Cambridge, 2000).
2. G. S. Jaeger, *Quantum Information: An overview* (Springer; New York, 2006).
3. J. Preskill, Proc. Roy. Soc. London A **454**, 385 (1998).
4. T. Yu and J. H. Eberly, Opt. Comm. **264**, 393 (2006).
5. T. Yu and J. H. Eberly, Phys. Rev. Lett. **97**, 140403 (2006).
6. M. Yonac, T. Yu, J. H. Eberly, J. Phys. B: At. Mol. Opt. Phys. **40**, S45 (2007).
7. K. Ann, G. S. Jaeger, Phys. Lett. A, to appear; doi:10.1016/j.physleta.2007.07.07.
8. G. S. Jaeger, K. Ann, J. Mod. Opt, too appear.
9. T. Yu and J. H. Eberly, Phys. Rev. B **68**, 165322 (2003).
10. K. Ann, G. S. Jaeger, Phys. Rev. B **75**, 115307 (2007).
11. P. J. Dodd, J. J. Halliwell, Phys. Rev. A **69**, 052105 (2004).
12. W. H. Zurek, Rev. Mod. Phys. **75**, 715 (2003).
13. R. Blume-Kohout, W. H. Zurek, Phys. Rev. A **73**, 062310 (2006).
14. M. P. Almeida, et al., Science **316**, 579 (2007).
15. J. Laurat, K. S. Choi, H. Deng, C. W. Chou, H. J. Kimble, arxiv:0706.0528.
16. W. F. Stinespring, Proc. Amer. Math. Soc. **6**, 211 (1955).
17. W. K. Wootters, Phys. Rev. Lett. **80**, 2245 (1998).
18. G. S. Jaeger and A. Shimony, Phys. Lett. A **197**, 83 (1995).
19. C. H. Bennett, S. Popescu, D. Rohrlich, J. A. Smolin, A. V. Thapliyal, Phys. Rev. A **63**, 012307 (2000).
20. C. H. Bennett, D. P. DiVincenzo, J. A. Smolin, W. K. Wootters, Phys. Rev. A. **54**, 3824 (1996).
21. V. Coffman, J. Kundu and W. K. Wootters, Phys. Rev. A **61**, 052306 (2000).
22. F. Verstraete, J. Dehaene, B. De Moor, H. Verschelde, Phys. Rev. A **65**, 052112 (2002).
23. D. Li, X. Li, H. Huang, X. Li, Phys. Lett. A **359**, 428 (2006).
24. G. Vidal, R.F. Werner, Phys. Rev. A 65, 032314 (2007).
25. A. Peres, Phys. Rev. Lett. 77, 1413 (1996).
26. M. Horodecki, P. Horodecki, R. Horodecki, Phys. Lett. A 223, 1 (1996).
27. P. Horodecki, Phys. Lett. A 232, 333-339 (1997).
28. M. O. T. Cunha, New J. Phys. **9**, S1367 (2007).
29. K. Ann, G. S. Jaeger, Phys. Rev. A, in press.
30. W. H. Zurek, Phys. Rev. D **26**, 1862 (1982).

Prequantum Classical Statistical Field Theory – PCSFT

Andrei Khrennikov

International Center for Mathematical Modeling in Physics, Engineering and Cognitive science MSI, Växjö University, S-35195, Sweden

Abstract. In this short note we present (without to go into mathematical details) the main notions of Prequantum Classical Statistical Field Theory (PCSFT): theory of classical fields which fluctuations produce averages approximatively coinciding with averages given by the quantum formalism. Thus the conventional quantum mechanics can be considered as a probabilistic approximation of a more detailed classical field model. Experimental predictions of PCSFT are discussed

Keywords: quantum averages, classical averages, ensembles of classical fields, random fluctuations, time scales of measuring device and field fluctuations, approximative calculation of averages
PACS: 03.50.-z

INTRODUCTION

In our model [1], [2] all quantum systems (not only photons) are described by classical vector fields:

$$\psi(x) = (q(x), p(x)).$$

In the complex representation such a vector field can be written as

$$\psi(x) = q(x) + ip(x)$$

(cf. Riemann-Silberstein representation for the classical electromagnetic field). These fields can be considered as prequantum hidden variables. The field amplitude is its physical characteristic. These are really waves and not "waves of probability." A "quantum state" is an ensemble of such fields, or in other words a random field (fluctuating with time).

In some sense PCSFT is a comeback to the wave mechanics. We can mention e.g. Lande [3], [4] and Lamb[5] or recently Kracklauer and Roychoudhuri (see papers in this volume) who denied the wave-particle duality for photons. However, PCSFT implies the illusoriness of the particle picture for all types of quantum systems. Thus it is closer to original views of Schrödinger[1] and late views of Einstein, see e.g. [6]–[9].

QUANTUM AVERAGES AS APPROXIMATIONS OF CLASSICAL FIELD AVERAGES

We recall the quantum formula for averages

$$< A >= \mathrm{Tr}\rho\ \mathrm{A},$$

where A is an observable (represented by a self-adjoint operator) and ρ is a statistical state (represented by a density operator [10]). This formula is a simple consequence of Born's probabilistic rule.

The main mathematical problem in the PCSFT framework was to derive this formula from this random field model. And it was derived [1], [2] as an approximation of the average with respect to an ensemble of classical ψ-fields (classical random field).

A quantum observable – a self-adjoint operator A – arises as Hessian (the second derivative) of a prequantum variable given by a function $f(\psi)$ of classical field ψ. We recall that the second derivative is always symmetric (even

---

[1] At the end of his life Schrödinger tried again to exclude completely particles from quantum theory. But this latest comeback to wave theory was toward quantum and not classical field theory, see [7].

CP962, *Quantum Theory, Reconsideration of Foundations – 4*,
edited by G. Adenier, A. Yu. Khrennikov, P. Lahti, V. I. Man'ko, and Th. M. Nieuwenhuizen

in the case functionals of fields). This mapping of classical variables onto quantum, $f(\psi) \to A \equiv f''(0)$ is not one-to-one. It has huge degeneracy.

A statistical state – a density operator ρ – is obtained as the normalization of the covariance operator of the classical random field. This correspondence is neither one-to-to one: different classical random fields can have the same covariance operator and hence they are represented by the same quantum state.

Let the distribution of random prequantum field is given by a probability measure μ (on the L_2-space of fields). Let cov μ be its covariance operator and let σ_μ^2 be its dispersion. Then this random field is represented in the conventional quantum formalism by the density matrix: $\rho = \frac{\text{cov}\,\mu}{\sigma_\mu^2}$.

DYNAMICS

The PCSFT-dynamical equation is the Hamilton system for field variables q and p. In the complex form this is a nonlinear Schrodinger equation. The usual linear Schrödinger equation arises as an approximation of the PCSFT-dynamical equation by expanding the Hamilton function for classical field variables $\mathscr{H}(q,p)$ into the Taylor series and preserving only the quadratic term. Thus quantum Hamiltonian H arises as the second derivative of the Hamilton function $\mathscr{H}(q,p)$:

$$H = \mathscr{H}''(0).$$

We recall that by using the Riemann-Silberstein representation we can write the system of Maxwell's equations in the form of Schrödinger equation. PCSFT explores this possibility for arbitrary vector fields (by determining the linearization procedure).

DEGREE OF APPROXIMATION AND POSSIBLE EXPERIMENTAL VERIFICATION

Averages with respect to classical fields do not coincide precisely with $< A >= \text{Tr}\rho$ A. The difference between the quantum average and the classical average are of the magnitude

$$\kappa = t/T,$$

where T is the scale of operation of modern detectors and t is the scale of the fluctuation of the classical prequantum field, see [11]. The main problem is that the latter is unknown.

If it is one attosecond (10^{-18} of a second) then we shall hope to find violation of statistical predictions of quantum mechanics – averages would deviate from $< A >= \text{Tr}\rho A$. We recall that modern detectors operates on the nanosecond scale, so T is about one nanosecond. One could expect deviation already with approaching the T of the magnitude of one femtosecond (10^{-15} of a second).

But, if $t =$ Planck time, we would practically never find deviations. In such a case even if PCSFT were correct, it would have no practical consequences, since deviation from the quantum predictions would be totally negligible.

Thus the main open question in PCSFT is the time scale of prequantum classical field fluctuations. And its main prediction deviation of experimental statistics from statistics predicted by quantum formalism.

REFERENCES

1. A. Yu. Khrennikov, *J. Phys. A: Math. Gen.* **38**, 9051-9073 (2005); *Found. Phys. Lett.*, **18**, 637-650 (2006); *Physics Letters A*, **357**, 171-176 (2006).
2. A. Yu. Khrennikov, "Quantum mechanics as an asymptotic projection of statistical mechanics of classical fields," in *Quantum theory: reconsideration of foundations—3*, edited by G. Adenier, A. Yu. Khrennikov and Th.M. Nieuwenhuizen, AIP Conference Proceedings Ser. **810**, Melville, New York, 2006, pp. 179–197.
3. A. Lande, *Foundations of quantum theory,* Yale Univ. Press, 1955.
4. A. Lande, *New foundations of quantum mechanics,* Cambridge Univ. Press, Cambridge, 1965.
5. W. E. Lamb, *The interpretation of quantum mechanics,* Rinton Press, Inc, Princeton, NJ, 2001.
6. E. Schrödinger, M. Planck, A. Einstein, H.A. Lorentz, *Briefe zur Wellenmechanik,* Springer Verlag, Wien, 1976.
7. D'Agostino, "Continuity and completeness in physical theory: Schrödinger's return to the wave interpretation of quantum mechanics in the 1950's", in *E. Schrödinger: Philosophy and the Birth of Quantum Mechanics,* edited by M. Bitbol, O. Darrigol, Editions Frontieres, Gif-sur-Yvette, 1992, pp. 339-360.

8. A. Einstein, *The Collected Papers of Albert Einstein,* Princeton Univ. Press, Princeton, 1993.
9. A. Einstein and L. Infeld, *Evolution of Physics: The Growth of Ideas from Early Concepts to Relativity and Quanta,* Simon and Schuster, 1961.
10. J. von Neumann, *Mathematical foundations of quantum mechanics,* Princeton Univ. Press, Princeton, N.J., 1955.
11. A. Yu. Khrennikov, "To quantum mechanics through random fluctuations at the Planck time scale, arXiv:hep-th/0604011v3."

Bell's Inequality: Nonlocalty, "Death of Reality", or Incompatibility of Random Variables?

Andrei Khrennikov

International Center for Mathematical Modeling in Physics, Engineering and Cognitive Science, Växjö University, S-35195, Sweden

Abstract. We remind the viewpoint that violation of Bell's inequality might be interpreted not only as an evidence of the alternative – either nonlocality or "death of reality" (under the assumption the quantum mechanics is incomplete). Violation of Bell's type inequalities is a well known sufficient condition of incompatibility of random variables – impossibility to realize them on a single probability space. Thus, in fact, we should take into account an additional interpretation of violation of Bell's inequality – a few pairs of random variables (two dimensional vector variables) involved in the EPR-Bohm experiment are incompatible. They could not be realized on a single Kolmogorov probability space. Thus one can choose between: a) completeness of quantum mechanics; b) nonlocality; c) " death of reality"; d) non-Kolmogorovness. In any event, violation of Bell's inequality has a variety of possible interpretations. Hence, it could not be used to derive the unique conclusion on the relation between quantum and classical models.

Keywords: EPR experiment, EPR-Bohm experiment, Bell's inequality, probability measure, probabilistic compatibility and incompatibility, Kolmogorov probability space, contextuality, detectors efficiency, fair and unfair sampling, negative probabilities, rejection of the photon hypothesis, frequency probabilities
PACS: 03.65.Ca, 03.65.Ta, 03.50.-z

INTRODUCTION

The aim of this report is to present to the physical community (especially, its quantum information part) results of purely probabilistic studies on the problem of *probabilistic compatibility of a family of random variables.* They were done during last hundred years. And they have the direct relation to Bell's inequality. Surprisingly, a priory studies on probabilistic compatibility have no direct relation to the well known fundamental problems which are typically discussed by physicists, namely, *realism and locality* [1]–[15], see e.g. [16]–[21] for recent debates. After a review on conditions of probabilistic compatibility, we shall try to find traces of this mathematical research in physics, e,g,. parameters of measurement devices, negative probabilities, detectors inefficiency, unfair sampling, rejection of photon.

We remark that our considerations would not imply that the conventional interpretation of Bell's inequality [1]–[15] should be rejected. In principle, Bell's conditions (nonlocality, "death of reality") could also be taken into account. Our aim is to show that Bell's conditions are only *sufficient, but not necessary* for violation of Bell's inequality. Therefore other interpretations of violation of this inequality are also possible. Bell's alternative – either quantum mechanics or local realism – can be extended: either existence of a single probability measure [1] for incompatible experimental contexts or quantum mechanics. We notice that existence of such a single probability was never assumed in classical (Kolmogorov) probability theory, but it was used by J. Bell to derive his inequality (it was denoted by ρ in Bell's derivation). Therefore if one wants to use Bell's inequality, he should find reasonable arguments supporting Bell's derivation. Roughly speaking: Why do we use such an assumption in quantum physics, although we have never used it in classical probability theory?

In classical probability theory researchers never try to put statistical data collected on the basis of different sampling experiments into one single probability space. However, in quantum mechanics we try to do this. From the point of view of "the probabilistic opposition" to the conventional interpretation of violation of Bell's inequality, the crucial point of Bell's considerations was placing statistical data collected in a few totally different experiments (corresponding to different setting of polarization beam splitters) in one probabilistic inequality. We remark that the same trick was done by Feynman [24] in his probabilistic analysis of the two slit experiment, However, Feynman

[1] By using the terminology of modern probability theory one should speak about existence of a single Kolmogorov probability space [22], [23].

CP962, *Quantum Theory, Reconsideration of Foundations – 4,*
edited by G. Adenier, A. Yu. Khrennikov, P. Lahti, V. I. Man'ko, and Th. M. Nieuwenhuizen

remarked that his analysis demonstrated violation of laws of classical probability theory. [2] And this is not surprising, because he proceeded against all rules of the conventional probabilistic practice. Kolmogorov emphasized from the very beginning [22] that any experimental context induces its own probability space. But Feynman considered three different contexts: C_{12} two slits are open, C_1 only the first slit is open, C_2 only the second slit is open. By some reason (I think simply because of lack of education in probability) he wanted to put such data collected under three different contexts into a single probability space. The impossibility to do this Feynman interpreted as astonishing violation of laws of classical probability theory. To solve this problem, he decided to assume (in accordance with fathers of quantum mechanics, see e.g. Dirac [25]) that each quantum particle interferes with itself. [3] Bell did more or less the same thing with the EPR-Bohm experiment. In contrast to Feynman, Bell did not even see the probabilistic inconsistency of his considerations. This story was presented in detail in my book [26] (which is definitely unreadable for physicists, because of too much probability).

This paper is based on the results of research on the probabilistic structure of Bell's inequality. We can call this (very inhomogeneous) group of researchers "probabilistic opposition" [4]: Accardi [27]–[29], Fine [30], Garola, Solombrino and Sozzo [31]–[34] (and their papers in this volume), Hess and Philipp [35], Khrennikov [26], [36]–[39], Kupczynski [40]- [42] (and his paper in this volume), Pitowsky [43], [44], Rastal [45]. On one hand, it is amazing that so many people came to the same conclusion practically independently. On the other hand, it is also amazing that this conclusion is not so much known by physicists (even for mathematically interested researchers working in quantum information theory). There is definitely a problem of communication. I hope that this review would inform physicists about some general mathematical ideas on Bell's inequality.

I remark that (to my knowledge) only one physicist (and moreover very good experimenter!), Klyshko [46]-[49], obtained similar conclusion – independently from mathematicians! (he did not know anything about mentioned research in probabilistic analysis of Bell's arguments). I also point out to a series of papers of Andreev and Manko et al. [50]– [54] who used quantum tomographic [55]– [59] interpretation Bell's inequality. Although they have never emphasized "non-Kolmogorovness" of their approach, it is evident that the quantum tomographic scheme for Bell's inequality is based on a family of probability measures associated with involved experimental contexts. It is impossible to construct a single probability serving for the total collection of statistical data.

Finally, we note that Volovich [60], [61] pointed out to the role of space variables in the EPR-Bohm studies (which was surprisingly missed from modern considerations). He also emphasized the crucial difference between the original EPR experiment for correlations of positions and momenta and its Bohm version for photon polarization or electron spin. Khrennikov and Volovich demonstrated in the rigorous probabilistic framework that the original EPR experiment for measurement of continuous variables for entangled systems (position and momentum) differs crucially from its EPR-Bohm version for measurement of discrete variables for entangled systems (projections of polarization or spin). We discuss this point in the present review.

Since in this paper we shall discuss Bell's proof of its inequality and its versions, we present (for reader's convenience) these proofs in the appendix.

STUDIES OF BOOLE AND VOROBJEV ON PROBABILISTIC COMPATIBILITY

Consider a system of three random variables $a_i, i = 1,2,3$. Suppose for simplicity that they take discrete values and moreover they are dichotomous: $a_i = \pm 1$. Suppose that these variables as well as their pairs can be measured and hence joint probabilities for pairs are well defined:

$$P_{a_i,a_j}(\alpha_i, \alpha_j) \geq 0 \tag{1}$$

[2] He had no idea about the rigorous axiomatics of modern classical probability theory (the Kolmogorov model). Therefore he wrote about violation of laws of classical Laplacian probability.

[3] Such a conclusion is in visible contradiction with his own path integral approach to quantum mechanics. Of course, one might say (as Feynman did) that these trajectories are virtual. However, in such a case one should explain such a funny coincidence of prediction of the quantum operator formalism with the result of integration with respect to these totally nonphysical trajectories. Feynman did not do this. It is clear that we do not know all historical details. It might be that at the beginning Feynman assigned to these trajectories more physical meaning. We only know that his path integral approach was terribly criticized by Bohr. Feynman was able to sell his path integral approach to Bohr (and hence to the quantum community) only through personal contacts with Pauli. It is clear that the conflict with Bohr was not possible to resolve in the framework of physical trajectories.

[4] Names are presented in the alphabetic order. In spite the evident fact that physicists totally ignore this probabilistic research, the "probabilistic opposition" is permanently disturbed by conflicts on priority.

and

$$\sum_{\alpha_i,\alpha_j=\pm 1} P_{a_i,a_j}(\alpha_i,\alpha_j) = 1. \tag{2}$$

Question: *Is it possible to construct the joint probability distribution, $P_{a_1,a_2,a_3}(\alpha_1,\alpha_2,\alpha_3)$, for any triple of random variables?*

Surprisingly this question was asked and answered for hundred years ago by Boole (who invented Boolean algebras). This was found by Itamar Pitowsky [63], [64], see also preface [18]. To study this problem, Boole derived inequality which coincides with the well known in physics Bell's inequality. Violation of this Boole-Bell inequality implies that for such a system of three random variables the joint probability distribution $P_{a_1,a_2,a_3}(\alpha_1,\alpha_2,\alpha_3)$ does not exist.

Thus Bell's inequality was known in probability theory. It was derived as a constraint which violation implies nonexistence of the joint probability distribution.

Different generalizations of this problem were studied in probability theory. The final solution (for a system of n random variable) was obtained by Soviet mathematician Vorobjev [65] (as was found by Hess and Philipp [35]). His result was applied in purely macroscopic situations – in game theory and optimization theory.

We emphasize that for mathematicians consideration of Bell's type inequalities did not induce revolutionary reconsideration of laws of nature. The joint probability distribution does not exist just because those observables could not be measured simultaneously.

THE EPR-BOHM EXPERIMENT, IMPOSSIBILITY TO MEASURE THREE POLARIZATION PROJECTIONS SIMULTANEOUSLY

We consider now one special application of Boole's theorem the EPR-Bohm experiment for measurements of spin projections for pairs of entangled photons.[5] Denote corresponding random variables by a_θ^1 and a_θ^2, respectively (the upper index $k = 1,2$ denotes observables for corresponding particles in a pair of entangled photons). Here θ is the angle parameter determining the setting of polarization beam splitter. For our purpose it is sufficient to consider three different angles: $\theta_1, \theta_2, \theta_3$. (In fact, for real experimental tests we should consider four angles, but it does not change anything in our considerations).

By using the condition of precise correlation for the singlet state we can identify observables

$$a_\theta(\lambda) \equiv a_\theta^1(\lambda) = a_\theta^2(\lambda). \tag{3}$$

The following discrete probability distributions are well defined: $P_{a_\theta}(\alpha)$ and $P_{a_{\theta_i},a_{\theta_j}}(\alpha,\beta)$. Here $\alpha,\beta = \pm 1$. We remark that in standard derivations of Bell's type inequality for probabilities (and not correlations), see appendix, there are typically used the following symbolic expressions of probabilities: $P(a_\theta(\lambda) = \alpha)$ and $P(a_{\theta_i}(\lambda) = \alpha, a_{\theta_j}(\lambda) = \beta)$. However, by starting with a single probability P (defined on a single space of "hidden variables" Λ) we repeat Bell's schema (which we would not like to repeat in this paper).

Thus we are precisely in the situation which was considered in probability theory. Boole (and Vorobjev) would ask: Do polarization-projections for any triple of angles have the joint probability distribution? Can one use a single probability measure P? The answer is negative – because the Boole-Bell inequality is violated (or because necessary condition of Vorobjev theorem is violated). Thus it is impossible to introduce the joint probability distribution for an arbitrary triple of angles.

On the other hand, Bell started his considerations with the assumption that such a single probability measure exists, see appendix. He represented all correlations as integrals with respect to the same probability measure ρ :

$$\langle a_{\theta_i}, a_{\theta_j}\rangle = \int_\Lambda a_{\theta_i}(\lambda) a_{\theta_j}(\lambda) dP(\lambda). \tag{4}$$

(We shall use the symbol P, instead of Bell's ρ to denote probability).

[5] Although both Boole's and Bell's theorems are based on the same inequality, the conclusions are totally different. These are "nonexistence of the joint probability distribution" and "either local realism or quantum mechanics", respectively. Thus we would like to analyze the EPR-Bohm experiment from the viewpoint of Boole (Vorobjev, Accardi, Fine, Pitowsky, Rastal, Hess and Philipp and the author).

In opposite to Bell, Boole would not be so much excited by evidence of violation of Bell's inequality in the EPR-Bohm experiment. The situation when pairwise probability distributions exist. but a single probability measure P could not be constructed is rather standard. What would be a reason for existence of P in the case when the simultaneous measurement of three projections of polarization is impossible?

A priory nonexistence of P has nothing to do with nonlocality or "death or reality." The main problem is not the assumption that polarization projections are represented in the "local form":

$$a^1_{\theta_i}(\lambda), a^2_{\theta_j}(\lambda) \tag{5}$$

and not in the "nonlocal form"

$$a^1_{\theta_i}(\lambda | a^2_{\theta_j} = \beta), a^2_{\theta_j}(\lambda | a^1_{\theta_i} = \alpha), \tag{6}$$

where $\alpha, \beta = \pm 1$. The problem is nor assigning to each λ the definite value of the random variable – "realism."

The problem is impossibility to realize three random variables

$$a_{\theta_1}(\lambda), a_{\theta_2}(\lambda), a_{\theta_3}(\lambda) \tag{7}$$

on the same space of parameters Λ with same probability measure P. By using the modern terminology we say that it is impossible to construct a Kolmogorov probability space for such three random variables.

In this situation it would be reasonable to find sources of nonexistence of a Kolmogorov probability space. We remark that up to now we work in purely classical framework– neither the ψ-function or noncommutative operators were considered. We have just seen [7], [14], [15] that experimental statistical data violates the necessary condition for the existence of a single probability P. Therefore it would be useful to try to proceed purely classically in the probabilistic analysis of the EPR-Bohm experiment. We shall do this in the next section.

CONTEXTUAL ANALYSIS OF THE EPR-BOHM EXPERIMENT

As was already emphasized in my book [26], [39] the crucial point is that in this experiment one combine statistical data collected on the basis of three different complexes of physical conditions (contexts). We consider context C_1 – setting θ_1, θ_2, context C_2 – setting θ_1, θ_3, and finally context C_3 – setting θ_2, θ_3. We recall that already in Kolmogorov's book [22] (where the modern axiomatics of probability theory was presented) it was pointed out that each experimental context determines its own probability space. By Kolmogorov in general three contexts $C_j, j = 1, 2, 3$, should generate three Kolmogorov spaces: with sets of parameters Ω_j and probabilities P_j.

The most natural way to see the source of appearance of such spaces is to pay attention to the fact that (as it was underlined by Bohr) the result of measurement is determined not only by the initial state of a system (before measurement), but also by the *whole measurement arrangement*. Thus states of measurement devices are definitely involved. We should introduce not only space Λ of states of a system (a pair of photons), but also spaces of states of polarization beam splitters – Λ_θ. (We proceed under the assumption that the state of polarization beam splitter depends only on the orientation θ. In principle, we should consider two spaces for each θ for the first and the second splitters. In reality they are not identical.) Thus, see [26], for the context C_1 the space of parameters ("hidden variables") is given by

$$\Lambda_1 = \Lambda \times \Lambda_{\theta_1} \times \Lambda_{\theta_2}, \tag{8}$$

for the context C_2 it is

$$\Lambda_2 = \Lambda \times \Lambda_{\theta_1} \times \Lambda_{\theta_3}, \tag{9}$$

for the context C_3 it is

$$\Lambda_3 = \Lambda \times \Lambda_{\theta_2} \times \Lambda_{\theta_3}. \tag{10}$$

And, of course, we should consider three probability measures

$$dP_1(\lambda, \lambda_{\theta_1}, \lambda_{\theta_2}), dP_2(\lambda, \lambda_{\theta_1}, \lambda_{\theta_3}), dP_3(\lambda, \lambda_{\theta_2}, \lambda_{\theta_3}). \tag{11}$$

Random variables are functions on corresponding spaces

$$a_{\theta_1}(\lambda, \lambda_{\theta_1}), a_{\theta_2}(\lambda, \lambda_{\theta_2}), a_{\theta_3}(\lambda, \lambda_{\theta_3}). \tag{12}$$

Of course, Bell's "condition of locality" is satisfied (otherwise we would have e.g. $a_{\theta_1}(\lambda,\lambda_{\theta_1},\lambda_{\theta_2}), a_{\theta_2}(\lambda,\lambda_{\theta_2},\lambda_{\theta_1})$ for the context C_1).

In this situation one should have strong arguments to assume that these three probability distributions could be obtained from a single probability measure

$$dP_1(\lambda,\lambda_{\theta_1},\lambda_{\theta_2},\lambda_{\theta_3}) \tag{13}$$

on the space

$$\Lambda = \Lambda \times \Lambda_{\theta_1} \times \times \Lambda_{\theta_2} \times \Lambda_{\theta_3}. \tag{14}$$

CONSEQUENCES FOR QUANTUM MECHANICS

Finally, we come to quantum mechanics. Our contextual analysis of the EPR-Bohm experiment implies that the most natural explanation of nonexistence of a single probability space is that *the wave function does not determine probability in quantum mechanics* (in contrast to Bell's assumption). We recall that Born's rule contains not only the ψ-function but also spectral families of commutative operators which are measured simultaneously. Hence, the probability distribution is determined by the ψ-function as well as spectral families, i.e., observables.

Such an interpretation of mathematical symbols of the quantum formalism does not imply neither nonlocality nor "death of reality."[6]

BELL'S INEQUALITY AND NEGATIVE AND P-ADIC PROBABILITIES

By looking for a trace in physics of the Boole-Vorobjev conclusion on nonexistence of probability one can find that this problem was intensively discussed, but in rather unusual form (at least from the mathematical viewpoint). During our conversations on the probabilistic structure of Bell's inequality Alain Aspect permanently pointed out to a probabilistic possibility to escape Bell's alternative: either local realism or quantum mechanics. This possibility mentioned by Alain Aaspect is consideration of negative valued probabilities. A complete review on solving "Bell's paradox" with the aid of negative probabilities was done by Muckenheim [67]. Although negative probabilities are meaningless from the mathematical viewpoint (however, see [68]- [75] for an attempt to define them mathematically by using p-adic analysis), there is some point in consideration of negative probabilities by physicists. In the light of our previous studies this activity can be interpreted as a sign of understanding that "normal probability distribution" does not exist. Surprisingly, but negative probability approach to Bell's inequality can be considered as a link to Boole-Vorobjev's viewpoint on violation of Bell's inequality.

DETECTORS EFFICIENCY

Another trace of nonexistence of a single probability space can be found in physical literature on detectors efficiency [84]– [86]. Theoreticians as well as experimenters are well aware about the fact that the real experiments induce huge losses of photons. Even if one associates (as Bell did) one fixed probability distribution with the source (the initial state), there are no reasons to assume that it is preserved by detectors.

FAIR SAMPLING

The assumption of fair sampling is typically misidentified with detectors efficiency. However, they are essentially different. Unfair sampling could take place even for detectors having 100% efficiency.

By the fair sampling assumption ensembles of pairs of output photons from two polarization beam splitters have the same probabilistic properties independently on orientations of splitters. Thus one can operate with a single probability distribution.

---

[6] One should not accuse the author in critique of J. Bell. J. Bell by himself did a similar thing with the von Neumann no-go theorem[66] , see [1], by pointing out that some assumptions of von Neumann were unphysical.

However, a priory there are no reasons to assume that ensembles of pairs of photons which pass polarization beam splitters for different choices of orientations have identical statistical properties, see e.g. [87],[88]. Unfair sampling implies that a single probability which would serve all orientations does not exist.

Finally, we remark that in general nonexistence of probability is not reduced to inefficiency of detectors or unfair sampling.

EXTENDED SEMANTIC REALISM

This is a generalization of quantum formalism, see Garola and Solombrino [31]–[33], Sozzo [34], by which each quantum observable gets an additional point to its spectrum, say a_0, denoting the event of nonregistration. In principle, one could consider extended semantic realism as a possible formalization of unfair sampling.

PHOTON HYPOTHESIS

All real experiments demonstrating violation of Bell's inequality and at the same time guaranteing locality have been done for photons. However, since first days of quantum mechanics, many prominent quantum physicists criticized Einstein's proposal of photon, e.g. Lande [89], [90] and Lamb [91] and recently Kracklauer and Roychoudhuri (see papers of the last two authors in this volume). For example, Lamb wrote in his "Anti-photon" [91] p. 221: "It is high time to give up the use of the word "photon", and of a bad concept which will shortly be a century old. Radiation does not consists of particles ..."

The idea about purely wave structure of classical as well as quantum light has interesting consequences in the EPR-Bohm experimental framework. If one rejects the conventional picture of detectors registering particles (photons) and if one considers detectors as devices integrating continuously (up to a certain threshold) electromagnetic radiation, then the whole Bell's representation loses its meaning. We could not associate hidden variables to clicks of detectors. We could not consider an ensemble of photons produced by a source and the corresponding probability distribution. This idea was explored in different versions by Santos, Thompson, Kracklauer [92], [93] (and his paper in this volume).

EBERHARD-BELL THEOREM

In quantum information community rather common opinion is that one could completely exclude probability distributions from derivation of Bell's inequality and proceed by operating with frequencies. One typically refers to the result of works [10]–[12] which we shall call the *Eberhard-Bell* theorem (in fact, the first frequency derivation of Bell's inequality was done by Stapp[9], thus it may be better to speak about Bell-Stapp-Eberhard theorem). By this theorem Bell's inequality can be obtain only under assumptions of realism – the maps $\lambda \longrightarrow a_\theta(\lambda)$ is well defined – and locality – the random variable $a_\theta(\lambda)$ does not depend on other variables which are measured simultaneously with it. Thus (in opposite to the original Bell derivation) existence of the probability measure P serving for all polarization (or spin) projections is not assumed.

At the first sight it seems that our previous considerations have no relation to the Eberhard-Bell theorem. One might say: "Yes, Bell proceeded wrongly, but his arguments are still true, because they were justified by Eberhard in the frequency framework."

As was shown by Khrennikov [26],[39] the use of frequencies, instead of probabilities, does not improve Bell's considerations, see also Hess and Philipp [35]. The contextual structure of the EPR-Bohm experiment plays again the crucial role. If we go into details of Eberhard's proof, we shall immediately see that he operated with statistical data obtained from three different experimental contexts, C_1, C_2, C_3, in such a way as it was obtained on the basis of a single context. He took results belonging to one experimental setup and add or subtract them from results belonging to another experimental setup. These are not proper manipulations from the viewpoint of statistics. One never performs algebraic mixing of data obtained for totally different sample. Thus if one wants to proceed in Eberhard's framework, he should find some strong reasons that the situation in the EPR-Bohm experiment differs crucially from the general situation in statistical experiments. I do not see such reasons. Moreover, the EPR-Bohm experimental setup is very common from the general statistical viewpoint.

Moreover, Eberhard's framework pointed to an additional source of nonexistence of a single probability distribution, see De Baere [76] and also [77]–[79]. Even if we ignore the contribution of measurement devices, then the ψ-function still need not determine a single probability distribution. In Eberhard's framework we should operate with results which are obtained in different runs. One could ask: Is it possible to guarantee that different runs of experiment produce the same probability distribution of hidden parameters? It seems that there are no reasons for such an assumption. We are not able to control the source on the level of hidden variables. It may be that the ψ-function is just a symbolic representation of the source, but it represents a huge ensemble of probability distributions of hidden variables. If e.g. hidden variables are given by classical fields, see e.g. [80]–[82], then a finite run of realizations (emissions of entangled photons) may be, but may be not representative for the ensemble of hidden variables produced by the source.

COMPARING OF THE EPR AND THE EPR-BOHM EXPERIMENTS

Typically the original EPR experiment [83] for correlations of coordinates and momenta and the EPR-Bohm experiment for spin (or polarization) projections are not sharply distinguished. People are almost sure that it is the same story, but the experimental setup was modified to move from "gedanken experiment" to real physical experiment. However, it was not the case! We should sharply distinguish these two experimental frameworks.

The crucial difference between the original EPR experiment and a new experiment which was proposed by Bohm is that these experiments are based on quantum states having essentially different properties. The original EPR state

$$\Psi(x_1,x_2) = \int_{-\infty}^{\infty} \exp\left\{\frac{i}{\hbar}(x_1 - x_2 + x_0)p\right\} dp, \tag{15}$$

and the singlet state

$$\psi = \frac{1}{\sqrt{2}}(|+>|->-|->|+>) \tag{16}$$

which is used in the EPR-Bohm experiment have in common only one thing: they describe correlated (or by using the modern terminology entangled) systems. But, in contrast to the EPR-Bohm state, one can really (as EPR claimed) associate with the original EPR state a single probability measure describing incompatible quantum observables (position and momentum). The rigorous prove in probabilistic terms was proposed by the author and Igor Volovich in [62]. On the other hand, as we have seen for the singlet state one could not construct a probabilistic model describing elements of reality corresponding to incompatible observables.

Thus the original EPR state is really exceptional from the general viewpoint of statistical analysis. But the EPR-Bohm state behaves "normally." In fact, there is no clear physical explanation why statistical data for incompatible contexts can be based on a single Kolmogorov space in one case and not in another. One possible explanation is that "nice probabilistic features of the original EPR-experiment" arise only due to the fact that it is "gedanken experiment."

APPENDIX: PROOFS

Bell's inequality

Let $\mathcal{P} = (\Lambda, \mathcal{F}, P)$ be a Kolmogorov probability space: Λ is the set of parameters, $\mathcal{F}$ is a σ-algebra of its subsets (used to define a probability measure), P is a probability measure. For any pair of random variables $u(\lambda), v(\lambda)$, their covariation is defined by

$$< u,v >= \mathrm{cov(u,v)} = \int_\Lambda \mathrm{u}(\lambda)\,\mathrm{v}(\lambda)\,\mathbf{dP}(\lambda). \tag{17}$$

We reproduce the proof of Bell's inequality in the measure-theoretic framework.

Theorem. (Bell inequality for covariations) *Let $a,b,c = \pm 1$ be random variables on $\mathcal{P}$. Then Bell's inequality*

$$|< a,b > - < c,b >| \leq 1 - < a,c > \tag{18}$$

holds.

Proof. Set $\Delta = < a,b > - < c,b >$. By linearity of Lebesgue integral we obtain

$$\Delta = \int_\Lambda a(\lambda)b(\lambda)d\mathbf{P}(\lambda) - \int_\Lambda c(\lambda)b(\lambda)dP(\lambda) \tag{19}$$

$$= \int_\Lambda [a(\lambda) - c(\lambda)] b(\lambda) dP(\lambda). \tag{20}$$

As

$$a(\lambda)^2 = 1, \tag{21}$$

we have:

$$|\Delta| = |\int_\Lambda [1 - a(\lambda)c(\lambda)] a(\lambda) b(\lambda) dP(\lambda)| \tag{22}$$

$$\leq \int_\Lambda [1 - a(\lambda)c(\lambda)] dP(\lambda). \tag{23}$$

It is evident that "hidden Bell's postulate" on the existence of a single probability measure P serving for three different experimental contexts (probabilistic compatibility of three random variables) plays the crucial role in derivation of Bell's inequality.

Wigner inequality

We recall the following simple mathematical result, see Wigner [6]:

Theorem 1.2. (Wigner inequality) *Let $a, b, c = \pm 1$ be arbitrary random variables on a Kolmogorov space $\mathscr{P}$. Then the following inequality holds:*

$$P(a = +1, b = +1) + P(b = -1, c = +1) \tag{24}$$

$$\geq \mathbf{P}(a = +1, c = +1). \tag{25}$$

Proof. We have:

$$P(a(\lambda) = +1, b(\lambda) = +1)$$

$$= P(a(\lambda) = +1, b(\lambda) = +1, c(\lambda) = +1) \tag{26}$$

$$+ P(a(\lambda) = +1, b(\lambda) = +1, c(\lambda) = -1),$$

$$P(b(\lambda) = -1, c(\lambda) = +1)$$

$$= P(a(\lambda) = +1, b(\lambda) = -1, c(\lambda) = +1) \tag{27}$$

$$+ P(\lambda \in \Lambda : a(\lambda) = -1, b(\lambda) = -1, c(\lambda) = +1),$$

and

$$P(a(\lambda) = +1, c(\lambda) = +1)$$

$$= P(a(\lambda) = +1, b(\lambda) = +1, c(\lambda) = +1) \tag{28}$$

$$+ P(a(\lambda) = +1, b(\lambda) = -1, c(\lambda) = +1).$$

If we add together the equations (26) and (27) we obtain

$$P(a(\lambda) = +1, b(\lambda) = +1) + P(b(\lambda) = -1, c(\lambda) = +1)$$

$$= P(a(\lambda) = +1, b(\lambda) = +1, c(\lambda) = +1)$$

$$+ P(a(\lambda) = +1, b(\lambda) = +1, c(\lambda) = -1) \tag{29}$$

$$+ P(a(\lambda) = +1, b(\lambda) = -1, c(\lambda) = +1)$$

$$+ P(a(\lambda) = -1, b(\lambda) = -1, c(\lambda) = +1).$$

But the first and the third terms on the right hand side of this equation are just those which when added together make up the term $P(a(\lambda)=+1, c(\lambda)=+1)$ (Kolmogorov probability is additive). It therefore follows that:

$$\begin{aligned} & P(a(\lambda)=+1, b(\lambda)=+1)+\mathbf{P}(b(\lambda)=-1, c(\lambda)=+1) \\ & =P(a(\lambda)=+1, c(\lambda)=+1) \\ & +P(a(\lambda)=+1, b(\lambda)=+1, c(\lambda)=-1) \\ & +P(a(\lambda)=-1, b(\lambda)=-1, c(\lambda)=+1) \end{aligned} \tag{30}$$

By using non negativity of probability we obtain the inequality:

$$\begin{aligned} & P(a(\lambda)=+1, b(\lambda)=+1)+\mathbf{P}(b(\lambda)=-1, c(\lambda)=+1) \\ & \geq P(a(\lambda)=+1, c(\lambda)=+1) \end{aligned} \tag{31}$$

It is evident that "hidden Bell's postulate" on the existence of a single probability measure P serving for three different experimental contexts (probabilistic compatibility of three random variables) plays the crucial role in derivation of Wigner's inequality.

CONCLUSION

In probability theory Bell's type inequalities were studied during last hundred years as constraints for probabilistic compatibility of families of random variables – possibility to realize them on a single probability space. In opposite to quantum physics, such arguments as nonlocality and "death of reality" were not involved in considerations. In particular, nonexistence of a single probability space does not imply that the realistic description (a map $\lambda \rightarrow a(\lambda)$) is impossible to construct. Bell's type inequalities were considered as signs (sufficient conditions) of impossibility to perform simultaneous measurement *all random variables* from a family under consideration. Such an interpretation can be used for statistical data obtained in the EPR-Bohm experiment for entangled photons.

In any event, Bell's inequality could not be used to obtain the definite conclusion on the relation between quantum and classical models.

ACKNOWLEDGMENT

I am thankful for discussions to L. Accardi, S. Albeverio, M. Appleby, L. E. Ballentine, V. Belavkin, L. Hardy, I. Bengtsson, M. Bozejko, C. Fuchs, S. Gudder, K. Gustafson, I. Helland, A. Holevo, R. Hudson, S. Kozyrev, E. Loubenets, M. Manko and V. Manko, D. Petz, R. Schack, A. Caticha, D. Aerts, S. Aerts, B. Coecke, A. Grib, P. Lahti, A. Aspect, D. Mermin, L. Accardi, G. Adenier, W. De Baere, W. De Muynck, K. Hess and W. Philipp, R. Gill, D. Greenberger, J-A. Larsson, A. Peres, I. Pitowsky, M. Scully, I. Volovich.

REFERENCES

1. J. S. Bell, *Speakable and unspeakable in quantum mechanics,* Cambridge Univ. Press, Cambridge, 1987.
2. B. d'Espagnat, *Veiled Reality. An anlysis of present-day quantum mechanical concepts,* Addison-Wesley, 1995.
3. A. Shimony, *Search for a naturalistic world view,* Cambridge Univ. Press, Cambridge, 1993.
4. J. F. Clauser, M. A. Horne, A. Shimony, R. A. Holt, *Phys. Rev. Letters* **49**, 1804-1806 (1969).
5. J. F. Clauser, A. Shimony, *Rep. Progr. Phys.* **41**, 1881-1901 (1978).
6. E. P. Wigner, *Am J. Phys.* **38**, 1005-1015 (1970).
7. A. Aspect, J. Dalibard, G. Roger, *Phys. Rev. Lett.* **49**, 1804-1807 (1982).
8. D. Home, F. Selleri, *Nuovo Cim. Rivista* **14**, 2–176 (1991).
9. H. P. Stapp, *Phys. Rev.* ser. D **3**, 1303-1320 (1971).
10. P. H. Eberhard, *Il Nuovo Cimento B* **38**, 75-80 (1977).
11. P. H. Eberhard, *Il Nuovo Cimento B* **46**, 392-419 (1978).

12. P. H. Eberhard, *Phys. Rev. Letters* **49**, 1474-1477 (1982).
13. A. Peres, *Am. J. of Physics* **46**, 745-750 (1978).
14. G. Weihs, T. Jennewein, C. Simon, H. Weinfurter, and A. Zeilinger, *Phys. Rev. Lett.* **81**, 5039-5042 (1998).
15. G, Weihs, "A test of Bell's inequality with spacelike separation," in *Proc. Conf. Foundations of Probability and Physics-4,* edited by G. Adenier, C. Fuchs, and A. Yu. Khrennikov, Ser. Conf. Proc. **889**, American Institute of Physics, Melville, NY, 2007, pp. 250-262.
16. A. Yu. Khrennikov (ed.), *Proc. Conf. Foundations of Probability and Physics,* Ser. Quantum Probability and White Noise Analysis **13**, WSP, Singapore, 2001.
17. A. Yu. Khrennikov (ed.), *Proc. Conf. Quantum Theory: Reconsideration of Foundations,* Ser. Math. Modeling **2**, Växjö Univ. Press, Växjö, 2002.
18. A. Yu. Khrennikov (ed.), *Proc. Conf. Foundations of Probability and Physics-2,* Ser. Math. Modeling **5**, Växjö Univ. Press, Växjö, 2003.
19. A. Yu. Khrennikov (ed.), *Proc. Conf. Foundations of Probability and Physics-3,* Ser. Conf. Proc. **750**, American Institute of Physics, Melville, NY, 2005.
20. G. Adenier, A. Yu. Khrennikov, and Th.M. Nieuwenhuizen (eds.), *Proc. Conf. Quantum Theory: Reconsideration of Foundations-3,* Ser. Conf. Proc. **810**, American Institute of Physics, Melville, NY, 2006.
21. G. Adenier, C. Fuchs, and A. Yu. Khrennikov, (eds), *Proc. Conf. Foundations of Probability and Physics-3,* Ser. Conf. Proc. **889**, American Institute of Physics, Melville, NY, 2007.
22. A. N. Kolmogoroff, *Grundbegriffe der Wahrscheinlichkeitsrech,* Springer Verlag, Berlin, 1933; reprinted : *Foundations of the Probability Theory,* Chelsea Publ. Comp., New York, 1956.
23. B. V. Gnedenko, *The theory of probability,* Chelsea Publ. Com., New-York, 1962.
24. R. P. Feynman and A. R. Hibs, *Quantum mechanics and path integrals,* Wiley, New York, 1965.
25. P. A. M. Dirac, *The Principles of Quantum Mechanics,* Oxford Univ. Press, 1930.
26. A. Yu. Khrennikov, *Interpretations of Probability,* VSP Int. Sc. Publishers, Utrecht/Tokyo, 1999 (second edition, 2004).
27. L. Accardi, "The probabilistic roots of the quantum mechanical paradoxes," in *Proc. Conf. The wave–particle dualism. A tribute to Louis de Broglie on his 90th Birthday,* edited by S. Diner, D. Fargue, G. Lochak and F. Selleri, Reidel Publ. Company, Dordrecht, 1970, pp. 47–55.
28. L. Accardi, *Urne e Camaleoni: Dialogo sulla realta, le leggi del caso e la teoria quantistica,* Il Saggiatore, Rome, 1997.
29. L. Accardi, "Could one now convince Einstein?" In *Proc. Conf. Quantum Theory: Reconsideration of Foundations-3,* edited by G. Adenier, A. Yu. Khrennikov, and Th.M. Nieuwenhuizen, Ser. Conf. Proc. **810**, American Institute of Physics, Melville, NY, 2006, pp. 3-18.
30. A. Fine, *Phys. Rev. Lett.* **48**, 291–295 (1982).
31. C. Garola and L. Solombrino, *Found. Phys.* **26**, 1121 (1996).
32. C. Garola and L. Solombrino, *Found. Phys.* **26**, 1329 (1996)
33. C. Garola, *Found. Phys.* **32**, 1597 (1996)
34. C. Garola and S. Sozzo, quant-ph/0703260 (2007).
35. K. Hess and W. Philipp, "Bell's theorem: critique of proofs with and without inequalities," in *Proc. Conf. Foundations of Probability and Physics-3,* edited by A. Yu. Khrennikov, Ser. Conf. Proc. **750**, American Institute of Physics, Melville, NY, 2005, pp. 150-155.
36. A. Yu. Khrennikov, *Il Nuovo Cimento B* **115**, 179-184 (1999).
37. A. Yu. Khrennikov, *J. of Math. Physics* **41**, 1768-1777 (2000).
38. A. Yu. Khrennikov, *J. of Math. Physics* **41**, 5934-5944 (2000).
39. A. Yu. Khrennikov, *Found. of Phys.* **32**, 1159-1174 (2002).
40. M. Kupczynski, *Phys. Lett A* **116**, 417-422 (1986).
41. M. Kupczynski, *Phys. Lett A* **121**, 51-56 (1987).
42. M. Kupczynski, *Phys. Lett A* **121**, 205-210 (1987).
43. I. Pitowsky, *Phys. Rev. Lett.* **48**, 1299-1302 (1982).
44. I. Pitowsky, *Phys. Rev.* ser. D **27**, 2316-2326 (1983).
45. P. Rastal, *Found. of Physics* **13**, 555-575 (1983).
46. D. N. Klyshko *Phys. Lett. A* **172**, 399-403 (1993); ibid **176**, 415-420 (1993); ibid **218**, 119-127 (1996); ibid **247**, 261-266 (1998).
47. D. N. Klyshko, *Laser Physics* **6**, 1056-1076 (1996).
48. D. N. Klyshko, *Annals of New York Academy of Science* **755**, 13-27 (1995).
49. D. N. Klyshko, *Uspehi Fizicheskih Nauk 168*, 975-1015 (1998).
50. V. A. Andreev, V. I. Man'ko, *JETP LETTERS* **72**, 93-96 (2000).
51. V. A. Andreev, V. I. Man'ko, *Theor. Math. Phys.* **140**, 1135-1145 (2004).
52. V. A. Andreev, V. I. Man'ko, *Phys. Lett. A* **281**, 278-288 (2001)
53. V. A. Andreev, V. I. Man'ko, O. V. Man'ko, et al., *Theor. Math. Phys.* **146**, 140-151 (2006).
54. V. A. Andreev, *J. Russian Laser Research* **27**, 327-331 (2006).
55. V. I. Manko, *J. of Russian Laser Research* **17** 579-584 (1996).
56. O. V. Manko and V.I. Manko, *J. Russian Laser Research* **25**, 477-492 (2004).
57. V. I. Manko and E. V. Shchukin, *J. Russian Laser Research* **22**, 545-560 (2001).
58. M. A. Manko, V. I. Manko, R. V. Mendes, *J. Russian Laser Research* **27**, 507-532 (2006).

59. Yu. M. Belousov, V.I. Manko, *Density matrix: representations and applications in statistical mechanics, v. 1, 2,* MFTI-Publ., Moscow, 2004
60. I. V. Volovich, "Quantum cryptography in space and Bell's theorem", in *Proc. Conf. Foundations of Probability and Physics,* edited by A. Yu. Khrennikov, Ser. Quantum Probability and White Noise Analysis **13**, WSP, Singapore, 2001, pp.364-372
61. I. V. Volovich, "Towards quantum information theory in space and time", *Proc. Conf. Quantum Theory: Reconsideration of Foundations,* edited by A. Yu. Khrennikov, Ser. Math. Modeling **2**, Växjö Univ. Press, Växjö, 2002, pp. 423-440.
62. A. Yu. Khrennikov and I. V. Volovich, *Soft Computing* **10**, 521 - 529 (2005).
63. I. Pitowsky, "From George Boole to John Bell: The Origins of Bell's Inequalities," in *Proc. Conf. Bell's Theorem, Quantum theory and Conceptions of the Universe,* Kluwer, Dordrecht, 1989, pp. 37-49.
64. I. Pitowsky, "Range Theorems for Quantum Probability and Entanglement," in *Proc. Conf. Quantum Theory: Reconsideration of Foundations,* edited by A. Yu. Khrennikov, Växjö University Press, Växjö, 2002, pp. 299-308.
65. N. N. Vorob'ev, *Theory of Probability and its Applications* **7**, 147-162 (1962).
66. J. von Neumann, *Mathematical foundations of quantum mechanics,* Princeton Univ. Press, Princeton, N.J., 1955.
67. W. Muckenheim, *Phys. Rep.* **133**, 338–401 (1986).
68. A. Yu. Khrennikov, *Dokl. Akad. Nauk SSSR, Matem.* **322**, 1075–1079 (1992).
69. A. Yu. Khrennikov, *J. Math. Phys.* **32**, 932–937 (1991).
70. A. Yu. Khrennikov, *Phys. Lett. A* **200**, 119–223 (1995).
71. A. Yu. Khrennikov, *Physica A* **215**, 577–587 (1995).
72. A. Yu. Khrennikov, *Int. J. Theor. Phys.* **34**, 2423–2434 (1995).
73. A. Yu. Khrennikov, *J. Math. Phys.* **36**, 6625–6632 (1995).
74. A. Yu. Khrennikov, *p-adic valued distributions in mathematical physics,* Kluwer Academic Publ., Dordrecht, 1994.
75. A.Yu. Khrennikov, *Non-Archimedean analysis: quantum paradoxes, dynamical systems and biological models,* Kluwer Academic Publ., Dordrecht, 1997.
76. W. De Baere, *Lett. Nuovo Cimento* **39**, 234-238 (1984).
77. W. De Muynck and W. De Baere, *Ann. Israel Phys. Soc.* **12**, 1-22 (1996).
78. W. De Muynck, W. De Baere and H. Marten, *Found. of Physics* **24**, 1589–1663 (1994).
79. W. De Muynck, J. T. Stekelenborg, *Annalen der Physik* **45**, 222-234 (1988).
80. A. Yu. Khrennikov, *J. Phys. A: Math. Gen.* **38**, 9051-9073 (2005).
81. A. Yu. Khrennikov, *Found. Phys. Lett.* **18**, 637-650 (2006).
82. A. Yu. Khrennikov, *Physics Letters A* **357**, 171-176 (2006).
83. A. Einstein, B. Podolsky, N. Rosen, *Phys. Rev.* **47**, 777–780 (1935).
84. P. Pearle, *Phys. Rev. D* **2**, 1418-1428 (1970).
85. N. Gisin and B. Gisin, *Phys. Lett. A* **260**, 323-333 (1999).
86. Jan-Åke Larsson, *Quantum Paradoxes, Probability Theory, and Change of Ensemble,* Linköping Univ. Press, Linköping, Sweden, 2000.
87. G. Adenier, A. Khrennikov, "Anomalies in EPR-Bell experiments," in *Proc. Conf. Quantum theory: reconsideration of foundations—3,* edited by G. Adenier, A. Yu. Khrennikov, and Th.M. Nieuwenhuizen, Ser. Conf. Proc. **810**, American Institute of Physics, Melville, NY, 2006, pp. 283–293.
88. G. Adenier, A. Khrennikov, *J. Phys. B: Atomic, Molecular and Optical Physics* **40** (1), 131-141 (2007).
89. A. Lande, *Foundations of quantum theory,* Yale Univ. Press, Yale, 1955.
90. A. Lande, *New foundations of quantum mechanics,* Cambridge Univ. Press, Cambridge, 1965.
91. W. E. Lamb, *The interpretation of quantum mechanics,* Rinton Press, Inc, Princeton, NJ, 2001.
92. A. F. Kracklauer, quant-ph/0610173 (2006).
93. A. F. Kracklauer, *Found. Phys. Lett.* **19** (6), 625-629 (2006).

Tomographic Entropy and New Entropic Uncertainty Relations

Margarita A. Man'ko

P. N. Lebedev Physical Institute, Leninskii Prospect 53, Moscow 119991, Russia

Abstract. Such entropies as Shannon entropy, Rènyi entropy, and von Neumann entropy are discussed for quantum states. New inequalities for quantum tomographic-probability densities are considered.

Keywords: Entropy, quantum tomography, uncertainty relations, symplectic tomogram.
PACS: 42.50.Ct, 42.50.Ex, 02.50.Cw

INTRODUCTION

Entropy and information are important characteristics of probability distributions. In any physical problem, which can be studied using a probability distribution, the Shannon entropy [1] can be introduced for characterization of the properties of the physical system under study.

The Shannon entropy is the real number associated with a probability distribution. The number corresponds to the degree of order/disorder of the system under consideration. The more disorder in the system, the larger value of the Shannon entropy. The finer properties of probability distributions are described by the Rènyi entropy [2] which is the function of a q-parameter; the limit case of the Rènyi entropy for a specific value of the parameter $q = 1$ yields the Shannon entropy.

In a conventional approach to quantum systems, the quantum states are associated to the wave functions [3] for pure states and the density matrix for mixed states [4, 5]. Entropy as a characteristics of the degree of order in quantum systems was introduced by von Neumann as the functional of the density matrix (see [6]). For pure states, the von Neumann entropy, by definition, is equal to zero.

There exist some inequalities for entropies related to the quantum states introduced in [7] and called entropic uncertainty relations; these inequalities were studied in [8]. Recently it was suggested to describe quantum states by standard probability distributions (quantum-state tomograms) containing the same information on the quantum states, that the density operator and the wave function have. In view of this approach (called the probability representation of quantum mechanics), one can employ elaborated mathematical schemes of entropies to apply these schemes to characterize the quantum states. The states of the systems with continuous degrees of freedom like photon quadratures are described by symplectic tomograms [9, 10], which are fair probability densities of random variable and, in addition, they are dependent on extra parameters. The symplectic entropy associated to symplectic tomogram of the quantum state was introduced in [11]. The symplectic entropy of signal depending on time was introduced in [12] using the known analogy (see, for example, [13]) of the wave function of the quantum state $\psi(x)$ with the analytic signal $f(t)$.

The different kinds of entropy, within the framework of the probability representation of quantum mechanics, were studied in [14]. For tomographic probability of spin states, the tomographic entropy was introduced in [15].

Our aim here is to review the new entropic uncertainty relations for Rènyi entropy considered in [16] and applied in [17–23] for studying the properties of optical [24] and symplectic tomogram.

The paper is organized as follows.

In Section 2, we present a review of the properties of Shannon entropy and entropic uncertainty relations. In Section 3, the symplectic tomography is considered and the notion of tomographic entropy is discussed. In Section 4, the q-deformed entropic uncertainty relations are studied, while in Section 5 tomographic entropy for spin states is considered and new uncertainty relations for this entropy in the case of tripartite spin system are reviewed. Conclusions are done in Section 6.

CP962, *Quantum Theory, Reconsideration of Foundations – 4*,
edited by G. Adenier, A. Yu. Khrennikov, P. Lahti, V. I. Man'ko, and Th. M. Nieuwenhuizen

ENTROPY AND ENTROPIC UNCERTAINTY RELATIONS

Below we consider the notion of entropy for the wave functions $\psi(x)$ (state vectors $|\psi\rangle$) following [18, 23]. In the information-theory context, entropy is related to an arbitrary probability-distribution function [1].

For example, given the probability distribution $P(n)$, where n is a discrete random variable, i.e.,

$$P(n) \geq 0, \tag{1}$$

and the normalization condition holds

$$\sum_n P(n) = 1, \tag{2}$$

one has, by definition, the Shannon entropy

$$S = -\sum_n P(n) \ln P(n) = -\langle \ln P(n) \rangle. \tag{3}$$

In quantum mechanics, for continuous variables the wave function $\psi(x)$ provides the probability-distribution density of position

$$P(x) = |\psi(x)|^2. \tag{4}$$

The corresponding entropy S_x reads (see, for example, [8])

$$S_x = -\int |\psi(x)|^2 \ln |\psi(x)|^2 \, dx. \tag{5}$$

In the momentum representation, the wave function reads

$$\widetilde{\psi}(p) = \frac{1}{\sqrt{2\pi}} \int \psi(x) e^{-ipx} \, dx \qquad (\hbar = 1). \tag{6}$$

The corresponding entropy S_p related to the momentum-probability density $|\widetilde{\psi}(p)|^2$ is

$$S_p = -\int |\widetilde{\psi}(p)|^2 \ln |\widetilde{\psi}(p)|^2 \, dp. \tag{7}$$

It is worthy noting that one can construct entropies S_x and S_p not only in quantum mechanics. If the function $\psi(x)$ is replaced by a signal function $f(t)$ depending on time t, the function $\widetilde{\psi}(p)$ is replaced by the function $\widetilde{f}(\omega)$ describing the signal spectrum. In this case, the entropy of the signal

$$S_t = -\int |f(t)|^2 \ln |f(t)|^2 \, dt \tag{8}$$

and the entropy of its frequency spectrum

$$S_\omega = -\int \left|\widetilde{f}(\omega)\right|^2 \ln \left|\widetilde{f}(\omega)\right|^2 \, d\omega \tag{9}$$

provide some information characteristics of the signal.

Analogous approach can be applied also for soliton solutions of nonlinear equations [18, 23].

From mathematical point of view, there exists correlation of entropies S_x and S_p (S_t and S_ω), since the function $\psi(x)$ $[f(t)]$ determines the Fourier component $\widetilde{\psi}(p)$ $[\widetilde{f}(\omega)]$. This means that entropies S_x and S_p have to obey some constrains.

These constrains are well-known entropic uncertainty relations (some inequalities). For one-mode system, the inequalities read (see, e.g., [8], p. 28)

$$S_x + S_p \geq \ln(\pi e), \qquad S_t + S_\omega \geq \ln(\pi e). \tag{10}$$

For multimode quantum systems (or multicomponent signals), the entropy uncertainty relation reads

$$S_{\vec{x}} + S_{\vec{p}} \geq N \ln(\pi e), \tag{11}$$

where N is the number of degrees of freedom of the system and

$$S_{\vec{x}} = -\int |\psi(\vec{x})|^2 \ln |\psi(\vec{x})|^2 \, d\vec{x},$$
$$S_{\vec{p}} = -\int |\widetilde{\psi}(\vec{p})|^2 \ln |\widetilde{\psi}(\vec{p})|^2 \, d\vec{p}. \tag{12}$$

The functions $\psi(\vec{x})$ and $\widetilde{\psi}(\vec{p})$ are connected by the Fourier transform

$$\widetilde{\psi}(\vec{p}) = (2\pi)^{-N/2} \int \psi(\vec{x}) e^{-i\vec{p}\vec{x}} \, d\vec{x}. \tag{13}$$

For the Gaussian wave function corresponding to factorized squeezed and noncorrelated state of several modes, we have [8]

$$S_{\vec{x}} + S_{\vec{p}} = N \ln(\pi e). \tag{14}$$

SYMPLECTIC TOMOGRAPHY AND TOMOGRAPHIC ENTROPIES

There exists an invertible map, which is the tomographic map of the wave function $\psi(x)$. In terms of the density operator (matrix) $\hat{\rho}$, the symplectic tomogram, which is the probability-density of random position X, can be presented in the form

$$w(X, \mu, \nu) = \mathrm{Tr}\hat{\rho}\ \delta(X - \mu\hat{q} - \nu\hat{p}), \tag{15}$$

where $\hat{\rho}$ is the density operator, μ and ν are real parameters, $\hat{q}$ and $\hat{p}$ are the position and momentum operators, respectively.

For the pure state $\hat{\rho}_\psi = | \psi\rangle\langle\psi |$, transform (15) yields

$$w(X, \mu, \nu) = \frac{1}{2\pi|\nu|} \left| \int \psi(y) \exp\left(\frac{i\mu}{2\nu} y^2 - \frac{iX}{\nu} y \right) dy \right|^2. \tag{16}$$

The function $w(X, \mu, \nu)$ (the state tomogram) is the position-probability density of the position X, i.e., it is nonnegative $w(X, \mu, \nu) \geq 0$ and normalized

$$\int w(X, \mu, \nu) \, dX = 1. \tag{17}$$

Since the symplectic tomogram is the standard probability distribution, one can introduce entropy associated with tomogram of quantum state [11] or with tomogram of analytic signal [20]. Thus one has entropy as the function of two real variables

$$S(\mu, \nu) = -\int w(X, \mu, \nu) \ln w(X, \mu, \nu) \, dX. \tag{18}$$

We call this entropy the symplectic entropy.

In view of the homogeneity and normalization conditions for tomogram, one has the additivity property for entropy

$$S(\lambda\mu, \lambda\nu) = S(\mu, \nu) + \ln|\lambda|. \tag{19}$$

For the pure state $| \psi\rangle$, in view of the relations

$$w(X, 1, 0) = |\psi(X)|^2 \qquad \text{and} \qquad w(X, 0, 1) = |\widetilde{\psi}(X)|^2,$$

one obtains the entropies S_x and S_p by posing in (18) $\mu = 1$ and $\nu = 0$

$$S(1, 0) = S_x, \tag{20}$$

or $\mu = 0$ and $\nu = 1$

$$S(0, 1) = S_p. \tag{21}$$

In view of inequality (10), one has the inequality for tomographic entropies

$$S(1, 0) + S(0, 1) \geq \ln(\pi e). \tag{22}$$

For the multimode system, the symplectic entropy reads

$$S(\vec{\mu},\vec{\nu}) = -\int w(\vec{X},\vec{\mu},\vec{\nu})\ln w(\vec{X},\vec{\mu},\vec{\nu})\,d\vec{X}. \tag{23}$$

Since the symplectic entropy is related to entropies $S_{\vec{x}}$ and $S_{\vec{p}}$ of the multimode-system state, one can use inequality (11) to obtain the entropic uncertainty relation in the form of inequality for symplectic entropies

$$S(\vec{1},\vec{0}) + S(\vec{0},\vec{1}) \geq N\ln(\pi e), \tag{24}$$

where $\vec{\mu} = \vec{1} = (1,1,\ldots,1)$ and $\vec{\nu} = \vec{1} = (1,1,\ldots,1)$.

For the optical tomogram [24], one has [25]

$$w(X,t) = \left|\int \psi(y)\exp\left[\frac{i}{2}\left(\cot t\,(y^2+X^2) - \frac{2X}{\sin t}y\right)\right]\frac{dy}{\sqrt{2\pi i\sin t}}\right|^2, \tag{25}$$

and entropy is defined by the formula

$$S(t) = -\int w(X,t)\ln w(X,t)\,dX. \tag{26}$$

For the pure state,

$$S(0) = S_x \tag{27}$$

$$S(\pi/2) = S_p. \tag{28}$$

In view of the expression of tomogram in terms of the wave function (25) and the relation

$$\psi(X,t) = \frac{1}{\sqrt{2\pi i\sin t}}\int \exp\left[\frac{i}{2}\left(\cot t\,(y^2+X^2) - \frac{2X}{\sin t}y\right)\right]\psi(y)\,dy, \tag{29}$$

one has the entropic uncertainty relation in the form

$$S(t) + S(t+\pi/2) \geq \ln \pi e. \tag{30}$$

In relation (29), the kernel of the integral transform coincides with the Green function of the quantum harmonic oscillator. Thus we extended the entropic uncertainty relation to an arbitrary value of the parameter t using the interpretation of symplectic tomogram as modulus squared [see (25)] of the wave function of an "artificial harmonic oscillator."

INEQUALITIES WITH EXTRA PARAMETERS FOR TOMOGRAMS

In recent paper [16], the new uncertainty relation was obtained for Rènyi entropy [2] related to the probability distributions for position and momentum of the quantum state with the density operator $\hat{\rho}$. Using the same argument that we employed to obtain inequality (30) for Shannon entropies, we arrive at the condition for optical tomogram [19, 23]

$$\begin{aligned}&(q-1)\ln\left\{\int_{-\infty}^{\infty} dX\,[w(X,\theta+\pi/2)]^{1/(1-q)}\right\} + (q+1)\ln\left\{\int_{-\infty}^{\infty} dX[w(X,\theta)]^{1/(1+q)}\right\}\\ &\geq (1/2)\Big\{(q-1)\ln[\pi(1-q)] + (q+1)\ln[\pi(1+q)]\Big\},\end{aligned} \tag{31}$$

where the parameter q is a real number.

We used in Eq. (31) the angle parameter θ instead of the notation t used in the previous sections, since the notation θ is the standard one for optical tomography.

Now we present a new inequality for symplectic tomogram in the one-mode case:

$$(q-1)\ln\left\{\int_{-\infty}^{\infty}dX\left[w\left(X,-\sqrt{\mu^2+\nu^2}\sin\theta,\sqrt{\mu^2+\nu^2}\cos\theta\right)\right]^{1/(1-q)}\right\}$$
$$+(q+1)\ln\left\{\int_{-\infty}^{\infty}dX\left[w\left(X,\sqrt{\mu^2+\nu^2}\cos\theta,\sqrt{\mu^2+\nu^2}\sin\theta\right)\right]^{1/(1+q)}\right\}$$
$$\geq\frac{1}{2}\Big\{(q-1)\ln[\pi(1-q)]+(q+1)\ln[\pi(1+q)]\Big\}. \tag{32}$$

This inequality can be interpreted as a generalization of inequality (31) extended from optical tomogram to symplectic tomogram.

In view of the inequality for Rènyi entropy adopted from [16], the condition (32) for tomogram of the quantum state can be generalized for the multimode case as well.

Then for symplectic tomogram of the quantum state with the density operator $\hat{\rho}$ defined as

$$w(X_1,X_2,\ldots,X_N,\mu_1,\mu_2,\ldots,\mu_N,\nu_1,\nu_2,\ldots,\nu_N)$$
$$=\mathrm{Tr}\left[\hat{\rho}\,\delta\left(X_1-\mu_1\hat{q}_1-\nu_1\hat{p}_1\right)\delta\left(X_2-\mu_2\hat{q}_2-\nu_2\hat{p}_2\right)\cdots\delta\left(X_N-\mu_N\hat{q}_N-\nu_N\hat{p}_N\right)\right], \tag{33}$$

where $\hat{q}_k$ and $\hat{p}_k$ $(k=1,2,\ldots,N)$ are the position and momentum operators, respectively, we obtain

$$(q-1)\ln\left\{\int_{-\infty}^{\infty}d\vec{X}\left[w\left(X_1,X_2,\ldots,X_N,-\sqrt{\mu_1^2+\nu_1^2}\sin\theta_1,\ldots,\right.\right.\right.$$
$$\left.\left.\left.-\sqrt{\mu_N^2+\nu_N^2}\sin\theta_N,\sqrt{\mu_1^2+\nu_1^2}\cos\theta_1,\ldots,\sqrt{\mu_N^2+\nu_N^2}\cos\theta_N\right)\right]^{1/(1-q)}\right\}$$
$$+(q+1)\ln\left\{\int_{-\infty}^{\infty}d\vec{X}\left[w\left(X_1,X_2,\ldots,X_N,\sqrt{\mu_1^2+\nu_1^2}\cos\theta_1,\ldots,\right.\right.\right.$$
$$\left.\left.\left.\sqrt{\mu_N^2+\nu_N^2}\cos\theta_N,\sqrt{\mu_1^2+\nu_1^2}\sin\theta_1,\ldots,\sqrt{\mu_N^2+\nu_N^2}\sin\theta_N\right)\right]^{1/(1+q)}\right\}$$
$$\geq\frac{N}{2}\Big\{(q-1)\ln[\pi(1-q)]+(q+1)\ln[\pi(1+q)]\Big\}. \tag{34}$$

For optical tomogram $w(X_1,\ldots,X_N,\theta_1,\ldots,\theta_N)$, the inequality reads

$$(q-1)\ln\left\{\int_{-\infty}^{\infty}d\vec{X}\,[w(X_1,X_2,\ldots,X_N,\theta_1,\ldots,\theta_N)]^{1/(1-q)}\right\}$$
$$+(q+1)\ln\left\{\int_{-\infty}^{\infty}d\vec{X}\,[w(X_1,X_2,\ldots,X_N,\theta_1+\pi/2,\ldots,\theta_N+\pi/2)]^{1/(1+q)}\right\}$$
$$\geq\frac{N}{2}\Big\{(q-1)\ln[\pi(1-q)]+(q+1)\ln[\pi(1+q)]\Big\}. \tag{35}$$

Inequalities (34) and (35) saturate for Gaussian tomograms of coherent states.

We introduce the function $R(\theta,q)$ which is equal to the difference of left- and right-hand sides of Eq. (31)

$$\begin{aligned} R(\theta,q) \;=\;& (q-1)\ln\left\{\int_{-\infty}^{\infty}dX\left[w\left(X,\theta+\frac{\pi}{2}\right)\right]^{1/(1-q)}\right\} \\ & +(q+1)\ln\left\{\int_{-\infty}^{\infty}dX\,[w(X,\theta)]^{1/(1+q)}\right\} \\ & -\frac{1}{2}\Big\{(q-1)\ln[\pi(1-q)]+(q+1)\ln[\pi(1+q)]\Big\}. \end{aligned} \tag{36}$$

This function must be nonnegative for any tomogram describing a quantum state.

The inequality $R(\theta,q)\geq 0$ was checked numerically for some tomograms [18, 23].

SPIN ENTROPIC INEQUALITIES

For any given quantum state of two spins with the density operator $\hat{\rho}$, the entropic characteristics is the von Neumann entropy

$$S = -\mathrm{Tr}\,\hat{\rho}\ln\hat{\rho}. \tag{37}$$

For the composite systems, there exist known inequalities for the von Neumann entropy.

For a two-partite system with the density operator $\hat{\rho}(1,2)$ depending on variables of the first subsystem 1 and the second subsystem 2, there exists the inequality called the subadditivity

$$S_{12} \leq S_1 + S_2, \tag{38}$$

where

$$S_{12} = -\mathrm{Tr}\,\hat{\rho}(1,2)\ln\hat{\rho}(1,2), \tag{39}$$
$$S_1 = -\mathrm{Tr}\,\hat{\rho}(1)\ln\hat{\rho}(1), \tag{40}$$
$$S_2 = -\mathrm{Tr}\,\hat{\rho}(2)\ln\hat{\rho}(2). \tag{41}$$

The operators $\hat{\rho}(1)$ and $\hat{\rho}(2)$ are the density operators of two subsystems, respectively.

Now we present new tomographic entropic inequalities.

Consider a two-partite system with the density matrix ρ_{12} and the tomographic-probability distribution for two spin projections m_1 and m_2

$$w(m_1,m_2,u) = \langle m_1m_2 \mid u^\dagger\rho_{12}u \mid m_1m_2\rangle, \tag{42}$$

u being a $(2j_1+1)(2j_2+1)\times(2j_1+1)(2j_2+1)$ unitary matrix. The Shannon entropy is defined by tomographic probability distributions

$$H_u(12) = -\sum_{m_1m_2} w(m_1,m_2,u)\ln w(m_1,m_2,u). \tag{43}$$

From the tomographic probability for the first and second spin projections

$$w(m_1,u) = \sum_{m_2} w(m_1,m_2,u), \qquad w(m_2,u) = \sum_{m_1} w(m_1,m_2,u), \tag{44}$$

one writes the tomographic entropies

$$H_u(1) = -\sum_{m_1} w(m_1,u)\ln w(m_1,u), \qquad H_u(2) = -\sum_{m_2} w(m_2,u)\ln w(m_2,u). \tag{45}$$

For each fixed u, tomographic probabilities are ordinary probability distributions.

For spin system with two spins j_1 and j_2, one can construct the spin tomogram [26, 27], which completely determines the density operator of the state. For two spins, the tomographic probability can be given as diagonal matrix elements in unitarily rotated basis.

In classical probability theory, Shannon entropies associated with joint probability distributions of bipartite and tripartite systems satisfy known inequalities called subadditivity conditions and strong subadditivity conditions. We can adopt these inequalities and write them for quantum spin systems.

Therefore, by the subadditivity of classical entropy, one has the inequality

$$H_u(12) \leq H_u(1) + H_u(2).$$

This new inequality is a tomographic counterpart of the subadditivity condition for the von Neumann entropy of the bipartite quantum system.

Strong subadditivity existing in classical probability theory also holds for any u for the tripartite system

$$H_u(123) + H_u(2) \leq H_u(12) + H_u(23), \tag{46}$$

where the Shannon entropy $H_u(123)$ is determined by the standard formula via the tomographic-probability distribution

$$w(m_1,m_2,m_3,u) = \langle m_1m_2m_3 \mid u\rho(1,2,3)u^\dagger \mid m_1m_2m_3\rangle,$$

with the density matrix $\rho(1,2,3)$ of three spin systems with spins j_1, j_2, and j_3.

The probability distribution depends on unitary matrix $u(N)$ with dimensionality $N = (2j_1+1)(2j_2+1)(2j_3+1)$. The other entropies $H_u(2)$, $H_u(1,2)$, and $H_u(2,3)$ are calculated via marginal distributions for spin subsystems.

Equation (46) is a tomographic counterpart of the known strong subadditivity which exists for the von Neumann entropy (37) of the tripartite spin system, i.e.,

$$S(123)+S(2) \leq S(12)+S(23). \quad (47)$$

Therefore, strong subadditivity for the von Neumann entropy (47) and the new strong subadditivity for the tomographic entropies (46) are the properties of all quantum tripartite spin states.

CONCLUSIONS

To conclude, we summarize the main results of our study.

The new uncertainty relations were reviewed within the framework of the probability representation of quantum mechanics. The entropic uncertainty relations have the form of integral condition for tomograms of quantum states which contain the complete information on the states. The new inequality with extra parameter was obtained for some integral expressions containing the quantum-state tomograms on the base of recently found uncertainty relations [16] for Rènyi entropy of quantum states.

The conditions for the one-mode and multimode optical tomograms are of particular interest, since these tomograms are directly measured in quantum-optics experiments [24]. Obtained tomographic entropic uncertainty relations written in the form of integral inequalities were discussed and their role for controlling the accuracy of homodyne measuring the quantum state based on optical tomography of the Wigner function was pointed out in [19].

The new subadditivity and strong subadditivity conditions for spin tomographic probability distributions have been formulated.

ACKNOWLEDGMENTS

This study was supported by the Russian Foundation for Basic Research under Project No. 07-02-00598. The author thanks the Organizers of the International conference "QTRF-4 – Quantum Theory: Reconsiderations and Foundations-4" (Vaxjo, Sweden, June 2007) for invitation and kind hospitality.

REFERENCES

1. C. E. Shannon, *B*ell. Tech. J., **27**, 379 (1948).
2. A. Rènyi, *Probability Theory*, North-Holland, Amsterdam, 1970.
3. E. Schrödinger, "Der stertige Übergang von der Mikro- zur Markomechanik," *Naturwissenschaften*, **14**, 664–666 (1926) ["The continuous transition from micro- to makromechanics," in E. Schrödinger, *Collected papers on Wave Mechanics*, the second edition, Chelsea Publ. Corp., New York, 1978, pp. 41–44].
4. L. Landau, "Das Dämpfungsproblem in der Wellenmechanik," *Z. Phys.*, **45**, 430–441 (1927) ["The damping problem in wave mechanics," in *Collected Papers of L. D. Landau*, edited by D. Ter Haar, Gordon & Breach, New York, 1965, pp. 8–18].
5. J. von Neumann, *Göttingenische Nachrichten*, **11**, S. 245 (November 1927).
6. J. von Neumann, *Mathematische Grundlagen der Quantummechanik*, Springer, Berlin, 1932.
7. I. I. Hirschman, "A note on entropy," *Amer. J. Math.*, **79**, 152–156 (1957).
8. V. V. Dodonov, and V. I. Man'ko, *Invariants and the Evolution of Nonstationary Quantum Systems, Proceedings of the P. N. Lebedev Physical Institute*, Nauka, Moscow, 1987, Vol. 183 [translated by Nova Science, New York, 1989].
9. S. Mancini, V. I. Man'ko, and P. Tombesi, "Wigner function and probability distribution for shifted and squeezed quadratures," *Quantum Semiclass. Opt.*, **7**, 615–623 (1995).
10. G. M. D'Ariano, S. Mancini, V. I. Man'ko, and P. Tombesi, "Reconstructing the density operator by using generalized field quadratures," *Quantum Semiclass. Opt.*, **8**, 1017–1028 (1996).
11. O. V. Man'ko, and V. I. Man'ko, "Quantum states in probability representation and tomography," *J. Russ. Laser Res.*, **18**, 407–444 (1997).
12. M. A. Man'ko, "Noncommutative tomography of an analytic signal and entanglement in the probability representation of quantum mechanics," *J. Russ. Laser Res.*, **22**, 168–174 (2001).
13. M. A. Man'ko, "Fractional Fourier transform in information processing, tomography of optical signal, and Green function of harmonic oscillator," *J. Russ. Laser Res.*, **21**, 411–437 (2000).

14. M. A. Man'ko, V. I. Man'ko, and R. Vilela Mendes, "A probabilistic operator symbol framework for quantum information," *J. Russ. Laser Res.*, **27**, 507–532 (2006) [ArXiv quant-ph/0602189 v1].
15. O. V. Man'ko, and V. I. Man'ko, "Probability-Representation Entropy for Spin-State Tomogram," *J. Russ. Laser Res.*, **25**, 115–122 (2004) [ArXiv quant-ph/041131].
16. I. Bialynicki-Birula, "Formulation of the uncertainty relations in terms of the Rényi entropies," *Phys. Rev. A*, **74**, 052101 (2006) [ArXiv quant-ph/0608116 v1].
17. M. A. Man'ko, O. V. Man'ko, and V. I. Man'ko, "Probabilic representation of optical signal and new entropy of quantum states," *Int. J. Mod. Phys. B*, **20**, 1399–1407 (2006).
18. S. De Nicola, R. Fedele, M. A. Man'ko, and V. I. Man'ko, "New uncertainty relation for tomographic entropy: Application to squeezed states and solitons," *Eur. Phys. J. B*, **52** 191–198 (2006) [ArXiv quant-ph/0607200v1].
19. S. De Nicola, R. Fedele, M. A. Man'ko, and V. I. Man'ko, "New inequalities for tomograms in the probability representation of quantum mechanics"*Teor. Mat. Fiz.*, **152**, 225–231 (2007) [*Theor. Math. Phys.*, **152**, 1069–1074 (2007)], [ArXiv quant-ph/0611114 v1].
20. M. A. Man'ko, "Entropy of an analytic signal," *J. Russ. Laser Res.*, **27**, 405–413 (2006).
21. M. A. Man'ko, V. I. Man'ko, S. De Nicola, and R. Fedele, "Probability representation and new entropic uncertainty relations for symplectic and optical tomograms" *Acta Phys. Hung. B*, **26**, 71–77 (2006).
22. M. A. Man'ko, "Linear and nonlinear optical signals in probability and phase-space representations," *J. Phys. Conf. Ser.*, **36**, 126-136 (2006).
23. S. De Nicola, R. Fedele, M. A. Man'ko, and V. I. Man'ko, "Symplectic entropy" *J. Phys. Conf. Ser.*, **70**, 012007 (2007).
24. D. T. Smithey, M. Beck, M. G. Raymer, and A. Faridani, "Measurement of the Wigner distribution and the density matrix of a light mode using optical homodyne tomography: Application to squeezed states and the vacuum," *Phys. Rev. Lett.*, **70**, 1244–1247 (1993).
25. V. I. Man'ko and R. V. Mendes, "Noncommutative time–frequency tomography," *Phys. Lett. A*, **263**, 53–56 (1999) [ArXiv Physica/9712022 Data Analysis, Statistics, and Probability].
26. V. V. Dodonov and V. I. Man'ko, "Positive distribution description for spin states," *Phys. Lett.* A, **239**, 335–339 (1997).
27. V. I. Man'ko and O. V. Man'ko, "Spin state tomography," *Zh. Éksp. Teor. Fiz*, **112**, 796–800 (1997) [*J. Exp. Theor. Phys.*, **85**, 430–434 (1997)].

Probability Instead of Wave Function and Bell Inequalities as Entanglement Criterion

Vladimir I. Man'ko

P. N. Lebedev Physical Institute, Leninskii Prospect 53, Moscow 119991, Russia

Abstract. Review of the probability representation of quantum mechanics, in which a fair probability distribution is used instead of the density matrix or wave function to describe quantum states is presented. Bell inequalities are formulated as properties of joint probability distributions describing the states of multipartite spin systems in the probability representation. Probability distributions depending on extra parameters are discussed in the connection with problems of classical and quantum statistical mechanics, where these distributions can be considered in the context of non-Kolmogorov probability theory.

Keywords: probability distributions, tomograms, Radon transform, Bell's inequalities, spin states, entanglement, wave function, density matrix.
PACS: 42.50.Ct, 42.50.Ex, 02.50.Cw

INTRODUCTION

The probability theory or Kolmogorov probability theory [1] (see [2]) is a basic mathematical tool to formulate many problems of both classical and quantum statistical mechanics. In classical statistics of systems with fluctuations, the probability-distribution density on the phase space of the system is the object associated with the notion of "state" of the system under consideration. The notion of possible correlations of different physical observables in multipartite systems is usually described in the standard probability theory by means of joint probability distributions associated to the states of classical systems. In conventional quantum mechanics and quantum statistical mechanics as well as kinetics, the system states are described by the wave function [3] or density operator [4–6]. The probability distribution appear either as modulus squared of the wave functions (for pure states) or as diagonal matrix elements of the density matrix of the system states (for mixed states). The observable means and correlations are calculated within the framework of quantum mechanics, in view of the rules dealing with the Hermitian operators – physical observables and density operators describing the quantum states; these rules are quite different from the rules of standard probability theory. In this connection, the approach with quantum probability was suggested [7] and consideration of non-Kolmogorov probability theory has been emphasized [8].

The probability approach used in quantum mechanics to describe the oscillator's quantum states [9] or spin system [10, 11] depend not only on a random variables but also on extra parameters. Such probability distributions are used also in classical statistics. For example, the Maxwell's velocity distribution of molecules in a gas at temperature T depends on the random fluctuating variable (velocity) and two extra parameters; one of them is temperature and the other one is the mass of the molecule. Thus one has the probability-distribution function

$$f(v,T,m) = \sqrt{\frac{m}{2\pi T}}\, e^{-mv^2/2T}$$

(we take the Boltzman constant $k = 1$). In fact, this function is the homogeneous function of the parameters since $f(v,\lambda T,\lambda m) = f(v,T,m)$.

Our aim here is to review the approach called the probability representation of quantum mechanics, in which the quantum states are described by the probability distributions depending on extra parameters analogously to the case of Maxwell's velocity distributions of classical statistics. The other goal of this study is to discuss the entanglement phenomenon and Bell's inequalities [12] within the framework of the probability representation of spin states [13, 14] using the standard probability distributions for spin projections, which contain the same information on the spin-system states that the density matrix or the wave function has.

The paper is organized as follows.

CP962, *Quantum Theory, Reconsideration of Foundations – 4*,
edited by G. Adenier, A. Yu. Khrennikov, P. Lahti, V. I. Man'ko, and Th. M. Nieuwenhuizen

In Sec. 2 we discuss the probability distribution in the phase space of classical particle and its Radon transform [15], which is called tomogram of the distribution. In Sec. 3 we consider Radon transform of the Wigner function [16] of quantum particle, which is a fair probability distribution; it is called symplectic tomogram. In Sec. 4 we study the spin tomography and discuss Bell's inequalities [17] and in Sec. 5 we present conclusions and perspectives.

CLASSICAL TOMOGRAMS

Let us have a classical particle with fluctuating position q and momentum p. The state of such a particle is described by the probability density $f(q,p) \geq 0$, which is normalized

$$\int f(q,p)\,dq\,dp = 1. \tag{1}$$

There exist two marginal probability distributions, one for the position

$$P(q) = \int f(q,p)\,dp \tag{2}$$

and the other one for the momentum of the particle

$$\mathscr{P}(p) = \int f(q,p)\,dq. \tag{3}$$

One can consider [18] the marginal probability distribution $w(X,\mu,\nu)$ of the position X measured in a rotated and scaled reference frame in the phase space labeled by two real parameters $\mu = s\cos\theta$ and $\nu = s^{-1}\sin\theta$ where angle θ determines the rotation and s is a scaling parameter. This probability distribution is expressed in terms of the function $f(q,p)$ as follows:

$$w(X,\mu,\nu) = \int f(q,p)\delta(X-\mu q-\nu p)\,dq\,dp. \tag{4}$$

By construction, (2) and (3) are the partial case of (4)

$$P(q) = w(X=q,1,0), \qquad \mathscr{P}(p) = w(X=p,0,1). \tag{5}$$

Formula (4) is the known Radon integral transform; its inverse reads

$$f(q,p) = \int w(X,\mu,\nu)e^{i(X-\mu q-\nu p)}\frac{dX\,d\mu\,d\nu}{(2\pi)^2}. \tag{6}$$

Thus the marginal probability distribution $w(X,\mu,\nu)$ is the "Radon component" in the decomposition of the probability density $f(q,p)$ into the Radon integral. The Radon component is the homogeneous function

$$w(\lambda X,\lambda\mu,\lambda\nu) = \frac{1}{|\lambda|}w(X,\mu,\nu). \tag{7}$$

The Radon component $w(X,\mu,\nu)$ (called also symplectic tomogram) contains the same information on the state of a particle with fluctuating position and momentum that the probability distribution function $f(q,p)$ on the phase space of the classical particle. On the other hand, tomogram is fair probability distribution of random position X; it is nonnegative function and it is normalized for any value of parameters μ and ν, i.e.,

$$\int w(X,\mu,\nu)\,dX = 1. \tag{8}$$

The Radon transform provides the invertible map

$$f(q,p) \Longleftrightarrow w(X,\mu,\nu) \tag{9}$$

of two probability distributions. One of the distributions (tomogram) depends on extra parameters μ and ν analogously to Maxwell's velocity distribution for molecules in a gas. Thus one has the description of the particle classical state either by a single probability distribution $f(q,p)$ or by the family of the distributions $w(X,\mu,\nu)$ labeled by the parameters μ and ν.

One can write the Liouville kinetic equation either for the probability density $f(q,p,t)$ in the phase space of a particle with mass $m=1$

$$\frac{\partial f(q,p,t)}{\partial t}+p\frac{\partial f(q,p,t)}{\partial q}-\frac{\partial U(q)}{\partial q}\frac{\partial f(q,p,t)}{\partial p}=0, \tag{10}$$

where $U(q)$ is the potential energy, or for the Radon component $w(X,\mu,\nu,t)$, as follows:

$$\frac{\partial w(X,\mu,\nu,t)}{\partial t}-\mu\frac{\partial w(X,\mu,\nu,t)}{\partial \nu}-\nu\frac{\partial}{\partial X}\left(\frac{\partial}{\partial \tilde{q}}U(\tilde{q})\right)w(X,\mu,\nu,t)=0, \tag{11}$$

with

$$\tilde{q}=-\left(\frac{\partial}{\partial X}\right)^{-1}\frac{\partial}{\partial \mu}. \tag{12}$$

Thus, one can write for a classical particle the Liouville kinetic equation in two equivalent forms — one is the standard evolution equation for the probability distribution $f(q,p,t)$ in the phase space and the other one, the evolution equation for the marginal probability distribution (Radon component of the distribution) $w(X,\mu,\nu,t)$ of random position X depending on parameters μ and ν labeling the reference frame in the phase space where the position is measured.

Equation (11) is integro-differential equation containing the inverse of derivative $(\partial/\partial X)^{-1}$, which is the standard integration defined as follows:

$$\left(\frac{\partial}{\partial X}\right)^{-1}f(X)=\int_{-\infty}^{X}f(y)\,dy. \tag{13}$$

One has

$$\frac{\partial}{\partial X}\left(\frac{\partial}{\partial X}\right)^{-1}=1. \tag{14}$$

QUANTUM PARTICLE

In the conventional representation, the quantum particle state is described by the density matrix $\rho(x,x',t)$ or the wave function $\psi(x,t)$.

The density matrix in the position representation satisfies the von Neumann evolution equation

$$i\frac{\partial \rho(x,x',t)}{\partial t}=-\frac{1}{2}\frac{\partial^2\rho(x,x',t)}{\partial x^2}+\frac{1}{2}\frac{\partial^2\rho(x,x',t)}{\partial x'^2}+U(x)\rho(x,x',t)-U(x')\rho(x,x',t), \tag{15}$$

where we take the Planck's constant $\hbar=1$ and the particle mass $m=1$.

The Wigner function is defined as Fourier transform of the density matrix

$$W(q,p,t)=\int\rho\left(q+\frac{u}{2},q-\frac{u}{2},t\right)e^{-ipu}du. \tag{16}$$

The Wigner function is known to be an analog of the probability distribution $f(q,p,t)$ of classical particle. The Radon component of the Wigner function

$$w(X,\mu,\nu,t)=\int W(q,p,t)\delta(X-\mu q-\nu p)\frac{dq\,dp}{2\pi} \tag{17}$$

called symplectic tomogram is the standard nonnegative probability density [19] which is normalized

$$\int w(X,\mu,\nu,t)dX=1 \tag{18}$$

for all values of parameters μ and ν.

In invariant form, the invertible relation of the density operator $\hat{\rho}$ of a particle state with symplectic tomogram $w(X,\mu,\nu)$ reads [20]

$$w(X,\mu,\nu) = \mathrm{Tr}\Big(\hat{\rho}\,\delta\left(X - \mu\hat{q} - \nu\hat{p}\right)\Big) \tag{19}$$

and [21]

$$\hat{\rho} = \frac{1}{2\pi}\int w(X,\mu,\nu)e^{i(X-\mu\hat{q}-\nu\hat{p})}, \tag{20}$$

where $\hat{q}$ and $\hat{p}$ are operators of position and momentum, respectively, and the notation X means $X\,\hat{1}$, with $\hat{1}$ being unity operator.

These relations were considered within the framework of star-product quantization scheme in [22]. Relations (19) and (20) are quantum generalization of the classical Radon transform of the probability-distribution function $f(q,p)$. In fact, Eq. (4) can be written in the form

$$w(X,\mu,\nu) = \langle\delta\left(X - \mu q - \nu p\right)\rangle, \tag{21}$$

where $\langle\rangle$ means mathematical expectation.

Comparing relations (19) and (20) with (21) and (6), one can see that (19) and (20) are obtained from classical Radon formulas by "quantization" i.e., replacing $q \rightarrow \hat{q}$ and $p \rightarrow \hat{p}$ in classical formulas. The quantum symplectic tomogram (19) satisfies the homogeneity property (7).

Thus we shown that in both classical and quantum domains the state of a particle with fluctuating position and momentum (including quantum fluctuations in quantum domain) is described by nonnegative probability distribution $w(X,\mu,\nu,t)$, which obeys the evolution equation (11) in the classical case and

$$\frac{\partial w(X,\mu,\nu,t)}{\partial t} - \mu\frac{\partial w(X,\mu,\nu,t)}{\partial \nu}$$
$$= -i\left[U\left(-\frac{\partial}{\partial\mu}\left(\frac{\partial}{\partial X}\right)^{-1} + \frac{i}{2}\nu\frac{\partial}{\partial X}\right) - U\left(-\frac{\partial}{\partial\mu}\left(\frac{\partial}{\partial X}\right)^{-1} - \frac{i}{2}\nu\frac{\partial}{\partial X}\right)\right]w(X,\mu,\nu,t) \tag{22}$$

in the quantum case.

The decomposition of the right-hand side of (22) into series in powers of parameters ν provides both the classical equation (11), if one keeps term ν only and all the quantum corrections to the classical equation. If one reconstructs dimensional units, this series is equivalent to the series in powers of the Planck's constant $\hbar$. In fact, the linear equation (22) is the evolution equation for Radon component of the Wigner function; it can be obtained by means of the Radon transform the well-known evolution equation for this function obtained by Moyal [23]. The essential difference with the Moyal equation is that Eq. (22) is written for the standard probability distribution similar to equations of classical kinetic theory. From the previous discussion of the properties of symplectic tomograms, it follows that solutions of integro-differential equation (22), if one chooses the initial functions $w(X,\mu,\nu,0)$ to be nonnegative symplectic tomograms for any time t are real nonnegative normalized functions satisfying the homogeneity condition (7).

SPIN STATES AND BELL INEQUALITIES IN THE PROBABILITY REPRESENTATION

Spinors and density matrices of spin states can be mapped onto standard probability distributions called spin tomograms and depending on some extra parameters; this map which can be considered as an analog of Radon transform for discrete spin variables is invertible [11]. Given the density matrix of j-spin particle (or qudit) the spin tomogram is defined as diagonal matrix elements of "rotated" density matrix in standard spin basis $\mid m\rangle$, where m is spin projection on the z axis, i.e.,

$$w(m,\vec{n}) = \langle m \mid u^{\dagger}(\vec{n})\rho u(\vec{n}) \mid m\rangle, \tag{23}$$

where unit vector $\vec{n} = (\sin\theta\cos\varphi, \sin\theta\sin\varphi, \cos\theta)$, angles θ and φ are the standard Euler angles, the third one is taken to be zero, and $u(\vec{n})$ is the standard unitary representation matrix of the group $SU(2)$ depending on the Euler angles.

By construction, the spin tomogram is normalized probability distribution satisfying the condition

$$\sum_{m=-j}^{j} w(m,\vec{n}) = 1. \tag{24}$$

Thus the spin state is associated with a family of probability distributions labeled by two parameters θ and φ which determine the reference frame where we measure random spin projection in complete analogy with symplectic tomogram considered in Secs. 1 and 2.

For a state with two spins j_1 and j_2 with the density matrix $\rho(1,2)$, tomogram

$$w(m_1,m_2,\vec{n}_1,\vec{n}_2) = \langle m_1m_2 \mid u_1^\dagger(\vec{n}_1) \otimes u_2^\dagger(\vec{n}_2)\rho(1,2)u_1(\vec{n}_1) \otimes u_2(\vec{n}_2) \mid m_1m_2\rangle \tag{25}$$

is joint probability distribution of two random spin projections m_1 and m_2 which depend on the Euler angles θ_1, φ_1, θ_2, and φ_2 determining the two directions $\vec{n}_1 = (\sin\theta_1\cos\varphi_1, \sin\theta_1\sin\varphi_1, \cos\theta_1)$ and $\vec{n}_2 = (\sin\theta_2\cos\varphi_2, \sin\theta_2\sin\varphi_2, \cos\theta_2)$. The matrices $u_1(\vec{n}_1)$ and $u_2(\vec{n}_2)$ are standard unitary matrices of irreducible representations of $SU(2)$ group for spin j_1 and j_2, respectively. There exist inverse formulas (see, for example, [11, 13]) for density matrices expressed in terms of spin tomograms in both cases; in view of this, we have invertible linear maps $\rho \leftrightarrow w(m,\vec{n})$ and $\rho(1,2) \leftrightarrow w(m_1,m_2,\vec{n}_1,\vec{n}_2)$ of density matrices of spin states onto standard probability distributions depending on extra parameters. For two-qudit state, the normalization condition reads

$$\sum_{m_1=-j_1}^{j_1}\sum_{m_2=-j_2}^{j_2} w(m_1,m_2,\vec{n}_1,\vec{n}_2) = 1. \tag{26}$$

The calculation of spin-projection correlations for two-qudits is given by a standard formula from probability theory

$$\langle m_1m_2\rangle = \sum_{m_1=-j_1}^{j_1}\sum_{m_2=-j_2}^{j_2} m_1m_2w(m_1,m_2,\vec{n}_1,\vec{n}_2), \tag{27}$$

in which the tomogram $w(m_1,m_2,\vec{n}_1,\vec{n}_2)$ describes the quantum state instead of the density matrix $\rho(1,2)$. The separability of quantum state $\rho(1,2)$ defined by the property of the density matrix to be presented as a convex sum of the form

$$\rho(1,2) = \sum_k p_k\rho^{(k)}(1) \otimes \rho^{(k)}(2), \qquad p_k \geq 0, \qquad \sum_k p_k = 1, \tag{28}$$

can be reformulated as the property of the joint probability distribution; namely, a state is called separable iff its tomogram can be presented in the form

$$w(m_1,m_2,\vec{n}_1,\vec{n}_2) = \sum_k p_kw_1^{(k)}(m_1,\vec{n}_1)w_2^{(k)}(m_2,\vec{n}_2), \tag{29}$$

where $w_1^{(k)}(m_1,\vec{n}_1)$ and $w_2^{(k)}(m_2,\vec{n}_2)$ are tomograms of qudits j_1 and j_2, respectively.

If the state tomogram [joint probability distribution $w(m_1,m_2,\vec{n}_1,\vec{n}_2)$] cannot be presented in form (29),the two-qudit state is called entangled. Thus in the probability representation of quantum mechanics, the problem of separability and entanglement takes the form of problem of properties of the joint-probability-distribution functions depending on extra parameters. It is worthy noting that exactly the same formulation of separability and entanglement can be given for bipartite state of the system with continuous variables. In fact, for two degrees of freedom symplectic tomogram reads

$$w(X_1,\mu_1,\nu_1,X_2,\mu_2,\nu_2) = \langle\delta(X_1-\mu_1\hat{q}_1-\nu_1\hat{p}_1)\,\delta(X_2-\mu_2\hat{q}_2-\nu_2\hat{p}_2)\rangle, \tag{30}$$

where mathematical expectation means $\langle(\ldots)\rangle = \mathrm{Tr}\left(\hat{\rho}(1,2)(\ldots)\right)$, $\hat{\rho}(1,2)$ is the density operator of bipartite system state, and μ_1,ν_1 and μ_2,ν_2 are real parameters.

Symplectic tomogram (30) is joint probability distribution of two continuous random position X_1 and X_2, which depends on real parameters μ_1,ν_1 and μ_2,ν_2. The system state is separable iff one has decomposition of the function in the form of convex sum

$$w(X_1,\mu_1,\nu_1,X_2,\mu_2,\nu_2) = \sum_k p_kw_1^{(k)}(X_1,\mu_1,\nu_1)w_2^{(k)}(X_2,\mu_2,\nu_2) \tag{31}$$

of the products of tomograms of the states of the subsystems.

Thus the problem of entanglement of the state of the system with two continuous degree of freedom is reduced to the problem of studying the properties of joint probability distribution of two random positions depending on extra parameters. One can immediately see that joint probability distributions of the separable form (29) have specific properties. The generic joint probability distribution can be considered as a point on simplex s. The product of two probabilities $w_1^{(k)}(m_1,\vec{n}_1)w_2^{(k)}(m_2,\vec{n}_2)$ is also a point on the simplex but, due to the obvious geometrical reasons, the domain occupied on the simplex s by points corresponding to the product of probabilities $w_1^{(k)}(m_1,\vec{n}_1)w_2^{(k)}(m_2,\vec{n}_2)$ and to convex sum of such products is quite different from the whole simplex. Thus, the problem of entanglement is reduced to the problem of simplex geometry. It can be formulated as follows:

Given the "small" simplex s_1 corresponding to the probability points (vectors) $w_1^{(k)}(m_1,\vec{n}_1)$ and the "small" simplex s_2 corresponding to the probability points (vectors) $w_2^{(k)}(m_2,\vec{n}_2)$. Let us construct direct product $s_1 \otimes s_2$. This product can be embedded into generic space containing probability points (vectors) $w(m_1,m_2,\vec{n}_1,\vec{n}_2)$. Thus, we address the question what is domain in s occupied by $s_1 \times s_2$? Also how one can numerically characterize the difference of the points, which belong to the rest part of the simplex s? Bell's inequalities give just such numerical characteristics. For example, the Bell–CHSH inequalities [17] can be presented as the following constrains onto the joint probability for two qubits $w(m_1,m_2,\vec{n}_1,\vec{n}_2)$ given in the invariant matrix form [24, 14]

$$\mathrm{Tr}(IM)| \leq 2, \tag{32}$$

where the matrix I reads

$$I = \begin{pmatrix} 1 & -1 & -1 & 1 \\ 1 & -1 & -1 & 1 \\ 1 & -1 & -1 & 1 \\ -1 & 1 & 1 & -1 \end{pmatrix}. \tag{33}$$

The stochastic matrix M belonging to semigroup has the form

$$I = \begin{pmatrix} w_1\left(\vec{a},\vec{b}\right) & w_1\left(\vec{a},\vec{c}\right) & w_1\left(\vec{d},\vec{b}\right) & w_1\left(\vec{d},\vec{c}\right) \\ w_2\left(\vec{a},\vec{b}\right) & w_2\left(\vec{a},\vec{c}\right) & w_2\left(\vec{d},\vec{b}\right) & w_2\left(\vec{d},\vec{c}\right) \\ w_3\left(\vec{a},\vec{b}\right) & w_3\left(\vec{a},\vec{c}\right) & w_3\left(\vec{d},\vec{b}\right) & w_3\left(\vec{d},\vec{c}\right) \\ w_4\left(\vec{a},\vec{b}\right) & w_4\left(\vec{a},\vec{c}\right) & w_4\left(\vec{d},\vec{b}\right) & w_4\left(\vec{d},\vec{c}\right) \end{pmatrix}. \tag{34}$$

The four probabilities in columns of the matrix M are probabilities $w(m_1,m_2,\vec{n}_1,\vec{n}_2)$ for system of two qubits

$$\begin{aligned} w_1(\vec{n}_1,\vec{n}_2) &= w\left(+\frac{1}{2},+\frac{1}{2},\vec{n}_1,\vec{n}_2\right), \\ w_2(\vec{n}_1,\vec{n}_2) &= w\left(+\frac{1}{2},-\frac{1}{2},\vec{n}_1,\vec{n}_2\right), \\ w_3(\vec{n}_1,\vec{n}_2) &= w\left(-\frac{1}{2},+\frac{1}{2},\vec{n}_1,\vec{n}_2\right), \\ w_4(\vec{n}_1,\vec{n}_2) &= w\left(-\frac{1}{2},-\frac{1}{2},\vec{n}_1,\vec{n}_2\right) \end{aligned}$$

and we take four pairs of different vectors $\vec{n}_1$ and $\vec{n}_2$ in each column of the matrix M. If the state is entangled, inequality (32) can be violated. The constrain (32) is one of numerical characteristics of difference between two sets $s_1 \times s_2$ and s. There are many other numerical characteristics and it could be possible that in conventional probability theory the answers to the questions formulated are known, either explicitly or implicitly. In this case, one could use such results to use Bell's inequalities which could be understood as all the set of numerical characteristics of the difference between two sets $s_1 \times s_2$ and s as a criterion of entanglement and separability of bipartite (and multipartite) quantum states. It is worthy noting that approach to Bell's inequality in the context of probability theory was discussed recently in [25].

CONCLUSIONS

To conclude, we summarize the main items of the article.

Both in classical and quantum statistical mechanics, one can describe the states of the system by families of the standard probability distributions depending on extra parameters. Since the classical or quantum states are associated not with a single probability distribution but with families of the probabilities, one can consider the state within the framework of non-Kolmogorov probability theory. The comparison of a classical particle with a quantum particle in this context shows that non-Kolmogorov probability approach is not connected with quantum physics only, but can be used also for the description of classical statistics as well.

There are essential differences in the tomographic probabilities describing the states in classical and quantum domains. The evolution equations in these two cases are different. Also the sets of admissible classical and quantum tomograms are different to satisfy the uncertainty relations (condition of positivity of density operators) in quantum domain and to satisfy the positivity of probability density in the phase space in classical domain. But from the point of view of Kolmogorov or non-Kolmogorov approaches in probability theory, the difference of classical and quantum domains is not that large.

One can see that Bell's inequalities can be naturally associated with the property of joint probability distributions describing the quantum states of spin systems without using quantum procedure of calculating mean values and correlations of random variables by means of the density-matrix formalism. The similarity of the description of classical and quantum states by means of tomographic probabilities makes obvious the possibility to use in classical statistics the quantum-like formalism for describing, for example, the state of classical harmonic oscillator by means of the density matrix or wave function (see [26].

We point out that the tomographic-probability formulation, using Radon-transform formalism, is completely equivalent to all other known formulations of quantum mechanics like Feynmann path integral approach or Moyal star-product, as well as the others.

ACKNOWLEDGMENTS

This study was supported by the Russian Foundation for Basic Research under Project No. 07-02-00598. The author thanks the Organizers of the International conference "QTRF-4 – Quantum Theory: Reconsiderations and Foundations-4" (Vaxjo, Sweden, June 2007) for invitation and kind hospitality.

REFERENCES

1. A. N. Kolmogorof, "Über die analytischen Methoden in der Wahrscheinlichtkeitsrechnung," it Math. Ann., **104**, 415–458 (1931).
2. A. N. Kolmogorof, "Logic basis for information theory and probability theory," *IEEE Trans.*, **IT-14**, 662–664 (1968).
3. E. Schrödinger, "Der stertige Übergang von der Mikro- zur Markomechanik," *Naturwissenschaften*, **14**, 664–666 (1926) ["The continuous transition from micro- to makromechanics," in E. Schrödinger, *Collected papers on Wave Mechanics*, the second edition, Chelsea Publ. Corp., New York, 1978, pp. 41–44].
4. L. Landau, "Das Dämpfungsproblem in der Wellenmechanik," *Z. Phys.*, **45**, 430–441 (1927) ["The damping problem in wave mechanics," in *Collected Papers of L. D. Landau*, edited by D. Ter Haar, Gordon & Breach, New York, 1965, pp. 8–18].
5. J. von Neumann, *Göttingenische Nachrichten*, **11**, S. 245 (November 1927).
6. J. von Neumann, *Mathematische Grundlagen der Quantummechanik*, Springer, Berlin, 1932.
7. L. Accardi, " The probabilistic roots of the quantum-mechanical paradoxes," in *The wave–particle dualism. A tribute to Louis de Broglie on his 90th Birthday*, edited by S. Dinner et al., D. Reidel Publ. Company, Dodrecht, 1970, pp. 47–55.
8. A. Yu. Khrennikov, *Non-Kolmogorov Probability Theory and Quantum Physics*, Nauka-Fizmatlit, Moscow, 2002 [in Russian].
9. S. Mancini, V. I. Man'ko, and P. Tombesi, "Symplectic tomography as classical approach to quantum systems," *Phys. Lett. A*, **213**, 1–6 (1996).
10. V. V. Dodonov, and V. I. Man'ko, "Positive distribution description for spin states," *Phys. Lett.* A, **239**, 335–339 (1997).
11. V. I. Man'ko, and O. V. Man'ko, "Spin state tomography," *Zh. Éksp. Teor. Fiz*, **112**, 796–800 (1997) [*J. Exp. Theor. Phys.*, **85**, 430–434 (1997)].
12. J. S. Bell, *Physica*, **1**, 195–200 (1964); *Speakable and Unspeakable in Quantum Mechanics*, Cambridge University Press, Cambridge (1987).
13. B. A. Andreev, O. V. Man'ko, V. I. Man'ko, and E. V. Shchukin, "Tomography of Spin States, the Entanglement Criterion, and Bell's Inequalities," *Teor. Mat. Fiz.*, **146**, 172–185 (2006) [*Theor. Math. Phys.*, **146**, 140–151 (2006)].
14. C. Lupo, V. I. Man'ko, and G. Marmo, "Qubit portrait of qudit states and quantum correlations," *J. Phys. A: Math. Theor.*, **40**, 13091–13100 (2007).
15. J. Radon, "Über die Bestimmung von Funktionen durch ihre Integralwerte längs gewisser Mannigfaltigkeiten", *Berichte Sachsische Akademie der Wissenschaften, Leipzig, Mathematische–Physikalische Klasse*, **69** (1917), S. 262.

16. E. Wigner, "On the quantum correlations for thermodynamic equilibrium," *Phys. Rev.*, **40**, 749–759 (1932).
17. J. F. Clauser, M. A. Horne, A. Shimony, and R. A. Holt, "Proposed Experiment to Test Local Hidden-Variable Theories," *Phys. Rev Lett.*, **23**, 880–884 (1969).
18. O. V. Man'ko, and V. I. Man'ko, "Quantum states in the probability representation and tomography," *J. Russ. Laser Res.*, **18**, 407–444 (1997).
19. S. Mancini, V. I. Man'ko, and P. Tombesi, "Wigner function and probability distribution for shifted and squeezed quadratures," *Quantum Semiclass. Opt.*, **7**, 615–623 (1995).
20. M. A. Man'ko, V. I. Man'ko, and R. Vilela Mendes, "Tomograms and other transforms: a unified view," *J. Phys. A: Math. Gen.*, **34**, 8321–8332 (2001).
21. G. M. D'Ariano, S. Mancini, V. I. Man'ko, and P. Tombesi, "Reconstructing the density operator by using generalized field quadratures," *Quantum Semiclass. Opt.*, **8**, 1017–1028 (1996).
22. O. V. Man'ko, V. I. Man'ko, and G. Marmo, "Alternative commutation relations, star products and tomography," *J. Phys. A: Math. Gen.*, **35**, 699–720 (2002).
23. J. E. Moyal, "Quantum mechanics as a statistical theory," *Proc. Camb. Phil. Soc.*, **45**, 99–124 (1949).
24. V. N. Chernega, and V. I. Man'ko, "Qubit portrait of qugit states and Bell inequalities," *J. Russ. Laser Res.*, **28**, 103–124 (2007).
25. A. Khrennikov, "Bell's inequality: Physics meets probability," ArXiv:0709.3909v.1 [quant-ph].
26. V. N. Chernega, and V. I. Man'ko, "Wave function of the harmonic oscillator in classical statistical mechanics," *J. Russ. Laser Res.*, **28**, iss. 6 (2007).

The Pullback Mechanism in Stochastic Electrodynamics

Th. M. Nieuwenhuizen

Institute for Theoretical Physics, Valckenierstraat 65, 1018 XE Amsterdam, The Netherlands

Abstract. An argument is given why the classical theory called Stochastic Electrodynamics may reproduce scattering and ionization experiments of electrons on atomic hydrogen.

Keywords: hydrogen atom, classical description, stochastic electrodynamics
PACS: 02.50.Ey, 03.50.DE, 03.65.Ta, 03.65.Ud

In our view, quantum theory describes an underlying reality, where particles are definite entities, subject to some still mysterious "quantum forces". Our view has arisen from insights from the quantum measurement process [1] and "intuition"– whatever that is. We consider Stochastic Electromagnetics (SED) [2] as the most promising option to deal with this underlying level. The proper connection with quantum mechanics may have been put forward by Cetto and de la Peña already some 15 years ago in their approach called "Linear Stochastic Electromagnetics" , see e.g. [2, 3].

In his opening lecture of the 2007 Växjö conference, Timothy Boyer recalled his finding that classical electrodynamics already describes the Aharonov-Casher effect, and that there is a measurable difference with the quantum predictions. [4] This reminded us of an argument that we conceived when reporting in these Proceedings about the SED approach to the relativistic hydrogen ground state problem. [5] We explain it here.

For scattering and ionization experiments of electrons on atomic hydrogen, [6] the energy loss of the incoming electron can only take one of the discrete values needed to bring the ground state electron to the one of the excited levels (Lyman lines). Indeed, in the recent work [7] transitions to $n = 2$, 3 and 4 are clearly observed, while the higher states form a quasi continuum, that matches the ionization continuum. When one sees this from the classical SED approach, all intermediate energy values can also occur, so it remains a question how the quantum predictions can emerge, even in principle, from SED. To answer this qualitatively is the purpose of the present note.

To excite the atom, the incoming electron must have an energy of Rydberg size, $E \sim \alpha^2 mc^2$, with $\alpha = e^2/\hbar c$ the fine structure constant. The typical distance of the bound electron to the nucleus is the Bohr radius $a_0 = \hbar/\alpha mc$; its typical speed is αc. The typical orbit time is $t = \hbar/\alpha^2 mc^2$. In the scattering process, we should expect the arisal of a bunch of virtual photons. Their typical energy is also $E_\gamma \sim \alpha^2 mc^2$, so their wavelength is $\lambda \sim \hbar/\alpha^2 mc = a_0/\alpha$. Since this is some 137 times larger than the Bohr radius, it is conceivable that after the incoming electron has encountered the H-atom, the virtual photons will readjust its energy and momentum, and with it, the structure of the atom, according to the rules of QM. We call this the pull back mechanism. To demonstrate it in practice will require more work.

A similar pull back mechanism may play a role in the experiment discussed by Boyer.

REFERENCES

1. A. E. Allahverdyan, R. Balian and Th. M. Nieuwenhuizen, *Curie-Weiss model of the quantum measurement process*, Eur. Phys. Lett., **61**, 452 (2003).
2. L. de la Peña and A. M. Cetto, *The Quantum Dice: An Introduction to Stochastic Electrodynamics*, (Kluwer, Dordrecht, 1996).
3. L. de la Peña and A. M. Cetto, *Contribution from stochastic electrodynamics to the understanding of quantum mechanics*, in *Quantum Theory: Reconsideration of Foundations – 3*, G. Adenier, A. Yu. Khrennikov and Th.M. Nieuwenhuizen, eds, (AIP Conference Proceedings 810, Melville, 2006), pp 131.
4. Timothy H. Boyer, *Proposed Aharonov-Casher effect: Another example of an Aharonov-Bohm effect arising from a classical lag*, Phys. Rev. A **36**, 5083 (1987); these proceedings.
5. Th. M. Nieuwenhuizen, *Classical Phase Space Density for the Relativistic Hydrogen Atom* in *Quantum Theory: Reconsideration of Foundations – 3*, G. Adenier, A. Yu. Khrennikov and Th.M. Nieuwenhuizen, eds, (AIP Conference Proceedings 810, Melville, 2006), pp 198.
6. N.F. Mott and H.S.W. Massey, *The Theory of Atomic Collisions*, (Clarendon, Oxford, 1965)
7. J.G. Childers, K.E. James, M. Hughes, I. Bray, M. Baertschy and M.A. Khakoo, Phys. Rev. A **28**, 030702(R) (2003).

CP962, *Quantum Theory, Reconsideration of Foundations – 4*,
edited by G. Adenier, A. Yu. Khrennikov, P. Lahti, V. I. Man'ko, and Th. M. Nieuwenhuizen

The Relativistic Theory of Gravitation and its Application to Cosmology and Macroscopic Quantum Black Holes

Th. M. Nieuwenhuizen

Institute for Theoretical Physics, Valckenierstraat 65, 1018 XE Amsterdam, The Netherlands

Abstract. In the Relativistic Theory of Gravitation, space-time is Minkowskian, and gravitation is a physical field, next to electromagnetism. The bimetric coupling of Riemann and Minkowski metrics allows to defines a unique, local gravitational energy density. The case of a positive cosmological constant is considered. For cosmology, the ΛCDM model is covered, while there appears a form of inflation at early times. Furthermore, both the total energy and the zero point energy vanish.

Within this theory an exact solution is presented for the interior of supermassive black holes ($M_\star \sim 10^8 M_\odot$). They are consistent with a Bose-Einstein condensates of densely packed H-atoms. The solution is regular everywhere with a specific shape at the origin. The redshift at the horizon is finite but large, $z \sim 10^{14} M_\star/M$. The problem is exactly and self-consistently solved within a quantum field theoretic approach for the matter. Its direct back-reaction on the metric is 100%. In the BEC ground state, the binding energy equals 71% of the rest energy of the constituting matter. Extension to finite temperatures reveals a $T^{3/2}$ decay of the entropy. At $T = 0$ the black hole does not radiate.

Keywords: Minkowski space-time, cosmological constant, bimetric coupling, bimetrism, black hole, supermassive black hole, cosmology, Bose-Einstein condensation
PACS: 04.20.Cv, 97.60.Lf, 04.20

1. INTRODUCTION

The first attempt to describe gravitation by a field theory was made by Maxwell. After his success with electromagnetism, he attacked gravitation, but, assuming a scalar field, he got stuck on the problem of negative masses – a tensor field is needed. It is said that in introducing the General Theory of Relativity (GTR), Einstein made the step that Lorentz and Poincaré had failed to make: to go from flat space to curved space. This arises from the equivalence principle, or, technically, from the group of general coordinate transformations [1, 2]. One fundamental difficulty is then how to deal with the physics of gravitation itself, since there is only a quasi energy-momentum tensor [3]. The correspondence of Einstein exhibits how he changed position from first requiring a gravitational energy-momentum tensor and later discarding it. For gravitational wave detection, however, this absence leaves open the question of how energy can be faithfully transferred from the wave to the detector. The proper energy momentum tensor of gravitation was derived only recently by Babak and Grishchuk [4], who actually show that the proper tensor is a Minkowski space tensor form of the Landau-Lifshitz quasi-tensor. They start with a field theoretic approach to gravitation, in terms of a tensor field $\phi^{\mu\nu}$ in a Minkowski background space-time. The metric of the latter, $\eta_{\mu\nu} = \mathrm{diag}(1,-1,-1,-1)$, is denoted in arbitrary coordinates by $\gamma_{\mu\nu} = (\gamma^{\mu\nu})^{-1}$. The Riemann metric tensor $g_{\mu\nu} = (g^{\mu\nu})^{-1}$, is then defined by

$$\sqrt{\frac{g}{\gamma}}g^{\mu\nu} = \gamma^{\mu\nu} + \phi^{\mu\nu}, \qquad g = \det(g_{\mu\nu}), \qquad \gamma = \det(\gamma_{\mu\nu}). \tag{1}$$

It is just a way to code the gravitational field, allowing to expresses distances by $\mathrm{d}s^2 = g_{\mu\nu}\mathrm{d}x^\mu \mathrm{d}x^\nu$.

Such a non-linear way to code distances in a flat space is not uncommon. For diffuse light transport through clouds, one may express distances in the optical thickness, the number of extinction lengths. If the cloud is not homogeneous, points at the same physical distance are described by a different optical distance and, vice versa. To see the kind of questions that arise then, let us ask: can one know the thickness of a cloud that dims the sunlight by a factor 2? (No, one observes the product of thickness and humidity.) So this is an incomplete description of the underlying reality.

The aim of the present paper is first to update our view expressed recently [5] on Logunov's Relativistic Theory of Gravitation [6, 7]. In particular, we shall discuss the material energy density in cosmology. Next we consider black holes. By studying macroscopic quantum black holes we shall put forward the case of a negative bimetric energy density (positive parameter ρ_{bi}), leading to a positive cosmological constant, in agreement with cosmological data of last decades.

CP962, *Quantum Theory, Reconsideration of Foundations – 4*,
edited by G. Adenier, A. Yu. Khrennikov, P. Lahti, V. I. Man'ko, and Th. M. Nieuwenhuizen

2. FIELD THEORETIC DESCRIPTION OF THE GENERAL THEORY OF RELATIVITY

The Maxwell view that gravitation is a field in flat space, was actually the starting point for Einstein, and reappeared regularly. Nathan "EPR" Rosen [8] considers a bimetric theory, involving the Minkowski metric and the Riemann metric. Bimetrism is quite natural, with $\eta_{\mu\nu}$ entering e.g. particle physics, and $g_{\mu\nu}$ e.g. cosmology. Rosen considers covariant derivatives D_μ of Minkowski space, with Christoffel symbols $\gamma^{\lambda}_{\cdot\mu\nu}$ vanishing in Cartesian coordinates. When replacing partial derivatives by Minkowski covariant ones in the Riemann Christoffel symbols,

$$\Gamma^{\lambda}_{\cdot\mu\nu}=\frac{1}{2}g^{\lambda\sigma}(\partial_\mu g_{\nu\sigma}+\partial_\nu g_{\mu\sigma}-\partial_\sigma g_{\mu\nu})\rightarrow G^{\lambda}_{\cdot\mu\nu}=\frac{1}{2}g^{\lambda\sigma}(D_\mu g_{\nu\sigma}+D_\nu g_{\mu\sigma}-D_\sigma g_{\mu\nu})=\Gamma^{\lambda}_{\cdot\mu\nu}-\gamma^{\lambda}_{\cdot\mu\nu}, \qquad (2)$$

the obtained Christoffel-type symbols $G^{\lambda}_{\cdot\mu\nu}$ are tensors in Minkowski space. Inspired by the Landau-Lifshitz and Babak-Grishchuk results, we may define the *acceleration tensor*

$$A^{\mu\nu} = \frac{1}{2}D_\alpha D_\beta(k^{\mu\nu}k^{\alpha\beta}-k^{\mu\alpha}k^{\nu\beta}), \qquad k^{\mu\nu}=\gamma^{\mu\nu}+\phi^{\mu\nu}, \qquad (3)$$

in which the $\gamma\gamma$ terms do not contribute. Then we can calculate the combination

$$\tau^{\mu\nu}=\frac{c^4}{8\pi G}\left[\frac{\gamma}{g}A^{\mu\nu}-(R^{\mu\nu}-\frac{1}{2}g^{\mu\nu}R)\right]. \qquad (4)$$

In doing so, we make use of Rosen's observation that $R^{\mu\nu}$ remain unchanged if one replaces all partial derivatives by covariant ones in Minkowski space [8]. It appears that all second order derivatives drop out from (4), leaving a bilinear form in first order covariant derivatives,

$$\begin{aligned}\tau^{\mu\nu} \;=\;& \frac{c^4\gamma}{8\pi G g}\Big(\frac{1}{2}\phi^{\mu\nu}_{\cdot\cdot:\lambda}\phi^{\lambda\rho}_{\cdot\cdot:\rho}-\frac{1}{2}\phi^{\mu\lambda}_{\cdot\cdot:\lambda}\phi^{\nu\rho}_{\cdot\cdot:\rho}+\frac{1}{2}\phi^{\mu\lambda:\rho}\phi^{\nu}_{\cdot\lambda:\rho}+\frac{1}{4}k^{\mu\nu}\phi^{\lambda\rho:\sigma}\phi_{\lambda\sigma:\rho}-\frac{1}{2}\phi^{\lambda\rho:\mu}\phi^{\nu}_{\cdot\lambda:\rho}\\ &-\frac{1}{2}\phi^{\mu\lambda:\rho}\phi_{\lambda\rho}^{\cdot\cdot:\nu}+\frac{1}{4}\phi^{\lambda\rho:\mu}\phi_{\lambda\rho}^{\cdot\cdot:\nu}-\frac{1}{8}\phi_{\lambda}^{\cdot\lambda:\mu}\phi_{\rho}^{\cdot\rho:\nu}-\frac{1}{8}k^{\mu\nu}\phi^{\lambda\rho:\sigma}\phi_{\lambda\rho:\sigma}+\frac{1}{16}k^{\mu\nu}\phi_{\rho}^{\cdot\rho:\lambda}\phi^{\sigma}_{\cdot\sigma:\lambda}\Big). \qquad (5)\end{aligned}$$

where $X_{:\mu}\equiv D_\mu X$ and raising (lowering) of indices of $\phi^{\mu\nu}_{\cdot\cdot:\rho}$ is performed with $k^{\mu\nu}$ ($k_{\mu\nu}$). $\tau^{\mu\nu}$ is a tensor in Minkowski space. For Cartesian coordinates, it coincides with the Landau-Lifshitz quasi-tensor. In general, it coincides with the Babak-Grishchuk tensor $\gamma t^{\mu\nu}/g$. Inclusion of matter is now much easier than in [4]. Inserting the Einstein equations in the right hand side of (4), we may write the Einstein equations in the Newton shape: acceleration=mass $^{-1}\times$force,

$$A^{\mu\nu}=\frac{8\pi G}{c^4}\Theta^{\mu\nu}, \qquad \Theta^{\mu\nu}=t^{\mu\nu}+\frac{g}{\gamma}T^{\mu\nu}=\frac{g}{\gamma}\theta^{\mu\nu}, \qquad \theta^{\mu\nu}\equiv\tau^{\mu\nu}+T^{\mu\nu}. \qquad (6)$$

$\Theta^{\mu\nu}$ is the total energy momentum tensor of gravitation and matter. It is conserved, $D_\nu\Theta^{\mu\nu}=0$, because Eq. (3) implies $D_\nu A^{\mu\nu}=0$, since covariant Minkowski derivatives commute.

As an application, let us consider cosmology, described by the Friedman-Lemaitre-Robertson-Walker (FLRW) metric,

$$\mathrm{d}s^2=U(t)c^2\mathrm{d}t^2-V(t)\left(\frac{\mathrm{d}r^2}{1-kr^2}+r^2\mathrm{d}\Omega^2\right), \qquad \mathrm{d}\Omega^2=\mathrm{d}\theta^2+\sin^2\theta\,\mathrm{d}\phi^2. \qquad (7)$$

Let us consider flat space, $k=0$, and $U=1$, $V(t)=a^2(t)$ with a the scale factor. Then $\mathrm{d}s^2=c^2\mathrm{d}t^2-a^2(t)\mathrm{d}\mathbf{r}^2$ is space-independent, implying that $A^{00}=0$, due to the shape (3). According to (6) it then follows that the total energy density is zero, because the gravitational energy density, $\tau^{00}=-3c^4\dot{a}^2/(8\pi G a^2)$, is negative and cancels the one of matter, $T^{00}=\rho$, due to the Friedman equation. In plain words, such a universe contains no overall energy.

3. RELATIVISTIC THEORY OF GRAVITATION

Let us move on to an extension of GTR, giving up general coordinate invariance at cosmological scales. Discarding a total derivative of the Hilbert-Einstein action, Rosen expresses the gravitational action $S_R = \int d^3x dt \sqrt{-g}\, L_R$ in terms of the Lagrangian [8]

$$L_R = \frac{c^4 g^{\mu\nu}}{16\pi G}(G^\lambda_{\mu\nu}G^\sigma_{\lambda\sigma} - G^\lambda_{\mu\sigma}G^\sigma_{\nu\lambda}) = \frac{c^4\sqrt{\gamma/g}}{128\pi G}(2\phi^{\mu\nu:\rho}\phi_{\mu\nu:\rho} - 4\phi^{\mu\nu:\rho}\phi_{\mu\rho:\nu} - \phi^{\nu}_{\cdot\nu:\mu}\phi_\rho^{\cdot\rho:\mu}). \tag{8}$$

Involving only Minkowski covariant first order derivatives, it is close to general approaches in field theory. Logunov and coworkers continue on this [6]. The subgroup of gauge transformations that transform $\phi^{\mu\nu}$ but leave coordinates invariant, allows three extra terms [6],

$$L_g = L_R - \rho_\Lambda + \frac{1}{2}\rho_{\rm bi}\gamma_{\mu\nu}g^{\mu\nu} - \rho_0\sqrt{\gamma/g}\,, \tag{9}$$

Here ρ_Λ is the familiar energy related to a cosmological constant. The ρ_0 term describes a harmless shift of the zero level of energy, $\delta S = -\int d^3x dt \sqrt{-\gamma}\rho_0$. The bimetric term $\rho_{\rm bi}$ couples the Minkowski and the Riemann metrics. It acts like a mass term, because it breaks general coordinate invariance, and has some analogy to a mass term in massive electrodynamics. Logunov then imposes the relation

$$\rho_\Lambda = \rho_{\rm bi} = \rho_0, \tag{10}$$

which, in the absence of matter, keeps space flat, $\phi^{\mu\nu} = 0$, $g^{\mu\nu} = \gamma^{\mu\nu}$ and also $L_g = 0$. Thus one free parameter remains. Logunov's choice $\rho_{\rm bi} \equiv -m^2c^4/(16\pi G) < 0$ leads to an inverse length m and, in quantum language, a graviton mass $\hbar m/c$. The negative cosmological constant can be counteracted by an inflaton field [7]. The obtained theory has some drawbacks, such as self-repulsive properties for matter falling onto a black hole, and a minimal and a maximal size of the scale factor in cosmology [6, 7]. For a related approach to finite range gravity, based on a generalized Fierz-Pauli coupling, see [9].

We shall focus on the opposite choice, a positive cosmological constant Λ,

$$\rho_\Lambda \equiv \frac{\Lambda c^4}{8\pi G} = \frac{3c^2}{8\pi G}\Omega_{v,0}H_0^2 = \frac{3c^2}{8\pi G}0.74\left(\frac{0.71}{9.78\text{Gyr}}\right)^2, \qquad \rho_{\rm bi} \equiv \frac{\Lambda_{\rm bi}c^4}{8\pi G} \equiv \frac{\mu^2c^4}{16\pi G}. \tag{11}$$

Taking $\rho_{\rm bi} = \rho_\Lambda$, we define the inverse length μ and the dimensionless small parameter $\bar\mu$, [5] ($c = G = \hbar = 1$)

$$\mu = \sqrt{16\pi\rho_{\rm bi}} = \sqrt{2\Lambda}, \qquad \bar\mu = \mu M = 2.38\,10^{-23}\frac{M}{M_\odot} = 1.03\,10^{-14}\frac{M}{M_\star}. \tag{12}$$

Now the graviton has an "imaginary mass" $i\hbar\mu/c$, it is a "tachyon": Gravitational waves are unstable at today's Hubble scale. But this is of no concern, since on that scale, not single gravitational waves but the whole Universe matters. Though in applications we take $\Lambda_{\rm bi} = \Lambda$, our further notation is valid for the general case $\rho_{\rm bi} \neq \rho_\Lambda$, $\Lambda_{\rm bi} \neq \Lambda$.

3.1. The Einstein equations

The Einstein equations couple the Riemann metric to matter. They now read

$$R^{\mu\nu} - \frac{1}{2}g^{\mu\nu}R = \frac{8\pi G}{c^4}T^{\mu\nu}_{\rm tot}, \qquad T^{\mu\nu}_{\rm tot} = T^{\mu\nu} + \rho_\Lambda g^{\mu\nu} + \rho_{\rm bi}\gamma_{\rho\sigma}(g^{\mu\rho}g^{\sigma\nu} - \frac{1}{2}g^{\mu\nu}g^{\rho\sigma}). \tag{13}$$

Conservation of energy momentum, $T^{\mu\nu}_{\rm tot;\nu} = 0$, imposes a constraint due to the $\rho_{\rm bi}$ terms, [6]

$$D_\nu\left(\sqrt{\frac{g}{\gamma}}g^{\mu\nu}\right) = 0, \qquad \text{or} \qquad D_\nu\phi^{\mu\nu} = 0. \tag{14}$$

For Cartesian coordinates this coincides with the GTR harmonic condition $\partial_\nu(\sqrt{-g}g^{\mu\nu}) = 0$ [2]. Thus the theory automatically demands the harmonic constraint for $g^{\mu\nu}$, or, equivalently, the Lorentz gauge for $\phi^{\mu\nu}$, thereby severely reducing the gauge invariance of GTR.

Changes of Einstein's GTR have mostly met deep troubles with one or another established property, though not all proposals are ruled out [1, 10]. The present one is rather subtle and promising. For most applications, the Hubble-size $\rho_\Lambda = \rho_{bi}$ terms in Eq. (11,13) are too small to be relevant, so known results from general relativity can be reproduced. Indeed, viewed from a GTR standpoint, Eq. (14) is only a particular gauge, and actually often considered, while the cosmological constant only plays a role in cosmology. Logunov checked a number of effects in the solar system: deflection of light rays by the sun, the delay of a radio signal, the shift of Mercury's perihelion, the precession of a gyroscope, and the gravitational shift of spectral lines. [6] For gravitational radiation, e.g. in binary pulsars, the situation is subtle, because the graviton field has a spin zero mode, that could induce strong scalar or dipolar radiation. [10] However, in GTR the spin zero mode cannot appear in the gravitational radiation, because if it did, it would be gauge dependent. Now in RTG one takes the same radiation field, coming from Eq. (5), in the harmonic gauge. So also here the spin-0 mode cannot appear, i.e., it does not radiate. This argument is supported by explicit calculation [11],[12].

Differences between GTR and RTG may arise, though, for large gravitational fields, that we consider now.

4. RTG COSMOLOGY

Starting from the FLRW metric, the harmonic condition brings two relations: $U \sim V^3$ and $k = 0$: Minkowski space filled homogeneously with matter remains flat [6]. We may thus put $U = a^6(t)/a_*^4$, $V = a^2(t)$. Going from cosmic time t to conformal time $\tau = \int a^3 a_*^{-2} dt$ yields the familiar Einstein equations, extended by Λ_{bi} terms,

$$\frac{\dot{a}^2}{a^2c^2} = \frac{8\pi G}{3c^4}\rho + \frac{\Lambda}{3} - \frac{\Lambda_{bi}}{2a^2} + \frac{\Lambda_{bi}a_*^4}{6a^6}, \qquad \frac{\ddot{a}}{ac^2} = -\frac{4\pi G}{3c^4}(\rho + 3p) + \frac{\Lambda}{3} - \frac{\Lambda_{bi}}{3}\frac{a_*^4}{a^6}. \tag{15}$$

The first is the modified Friedman equation, the second corresponds to the first law $d(\rho_{tot}a^3) = -p_{tot}da^3$ provided we define $\rho_{tot} = \rho + \rho_\Lambda + \rho_2 + \rho_6$ and $p_{tot} = p - \rho_\Lambda - \frac{1}{3}\rho_2 + \rho_6$, with $\rho_2 = -3\rho_{bi}/2a^2$ and $\rho_6 = \rho_{bi}a_*^4/2a^6$. Note that ρ_2 acts as a positive curvature term.

The scale factor has an absolute meaning. If we assume that $a \gg 1$ and $a \gg a_*^{2/2}$, Eq. (15) just coincides with the ΛCDM model (cosmological constant plus cold dark matter), that gives the best fit of the observations. The ρ_2 term allows a positive curvature-type contribution. At large times, there is the exponential growth $a(\tau) = C\exp(H_\infty\tau)$ with $H_\infty = c\sqrt{\Lambda/3}$. In cosmic time this reads $a(t) = a_*^{2/3}[3H_\infty(t_0 - t)]^{-1/3}$, where t_0 is "the end of time", the moment where the scale factor has become infinite. The minimal scale factor is zero: in this theory a big bang can occur since $\rho_{bi} > 0$. Without including an inflaton field, Eq. (15) yields an initial growth of the expansion $a = (a_*^2 c\tau\sqrt{3\Lambda_{bi}/2})^{1/3}$. In cosmic time this reads $a = a_1\exp(ct\sqrt{\Lambda_{bi}/6})$, i. e., a certain inflation scenario starting at $t = -\infty$.

Also in RTG the gravitational energy precisely compensates the other energy contributions at all times. The vacuum energy vanishes: In empty space, the cosmological constant energy ρ_Λ cancels the ρ_{bi} terms, due to Eq. (10). See Eq. (13) for $g_{\mu\nu} = \gamma_{\mu\nu}$.

4.1. Material energy in RTG Cosmology

In a field theoretic approach it appears not to be easy to get a definition of cosmic energy density that matches our current understanding of this quantity. In conformal time $\tau = a_*^{-2}\int a^3 dt$ the metric takes the standard from $ds^2 = c^2d\tau^2 - a^2(\tau)dr^2$. The energy density $T^0_{\ 0} = \rho$ brings $T^{00} = a_*^4 a^{-6}\rho$. The material energy density in Riemann space becomes $dU_m^{(R)} = T^{00}\sqrt{-g}d^3x = a_*^2\rho d^3x$. In view of standard understanding, a factor a^3 is missing. Going from cosmic time t to conformal time τ, however, brings $d\tau = a^3 a_*^{-2}dt$ and $g'_{00} = 1$, $g' = -a^6$, $\gamma'_{00} = -\gamma' = a_*^4 a^{-6}$, implying indeed

$$dU_m^{(R)\prime} = (g^{00})'\rho\sqrt{-g'}d^3x = dU_m^{(R)}\frac{d\tau}{dt} = \sqrt{g_{00}}dU_m^{(R)} = a^3\rho d^3x. \tag{16}$$

The material energy in Minkowski space reads

$$dU_m^{(M)} = \frac{g}{\gamma}T^{00}d^3x = \frac{a^{12}}{a_*^4}\frac{a_*^4}{a^6}\rho = a^6\rho d^3x. \tag{17}$$

Let us try to find within RTG a physical interpretation for its a^6 factor. In conformal time we get

$$\mathrm{d}U_m^{(M)\prime} = \frac{g'}{\gamma' g'_{00}}\rho\sqrt{-\gamma'}\mathrm{d}^3x = \frac{a^9}{a_*^2}\rho\mathrm{d}^3x. \tag{18}$$

The same follows because $U_m^{(M)}$ behaves as the time-component of the momentum vector in Minkowski space, $\mathrm{d}U_m^{(M)\prime} = [\partial(x^{0\prime})/\partial x^0]\mathrm{d}U_m^{(M)} = (\partial\tau/\partial t)\mathrm{d}U_m^{(M)}$. Lowering its index with $\gamma'_{\mu\nu}$ brings

$$\mathrm{d}u_m^{(M)\prime} = \gamma'_{00}\mathrm{d}U_m^{(M)\prime} = \frac{a_*^4}{a^6}\frac{a^9}{a_*^2}\rho\mathrm{d}^3x = a_*^2a^3\rho\mathrm{d}^3x. \tag{19}$$

Up to a factor a_*^2 this coincides with $\mathrm{d}U'_R$ and $\mathrm{d}u'_R \equiv g'_{00}\mathrm{d}U'_R$. So there is a natural interpretation for the shape of energy density $\mathrm{d}u_m^{(M)\prime}/\mathrm{d}^3x$.

5. SUPERMASSIVE BLACK HOLES

Although it is tempting to fix the bimetric coupling constant ρ_{bi} to the observed positive cosmological constant, [5], for the opposite choice the cosmological data can be described by adding an inflaton field. [7] So the sign of ρ_{bi} must be obtained from another approach, e.g. by solving a realistic black hole in RTG.

On the basis of the Schwarzschild, Kerr and Kerr-Newman metrics, it generally believed that black holes (BHs) are singular objects with all matter localized in the center. Recent approaches challenge this assumption and consider matter just spread through out the interior. [13, 14] Here we shall follow this line of thought.

To start, let us just look at some orders of magnitude. For solar mass neutron stars it is known that their density is about the nuclear density. Solar mass black holes are about 10 times smaller, thus 1000 times more compact. Clearly, they beg for a description on the basis of Quantum Chromodynamics in curved space.

The other extreme concerns superheavy BHs. They occur in the center of each galaxy and weigh about one per mille of the bulge mass, $M_{\mathrm{BH}} = 0.0012M_{\mathrm{bulge}}$. [15] Let assume that they consist of Hydrogen atoms and that mass and particle number are related as $M \equiv \nu N m_H$ with some $\nu < 1$. The radius being $R = GM/c^2$ (see below), we may compare the BH number density $3N/4\pi R^3$ with the one of densely packed, non-overlapping H-atoms, that is, with the Bohr density $n_{\mathrm{B}} \equiv 3/4\pi a_0^3$, where $a_0 = 5.29\,10^{-11}$m is the Bohr radius. This yields a mass $M_\star = c^3(a_0^3/G^3m_H\nu)^{1/2}$, which appears to lie in the range of supermassive black holes, $M_\star = 8.60\,10^{38}$kg $= 4.32\,10^8 M_\odot$ (we take $\nu = \frac{1}{6}\sqrt{3}$, as we demonstrate below), and the induced bulge mass, $3.60\,10^{11}M_\odot$, lies in the range of observed bulges.

Next comes the question how matter can ever withstand the enormous pressure normally associated with such high densities. It was proposed originally by Sacharov that the vacuum equation of state $p = -\rho$ could describe matter at superhigh densities. [16] Laughlin and coworkers assumed that matter near the horizon could be in its Bose-Einstein condensed (BEC) phase, thus obeying the vacuum equation of state. [17] Dymnikova considered BHs obeying it in the interior, which, however, have one or two horizons. [18] Mazur and Mottola took the BEC idea over to the interior, and investigate a "gravo-star", of which the interior obeys the vacuum equation of state, and which is surrounded by a thin shell of normal matter having the stiff equation of state $p = +\rho$. This a solution is regular everywhere [13]

We shall demonstrate that an interior can exist in a homogeneous Bose-Einstein condensed phase, which, however, implies a non-constant energy density and pressure. Our first task is thus to show that gravitation can keep such matter together. We shall first consider a BEC condensed interior within the Relativistic Theory of Gravitation. Then we shall consider it in a self-consistent quantum field theoretic approach.

5.1. Static spherical shape of Einstein equations

Static spherical bodies, such as stars and non-rotating black holes have a metric

$$\mathrm{d}s^2 = U(r)c^2\mathrm{d}t^2 - V(r)\mathrm{d}r^2 - W^2(r)\mathrm{d}\Omega^2, \tag{20}$$

where $d\Omega^2 = d\theta^2 + \sin^2\theta d\phi^2$. In units where $c = G = 1$ they lead to the 00 and 11 Einstein equations

$$\frac{1}{W^2} - \frac{W'^2}{VW^2} - \frac{2W''}{VW} + \frac{V'W'}{V^2W} = 8\pi\rho_{\rm tot}, \qquad -\frac{1}{W^2} + \frac{W'^2}{VW^2} + \frac{U'W'}{UVW} = \frac{8\pi G}{c^4} p_{\rm tot}, \tag{21}$$

where the total density and pressure have the form

$$\rho_{\rm tot} = \rho + \rho_\Lambda + \frac{\rho_{\rm bi}}{2U} - \frac{\rho_{\rm bi}}{2V} - \frac{\rho_{\rm bi} r^2}{W^2}, \qquad p_{\rm tot} = p - \rho_\Lambda + \frac{\rho_{\rm bi}}{2U} - \frac{\rho_{\rm bi}}{2V} + \frac{\rho_{\rm bi} r^2}{W^2}. \tag{22}$$

In GTR one would just have $\rho_{\rm bi} = 0$. The harmonic constraint reads

$$\frac{U'}{U} - \frac{V'}{V} + 4\frac{W'}{W} = \frac{4rV}{W^2}. \tag{23}$$

In terms of the mass function $\mathcal{M}(r)$, defined by

$$V = \frac{W'^2}{1 - 2\mathcal{M}/W}, \tag{24}$$

the Einstein equations can be written as

$$\mathcal{M}' = 4\pi W' W^2 \rho_{\rm tot}, \qquad \frac{W - 2\mathcal{M}}{2UW^2}\frac{U'}{W'} - \frac{\mathcal{M}}{W^3} = 4\pi\, p_{\rm tot}. \tag{25}$$

With $\rho_\Lambda = \rho_{\rm bi}$ being very small, we shall only keep the $\rho_{\rm bi}/U$ terms in Eq. (22). Indeed, they cause the deformation of the Schwarzschild metric near the horizon [6, 5] and for a BH formed of ordinary matter describe the interior. [12] The key point of the present approach is that they can also describe the interior of macroscopic quantum BHs.

5.2. Black hole solution

Suppose that the condensed phase has a constant, isotropic pressure $p_0 = -\rho_0$, on top of a component with an isotropic pressure, obeying the stiff equation of state $p_n = +\rho_n - \rho_c$. The energy conservation $(\rho + p)U' + 2p'U = 0$ implies

$$\rho_n = \frac{\rho_1}{2U} + \frac{1}{2}\rho_c, \qquad p_n = \frac{\rho_1}{2U} - \frac{1}{2}\rho_c \tag{26}$$

There is a solution of Eqs. (25), [12]

$$U = 8\pi(\rho_1 + \rho_{\rm bi})W^2, \qquad \mathcal{M} = \frac{1}{4}W + \frac{4\pi}{3}(\rho_0 + \frac{1}{2}\rho_c)W^3 \equiv \frac{1}{4}W(1 + x^2), \tag{27}$$

where we introduce

$$x = \frac{W}{W_1}, \qquad W_1 = \sqrt{\frac{3}{8\pi(2\rho_0 + \rho_c)}}. \tag{28}$$

The locus follows by solving the harmonic constraint, which can be done analytically, [14]

$$r = r_1(1 + \frac{y}{\sqrt{5}})x^{\sqrt{5}-1}(1+y)^{-\sqrt{5}}, \qquad y = \sqrt{1-x^2}. \tag{29}$$

This determines W' and V,

$$W' = \frac{W_1\sqrt{5}}{4r_1}x^{2-\sqrt{5}}y(1+y)^{\sqrt{5}}, \qquad V = \frac{2W'^2}{y^2} = \frac{5W_1^2}{8r_1^2}x^{4-2\sqrt{5}}(1+y)^{2\sqrt{5}}. \tag{30}$$

In our model we may require that $p_n = p_1$ at $r = R$, where p_1 should follow from properties of the fluid, as will happen in our subsequent quantum approach. So far we have

$$\rho_1 = \frac{3(\rho_c + 2p_1)}{2(\rho_0 - \rho_c - 3p_1)}\rho_{\rm bi}, \tag{31}$$

and we require this to be positive. The important question whether the solution holds in case of positive or negative $\rho_{\rm bi}$, depends on p_1, and cannot be settled now. Let us introduce υ ('upsilon') by

$$U = \upsilon^2\bar{\mu}^2 x^2, \qquad \upsilon = \sqrt{\frac{2\rho_0 + \rho_c}{\rho_0 - \rho_c - 3p_1}}, \tag{32}$$

$$\rho = \rho_n + \rho_0, \quad p = p_n - \rho_0; \qquad \rho_n = \frac{1}{2}\rho_c + \frac{\rho_c + 2p_1}{2x^2}, \qquad p_n = -\frac{1}{2}\rho_c + \frac{\rho_c + 2p_1}{2x^2}. \tag{33}$$

At $r = M$, $y = y_c \ll 1$ this brings to leading order

$$U = \upsilon^2\bar{\mu}^2, \quad V = \frac{5}{2}, \quad W = 2M, \quad W' = \frac{\sqrt{5}}{2}y_c. \tag{34}$$

Well away from matter, the harmonic constraint brings for W the Schwarzschild shape $W_S = r + M$, where $M \equiv \mathscr{M}(R)$ is the mass, essentially as observed at infinity. This implies via Eq. (24)

$$V_S = \frac{1}{1 - 2M/W_S} = \frac{r+M}{r-M} = \frac{1}{U_S}. \tag{35}$$

Notice that in the harmonic gauge the Schwarzschild singularity occurs at $r = M$.

The values (34) are far from Schwarzschild's, even when r is near M (e.g., $W_S' = 1$). The problem nevertheless appears to be consistent. Beyond R we need the deformation of the Schwarzschild metric near the horizon, caused by the bimetric coupling [6, 7]. An elegant scaling form for small $\bar{\mu}$ was presented by us, [5]

$$r = M\frac{1 + \eta(e^{\xi} + \xi + r_0)}{1 - \eta(e^{\xi} + \xi + r_0)}, \qquad U = \eta e^{\xi}, \qquad V = \frac{e^{\xi}}{\eta(1 + e^{\xi})^2}, \qquad W = \frac{2M}{1 - \eta(e^{\xi} + w_0) - \bar{\mu}^2\xi}. \tag{36}$$

Here ξ is the running variable, η a small scale and $r_0 = \log\eta + 2$ and $w_0 = \mathscr{O}(\eta)$ are parameters determined by the behavior at finite distance beyond the horizon [12]. In this regime Eq. (36) yields $W'(r) = (e^{\xi} + \bar{\mu}^2/\eta)/(e^{\xi} + 1)$. Matching with the interior solution appears to be possible,

$$e^{\xi} = \upsilon\bar{\mu}\sqrt{\frac{5}{2}}, \qquad \eta = \upsilon\bar{\mu}\sqrt{\frac{2}{5}}, \qquad y_c = \upsilon\bar{\mu}\frac{3\sqrt{2}\rho_0}{2\rho_0 + \rho_c}. \tag{37}$$

Since to leading order $W_1 = 2M$, it holds that

$$\rho_0 + \frac{1}{2}\rho_c = \frac{3}{64\pi M^2}. \tag{38}$$

So this provides an exact solution of the problem. At the origin it exhibits the singularities

$$U = \bar{U}_1 r^{\gamma_\mu}, \quad V = \frac{1}{2}\gamma_\mu^2 \bar{W}_1^2 r^{\gamma_\mu - 2}, \quad W = \bar{W}_1 r^{\frac{1}{2}\gamma_\mu}, \tag{39}$$

where $\gamma_\mu = \frac{1}{2}(\sqrt{5}+1)$ is the golden mean. But if we take W as the coordinate, we have the shape

$$ds^2 = \frac{\upsilon^2\bar{\mu}^2}{4M^2}W^2 dt^2 - \frac{2dW^2}{1 - W^2/4M^2} - W^2 d\Omega^2, \tag{40}$$

that is regular at its origin. But the factor 2 codes the above singularities.

In the GTR limit $\rho_{\rm bi} \to 0$ one may attempt to derive a similar solution, starting with $U = 8\pi\rho_1 W^2$. This approach, however, will not fix all parameters, leaving, instead a one parameter family of solutions, set by either ρ_1 or y_c.

6. QUANTUM FIELD THEORY OF BOSE-EINSTEIN CONDENSED BLACK HOLES

Let our H-atoms be described by a field of creation and annihilation operators $[a_i, a_j^\dagger] = \delta_{ij}$,

$$\psi(\mathbf{r},t) = \sum_i a_i \psi_i(\mathbf{r}) e^{-iE_i t}, \qquad \psi^\dagger(\mathbf{r},t) = \sum_i a_i^\dagger \psi_i^*(\mathbf{r}) e^{iE_i t}. \tag{41}$$

where $i = \{n, \ell, m\}$ and $\psi_i(\mathbf{r}) = \phi_n(r) Y_{\ell m}(\theta, \phi)$. For a field theory in curved space, renormalization group arguments put forward that the field is coupled to the curvature scalar. [20] In the rotating wave approximation one keeps only the crossterms, $:(\psi^\dagger + \psi)^2 :\to 2\psi^\dagger\psi$ etc., giving the total Lagrangian

$$L = -\frac{R}{16\pi} - \rho_\Lambda + \frac{1}{2}\rho_{\rm bi}\gamma_{\mu\nu}g^{\mu\nu} - \rho_0\sqrt{\gamma/g} + g^{\mu\nu}\partial_\mu\hat{\psi}^\dagger\partial_\nu\hat{\psi} - (m^2 + \xi R)\psi^\dagger\psi - \frac{\lambda}{4}\psi^{\dagger 2}\psi^2. \tag{42}$$

The dimensionless coupling $\lambda = 8m^2 g/\hbar^3$ with $g = 4\pi\hbar^2 a_s/m$ codes the two particle interaction by the scattering length

$$a_s = 0.32 a_0 \quad \text{singlet state}, \qquad a_s = 1.34 a_0 \quad \text{triplet}. \tag{43}$$

For the singlet, it is large, $\lambda = 8.078\,10^6$. R follows from the Einstein equation in terms of the radial and transversal pressures,

$$R = -8\pi T_{\rm tot} = 8\pi(-\rho + p_r + 2p_\perp + \frac{\rho_{\rm bi}}{U}). \tag{44}$$

With $\Psi_0 = \sqrt{2E_0N_0}\psi_0 e^{-iE_0t}$ for N_0 atoms in the groundstate, the relativistic Gross-Pitaevskii equation reads

$$\left(\frac{1}{\sqrt{-g}}\partial_\mu\sqrt{-g}g^{\mu\nu}\partial_\nu + m^2 + \xi R + \frac{\lambda}{4E_0}|\Psi_0^2|\right)\Psi_0 = 0. \tag{45}$$

A homogeneous ground state, $\Psi_0(t) = |\Psi_0|e^{-iE_0t}$ requires

$$-\frac{E_0^2}{U} + m^2 + \xi R + \frac{\lambda}{4E_0}|\Psi_0|^2 = 0. \tag{46}$$

This has a solution due to the ξR-term and leads to

$$\xi = \xi_0(1 + \frac{\lambda|\Psi_0|^2}{4E_0 m^2}), \qquad E_0^2 = 8\pi\xi(\upsilon^2\bar{\mu}^2\rho_c + \rho_{\rm bi}) = \frac{1}{2}\xi(\upsilon^2\rho_c M^2 + 16\pi)\rho_{\rm bi}. \tag{47}$$

Inserting (38), the dimensionless parameter

$$\xi_0 = \frac{2}{3}m^2M^2 = 6.20\,10^{54}\left(\frac{M}{M_\star}\right)^2, \tag{48}$$

appears to be pretty large, but since $R \sim 1/M^2$ the combination ξR is just of order m^2.

Notice that $E_0^2 > 0$ requires that $\rho_{\rm bi} > 0$, which is the RTG with a tachyonic graviton, first considered in [5].

To determine $|\Psi_0|$, we need the $3d$ volume element in the future time direction, $\mathrm{d}\Sigma^\mu = \mathrm{d}r\mathrm{d}\theta\mathrm{d}\phi n^\mu\sqrt{-g_3} \equiv \delta_0^\mu \mathrm{d}\mathcal{V}$, set by the timelike unit vector $n^\mu = \delta_0^\mu/\sqrt{U}$ and $g_3 = -VW^4\sin^2\theta$. This yields $\mathrm{d}\mathcal{V} = \mathrm{d}r\mathrm{d}\Omega\sqrt{V/U}\,W^2$. The inner product [20]

$$(\psi_1, \psi_2) = -i\int \mathrm{d}\Sigma^\mu(\psi_1\partial_\mu\psi_2^* - \psi_2^*\partial_\mu\psi_1), \tag{49}$$

defines for eigenfunctions $(\psi_i, \psi_j^*) = 0$, and

$$(\psi_i, \psi_j) = (E_i + E_j)\int \mathrm{d}\mathcal{V}\,\psi_i\psi_j^* \equiv \delta_{ij}. \tag{50}$$

With $\mathrm{d}\mathscr{V} = \mathrm{d}x\mathrm{d}\Omega 8\sqrt{2}M^3x/(\upsilon\bar{\mu}y)$ it yields

$$|\Psi_0^2| = 2E_0N_0|\psi_0^2| = \frac{N_0\upsilon\bar{\mu}}{32\pi\sqrt{2}M^3} = \frac{\upsilon\bar{\mu}}{32\pi\sqrt{2}mM^3\nu}, \tag{51}$$

consistent with $\int \mathrm{d}\mathscr{V}|\Psi_0|^2 = N_0$. This allows to return to Eq. (47) and consider the correction

$$\frac{\lambda|\Psi_0|^2}{4E_0m^2} = \frac{\lambda\upsilon}{96\pi\nu m^4M^2} \approx 7.7\,10^{-13}\frac{M_\star^2}{M^2}. \tag{52}$$

These corrections seem relevant for BH's with masses $M \sim 8.2\,10^{-7}M_* \sim 350M_\odot$. Much less below $M_\star$ the Hydrogen atoms will be ionized, calling for fermionic fields for protons and electrons, which by a BCS pairing can again undergo a BEC transition. This scenario is beyond the aim of the present paper. For now we set $\lambda = 0$.

6.1. Self-consistent field theoretic approach

Rather than imposing an equation of state, the material energy-momentum tensor should be derived from first principles, i.e. from the bosonic quantum field theory for the H-atoms. The energy density induced by this field reads

$$\rho \quad = \quad \frac{\langle\partial_t\psi^\dagger\partial_t\psi\rangle}{U} + \frac{\langle\partial_r\psi^\dagger\partial_r\psi\rangle}{V} + \frac{\langle\partial_\theta\psi^\dagger\partial_\theta\psi\rangle}{W^2} + \frac{\langle\partial_\phi\psi^\dagger\partial_\phi\psi\rangle}{W^2\sin^2\theta} + m^2\langle\psi^\dagger\psi\rangle. \tag{53}$$

The pressures (p_r, p_θ, p_ϕ) have this shape with signatures $(++---)$, $(+-+--)$, and $(+--+-)$, respectively. Because of symmetry, it will hold that $p_\theta = p_\phi \equiv p_\perp$. In the homogeneous ground state p is isotropic,

$$\rho = \frac{1}{2}(\frac{E_0}{U} + \frac{m^2}{E_0})|\Psi_0^2|, \qquad p = \frac{1}{2}(\frac{E_0}{U} - \frac{m^2}{E_0})|\Psi_0^2|. \tag{54}$$

They indeed consist of a "vacuum" part $p = -\rho =$ const. and a "stiff" part $p = +\rho \sim 1/U$. As desired, in the non-relativistic ($E_0 = m$) and flat space ($U = 1$) limit, these expressions reduce to $\rho = mc^2|\Psi_0^2|$ and $p = 0$.

Because of the ξR-term, curvature couples directly to matter, and the Einstein equations embody a direct back reaction of matter on curvature. In the interior where $|\Psi_0|$ is constant, they take the form

$$R^{\mu\nu} - \frac{1}{2}g^{\mu\nu}R = 8\pi T_{\mathrm{eff}}^{\mu\nu}, \quad T_{\mathrm{eff}}^{\mu\nu} = \frac{T_{\mathrm{tot}}^{\mu\nu}}{1+8\pi\xi|\Psi_0^2|/E_0}. \tag{55}$$

The curvature scalar thus becomes

$$R = \frac{8\pi}{1+8\pi\xi|\Psi_0^2|/E_0}\left(\frac{E_0|\Psi_0^2| + \rho_{\mathrm{bi}}}{U} - \frac{2m^2}{E_0}|\Psi_0^2|\right). \tag{56}$$

In our RTG approach we can solve Eq. (46), to find two relations and a consequence,

$$\frac{8\pi\xi|\Psi_0^2|}{E_0} = 1, \qquad \xi\mu^2 = 2E_0^2, \qquad |\Psi_0^2| = \frac{E_0}{8\pi\xi} = \frac{\rho_{\mathrm{bi}}}{E_0}. \tag{57}$$

(The GTR situation, reached by taking $\mu^2 \sim \rho_{\mathrm{bi}} \to 0$ first, would not allow a meaningful solution of these equations.) The first identity expresses a 100% direct backreaction of matter on the metric. Replacing $\rho_{\mathrm{tot}} \to \rho_{\mathrm{eff}} = \frac{1}{2}\rho_{\mathrm{tot}}$, the mass function follows from the first relation of Eq. (25) as

$$\mathscr{M} = 8\pi M^3\left(\frac{E_0|\Psi_0^2| + \rho_{\mathrm{bi}}}{\bar{\mu}^2\upsilon^2}x + \frac{m^2|\Psi_0^2|}{3E_0}x^3\right). \tag{58}$$

The second relation demands that the linear term equals $\frac{1}{2}Mx$, implying $\upsilon = \sqrt{2}$. To have a horizon at $r = M$, it is needed that the cubic term equals $\frac{1}{2}Mx^3$, which yields

$$E_0 = \frac{1}{\sqrt{3}}\bar{\mu}m, \qquad \xi = \frac{2}{3}m^2M^2, \qquad \upsilon = \sqrt{2}. \tag{59}$$

Finally, comparing with (51) gives

$$\frac{M}{N_0} = \nu m, \qquad \nu = \frac{1}{2\sqrt{3}} = 0.288675135, \tag{60}$$

as announced before. Consequently, the binding energy in the groundstate is quite large, 71.13 % of the rest mass.

Putting these results together, it now follows that

$$\rho = \frac{1+6x^2}{64\pi M^2x^2}, \quad p = \frac{1-6x^2}{64\pi M^2x^2}, \quad U = 2\bar{\mu}^2x^2. \tag{61}$$

For x near 1, the pressure is negative, but it is positive below $x = 1/\sqrt{6}$. The bimetric energy density equals $\rho_{\rm bi}/2U = 1/64\pi M^2x^2$. The functions V and r are still given by (29) and (30), respectively, with $r_1 = M$, $W_1 = 2M$.

6.2. The exterior

The groundstate wavefunction is constant in the interior, but will decay in the outside on the Compton scale of one nucleon, $r - GM/c^2 \sim \hbar/m_Hc$. On a much larger scale $\ell_{\rm grav}$ there is discontinuity in slopes of interior and exterior gravitational solutions. We have to match the left and right values of U and V as before, yielding

$$\eta = \frac{2}{\sqrt{5}}\bar{\mu}, \qquad e^{\xi} = \sqrt{5}\,\bar{\mu}, \qquad y_c = 3\bar{\mu}. \tag{62}$$

The near-horizon growth of the metric functions $U/\eta \approx \eta V \approx e^{\xi}$ corresponds to a gravitational lengthscale in the micron range,

$$\ell_{\rm grav} = \frac{{\rm d}r}{{\rm d}\xi} = 2\eta M = \frac{4}{\sqrt{5}}\bar{\mu}M = 1.6\,10^{-6}\frac{M^2}{M_c^2}\,{\rm m}. \tag{63}$$

It is realistic, namely small compared to the size of the supermassive BH, but still large compared to the Bohr radius.

All by all, we may now rewrite the scaling form by eliminating ξ in favor of U,

$$r = M\frac{1+U+(2\bar{\mu}/\sqrt{5})(\log U+2)}{1-U-(2\bar{\mu}/\sqrt{5})(\log U+2)}, \qquad V = \frac{U}{(U+2\bar{\mu}/\sqrt{5})^2}, \qquad W = \frac{2M}{1-U+2\bar{\mu}^2-\bar{\mu}^2\log(U/2\bar{\mu}^2)}. \tag{64}$$

This describes the free space region $r \geq M$, where $2\bar{\mu}^2 \leq U \leq 1+\mathcal{O}(\bar{\mu})$. At scale $r \sim 1/\mu$ Newton's law picks up the Yukawa-type factor $\cos \mu r$, due to the tachyonic nature of gravitation in RTG with $\rho_{\rm bi} > 0$. [5]

The interior shape can also be expressed in U as running variable, where it lies in the range $(0, 2\,\bar{\mu}^2)$. Due to Eq. (29) it also holds that

$$x = \frac{1}{\bar{\mu}}\sqrt{\frac{1}{2}U}, \qquad y = \sqrt{1-\frac{U}{2\bar{\mu}^2}}. \tag{65}$$

The locus and the other two metric functions read

$$r = M(1+\frac{y}{\sqrt{5}})x^{\sqrt{5}-1}(1+y)^{-\sqrt{5}}, \qquad V = \frac{5}{2}x^{4-2\sqrt{5}}(1+y)^{2\sqrt{5}}, \quad W = 2Mx. \tag{66}$$

6.3. Properties of the solution

With the weight $\mathrm{d}\mathcal{V}$ given below (50), the standard expression for the energy, $\int \mathrm{d}\mathcal{V}\rho$ scales as $1/\bar{\mu}$ and even diverges logarithmically at $r = 0$. As discussed already, in RTG the energy is determined in a different way, set by the Landau-Lifshitz quasi-tensor, when properly generalized to a tensor in Minkowski space. The gravitational energy density takes for the metric (20) the form [1]

$$t^{00} = \frac{c^4W^2}{8\pi G r^6}\left(-\frac{r^2V'WW'}{V} + r^3V' - 5r^2W'^2 + \frac{2r^3VW'}{W} + 8rWW' - 2r^2V - 3W^2\right). \tag{67}$$

At the origin it diverges as $r^{\sqrt{5}-5}$, which is integrable with volume element $\mathrm{d}r\mathrm{d}\Omega r^2$. The gravitational energy inside the BH is

$$U_{\mathrm{grav,int}} = 4\pi \int_0^M \mathrm{d}r\, r^2 t^{00} = -842.898079M. \tag{68}$$

The total energy density reads $\Theta^{00} = t^{00} + VW^4\rho_{\mathrm{tot}}/r^4$. We can calculate the material energy, using

$$4\pi\int_0^M r^2\mathrm{d}r\frac{VW^4}{r^4}\frac{3}{64\pi} = 169.4316459M^3, \qquad 4\pi\int_0^M r^2\mathrm{d}r\frac{VW^4}{r^4}\frac{1}{64\pi x^2} = 686.4664330M^3. \tag{69}$$

We get

$$U_{\mathrm{mat}} = 4\pi\int_0^M \mathrm{d}r r^2\frac{VW^4}{r^4}\rho = 1025.32973M, \tag{70}$$

which is much larger than M. The bimetric term in ρ_{tot} contributes as

$$U_{\mathrm{bi}} = 4\pi\int_0^M \mathrm{d}r r^2\frac{VW^4}{r^4}\frac{\rho_{\mathrm{bi}}}{2U} = 686.4664M. \tag{71}$$

Taking into account the direct back reaction of matter on the metric, together they make up for

$$U_{\mathrm{interior}} = U_{\mathrm{grav,int}} + \frac{1}{2}U_{\mathrm{mat}} + \frac{1}{2}U_{\mathrm{bi}} = 13M. \tag{72}$$

The energy density in the skin layer first has a large positive and then a large negative part, due to the term r^3V'. The integrated effect is obtained easily since, in the formulation of the Einstein equations in Minkowski space, the total energy density is a total derivative,

$$\Theta^{00} = \frac{1}{4\pi r^2}\frac{\mathrm{d}}{\mathrm{d}r}\left(\frac{VW^2}{2r} + \frac{W^4}{2r^3} - \frac{W^3W'}{r^2}\right) \mapsto \frac{1}{4\pi r^2}\frac{\mathrm{d}}{\mathrm{d}r}\frac{M(r+M)^3(2r+M)}{2r^3(r-M)}, \tag{73}$$

the latter holding in the Schwarzschild regime. The region $M < r < \infty$ yields $U_{\mathrm{exterior}} = -12M$. Together with the interior it makes up the total BH energy $U = Mc^2$.

6.4. Finite temperature

We now assume that the BH is embedded in a thermal bath at temperature $T = 1/\beta$. The excited states have a Bose-Einstein occupation

$$n_{\mathrm{BE}}(E_i) = \frac{1}{\exp[\beta(E_i - \mu)] - 1}, \tag{74}$$

[1] In static, spherically symmetric RTG the energy density is unique, "energy can be localized". This is implied by the harmonic constraint. The residual gauges transformations allow only shapes that are unbounded at 0 or ∞. As they would make the energy infinite, they have to be discarded. This holds both for regular solutions and for our singular solution.

with $\mu = E_0$ in the condensed phase. The combination $\beta E_0 = \beta \bar{\mu} m/\sqrt{3} \equiv \beta_{\rm bs} m$ expresses the blue shift of temperature inside the BH, $T_{\rm bs} = T\sqrt{3}/\bar{\mu}$.

For eigenmodes $\psi_i(\mathbf{r}) = \phi_n(r) Y_{\ell m}(\theta, \phi) e^{-iE_i t}$ we obtain in the interior, if we simplify the ϕ' term using (23),

$$-\frac{1}{V}\phi'' - \frac{2r}{W^2}\phi' + \frac{\ell(\ell+1)}{W^2}\phi + \frac{E_0^2 - E^2}{U}\phi = 0. \tag{75}$$

The right hand side has vanished due to our requirement that $\phi =$ const. appears as groundstate at $E = E_0$, $\ell = 0$.

Let us define the dimensionless wavenumber k by

$$E = E_0 \left\{ 1 + \frac{1}{\xi} \left[\frac{k^2+1}{2} + \ell(\ell+1) \right] \right\}^{1/2}. \tag{76}$$

The transformation $x = 1/\cosh z$, $y = -\tanh z$, maps the interior $(0 \le x \le 1)$ onto the half-axis $(-\infty < z \le 0)$ and yields for $f = x\phi$ the well known eigenvalue problem

$$-f'' + \frac{2}{\cosh^2 z} f = k^2 f. \tag{77}$$

The solutions are explicit and form a complete set with the flat measure $\mathrm{d}\mathscr{V}/x^2 = \mathrm{d}z\mathrm{d}\Omega\, 8M^3/\bar{\mu}$,

$$f_n = C_n \left(1 + \frac{i}{k_n} \tanh z\right) e^{ik_n z}, \tag{78}$$

while the ground state $\phi_0 = C_0$ occurs for $k_0 = i$, $\ell = 0$. We can deal with the excited states as with plane waves, by requiring that $e^{ikL} = 1$ at some large $z = -L$ and take $L \to \infty$ at the end. This gives the quantization condition $k_n = 2\pi n/L$ for $n = 1, 2, \cdots$. These modes have normalization $C_n = \sqrt{\bar{\mu}/16 E_i L M^3}$ and thermal occupation

$$\langle \psi^\dagger \psi \rangle = \frac{1}{4\pi x^2} \sum_i C_i^2 n_{\rm BE}(E_i) \equiv \frac{\tau}{8\pi m^2 M^2 x^2} \tag{79}$$

where put $\sum_n \to (L/2\pi) \int \mathrm{d}k$, after which L drops out and the limit $L \to \infty$ can be taken trivially. In the non-relativistic approximation, it follows after a few steps that

$$\tau = \frac{m^4 M^2}{6\sqrt{2\pi}} \zeta\left(\frac{3}{2}\right) \left(\frac{T}{E_0}\right)^{3/2} = 9.588\, 10^{15} \frac{M^2}{M_\star^2} \left(\frac{T}{E_0}\right)^{3/2}. \tag{80}$$

The $1/x^2$ prefactor in (79) will bring a $1/x^4$ term in ρ and $\rho_{\rm eff}$, which changes the structure of the solution. However, at the leading order in τ, it drops out from $\rho_{\rm eff}$ and the still isotropic $p_{\rm eff}$, so we can pursue our present analysis. Extending the above approach, the results are for small τ:

$$\begin{aligned} \upsilon &= (1-\tau)\sqrt{2}, \quad \Psi_0^2 = \frac{\bar{\mu}\sqrt{3}(1-4\tau)}{16\pi m M^2} \quad E_0 = (1-4\tau)\frac{\bar{\mu} m}{\sqrt{3}} \quad \xi = \frac{2}{3} m^2 M^2 + \mathscr{O}(\tau^2), \\ \rho &= \frac{1+6x^2}{64\pi M^2 x^2} + \frac{2+3x^2}{96\pi M^2 x^4}\tau, \quad p = \frac{1-6x^2}{64\pi M^2 x^2} + \frac{2-21x^2}{96\pi M^2 x^4}\tau, \quad \rho_{\rm eff} = \frac{1+3x^2}{64\pi M^2 x^2}, \quad p_{\rm eff} = \frac{1-4\tau-3x^2}{64\pi M^2 x^2}. \end{aligned} \tag{81}$$

The number of ground state atoms comes from the integral of $|\Psi_0^2|$, and, because $M(T) = M(0) + \mathscr{O}(T^{5/2})$, it determines the number $N_n = N - N_0$ of atoms in the normal state,

$$N_0 = (1-3\tau)N, \qquad N_n = 3\tau N. \tag{82}$$

Clearly, we should expect the Bose-Einstein transition at some $\tau_c = \mathscr{O}(1)$.

If we assume that the entropy scales as usual with the particle number, $S_n/N_n = 2.972/2.315 = 1.283$, then $S_n = 3.85 N\tau \sim T^{3/2}$. With $N \sim M/m_H \sim 10^{65}\tau(M/M_\star)$, this is even at $\tau \sim 1$ much less than the Bekenstein-Hawking entropy $S_{\rm BH} = 4\pi M^2/m_P^2 = 1.971\, 10^{94}(M/M_\star)^2$.

In the quantum groundstate at $T = \tau = 0$, the BH does not radiate, also not by the Hawking mechanism.

7. CONCLUSION

We have first given motivation for the Relativistic Theory of Gravitation, RTG. This theory deals with the Hilbert-Einstein action extended with a bimetric coupling. We have discussed the shape of the gravitational energy momentum tensor, which is unique and thus solves the old paradox that "gravitational energy cannot be localized": In RTG it can be localized. We have shown that the related cosmological energy density can be understood as usual.

If we consider a local energy momentum density as a sine-qua-non property, then we are led to consider Minkowski space as a fixed "pre-space", that exist already without matter. Also for cosmology there is a different interpretation. In GTR coordinates are fixed to galaxies, this is called "coordinate space", but due to the increasing scale factor galaxies are said to move away from each other: physical space (i.e. Riemann space) is said to expand. In RTG we are led to another view: Coordinate space is physical space, and galaxies do not move in time. [6] However, the cosmic speed of light becomes a function of time. For GTR it is $c_{\rm cosmic} = {\rm d}r/{\rm d}t = c/a(t)$, very large at early times, while in RTG it reads $c_{\rm cosmic} = [a^2(t)/a_*^2]\,c$, very small at early times but exceeding c at large times. Logunov rejected that possibility by requiring that RTG light cones stay inside the Minkowski lightcones. [6] However, we see no reason to compare to such a would-be empty space situation. Nevertheless, this problem may be absent in a non-uniform cosmology, such as describing our Universe as a growing bubble in an otherwise empty Minkowski space time. This might connect the cosmic axis of evil, a preferred direction in the microwave background radiation, to the direction towards the Center.

Next we have considered a black hole in RTG. We have questioned the general wisdom that black holes have all their mass in the center and cannot be understood by present theories. We have recalled numerical estimates showing that a picture of closely packed hydrogen atoms naturally applies to the supermassive BH's in the center of galaxies, $M \sim 10^8 M_\odot$. Our previous finding that the bimetric coupling regulates the divergencies of the Schwarzschild singularity, continues to the behavior in the interior. We presented within RTG an exact solution for a black hole, of which the interior is governed by quantum matter in its Bose-Einstein condensed phase. The interior solution is matched with the Schwarzschild metric, which is deformed in RTG due to the bimetric coupling.

Powerlaw singularities occur at the origin. This has been a reason to discard the problem, [7] but they disappear in the Riemann description of the metric. Elsewhere, the solution is regular. The redshift at the horizon is finite, though of the order $1/\bar{\mu} \geq 10^{23} M_\odot/M$. In the interior, time keeps it standard role.

The large binding energy, $(1 - \frac{1}{6}\sqrt{3}) N m_H c^2 \approx 0.71 N m_H c^2$, may explain the enormous jets and energy output of quasars. Hawking radiation is absent, and Bekenstein-Hawking entropy has no bearing on the problem.

We failed to carry over our approach to GTR. This may imply that GTR must be abandoned and replaced by RTG, a step that will have far reaching consequences for classical and quantum approaches to gravitation and space-time.

REFERENCES

1. C.W. Misner, K.S. Thorne and J.A. Wheeler, *Gravitation*, (Freeman, San Francisco, 1973).
2. S. Weinberg, *Gravitation and Cosmology* , (Wiley, New York, 1972).
3. L. D. Landau and E. M. Lifshitz, *The Classical Theory of Fields*, (Pergamon, Oxford, U.K., 1951; revised 1979).
4. S. V. Babak and L. P. Grishchuk, Phys. Rev. D**61**, 024038 (1999).
5. Th. M. Nieuwenhuizen, Europhys. Lett. **78**, 10010 (2007).
6. A.A. Logunov, *The Theory of Gravity*, (Nauka, Moscow, 2001).
7. S.S. Gershtein, A.A. Logunov and M.A.Mestvirishvili, gr-qc/0602029.
8. N. Rosen, Phys. Rev. **57**, 147 (1940); ibid 150; Ann. Phys. **22**, 11 (1963).
9. S. V. Babak and L. P. Grishchuk, Int. J. Mod. Phys. D**12**, 1905 (2003).
10. C.M. Will, *Theory and Experiment in Gravitational Physics*, (Cambridge Univ. Press, New York, 1993), chapter 12.3, discusses that gravitational radiation of binary pulsars rules out a more recent bimetric theory of Rosen, in which black holes do not have the Schwarzschild shape.
11. Loskutov Yu.M., Vest. MGU, Series 3, Fizika, astronomia, **32** 1991, p.49; Proc. of the VI Marcel Grossman meeting on gen.relativ. 1991. Pt B. pp. 1.
12. Th.M. Nieuwenhuizen, unpublished.
13. P. O. Mazur and E. Mottola, Proc. Nat. Acad. Sci. **101**, 9545 (2004).
14. Th.M. Nieuwenhuizen, Europhysics Letters, 2007, submitted.
15. R. J. McLure and J. S. Dunlop, Mon. Not. Roy. Astron. Soc. **331**, 795 (2002).
16. A. D. Sacharov, Sov. Phys. JRTP **22**, 241 (1966).
17. G. Chapline, E. Hohlfield, R. B. Laughlin and D. I. Santiago, Philos. Mag. B **81**, 235 (2001).
18. I. Dymnikova, Int. J. Mod. Phys. D **12**, 1015 (2003).
19. C. Alcock et al., Astrophys. J. **550**, L169 (2001).
20. N. D. Birrell and P. C. W. Davies, *Quantum Fields in Curved Space*, (Cambridge Univ. Press, New York, 1994).

Classical Mechanics as Nonlinear Quantum Mechanics

Hrvoje Nikolić

Theoretical Physics Division, Rudjer Bošković Institute, P.O.B. 180, HR-10002 Zagreb, Croatia

Abstract. All measurable predictions of classical mechanics can be reproduced from a quantum-like interpretation of a nonlinear Schrödinger equation. The key observation leading to classical physics is the fact that a wave function that satisfies a linear equation is real and positive, rather than complex. This has profound implications on the role of the Bohmian classical-like interpretation of linear quantum mechanics, as well as on the possibilities to find a consistent interpretation of arbitrary nonlinear generalizations of quantum mechanics.

Keywords: classical mechanics, nonlinear quantum mechanics, Bohmian mechanics
PACS: 03.65.Ta, 05.20.-y

INTRODUCTION

In physics, linear equations are often approximations to nonlinear equations. On the other hand, we know that the Schrödinger equation is a linear equation. This raises an interesting question: Does it mean that the Schrödinger equation could be an approximation to a nonlinear equation?

We also know that the interpretation of the Schrödinger equation strongly rests on linearity. Therefore, the possibility of a nonlinear generalization of the Schrödinger equation raises further questions, such as: Does it mean that we have to modify the interpretation of quantum mechanics (QM)? Can nonlinearities teach us something new about the interpretation of QM?

Some results in that direction have already been found. In [1, 2, 3] the following result has been obtained: *If (nonlinear) wave function collapses, then EPR correlations can be used to transmit information instantaneously.* (For a comparison, it is well known that in ordinary linear QM the EPR correlations cannot be used to transmit information instantaneously.) This suggests that the concept of a true wave-function collapse is problematic. An alternative is to adopt an interpretation that does not rest on a true wave-function collapse. An example of such an interpretation is the many-world interpretation. However, when the many-world interpretation is applied to the nonlinear case, it is found that then different branches (worlds) may communicate [2]. But if they communicate, then it does not seem meaningful to interpret them as different worlds.

We see that the properties of nonlinear QM seem rather pathological. This suggests that a radically different interpretation should be adopted. In particular, interpretations in which the wave function is an objective-realistic entity are not expected to lead to such pathologies, as such interpretions are more similar to classical nonlinear waves (which, of course, do not lead to interpretational pathologies).

To resolve the interpretational problems, it would be desirable to have at least one example of nonlinear QM which we know how to interpret. On the other hand, it is widely believed that no such example of nonlinear QM exists. Nevertheless, we show that such an example exists, i.e., that *classical mechanics* (CM) represents an example of nonlinear QM. More precisely, we review our results of [4] which show that classical mechanics can be represented by a nonlinear Schrödinger equation

$$\left[\frac{\hat{\mathbf{p}}^2}{2m}+V-Q\right]\psi = i\hbar\partial_t\psi, \tag{1}$$

where

$$Q \equiv -\frac{\hbar^2}{2m}\frac{\nabla^2|\psi|}{|\psi|}. \tag{2}$$

We see that Q in (2) explicitly depends on ψ, which makes (1) nonlinear in ψ. We also note that the classical Schrödinger equation (1) has also been discussed earlier (see, e.g., [5]), but that before [4] it has not been shown that this equation alone is sufficient to reproduce *all* measurable predictions of the usual formulation of CM.

Another motivation for viewing CM as nonlinear QM is a hope that it could help in a better understanding of the big conceptual difference between CM and QM. For example, where does this difference comes from? A frequent answer

CP962, *Quantum Theory, Reconsideration of Foundations – 4*,
edited by G. Adenier, A. Yu. Khrennikov, P. Lahti, V. I. Man'ko, and Th. M. Nieuwenhuizen

is that this difference comes from the fact that CM is deterministic, whereas QM is probabilistic. Nevertheless, there is a deterministic interpretation of QM very similar to CM, but with the same probabilistic predictions as standard QM – the Bohmian interpretation [6]. Then why the Bohmian interpretation is not widely accepted? A frequent answer is: Because the purely probabilistic interpretation is simpler. (Or more precisely, because the Bohmian trajectories are unobservable, and thus unnecessary, hidden variables.) If we accept this argument against the Bohmian interpretation, then the following natural question arises: Can CM be made simpler by adopting a probabilistic interpretation? As we shall see in this paper, the answer will turn out to be – yes. We shall see that the probabilistic QM-like interpretation of the classical nonlinear Schrodinger equation reproduces all measurable predictions of standard CM. We shall also see that classical trajectories existing even without measurements play a role of unobservable hidden variables analogous to the Bohmian trajectories in QM. Thus, if it is (un)natural to accept the Bohmian interpretation of QM, then it is equally (un)natural to accept classical trajectories of CM.

QM AND THE BOHMIAN INTERPRETATION

Consider the standard Schrödinger equation

$$\left[\frac{\hat{\mathbf{p}}^2}{2m}+V(\mathbf{x},t)\right]\psi(\mathbf{x},t)=i\hbar\partial_t\psi(\mathbf{x},t), \tag{3}$$

where

$$\hat{\mathbf{p}}=-i\hbar\nabla. \tag{4}$$

We write ψ in the polar form

$$\psi(\mathbf{x},t)=R(\mathbf{x},t)e^{iS(\mathbf{x},t)/\hbar}, \tag{5}$$

where R and S are real functions and

$$R(\mathbf{x},t)\geq 0. \tag{6}$$

The complex Schrödinger equation is equivalent to a set of 2 real equations. These are the quantum Hamilton-Jacobi equation

$$\frac{(\nabla S)^2}{2m}+V+Q=-\partial_t S, \tag{7}$$

and the conservation equation

$$\partial_t\rho+\nabla\left(\rho\frac{\nabla S}{m}\right)=0, \tag{8}$$

where

$$Q\equiv-\frac{\hbar^2}{2m}\frac{\nabla^2 R}{R}. \tag{9}$$

Similarity with the classical Hamilton-Jacobi equation suggests the Bohmian interpretation [6]. In this interpretation, the particle has a trajectory satisfying

$$\frac{d\mathbf{x}}{dt}=\frac{\nabla S}{m}, \tag{10}$$

which is the same as an analogous equation in the classical Hamilton-Jacobi theory. The conservation equation provides that particles in a statistical ensemble are always distributed as in QM, with the probability density $\rho(\mathbf{x},t)$. Thus, in the Bohmian interpretation, all QM uncertainties emerge from the lack of knowledge of the actual initial particle position $\mathbf{x}(t_0)$.

CLASSICAL SCHRÖDINGER EQUATION

Now consider a classical statistical ensemble in the configuration space. The conservation equation is the same as in QM, which can be written as a linear equation for R

$$\left[\partial_t + \left(\frac{\nabla S}{m}\right)\nabla + \left(\frac{\nabla^2 S}{2m}\right)\right] R = 0, \tag{11}$$

where $R \equiv \sqrt{\rho}$. The classical Hamilton-Jacobi equation reads

$$\frac{(\nabla S)^2}{2m} + V = -\partial_t S. \tag{12}$$

Defining the classical wave function as

$$\psi(\mathbf{x},t) = R(\mathbf{x},t)e^{iS(\mathbf{x},t)/\hbar}, \tag{13}$$

the conservation equation and the Hamilton-Jacobi equation together turn out to be equivalent to the nonlinear classical Schrödinger equation

$$\left[\frac{\hat{\mathbf{p}}^2}{2m} + V - Q\right]\psi = i\hbar\partial_t\psi. \tag{14}$$

If, in addition, the wave function is required to be single-valued, then CM improves by including the Bohr quantization condition [4]

$$mvr = n\hbar. \tag{15}$$

MEASUREMENT IN NONLINEAR QM

We start from the observation that any function $\psi(\mathbf{x},t)$ representing a solution of some (not necessarily linear) equation can be expanded in terms of some other functions as

$$\psi(\mathbf{x},t) = \sum_a c_a\psi_a(\mathbf{x},t). \tag{16}$$

The linear case is special by having the property that the functions ψ_a can be chosen such that $c_a\psi_a$ and ψ_a are also solutions. Now, what is a measurement? Typically, a measurement is a process in which we obtain knowledge that the actual state is $c_a\psi_a$. Thus, to determine the subsequent post-measurement properties of the system, it is sufficient to know only that component. However, in the nonlinear case, it is not a solution, so to know that component one actually needs to know the *whole* solution. Only in the linear case the measured component evolves independently of other components. This explains the *effective* collapse in the linear case.

Now consider a measuring apparatus. Let the eigenstates of a measured hermitian operator be $\psi_a(\mathbf{x},t)$. A measurement that does not disturb the wave function requires entanglement with the measuring apparatus, such that the total wave function takes a form

$$\Psi(\mathbf{x},\mathbf{y},t) = \sum_a c_a\psi_a(\mathbf{x},t)\phi_a(\mathbf{y},t), \tag{17}$$

where the coefficients c_a are the same as those in (16) and $\phi_a(\mathbf{y},t)$ are some orthonormal states of the measuring apparatus. However, in the general nonlinear case, such a solution does not exist. From this, we conclude that *it is much more difficult to measure a quantity in nonlinear QM than in linear QM.*

MEASUREMENT FOR THE CLASSICAL SCHRÖDINGER EQUATION

The final conclusion of the preceding paragraph raises the following question: Does it mean that it is very difficult to measure anything in the quantum theory described by the classical Schrödinger equation? Fortunately, the answer is – no! Instead, as we show below, the classical Schrödinger equation has some specific properties that allow measurements that reproduce standard CM.

Although ψ does not satisfy a linear equation, there is a quantity that *does* satisfy a linear equation. This quantity is $R(\mathbf{x},t)$. (Note that for the linear Schrödinger equation R is not determined by a linear equation because R appears also in the quantum Hamilton-Jacobi equation.) Thus, *for measurement and effective collapse, the relevant "wave function" is* $R(\mathbf{x},t)$. The positivity of this wave function will turn out to be the source of classical properties that emerge from the classical Schrödinger equation.

First, in analogy with standard QM, we introduce the notation

$$R(\mathbf{x}) = \langle \mathbf{x}|R\rangle = \langle R|\mathbf{x}\rangle, \tag{18}$$

where the last equality is a consequence of reality. The scalar product is

$$\langle R_1|R_2\rangle \equiv \int d^3x \langle R_1|\mathbf{x}\rangle\langle \mathbf{x}|R_2\rangle \equiv \int d^3x R_1(\mathbf{x})R_2(\mathbf{x}), \tag{19}$$

which is *positive* (i.e., real and nonnegative). The only complete orthogonal basis $\{|R_i\rangle\}$ consistent with the positivity requirement $\langle \mathbf{x}|R_i\rangle \geq 0$ is the position basis $\{|\mathbf{x}\rangle\}$. Thus, *the position basis is the preferred basis.* The most general state consistent with the positivity requirement is

$$|R\rangle = \int d^3x c(\mathbf{x})|\mathbf{x}\rangle, \tag{20}$$

where $c(\mathbf{x}) \geq 0$.

We see that no real state $R(\mathbf{x})$ is an eigenstate of the momentum operator. Consequently, the state cannot collapse to the momentum eigenstate. This implies that the Heisenberg uncertainty relation $\Delta x \Delta p \geq \hbar/2$ cannot be revealed by an experiment, in agreement with the fact that there is no Heisenberg uncertainty relation in classical mechanics. Still, it does not mean that momentum cannot be measured. Instead, momentum can be measured *indirectly* by measuring two subsequent positions $\mathbf{x}_1$, $\mathbf{x}_2$ at times t_1, t_2, respectively. The momentum is then defined as

$$\mathbf{p} = (\mathbf{x}_2 - \mathbf{x}_1)/m(t_2 - t_1). \tag{21}$$

This, indeed, is how momentum is measured in classical mechanics.

EMERGENCE OF CLASSICAL STATISTICS

The origin of all nonclassical (i.e., typically quantum) probabilistic phenomena (e.g. destructive interference, EPR correlations, violation of Bell inequalities, ...) can be traced back to the fact that the scalar product $\langle\psi_1|\psi_2\rangle$ between the probability amplitudes does not need to be positive. Therefore, the positivity implies that there is no such nonclassical probabilistic phenomena in CM.

Furthermore, particles can always be distinguished. To see this, consider a 2-particle state

$$R(\mathbf{x},\mathbf{y}) = R_1(\mathbf{x})R_2(\mathbf{y}) + R_2(\mathbf{x})R_1(\mathbf{y}), \tag{22}$$

where R_1 and R_2 are *orthogonal*. Consequently, in the probability density R^2, the exchange term vanishes

$$2R_1(\mathbf{x})R_2(\mathbf{x})R_1(\mathbf{y})R_2(\mathbf{y}) = 0. \tag{23}$$

Just as in standard QM, this means that two particles can be regarded as distinguishable.

Now consider the density matrices. A pure state

$$|R\rangle = \sum_i \sqrt{w_i}|R_i\rangle, \tag{24}$$

where $w_i \geq 0$, can be represented by a density matrix

$$\hat{\rho}_{\text{pure}} = |R\rangle\langle R|. \tag{25}$$

The associated mixed state is a diagonal state

$$\hat{\rho}_{\text{mix}} = \sum_i w_i|R_i\rangle\langle R_i|. \tag{26}$$

The pure and the mixed state are related as

$$\hat{\rho}_{\text{pure}} = \hat{\rho}_{\text{mix}} + \sum_{i \neq j} \sqrt{w_i w_j} |R_i\rangle\langle R_j|. \tag{27}$$

Is there a measurable difference between pure and mixed states? Since $\hat{\mathbf{p}}$ is not measurable, the most general measurable operator is $\hat{A} = A(\mathbf{x})$, which is diagonal in the preferred basis. Consequently

$$\langle \hat{A} \rangle = \mathrm{Tr}(\hat{\rho}_{\text{mix}} \hat{A}) = \mathrm{Tr}(\hat{\rho}_{\text{pure}} \hat{A}) = \int d^3x \, \rho(\mathbf{x}) A(\mathbf{x}), \tag{28}$$

where $\rho(\mathbf{x}) = R^2(\mathbf{x})$ and $R(\mathbf{x}) = \sum_i \sqrt{w_i} R_i(\mathbf{x})$. This means that *there is no measurable difference between pure states and the associated mixed states*. The off-diagonal part plays no measurable role. Effectively it does not appear, which is a property of classical statistical mechanics. (Note a similarity with decoherence in ordinary QM).

EMERGENCE OF CLASSICAL TRAJECTORIES

The next question is: Why particles appear to move along classical trajectories? A partial answer is provided by the Ehrenfest theorem valid for the classical Schrödinger equation. One can show that

$$\langle \mathbf{x} \rangle = \int d^3x \, \psi^*(\mathbf{x},t) \mathbf{x} \psi(\mathbf{x},t) = \int d^3x \, \rho(\mathbf{x},t) \mathbf{x}, \tag{29}$$

$$\frac{d\langle \mathbf{x} \rangle}{dt} = \int d^3x \, \rho \frac{\nabla S}{m} = \int d^3x \, \psi^* \frac{\hat{\mathbf{p}}}{m} \psi, \tag{30}$$

$$m \frac{d^2 \langle \mathbf{x} \rangle}{dt^2} = \int d^3x \, \rho(-\nabla V) = \int d^3x \, \psi^*(-\nabla V)\psi. \tag{31}$$

Thus, as in ordinary QM, the Ehrenfest theorem says that the average position satisfies classical equations of motion, but that the actual position may be uncertain. But why particles appear as pointlike? As in ordinary QM, a measurement of the position induces a collapse of $R(\mathbf{x})$ to an arbitrarily narrow wave packet. However, in linear QM, we know that the wave packet suffers dispersion, i.e., that narrow wave packets are not stable. By contrast, the classical nonlinear Schrödinger equation contains arbitrarily narrow stable soliton solutions. The point-particle soliton solution is [4]

$$\psi_{\text{sol}}(\mathbf{x},t) = \sqrt{\delta^3(\mathbf{x} - \mathbf{y}(t))} \, e^{iS(\mathbf{x},t)/\hbar}, \tag{32}$$

where

$$\frac{d\mathbf{y}(t)}{dt} = \frac{\nabla S(\mathbf{y}(t),t)}{m}. \tag{33}$$

Note that (33) does not describe the motion of a pointlike particle associated with any solution of the classical Schrödinger equation, but the motion of the crest of the wave packet.

The results above can be interpreted as follows: *If a particle is measured to have a definite position at some time, then it will remain to have a definite position at later times and this position will change with time according to classical equations of motion.* However, *if the particle is not measured to have a definite position, then one is not allowed to claim that the particle has a definite position.* While this interpretation contradicts the usual interpretation of CM, it is analogous to the reasoning in the usual interpretation of QM and does not contradict any measurable result of CM.

We also note that, at each time, both a position and a momentum can be associated with such soliton solutions. This allows to introduce a classical phase space for this nonlinear QM [4].

DISCUSSION

Now we see that there are 4 consistent ways to interpret CM and QM:

1. Traditional: CM is deterministic, QM is probabilistic.
2. Bohmian: Both CM and QM are deterministic.
3. Anti-Bohmian: Both CM and QM are probabilistic.
4. Anti-traditional: CM is probabilistic, QM is deterministic.

The natural question is: How to know what is the correct way to interpret CM and QM? The Occam's razor says – the simplest one! But what does it mean "the simplest"? We have to chose some criterion of simplicity. With two different criteria we obtain two different answers:

1. Technical simplicity: no guiding equation for a particle trajectory – implies that both QM and CM are probabilistic.
2. Conceptual simplicity: particle positions exist even if we do not measure them – implies that both QM and CM are deterministic.

This suggests that we should either:

1. reject determinism of CM, or
2. accept the Bohmian deterministic interpretation of QM.

The interesting thing is that both possibilities seem rather heretic.

The next question we study is the following: Is there a consistent interpretation of nonlinear QM (in general)? From our results, we can conclude that if the consistent theory of measurement is the crucial consistency requirement, then the answer is – no! This is because we have seen that if there is no "wave function" (either complex or real) that satisfies a linear equation, then the theory of measurement based on wave-function collapse is inconsistent. The possible interpretations of this result are:

1. Only CM and linear QM are meaningful physical theories, or
2. More general nonlinear QM is physically meaningful, but a measurement cannot be performed, which is why we cannot measure effects of such theories, or
3. The wave-function collapse is not an essential part of measurement at all, so that all nonlinear generalizations of QM can be interpreted in the same way.

We note that the only known interpretation that allows consistent interpretation for any nonlinear generalization of QM by satisfying 3. is – the Bohmian interpretation.

It is also possible that the conclusions above suffer from our lack of imagination. We do not exclude a possibility that someone may think to have a different (and better) general interpretation of nonlinear QM. What we propose is a simple test of the consistency of any such different interpretation: one should test it on the classical nonlinear Schrödinger equation. If one does not reproduce the predictions of classical mechanics, then this interpretation of nonlinear QM is wrong!

ACKNOWLEDGMENTS

This work was supported by the Ministry of Science and Technology of the Republic of Croatia.

REFERENCES

1. N. Gisin, *Phys. Lett. A* **143**, 1 (1990).
2. J. Polchinski, *Phys. Rev. Lett.* **66**, 397 (1991).
3. M. Czachor, *Phys. Rev. A* **53**, 1310 (1996).
4. H. Nikolić, *Found. Phys. Lett.* **19**, 553 (2006).
5. N. Rosen, *Am. J. Phys.* **32**, 597 (1964); **33**, 146 (1965).
6. D. Bohm, *Phys. Rev.* **85**, 166, 180 (1952).

Algorithmic Container for Physics

Yu.I.Ozhigov and Al.Y.Ozhigov

Moscow State University, chair of quantum informatics, MSU, VMK, Vorobievi Gori, GSP-1,119992, Moscow, Russia

Abstract. We represent some arguments for the possibility to embed quantum physics to the special algorithmic container which permits to extend its methods to the dynamics of thousands of particles. The key role is played by the absolute decoherence model in which the limitation of classical computational resources in the simulation is treated as the fundamental source of decoherence. We show how quantum dynamics could be visualized by means of the method of collective behavior.

Keywords: simulation, quantum systems, decoherence
PACS: 07.05.Tp, 03,67,Lx

INTRODUCTION AND BACKGROUND

The problem of simulation of many body dynamics represents one of the oldest scientific problems. We treat it from the computational viewpoint, and mean by the simulation the computer program written on some of the algorithmic languages and designed for the implementation on the modern personal stations and supercomputers. We call this direction algorithmic approach. This approach is factually the plan of the extension of rigorous methods of theoretical physics to the large systems which treatment traditionally belongs to chemistry and biology. The actuality of this extension is confirmed by the activity in the simulation of such biological structures as a living cell.

We now analyze only the first step of this program: the method of extension of the rigorous methods to thousands of particles. Why we are not satisfied with the conventional statistical methods? The point is that the statistical method uses the averaging of magnitudes on large ensembles, whereas it is important to separate the influence of a separate single particle to the evolution of the large system. The conventional statistical methods cannot thus serve as the theoretical core of studying of factually complex systems as the biological objects where just the role of single particles can be crucial. The algorithmic approach assumes the application of the effective classical algorithms and at the same time it must be based on quantum mechanics.

These two requirements as applied to large systems immediately lead us to the necessity of the inbuilt decoherence model. Indeed, the dimensionality of the n particle quantum states space grows as exponential of n, which means the impossibility to simulate unitary dynamics in Hilbert formalism for more than 3 particles. This was the reason that Feynman ([1], and also Benioff ([2]), Deutsch and others put forward the idea of quantum computer. The attempts to build QC practically meets the serious problem of the decay of quantum states that is treated as decoherence. Decoherence is the fundamental factor which lies beyond the framework of quantum theory. It establishes the limitations for the applicability of Hilbert formalism for many particles to the real systems. But for algorithmic approach decoherence is helpful because it opens door to embed the quantum evolutions to the computational container based on the effective classical algorithms. The evident severe! limitation of classical computational resources which erects the insurmountable obstacle for the simulation of unitary dynamics is here treated as the decoherence factor. We call this model of decoherence the absolute model. All that is done up to nowadays in physics factually used just this model of decoherence. Our approach is thus the creation of the computational container for applications of quantum theory to large systems. We call so transformed quantum theory the reduced quantum theory; it is obtained from standard Copenhagen quantum physics by the special cut off that makes it embeddable into the classical computational container (realized by some set of super computers). We temporarily leave the important question how this cut off must be fulfilled (one possible way is shown in the next section, but it cannot be treated as the universal solution). With the help of this container one could hope to dis-

CP962, *Quantum Theory, Reconsideration of Foundations – 4*,
edited by G. Adenier, A. Yu. Khrennikov, P. Lahti, V. I. Man'ko, and Th. M. Nieuwenhuizen

cover the new collective effects in large systems induced by the ! quantum character of its particles, for example, to elucidate ! the role of complex entanglement in chemistry. We underline that this reduction of quantum theory touches the perspectives of quantum computing only, that is it makes no influence to the already confirmed in experiments part of quantum physics.

Algorithmic approach (see [3], [4]) should be considered in the context of the well known direction in physics that is called the reconsideration of the foundations of quantum theory. This direction initially had non algorithmic issues; it is induced by the non intuitiveness of quantum theory and has such a long history as this theory itself. It includes: the building of more intuitive formalism for quantum theory that must account time and space as the main categories, the hypothesis of grain for time and space (see the work [5]), etc. This activity gets rise to the attempts of de Broil, Rohm, Madelung and others to find some deterministic pre quantum field which gives the quantum dynamics of real particles. The modern results in this direction can be found in the works of Khrennikov (see [6]). This activity has the serious motivation because it is aimed to inbuilt quantum theory in the framework of the more traditional methodology, which is necessary f! or the further investigations, especially in the area of complex systems. The cost of the well known mathematical completeness of quantum theory is very high. This cost is the absence and even the refusal to obtain whenever the visual representation of quantum evolution (see the discussion in [7]). This fact, of course, is not the disadvantage of quantum theory itself, because its methods of computation of the probability in the results of simple experiments (and also the application to some complex systems which complexity can be treated as almost homogeneous, like crystals) give the excellent coincidence with all known experiments.

The disadvantage of quantum theory lies in the problems of its applicability to complex systems. Quantum theory predicts the existence of a scalable quantum computer. There is the exact criterion of such scalability: the ability to fulfill the so called fast quantum algorithms, e.g. algorithms which radically outperform its classical counterparts ([8]). The problems in the building of a scalable quantum computer seem now not only technological due to the role of decoherence which increases with the adding of each new particle to the considered system. We cannot now answer to the question: to what extend QC is scalable? Just this disadvantage of quantum theory in its application to complex systems raises the role of algorithmic approach. It is worth noting that even if QC could ever outperform classical computers, algorithmic approach will not lose its importance. In this case we would simply include QC as the subroutine for the fast solution of the scattering problems f! or complex ensembles.

In any case, the main aim of algorithmic approach: the simulation of the complex systems dynamics requires the solution of the so called measurement problem that is needed to compose the uniform simulating algorithms for evolutions of quantum systems without its division to unitary evolutions and measurements. We will show how to manage with this problem in the framework of algorithmic approach. The solution is based on the idea of grained amplitude. This solution of measurement problem as well as the further algorithmic methods utilizes the ideas of de Broil, Maxwell, Boltsman and others who attempted to create some intuitive mechanisms of quantum dynamics. The difference of those earlier attempts from algorithmic approach lies in that the last one is the integer mathematical formalism aimed to the introduction of quantum methods to the area of many body systems.

Algorithmic approach assumes the classical algorithms as the main mathematical tool. The power and flexibility of this tool are so that it can serve as the basement of new mathematical formalism for quantum theory of many body systems. This is real mathematical formalism though it differs radically from the language of formulas. Algorithmic approach means that we create the special video films representing the dynamics of the considered system. This video film is built by the simulating program which computes the evolution of the system accordingly to the reduced quantum theory, e.g., in the framework of the absolute model of decoherence. Only the film is the main tool of evaluation of the simulating program and of the whole container. It follows from the well known theorem of Turing Ű Riss in the algorithm theory: there is no mean to learn how a given algorithm works but to launch it and to observe the result. Hence, the role of visualization in algorithmic approach is cruc! ial. The traditional problems successfully solved by quantum theory, as the scattering problem, will here serve as the tools of tuning of the main program creating video film. This feature of algorithmic approach makes it substantially different from the existing theoretical physics, though it does not contradict to any existing method or even to any existing tradition. This feature requires the detailed workout and discussions; we leave it for the future.

In the next sections we represent the possibility to account the decoherence in the absolute model based on the assumption of the grain of amplitude, represent the method of collective behavior, and show the result of visualization of the simple scattering process made by the method of collective behavior.

GRAIN OF AMPLITUDE AND UNIFORM DESCRIPTION OF QUANTUM DYNAMICS

The key question in quantum theory is the model of decoherence. It determines the correspondence between quantum and classical worlds that is necessary for the model of dynamics of the real systems. The single exact form of this correspondence is BornŠs rule of quantum probability. We represent the explanation of Born rule through the nature element of algorithmic approach Ű the notion of amplitude quantum, or grain of amplitude.

Consider the general form of quantum state

$$|\Psi\rangle = \sum_j \lambda_j |e_j\rangle, \tag{1}$$

where the summing extends to the unlimited set of basic states of the system $|e_j\rangle$. Algorithmic approach requires to cut off this row to the finite sum by elimination of all summands with coefficients λ_j, which modules are less than some constant threshold ε. We call this procedure the reduction, and agree to fulfill the reduction on any state which we meet in evolution. Such reduced sum contains no more than $1/\varepsilon^2$ summands. Let N be the number of basic states for one particle. We then can agree that $\varepsilon = \frac{1}{\sqrt{N}}$. The state thus will have the form

$$|\Psi\rangle = \sum_{j=1}^{N} \lambda_j |e_j\rangle, \tag{2}$$

where some summands can be zeroes.

We call the constant ε amplitude quantum.

We now show how the reduction leads to the Born rule for the finding of quantum probabilities. For this we reduce the finding of probability to obtain some certain basic state A in the measurement of state Ψ to the application of classical rule

$$p(A) = \frac{N_{suc}}{N_{tot}}$$

where N_{suc} is the total number of successful events (e.g., such elementary events for which the event A is realized), N_{tot} the total number of elementary events. We define the set of elementary events and establish the correspondence between elementary events and the basic states of the measured system. We call an elementary event such a state of extended system (measured system + measuring apparatus) which module of amplitude in the given quantum state equals ε. The set of elementary events will then depend on a quantum state of extended system.

Let $|\Psi_j\rangle$ denote basic states of measured system, and $|\Phi_j\rangle$ denote basic states of measuring device (which can be, for example, the eye of observer), we obtain in the instant of contact between these two subsystems the state of the form

$$\sum_j \lambda_j |\Psi_j\rangle \bigotimes |\Phi_j\rangle \tag{3}$$

Now, in view that the measuring device is very massive in comparison with the measured object, when trying to describe its states we must divide states from (3) to the sum of l_j basic states (we must account states of all nuclei and electrons containing in the measuring device, all photons emitted and absorbed by it, etc.). If even in the instant of contact we have the state $|\Phi_j\rangle$, the evolution transforms it very fast to the state $|\Phi'_j\rangle = \sum_{k=1}^{l_j} \mu_{j,k} |\phi_{j,k}\rangle$, where the numbers l_j increase rapidly in time until the amplitudes modules achieve ε - when they will become nulls. Consequently, all modules of amplitudes $\mu_{j,k}$ must be considered as approximately equal. If we substitute the expression for $|\Phi'_j\rangle$ instead of $|\Phi_j\rangle$ in (3), the amplitudes of states $\phi_{j,k}$ become equal to $\frac{\lambda_j}{\sqrt{l_j}}$ because the quantum evolution is unitary.

We agreed to fulfill the reduction, that is the elimination of summands $\phi_{j,k}$ which module of amplitude is too small. Since the time frame when the division to the huge number of summands is very short, in the computations it means that we divide each summand in (3) to l_j new summands so that all newly arisen amplitudes have modules close to amplitude quantum and approximately equal, that makes all the states equivalent before the reduction and it makes possible to apply the classical urn scheme. The quantity l_j of summands with the first multiplier $|\Psi_j\rangle$, that is the total number of successful elementary events, is proportional to $|\lambda_j|^2$, and if only one summands survive in the reduction, we obtain Born rule for quantum probability.

The reduction of small amplitudes always takes place in the numerical simulation of quantum evolution. We thus obtain the important and general form of quantum evolution with the inbuilt model of decoherence. This model has the absolute form: decoherence is treated as the deficiency of classical memory needed for the storage of quantum states. The practical simulation based on this model gives the right general form of quantum dynamics for simple cases of 2-3 particles.

Representation of unitary quantum dynamics via method of collective behavior

We now show how the quantum unitary evolution determined by Shoedinger equation can be approximately represented in the form of diffusion dynamics. The attempts to do this get rise to de Broil, Madelung and some others. We now represent the new method which solves this problem. But the reduction of unitary evolution to classical dynamics has one substantial feature: we cannot write the single system of differential equations for classical dynamics instead of one Shroedinger equation. Adequate representation of quantum dynamics in terms of collective behavior gives the various differential equations depending on the chosen grain of spatial resolution. This is the cost of representation of quantum dynamics in terms of point wise particles.

The well known diffusion process can be described by Liouville equation: $\frac{\partial \rho}{\partial t} = a\,[H,\rho]$. This equation contains the scalar function $\rho(\bar{r},t)$ which is the density of ensemble of moving equivalent particles. Such equation cannot give the adequate picture of quantum dynamics with the basic quantum effects induced by interference.

The method of collective behavior is based on the representation of one quantum particle by the ensemble of point wise particles whose behavior obeys simple rules. We call this ensemble the swarm, and its members the samples of the initial quantum particle. One quantum particle is thus represented by the set consisting of point wise objects interacting with each other. This representation bears the resemblance with de Broil wave but is more robust for the computer simulation. Each sample has the mass and charge $m = M/n,\ c = C/n$, where M and C are the mass and charge of the initial particle, n is the total number of samples in the swarm. The density of swarm is given by the formula $\rho(\bar{r}) = \lim_{\delta x ar 0} N(a(\bar{r})/(\delta x)^3$, where $N(a(\bar{r}))$ is the total number of samples containing in the cube with the side δx with the center in the point $\bar{r}$. In each time instant t and point $\bar{r}$ it must obey the equation

$$|\Psi(\bar{r},t)|^2 = \rho(\bar{r},t) \tag{4}$$

which determines the adequateness of swarm approximation of the wave function Ψ. The density $\rho(\bar{r},t)$ of swarm cannot serve as the complete characteristic of quantum dynamics. The knowledge of wave function module is not sufficient for the description of its dynamics. Hence we introduce the total impulse of samples in the swarm in the point $\bar{r}$ by the analogous equation $p(\bar{r},t) = \lim_{\delta x ar 0} P(a(\bar{r})/(\delta x)^3$, where P denotes the sum of impulses of all samples containing in the cube a. It is shown in [3] that the quantum dynamics is completely determined by the pair of functions $\rho(\bar{r},t),\ p(\bar{r},t)$ provided the dynamics of swarm is given by the mechanism of dynamical diffusion. We suppose that each sample can either move along one of coordinate axes with some limited speed c or have the zero speed. The dynamical diffusion mechanism contains only one rule: two close samples with the opposite speed either obtain simultaneously the zero speed (if their speeds were not zero) or they obtain the opposite speeds (if initially their speed was zero). This simple rule ensures the exchange of impulses b! etween the neighboring samples. It is easy to trace qualitatively how this mechanism ensures the famous picture of the interference of the slow particle on two slits.

It is worth noting that the dynamical swarm determined by this mechanism of impulse exchange differs from the ensemble from diffusion Monte Carlo method. The total impulse in the spatial cell for the dynamical swarm can be nonzero whereas for DMC ensemble it is always zero. The reason is that the dynamical swarm obeys the law of impulse conservation on the level of separate samples whereas DMC ensemble does not. For a sample in DMC ensemble the impulse in the next instant is chosen randomly and it does not depend on the previous impulses of this sample or any other sample. Physically it means that DMC ensemble is the swarm with friction. Samples of DMC ensemble transmit their impulses to some media which does not contain in the ensemble and receive new impulses from this media (like particles in Brownian motion which interact with invisible molecules). In contrast, in the dynamical swarm the samples always obtain their speed via interaction with the certain other samples suc! h that the total impulse of the interacting pairs remains constant. The exchange of impulses between the close samples follows from this rule; it ensures that the changing of density ρ is accompanied by the changing of total impulses in the spatial cells.

If we fix some size δx of the linear spatial grain, we obtain the system of two differential equations on these two functions which part of solution $\rho(\bar{r},t)$ satisfies (4) with the accuracy $O(\delta x)$). The relative phase (only it has the physical sense) of wave function Ψ can be restored from the field of impulses $v(\bar{r},t)$ by the formula

$$\phi(r_1) = \int_{\gamma} k(\delta x)^2 \bar{v}\, d\bar{\gamma} \tag{5}$$

where the contour γ goes from r to r_1. The coefficients of the system of equations on ρ and $\bar{v} = \bar{p}/m$ approximating quantum dynamics depend on the grain δx, which makes impossible to launch it to zero as in the approximation of differential equations by the difference equations. It means that the intensiveness of dynamical diffusion (probability of that two samples exchange their impulses) depends on the spatial grain; factually it must be proportional to $1/(\delta x)^3$.

Factually, if we need to find the wave function with the accuracy $O(\delta x)$) we must fix δx and simulate the dynamics of swarm by the corresponding diffusion process. If we want to increase the accuracy of determining Ψ, we need to redefine the swarm and simulate the dynamics again. This looks like the situation with the statistical restoration of the wave function, when to find Ψ we need to collect the statistics over many independent measurements on the sequence of identically prefabricated state Ψ.

REPRESENTATION OF ENTANGLED STATES

The single mean for the description of entanglement in the collective behavior method is the forming of new particles from the existing ones by the operation of gluing, similar to operation of photon absorption. That is we treat the method of swarms as the universal form of representation of complex quantum systems in the algorithmic approach.

Let $S_1, S_2, \ldots, S_n$ be the swarms corresponding to particles $P_1, P_2, \ldots, P_n$. The main deposit to the energy (and dynamics) belongs to the sum $\bar{S}$ of these not entangled swarms S_j. Unitary evolution $U_j(t)$ of each swarm S_j is induced by the collective field of the others. It terms of Hilbert formalism: we treat the wave function as $\Psi_{not\ entangled} = \Psi_j \otimes \Psi_2 \otimes \ldots \otimes \Psi_n$. Here each sample of S_2 has the equal possibility to couple with each sample from S_2, end so on. But the small portion of samples from S_1 can obtain the exact partner for coupling from S_2. These pairs form the special matter inside of the whole ensemble. This matter consists of samples of the new particle Ű pair of the formers. It has its own wave function. The coupling is established if it is profitable on energy. Such a pairs can exist in two electrons systems like helium atom. When computing energy of ground state the main deposit b! elongs to not entangled states $\Psi_{not\ entangled}$. The small difference of its energy and the experimental value comes due to coupling of electrons and forming new particles. This process can be generalized to forming corteges of $3, 4, \ldots$ samples of the different real particles. It bears the resemblance to forming of ice crystals when it is energetically profitable. This is the way of description of entanglement in the algorithmic model. We can consider Hilbert space of states only for the separate swarms. The entanglement is thus exists exclusively in form of coupling of samples (or forming corteges). The total number of cortege like samples of new particles cannot exceed the total number of samples of each of the initial particles. This is why the total number N of all samples is limited and amplitude lower border is $1/sqrtN$. This is the grain of amplitude.

We consider Hilbert representation of the system of particles which in turn consist of more small particles, etc. We agree that every particle of the level $k-1$ is located in the center of mass of forming it particles of the level k, and in the group of these particles the determining of coordinates of all but one determines the coordinates of the last one (in the system of center of mass). Let $r_1, r_2, \ldots, r_n$ be coordinates of all particles partially ordered by nesting so that the particles of the more depth level of nesting has the much number, and the particles of the same level of nesting follow in the random order.

A wave function $|\Psi(\bar{r})\rangle$ can be represented in the form

$$\sum_{\bar{r}} \lambda_{\bar{r}} |\bar{r}\rangle \tag{6}$$

where $\bar{r}$ is the binary arithmetic notation of the values of coordinates of all particles containing in the considered system. Let n be the length of this list. Here the value of wave function $\Psi(\bar{r})$ is proportional to $\lambda_{\bar{r}}$. We consider the natural lexicographic order on the list $\bar{r}$, corresponding to the one particle case, but our consideration will be the general. We denote by $\bar{r}_k$ the initial segment of the sequence $\bar{r}$ of the length k, where r_k is the k-th element of this sequence. Any wave function of the form (6) can be represented as

$$|\Psi\rangle = \sum_{r_1} \left(\lambda_{\bar{r}_1} |r_1\rangle \bigotimes \sum_{r_2} \left(\lambda_{\bar{r}_2} |r_2\rangle \bigotimes \ldots \bigotimes \sum_{r_n} \lambda_{\bar{r}_n} |\bar{r}_n\rangle \right) \ldots \right) \tag{7}$$

If we consider the fixed value of j, the amplitude distribution $\lambda_{\bar{r}_j}$ can be treated as some wave function. Let all the functions $\lambda_{\bar{r}_j}$ really depend not on all the list $\bar{r}_j$, but on the coordinates $r_{j-p}, r_{j-p+1}, \ldots, r_j$ only. Then we call the set of such states the class of the depth p. The class $\mathscr{P}_0$ of the states of zero depth in the simple generalization of Schmidt states to the case of hierarchical system of nested particles. The generalization of this method gives the general form of swarm representation of any form of entanglement.

VISUAL REPRESENTATION OF RECHARGE CHANNEL OF SCATTERING OF A PROTON ON HYDROGEN ATOM

We represent in Fig. 1 the results of the qualitative simulation of scattering of a proton on a hydrogen atom. We can treat protons as the different classical particles because the dispersions of their spatial positions are substantially smaller than for electrons. There are three channels of this reaction: flying proton and hydrogen atom are conserved, they join and form the molecular ion of hydrogen, or electron passes from atom to flying proton. The last reaction is called the recharge. We show the sequential cadres of the video film simulating this channel of scattering.

CONCLUSION

We have demonstrated the potential of algorithmic approach to quantum theory on three items: the description of decoherence as the consequence of the limitation of classical memory, the possibility to build scalable models by the method of collective behavior, and the importance of visual representation of quantum evolutions. These advantages of algorithmic approach serve the evidence that this is the right way for the reconsideration of the foundations of quantum theory.

ACKNOWLEDGMENTS

The work is supported by Fond of NIX Computer Company grant # F793/8-05, INTAS grant 04-77-7289 and Russian Foundation for Basic Research grant 06-01-00494-a.

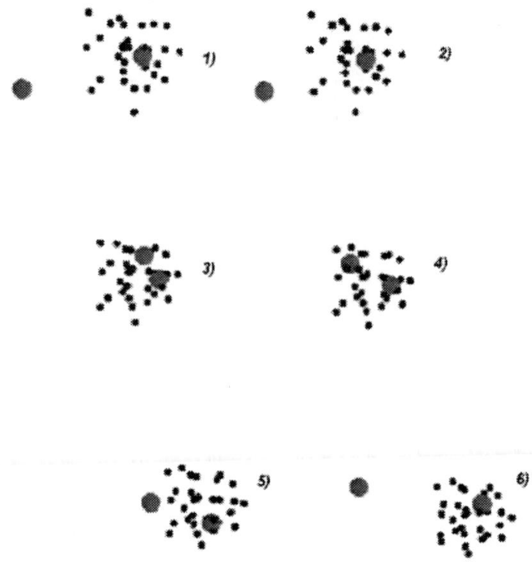

FIGURE 1. Scattering of proton of hydrogen atom. Recharge channel

REFERENCES

1. R.Feynman, Simulating physics with computers, J. Theoret. hys., 1982, 21, pp. 467-488.
2. P.Benioff, Quantum mechanical hamiltonian models of Turing machines, J. Stat. Phys. 29, 1982, 515-546.
3. Y.Ozhigov, Amplitude quanta in multi particle system simulation, Russian Microelectronics, 2006, vol. 35, 1, 53-65.
4. A.Ozhigov, K.Arakelov, Y.Ozhigov, Principles of the numerical simulation of many body quantum dynamics, Quantum computers and computing, 2006, vol. 6 N1, pp. 137-148.
5. A.Khrennikov, Y.Volovich, Energy levels of "hydrogen atom" in discrete dynamics, Open Sys. andInformation Dyn. , 2006, 13, pp. 119-132.
6. A.Khrennikov, Quantum mechanics from time scaling and random fluctuations at the quick time scale, IL NUOVO CIMENTO, vol. 121 B, 9, 2006.
7. J.L. Andrade E Silva, J. Loshak, Fields, Particles, Quanta. Moscow, Science, 1972.
8. L.Grover, A fast quantum mechanical algorithm for database search, Proceedings, 28th Annual ACM Symposium on the Theory of Computing (STOC), May 1996, pages 212-219. Proceedings, Melville, NY, 2006, vol. 810, electronic version: xxx.lanl.gov, quant-ph/0610052.

de Sitter Relativity and Quantum Physics

R. Aldrovandi, J. P. Beltrán Almeida, C. S. O. Mayor and J. G. Pereira

Instituto de Física Teórica, Universidade Estadual Paulista
Rua Pamplona 145, 01405-900 São Paulo, Brazil

Abstract. In the presence of a cosmological constant, interpreted as a purely geometric entity, absence of matter is represented by a de Sitter spacetime. As a consequence, ordinary Poincaré special relativity is no longer valid and must be replaced by a de Sitter special relativity. By considering the kinematics of a spinless particle in a de Sitter spacetime, we study the geodesics of this spacetime, the ensuing definitions of canonical momenta, and explore possible implications for quantum mechanics.

Keywords: de Sitter relativity, geodesics, quantum mechanics
PACS: 04.20.Cv; 04.50.+h

1. INTRODUCTION

When the cosmological constant Λ vanishes, absence of gravitation is represented by Minkowski spacetime, a solution of the sourceless Einstein's equation. Its isometry transformations are those of the Poincaré group, which is the group governing the kinematics of special relativity. For a non-vanishing Λ, however, Minkowski is no longer a solution of the corresponding Λ–modified Einstein equation and becomes, in this sense, physically meaningless. In this case, if we interpret Λ as a purely geometric entity, absence of gravitation turns out to be represented by the de Sitter spacetime. Now, the group governing the kinematics in a de Sitter spacetime is not the Poincaré, but the de Sitter group. This means that, in the presence of a non-vanishing Λ, ordinary Poincaré special relativity will no longer be valid, and must be replaced by a de Sitter special relativity [1].[1]

An important point of this theory is that it retains the quotient character of spacetime and, consequently, a notion of homogeneity. As in ordinary special relativity, whose underlying Minkowski spacetime M is the quotient space of the Poincaré by the Lorentz groups, the underlying spacetime of the de Sitter relativity will be the quotient space of the de Sitter and the Lorentz groups. Similarly to ordinary special relativity, therefore, in a de Sitter special relativity the Lorentz subgroup remains responsible for both the isotropy of space (rotation group) and the equivalence of inertial frames (boosts) [3]. The four additional transformations, given by a combination of translations and proper conformal transformations, define the homogeneity of spacetime.

A space is said to be transitive under a set of transformations — or homogeneous under them — when any two points of it can be attained from each other by a transformation belonging to the set. For example, Minkowski spacetime is transitive under spacetime translations. However, the de Sitter spacetime is transitive, not under translations, but under a combination of translations and proper conformal transformations, with the relative importance of these contributions being determined by the value of the cosmological constant. An immediate consequence of this property is that the ordinary notions of energy and momentum will change [4]. In fact, the conserved momentum, for example, will now be obtained from the invariance of the physical system, not under translations, but under a combination of translations and proper conformal transformations. The conserved momentum, therefore, will be a combination of ordinary and proper conformal momenta [5].

Due to the smallness of the observed cosmological constant, the difference between ordinary and de Sitter relativities will be very small. However, there are situations where this difference could become significant. For example, according to our current theories on fields and particles, the phase transitions associated to the spontaneously broken symmetries can be considered the primary source for a non-vanishing Λ [6]. According to this view, a high energy experiment modifies the *local structure of space-time for a short period of time*, in such a way that the immediate neighborhood of a high energy collision departs from Minkowski and becomes a de Sitter — or anti-de Sitter — spacetime [7]. There would then exist a connection between the energy scale of the experiment and the local value of

[1] Similar ideas have been explored in Ref. [2].

CP962, *Quantum Theory, Reconsideration of Foundations – 4,*
edited by G. Adenier, A. Yu. Khrennikov, P. Lahti, V. I. Man'ko, and Th. M. Nieuwenhuizen

Λ. The higher the energy, the larger the local value of Λ, and consequently the large the importance of the conformal symmetry. This is consistent with the idea that, at high energies the masses become negligible, thence the importance of the conformal symmetry. For a high enough energy, the local kinematics would be ruled preponderantly by a de Sitter special relativity.

The basic purpose of these notes is to study the kinematics subjacent to a de Sitter special relativity. In particular, we are going to study the equation of motion of spinless particles, which defines the geodesics of the de Sitter spacetime. Then, through an usual analysis of the action functional variation, we obtain the canonical momentum, and explore the consequences of this definition for quantum mechanics. We begin by introducing first the basic notions associated to the de Sitter spacetimes and groups.

2. DE SITTER SPACETIMES AND GROUPS

Spacetimes with constant scalar curvature R are maximally symmetric: they can lodge the highest possible number of Killing vectors. Given a metric signature, this spacetime is unique [8] for each value of R. Minkowski spacetime M, with $R = 0$, is the simplest one. Its group of motions is the Poincaré group $\mathscr{P} = \mathscr{L} \oslash \mathscr{T}$, the semi- direct product of the Lorentz $\mathscr{L} = SO(3,1)$ and the translation group $\mathscr{T}$. The latter acts transitively on M and its group manifold can be identified with M. Indeed, Minkowski spacetime is a homogeneous space under $\mathscr{P}$:

$$M = \mathscr{P}/\mathscr{L}.$$

Amongst curved spacetimes, the de Sitter and anti-de Sitter spaces are the only possibilities. One of them has negative, and the other has positive scalar curvature. They are hyper-surfaces in the "host" pseudo-Euclidean spaces $\mathbf{E}^{4,1}$ and $\mathbf{E}^{3,2}$, inclusions whose points in Cartesian coordinates $(\chi^A) = (\chi^0, \chi^1, \chi^2, \chi^3, \chi^4)$ satisfy respectively [9]

$$\eta_{AB}\chi^A\chi^B \equiv (\chi^0)^2 - (\chi^1)^2 - (\chi^2)^2 - (\chi^3)^2 - (\chi^4)^2 = -l^2$$

and

$$\eta_{AB}\chi^A\chi^B \equiv (\chi^0)^2 - (\chi^1)^2 - (\chi^2)^2 - (\chi^3)^2 + (\chi^4)^2 = l^2,$$

where l is the so-called de Sitter length-parameter. Using the Latin alphabet $(a, b, c \ldots = 0, 1, 2, 3)$ to denote the four-dimensional algebra and tangent space indices, whose metric tensor is $\eta_{ab} = \text{diag}\,(1, -1, -1, -1)$, and writing $\eta_{44} = \mathsf{s}$, the above conditions can be put together as

$$\eta_{ab}\,\chi^a\chi^b + \mathsf{s}\,(\chi^4)^2 = \mathsf{s}\,l^2. \tag{1}$$

For $\mathsf{s} = -1$, we have the de Sitter space $dS(4,1)$, whose metric is induced from the pseudo-Euclidean metric $\eta_{AB} = (+1, -1, -1, -1, -1)$. It has the pseudo-orthogonal group $SO(4,1)$ as group of motions. Sign $\mathsf{s} = +1$ corresponds to anti-de Sitter space, denoted by $dS(3,2)$. It comes from $\eta_{AB} = (+1, -1, -1, -1, +1)$, and has $SO(3,2)$ as its group of motions. Both spaces are homogeneous [10]:

$$dS(4,1) = SO(4,1)/\mathscr{L} \quad \text{and} \quad dS(3,2) = SO(3,2)/\mathscr{L}.$$

Furthermore, they are solutions of the sourceless Einstein's equation, provided the cosmological constant Λ and the de Sitter length-parameter l are related by

$$\Lambda = -\frac{3\mathsf{s}}{l^2}. \tag{2}$$

We consider from now on the dS spacetime, for which $\Lambda > 0$. In this case, the four-dimensional stereographic coordinates $\{x^a\}$ — which are obtained through a stereographic projection from the de Sitter hyper–surface into a target Minkowski spacetime — is given by [11]

$$\chi^a = \Omega(x)\,x^a \quad \text{and} \quad \chi^4 = -l\,\Omega(x)\left(1 + \frac{\sigma^2}{4l^2}\right), \tag{3}$$

where

$$\Omega(x) = \left(1 - \frac{\sigma^2}{4l^2}\right)^{-1}, \tag{4}$$

with $\sigma^2 = \eta_{ab} x^a x^b$ a Lorentz invariant quadratic interval. The inverse relations are

$$x^a = \Omega^{-1}(\chi)\,\chi^a \quad \text{and} \quad \frac{\sigma^2}{4l^2} = \frac{(\chi^4/l)+1}{(\chi^4/l)-1}, \tag{5}$$

where now

$$\Omega(\chi) = \frac{1}{2}\left(1 - \frac{\chi^4}{l}\right). \tag{6}$$

In terms of the stereographic coordinates, the five–dimensional line element

$$d\tau^2 = \eta_{AB}\, d\chi^A d\chi^B \tag{7}$$

reduces to the de Sitter line element

$$d\tau^2 = g_{ab}\, dx^a dx^b, \tag{8}$$

with

$$g_{ab} = \Omega^2\, \eta_{ab} \tag{9}$$

the metric of de Sitter spacetime.

The generators of infinitesimal de Sitter transformations are

$$\mathscr{L}_{AB} = \eta_{AC}\, \chi^C \frac{\partial}{\partial \chi^B} - \eta_{BC}\, \chi^C \frac{\partial}{\partial \chi^A}. \tag{10}$$

In terms of the stereographic coordinates $\{x^a\}$, these generators are written as

$$\mathscr{L}_{ab} = \eta_{ac}\, x^c\, \mathscr{P}_b - \eta_{bc}\, x^c\, \mathscr{P}_a \tag{11}$$

and

$$\mathscr{L}_{a4} = l\mathscr{P}_a - (4l)^{-1}\mathscr{K}_a, \tag{12}$$

where

$$\mathscr{P}_a = \partial/\partial x^a \tag{13}$$

are the translation generators (with dimension of $length^{-1}$), and

$$\mathscr{K}_a = \left(2\eta_{ab} x^b x^c - \sigma^2 \delta_a{}^c\right)\partial/\partial x^c \tag{14}$$

are the generators of *proper* conformal transformations (with dimension of *length*). Generators $\mathscr{L}_{ab}$ refer to the Lorentz subgroup of de Sitter, whereas $\mathscr{L}_{a4}$ define transitivity on the corresponding de Sitter space. For this reason, they are usually called the de Sitter "translation" generators. As implied by Eq. (12), the de Sitter spacetime is seen to be transitive under a combination of translations and proper conformal transformations. The relative importance of each one of these transformations is determined by the value of the length parameter l, that is, by the value of the cosmological constant.

The group contraction procedure requires that, before each limit is taken, the generators be modified through an appropriate insertion of parameters. These alterations are frequently guided by dimensional considerations and are different for different limits [12]. For this reason, the limits of small and large cosmological constant must be considered separately. In particular, for a small cosmological constant, it is convenient to rewrite the de Sitter generators in the forms

$$\mathscr{L}_{ab} = \eta_{ac}\, x^c\, \mathscr{P}_b - \eta_{bc}\, x^c\, \mathscr{P}_a \tag{15}$$

and

$$\Pi_a \equiv \frac{\mathscr{L}_{a4}}{l} = \mathscr{P}_a - \frac{1}{4l^2}\mathscr{K}_a. \tag{16}$$

For $l \to \infty$, the generators Π_a reduce to ordinary translations, and the de Sitter group contracts to the Poincaré group $\mathscr{P} = \mathscr{L} \oslash \mathscr{T}$. Concomitant with the algebra and group deformations, the de Sitter space $dS(4,1)$ reduces to the Minkowski spacetime $M = \mathscr{P}/\mathscr{L}$, which is transitive under ordinary translations only.

3. KINEMATICS IN MINKOWSKI SPACETIME REVISITED

The action functional describing a free particle of mass m moving in a Minkowski spacetime is

$$S = -mc \int_a^b ds, \tag{17}$$

where

$$ds = (\eta_{ab}\, dx^a dx^b)^{1/2} \tag{18}$$

is the Lorentz invariant interval. Now, the kinematic group of Minkowski is the Poincaré group $\mathscr{P} = \mathscr{L} \oslash \mathscr{T}$, the semi-direct product of the Lorentz $\mathscr{L}$ and the translation group $\mathscr{T}$. The first Casimir invariant of the Poincaré group, on the other hand, is

$$\mathscr{C}_P = \eta_{ab}\, p^a p^b = m^2 c^2 \tag{19}$$

where $p^a = mcu^a$ is the particle four-momentum, with $u^a = dx^a/ds$ the four–velocity. Considering that the action S and the Lagrangian L are related by

$$S = \frac{1}{c} \int_a^b L\, ds, \tag{20}$$

the corresponding Lagrangian can then be written in the form

$$L = -c\,(\eta_{ab}\, p^a p^b)^{1/2} = -c\sqrt{\mathscr{C}_P}. \tag{21}$$

The identity $\eta_{ab}\, p^a p^b = m^2 c^2$ is a weak constraint in the sense that it can be used only after the variational calculus is performed. The resulting equation of motion is

$$\frac{dp^a}{ds} = 0. \tag{22}$$

The equation of motion, therefore, coincides with the conservation of the particle four-momentum, which follows from the invariance of the system under spacetime translation. Its solution determines the geodesics of the Minkowski spacetime. The invariance of the system under Lorentz transformations yields the conservation of the particle angular momentum $\lambda^{ab} = x^a p^b - x^b p^a$, that is,

$$\frac{d\lambda^{ab}}{ds} = 0. \tag{23}$$

4. KINEMATICS IN DE SITTER SPACETIME

4.1. Casimir Invariant

For a spinless particle of mass m, the first Casimir invariant of the de Sitter group is given by [11]

$$\mathscr{C}_{dS} = -\frac{1}{2l^2}\, \eta_{AC}\, \eta_{BD}\, \lambda^{AB} \lambda^{CD}, \tag{24}$$

where

$$\lambda^{AB} = mc\left(\chi^A \frac{d\chi^B}{d\tau} - \chi^B \frac{d\chi^A}{d\tau}\right) \tag{25}$$

is the conserved five-dimensional angular momentum. In terms of the stereographic coordinates $\{x^a\}$, the Casimir invariant (24) assumes the form

$$\mathscr{C}_{dS} = \eta_{ab}\, \pi^a \pi^b - \frac{1}{2l^2}\, \eta_{ac}\, \eta_{bd}\, \lambda^{ab} \lambda^{cd}, \tag{26}$$

where[2]

$$\pi^a \equiv \frac{\lambda^{a4}}{l} = \Omega^2\left(p^a - \frac{k^a}{4l^2}\right), \qquad \pi_a \equiv \frac{\lambda_a{}^4}{l} = p_a - \frac{k_a}{4l^2} \tag{27}$$

[2] Analogously to the generators, we use a parameterization appropriate for a small cosmological constant.

represents the de Sitter momentum, with

$$p^a = mc\,\frac{dx^a}{d\tau} \qquad \text{and} \qquad k^a \equiv \bar{\delta}^a{}_b\, p^b = (2\eta_{cb}x^c x^a - \sigma^2\,\delta^a{}_b)\,p^b, \tag{28}$$

respectively, the linear and the conformal momentum,[3] and

$$\lambda^{ab} = \Omega^2\left(x^a p^b - x^b p^a\right), \qquad \lambda_{ab} = \eta_{ac}x^c p_b - \eta_{bc}x^c p_a \tag{29}$$

represents the orbital angular momentum. In the above expression,

$$\bar{\delta}^a{}_b = 2\eta_{bc}x^c x^a - \sigma^2\,\delta^a{}_b \tag{30}$$

is a kind of conformal Kroenecker delta. Since λ^{AB} is conserved, we have also

$$\frac{d\lambda^{ab}}{d\tau} = 0 \qquad \text{and} \qquad \frac{d\pi^a}{d\tau} = 0. \tag{31}$$

We remark that λ^{ab} is the Noether conserved momentum related to the invariance of the system under the transformations generated by $\mathscr{L}_{ab}$, whereas π^a is the Noether conserved momentum related to the invariance of the system under the transformations generated by Π_a.

4.2. Equations of Motion

Relying on the Minkowski case, the Lagrangian of a spinless particle of mass m in de Sitter spacetime can be assumed to be given by $-c\sqrt{\mathscr{C}_{dS}}$. In the five–dimensional spacetime, however, it is necessary to add a constraint restricting the movement to the de Sitter hyperboloid. In this case, therefore, the Lagrangian turns out to be

$$L = -c\left[\left(\mathscr{C}_{dS}\right)^{1/2} + \beta\left(\eta_{AB}\,\chi^A\chi^B + l^2\right)\right], \tag{32}$$

where β is a Lagrange multiplier. Using Eq. (24), the corresponding action is written as

$$S = -\int_a^b \left[\left(-\frac{1}{2l^2}\,\eta_{AC}\,\eta_{BD}\,\lambda^{AB}\,\lambda^{CD}\right)^{1/2} + \beta\left(\eta_{AB}\,\chi^A\chi^B + l^2\right)\right] d\tau, \tag{33}$$

with $d\tau$ the de Sitter invariant interval (7). Performing a functional variation, and neglecting the surface term coming from an integration by parts, the invariance of the action yields the equation of motion

$$\frac{d^2\chi^A}{d\tau^2} + \left(\frac{1}{l^2} - 2\beta\right)\chi^A = 0. \tag{34}$$

Using the constraints

$$\eta_{AB}\,\chi^A\chi^B = -\,l^2 \qquad \text{and} \qquad \eta_{AB}\,\frac{d\chi^A}{d\tau}\,\frac{d\chi^B}{d\tau} = 1, \tag{35}$$

the value of the Lagrange multiplier is found to be

$$\beta = \frac{1}{l^2}, \tag{36}$$

and the equation of motion reduces to to

$$\frac{d^2\chi^A}{d\tau^2} - \frac{\chi^A}{l^2} = 0. \tag{37}$$

---

[3] Similarly to the identification $p^a = T^{a0}$, with T^{ab} the energy-momentum tensor, the conformal momentum k^a is defined by $k^a = K^{a0}$, with K^{ab} the conformal current [13].

In terms of the stereographic coordinates, it becomes

$$\frac{d}{d\tau}\left(p^a + \frac{1}{4l^2}k^a\right) + \frac{x^c u_c}{l^2\Omega}\left(p^a + \frac{1}{4l^2}k^a\right) - \frac{mc}{l^2\Omega}x^a = 0. \tag{38}$$

The corresponding equation for the covariant components of the momentum is

$$\frac{d}{d\tau}\left(p_a + \frac{1}{4l^2}k_a\right) - \frac{mc\Omega}{l^2}\eta_{ab}x^b = 0. \tag{39}$$

Of course, due to the universality of gravitation, this equation is independent of the mass when written in terms of the four velocity. In fact, it is the same as

$$\frac{d}{d\tau}\left(u_a + \frac{1}{4l^2}\bar{\delta}_a{}^c u_c\right) - \frac{\Omega}{l^2}\eta_{ab}x^b = 0, \tag{40}$$

with $\bar{\delta}_a{}^c$ given by Eq. (30). This is the equation of motion of a spinless particle of mass m in a de Sitter spacetime. Its solutions determine the geodesics of this spacetime.

Using the second of the conservation laws (31), it is possible to obtain separate evolution equations for p^a and k^a. For example, the equation of motion for the linear momentum p^a is found to be

$$\frac{dp_a}{d\tau} - \frac{mc\Omega}{2l^2}\eta_{ab}x^b = 0, \tag{41}$$

or equivalently,

$$\frac{dp^a}{d\tau} + \frac{x^c u_c}{l^2\Omega}p^a - \frac{mc}{2l^2\Omega}x^a = 0. \tag{42}$$

This equation is nothing but the geodesic equation

$$\frac{dp^a}{d\tau} + \Gamma^a{}_{bc}p^b u^c = 0, \tag{43}$$

with $\Gamma^a{}_{bc}$ the Levi–Civita connection of the de Sitter metric (9). On the other hand, the equation of motion for the conformal momentum k^a assumes the form

$$\frac{dk^a}{d\tau} + \frac{x^c u_c}{l^2\Omega}k^a - \frac{2mc}{\Omega}x^a = 0. \tag{44}$$

Differently from the ordinary momentum p^a, which is conserved with a covariant derivative, the conformal momentum k^a is not covariantly conserved. In fact, it is found to satisfy

$$\frac{dk^a}{d\tau} + \Gamma^a{}_{bc}k^b u^c = \frac{2mc}{\Omega}\left[1 - \frac{1}{4l^2}\left(\frac{2}{\Omega^2}u_b u_c x^b x^c - \sigma^2\right)\right]x^a. \tag{45}$$

Put together, however, these two momenta yield the truly conserved total momentum π^a.

5. THE SCALAR FIELD

Let us consider now a de Sitter scalar field ϕ, that is, a field invariant under de Sitter transformations. We begin by considering the Casimir invariant (26), but now written as an operator, that is,

$$\mathscr{C}_{dS} = \eta^{ab}\Pi_a\Pi_b - \frac{1}{2l^2}\eta^{ac}\eta^{bd}\mathscr{L}_{ab}\mathscr{L}_{cd}, \tag{46}$$

with Π_a and $\mathscr{L}_{ab}$ the de Sitter generators, given respectively by Eqs. (15) and (16). From the theory of group representations, one finds [14]

$$\mathscr{C}_{dS} = m^2c^2 + \frac{1}{l^2}[s(s+1) - 2], \tag{47}$$

with m the mass and s the spin of the field ($\hbar = 1$). For a scalar field, $s = 0$, and we get[4]

$$\mathscr{C}_{dS} = \left(m^2 - \frac{2}{l^2 c^2} \right) c^2 \equiv \alpha^2(m) c^2. \tag{48}$$

Now, for a scalar field, we can identify

$$\mathscr{C}_{dS} = -\Box, \tag{49}$$

where

$$\Box = -g_{ab} \nabla^a \nabla^b \tag{50}$$

is the d'Alembertian, with ∇^a the covariant derivative in the Levi–Civita connection of the de Sitter metric (9). Accordingly, the field equation for ϕ is found to be

$$\Box\phi + m^2 c^2 \phi - \frac{R}{6} \phi = 0, \tag{51}$$

where $R = 12/l^2$ is the scalar curvature of the de Sitter spacetime. For a massless field, it reduces to the conformal invariant field equation for ϕ. This shows in a simple manner that the cosmological constant naturally introduces the conformal symmetry in any physical problem. This symmetry is enforced by the second term of the Casimir operator (47). In fact, it is responsible for the curvature term in the Klein–Gordon equation (51), which is essential for the conformal invariance of the equation. Observe that, for the case of the electromagnetic field ($s = 1$), which is naturally conformal invariant in the absence of sources, this term does not contribute to the corresponding field equation.

Equation (51), with zero mass and no self–interaction, has already been obtained in the study of conformally invariant equations for massless particles [15, 16]. It also appeared in connection with the so called improved energy–momentum tensor for the scalar field [17]. Here, it has been obtained simply by considering that, instead of a Lorentz scalar, the field is in a singlet representation of the de Sitter group.

6. QUANTUM PHYSICS

Let us consider again the action (33). Its total variation is given by

$$\delta S = mc \int_a^b \left[\frac{d^2\chi_A}{d\tau^2} - \frac{\chi_A}{l^2} \right] \delta\chi^A d\tau - \int_a^b \left[mc \frac{d\chi_A}{d\tau} \delta\chi^A \right], \tag{52}$$

where we have already used Eq. (36). If we admit now only possible trajectories, the first term in the variation vanishes identically. Then, the second term, with the upper limit considered as variable, gives the differential of the action as a function of the coordinates:

$$\delta S = -mc \frac{d\chi_A}{d\tau} \delta\chi^A. \tag{53}$$

In terms of the stereographic coordinates, it reduces to

$$\delta S = -p_b \delta x^b. \tag{54}$$

Consequently, we can write

$$\frac{\delta S}{\delta x^b} = -p_b, \tag{55}$$

where

$$p_b = mc u_b \tag{56}$$

is the canonical momentum conjugate to the coordinate x^b.

We define now the quantum mechanical operator

$$\hat{p}_b = -i\hbar \partial_b. \tag{57}$$

---

[4] Sometimes $\alpha(m)$ is called "de Sitter mass". However, we prefer not to use this terminology because, strictly speaking, it does not represent a mass [18]. The only mass present is the physical mass m.

As is well known, it satisfies the commutation relation

$$[\hat{x}^a, \hat{p}_b] = -i\hbar\,\delta^a{}_b. \tag{58}$$

Analogously to the momentum, we define the operator

$$\hat{k}_b = \bar{\delta}_b{}^c\,\hat{p}_c, \tag{59}$$

with $\bar{\delta}_b{}^c$ given by Eq. (30). Using the fundamental commutation relation (58), it is easy to see that

$$[\hat{x}^a, \hat{k}_b] = -i\hbar\,\bar{\delta}^a{}_b. \tag{60}$$

Now, as is well known, under a spacetime inversion

$$x^a \to y^a = -\frac{x^a}{\sigma^2}, \tag{61}$$

the translation generators are transformed into the proper conformal generators, and *vice–versa* [13]. Using this property, we obtain

$$[\hat{y}^a, \hat{k}_b] = -i\hbar\,\delta^a{}_b. \tag{62}$$

This means that the conformal momentum k_b is the canonical momentum conjugate to the coordinate y^a.

It is important to observe that the total momentum of the particle is given by $\pi^a = p^a - (1/4l^2)k^a$, with p^a representing the part related to translations, and k^a representing the part related to the proper conformal transformations. This follows from the fact that the de Sitter spacetime is transitive under a combination of translations and proper conformal transformations. The total energy of the particle, therefore, will be given by the time component of π^a, that is,

$$\frac{E}{c} \equiv \pi^0 = p^0 - \frac{1}{4l^2}\,k^0. \tag{63}$$

In other words,

$$E = E_p - \frac{1}{4l^2}\,E_k, \tag{64}$$

with $E_p = p^0$ and $E_k = k^0$. Accordingly, the total momentum operator is

$$\hat{\pi}_b = \hat{p}_b - \frac{1}{4l^2}\,\hat{k}_b, \tag{65}$$

which satisfies the commutation relation

$$[\hat{x}^a, \hat{\pi}_b] = -i\hbar\left(\delta^a{}_b - \frac{1}{4l^2}\,\bar{\delta}^a{}_b\right). \tag{66}$$

This is the quantization rule in de Sitter special relativity. Of course, since π_b is not the conjugate momentum to the coordinate x^a, the right–hand side is not a Kroenecker delta.

The commutation rules, as is well known, are used to construct the uncertainty relations of quantum mechanics. For example, the commutation relation (58) implies that

$$\Delta x^a\,\Delta p_b \geq \frac{1}{2}\,|\langle[\hat{x}^a, \hat{p}_b]\rangle| = \frac{\hbar}{2}\,\delta^a{}_b. \tag{67}$$

Analogously, the commutation relation (60) implies that

$$\Delta x^a\,\Delta k_b \geq \frac{1}{2}\,\left|\langle[\hat{x}^a, \hat{k}_b]\rangle\right| = \frac{\hbar}{2}\,\bar{\delta}^a{}_b. \tag{68}$$

The uncertainty relation for the total momentum, consequently, is

$$\Delta x^a\,\Delta\pi_b \geq \frac{1}{2}\,|\langle[\hat{x}^a, \hat{\pi}_b]\rangle| = \frac{\hbar}{2}\left(\delta^a{}_b - \frac{1}{4l^2}\,\bar{\delta}^a{}_b\right). \tag{69}$$

For small values of the cosmological term Λ, the corrections to ordinary quantum mechanics will be very small. However, for large values of Λ, the corrections coming from the conformal momentum will become important, giving rise to a new quantum mechanics.

7. FINAL REMARKS

A non-vanishing cosmological term Λ introduces the conformal generators in the definition of spacetime transitivity. As a consequence, the conformal transformations will naturally be incorporated in the kinematics of spacetime, and the corresponding conformal current will appear as part of the Noether conserved current. Of course, for a small cosmological term, the conformal modifications become negligible and ordinary physics remains valid. For large values of Λ, however, the conformal contributions to the physical quantities cannot be neglected, and these contributions will give rise to deep conceptual changes. For example, ordinary special relativity, which is based on the Poincaré group, will no longer be true, and must be replaced by a new special relativity based on the de Sitter group. As a consequence, the ordinary notions of energy and momentum will change [4]. The conserved momentum, for example, will now be obtained from the invariance of the physical system, not under translations, but under a combination of translations and proper conformal transformations. It will consequently be given by a combination of ordinary and proper conformal momenta. Energy, which is the time component of the momentum, will change accordingly. Due to the fundamental role played by energy and momentum, these modifications will affect all branches of physics, including of course quantum mechanics. Although these effects may be negligible for small values of the cosmological term Λ, there are situations where this difference could become significant.

As an example of such situation, let us consider the following hypotheses. Taking into account that conformal symmetry has a relevant role at high energies, it is conceivable to assume that a high–energy phenomenon could modify the *local structure of space-time for a short period of time*, in such a way that the immediate neighborhood of a high energy phenomenon would depart from Minkowski and become a de Sitter — or anti-de Sitter — spacetime [7]. According to this hypotheses, around a high–energy experiment there would exist a large Λ, and the local kinematics would consequently be ruled by the de Sitter special relativity. Concomitantly, the conformal symmetry would naturally acquire a relevant role. This scenario fits quite reasonably with the idea that a high–energy experiment should modify the local structure (texture) of spacetime. The important point is that the de Sitter special relativity gives a precise meaning to this change, opening up the door for a possible experimental verification [3]. For an experiment with energy of the order of the Planck energy, the local value of Λ would be of the order $\Lambda \sim 10^{66}$ cm^{-2}, which differs from the observed [19] cosmological constant $\Lambda \sim 10^{-56}$ cm^{-2} by roughly 120 orders of magnitude. The underlying spacetime in this case would approach a cone spacetime [5], which is transitive under proper conformal transformations only. In such extreme situation, the de Sitter special relativity would reduce to a conformal relativity, in which only the conformal notions of momentum and energy would survive. A very peculiar new quantum world would then emerge, whose physics has yet to be developed.

ACKNOWLEDGMENTS

The authors would like to thank FAPESP, CNPq and CAPES for financial support.

REFERENCES

1. R. Aldrovandi, J. P. Beltrán Almeida and J. G. Pereira, *Class. Quant. Grav.* **24**, 1385 (2007) (*Preprint* gr-qc/0606122).
2. H. Y. Guo, C. G. Huang, Z. Xu and B. Zhuo, *Phys. Lett.* A **331**, 1 (2204) (*Preprint* hep-th/0403171).
3. R. Aldrovandi, J. P. Beltrán Almeida, C. S. O. Mayor and J. G. Pereira, *Lorentz Transformations in de Sitter Relativity* (*Preprint* gr-qc/0709.3947).
4. R. Aldrovandi, J. P. Beltrán Almeida and J. G. Pereira, *Int. J. Mod. Phys.* D **13**, 2241 (2004) (*Preprint* gr-qc/0405104).
5. R. Aldrovandi, J. P. Beltrán Almeida and J. G. Pereira, *J. Geom. Phys* **56**, 1042 (2006) (*Preprint* gr-qc/0403099).
6. S. M. Carroll, *Living Rev. Rel.* **4**, 1 (2001) (*Preprint* astro-ph/0004075v2)
7. F. Mansouri, *Phys. Lett.* B **538** 239 (2002) (*Preprint* hep-th/0203150)
8. S. Weinberg, *Gravitation and Cosmology*, Wiley, New York, 1972.
9. S. W. Hawking and G. F. R Ellis, *The Large Scale Structure of Space-Time*, Cambridge University Press, Cambridge, 1973.
10. R. Aldrovandi and J. G. Pereira, *An Introduction to Geometrical Physics*, World Scientific, Singapore, 1995.
11. F. Gürsey, in *Group Theoretical Concepts and Methods in Elementary Particle Physics*, ed. by F. Gürsey, Istanbul Summer School of Theoretical Physics, Gordon and Breach, New York, 1962.
12. E. Inönü, in *Group Theoretical Concepts and Methods in Elementary Particle Physics*, ed. by F. Gürsey, Istanbul Summer School of Theoretical Physics, Gordon and Breach, New York, 1962; E. Inönü and E. P. Wigner, *Proc. Natl. Acad. Scien.* **39**, 510 (1953).
13. S. Coleman, *Aspects of Symmetry*, Cambridge University Press, Cambridge, 1985.

14. J. Dixmier, *Bulletin de la S. M. F.* **89**, 9 (1961) (available at http://www.numdam.org).
15. R. Penrose, *Proc. Roy. Soc. (London)* **284 A**, 204 (1965).
16. S. Deser and A. Waldron, *Phys. Lett.* **B 603**, 30 (2004) (*Preprint* hep-th/0408155).
17. C. G. Callan, S. Coleman and R. Jackiw, *Ann. Phys. (NY)* **59**, 42 (1970).
18. V. Faraoni and F. I. Cooperstock, *Eur. J. Phys.* **19**, 419 (1998) (*Preprint* physics/9807056).
19. A. G. Riess *et al*, *Ap. J.* **116**, 1009 (1998); S. Perlmutter *et al*, *Ap. J.* **517**, 565 (1999); P. de Bernardis *et al*, *Nature* **404**, 955 (2000); S. Hanany *et al*, *Ap. J. Letters* **545**, 5 (2000).

From Como to Copenhagen: The Beginnings and Ends of Complementarity

Arkady Plotnitsky

Theory and Cultural Studies Program, Department of English
Purdue University, West Lafayette, IN 47907, USA

Abstract. This paper examines the transformation of Bohr's interpretation of quantum phenomena and quantum mechanics, known as complementarity, from the version presented in his famous Como lecture (1927), which introduced complementarity, to a new version that was developed by Bohr shortly thereafter, in part under the impact of his intervening exchanges with Einstein. While the existence of more than one version of Bohr's complementarity (and of still different versions of the Copenhagen interpretation by others) has been noted, Bohr's rethinking itself of complementarity has rarely been adequately addressed. Apart from refining and indeed correcting the Como argument, the transformation in question was, I argue, defined by a shift from a more philosophical to a more experimental argument. My aim is to explore both this shift and the complexity of the relationships between physics and philosophy in Bohr's thinking in order to understand better the beginnings and the *ends*, also in the sense of *aims*, of Bohr's complementarity.

Keywords: complementarity, epistemology, measurement instruments, quantum objects, quantum phenomena
PACS: 01.65.+g, 01.70.+w , 03.65.-w , 03.65.Ud , 03.65.Ta

1. INTRODUCTION

This paper considers Niels Bohr's thinking about quantum phenomena and quantum theory that emerged in the wake of Werner Heisenberg's and Erwin Schrödinger's discovery of quantum mechanics and then Heisenberg's discovery of the uncertainty relations. This thinking led Bohr to his idea of complementarity, which was introduced in the so-called Como lecture of 1927 and gave rise to what we now call the Copenhagen interpretation of quantum mechanics or, more accurately, to "the spirit of Copenhagen," as Heisenberg called [1, p. iv]. There is a spectrum of interpretations, clustered (not always with due discrimination) under the rubric of the Copenhagen interpretation. Some of them were influenced by Bohr's ideas and, thus, may be said to be in "the spirit of Copenhagen," but many of them, even among those in the spirit of Copenhagen, diverge (often significantly) from Bohr's views. As in the Roman Empire all roads led to Rome, in the empire of quantum theory all roads seem to start at and move away from Copenhagen, which may be seen as a point of departure for our thinking concerning quantum theory mostly in the sense of moving away from it in directions different from those envisioned by Bohr. Indeed, one can speak of more than one Copenhagen interpretation even in Bohr's case, since over the years Bohr refined and even changed his interpretation of (jointly) both quantum phenomena and quantum mechanics, which he designated as 'complementarity' as well, the term that, unlike 'the Copenhagen interpretation,' is usually associated with Bohr's own interpretation. As it happened, the Como interpretation of quantum phenomena and quantum mechanics (to the degree one could speak of a full-fledged interpretation offered in the lecture) proved to be short lived. It appeared to have dissolved into the mountain air over Lake Como, as it were, as Bohr's train took off and Bohr was on his way from Como to Copenhagen. This last claim may be an exaggeration, especially given that a version of the Como lecture, "The Quantum Postulate and the Recent Development of Quantum Theory," was published later in 1927, and that Bohr, accordingly, continued to think along the same lines at least for a few months ([2], v. 1, pp. 92-101). Apart, however, from the fact that Bohr was never satisfied with the Como argument, by the time of this publication his thinking was on its way to a different version of his interpretation, which was developed within barely a year, in part under the impact of Bohr's exchanges with Einstein in 1927-1928. This version was eventually replaced as

CP962, *Quantum Theory, Reconsideration of Foundations – 4*,
edited by G. Adenier, A. Yu. Khrennikov, P. Lahti, V. I. Man'ko, and Th. M. Nieuwenhuizen

well, sometime in the late 1930s, by yet another version, following Bohr's exchange with Einstein concerning the EPR experiment. One can also locate certain nuances and modifications in between as well, but, I would argue, they are of a more incremental nature.

Bohr's post-EPR version of complementarity is his most rigorous interpretation of quantum phenomena and quantum mechanics, and it may, accordingly, be seen as most significant for the development of Bohr's thought and for the history of quantum theory. One might also argue, however, as I shall do here, that Bohr's transformation of his thinking and the new version of complementarity developed in the immediate wake of his Como argument were crucial in both of these contexts as well. This version of complementarity first appeared in print in Bohr's lesser known and rarely discussed 1929 article "The Quantum of Action and the Description of Nature." This article, in addition to the Como lecture, is my primary concern here, and my main aim is to explore how Bohr begins his journey toward his ultimate epistemology of quantum mechanics, developed in the wake of EPR's argument.[1]

First, however, I would like to comment on the idea and language of complementarity. Although an umbrella term for an interconnected set of concepts, 'complementarity' may be said to have two primary meanings in Bohr's work. Both of these meanings are retained throughout his writings, even though some aspects of complementarity (pertaining to either meaning) underwent , sometimes significant, refinements and revisions.

The first meaning of complementarity may be designated as *narrow*. It is defined by a *particular concept* of the mutual exclusivity of certain entities, phenomena, or conceptions. I stress "a particular concept" because this concept is not restricted to this mutual exclusivity alone and has certain other features that are equally crucial. More rigorously, complementarity in the narrow sense is defined as follows. It designates a) the mutual exclusivity of certain phenomena, entities, or conceptions (and hence, the impossibility of applying them at the same time); b) the possibility of applying each one of them separately at any given point; and, finally, c) the necessity of using all of them at different points for a comprehensive account of the totality of phenomena that we must consider.

The concept itself thus defined is clearly general in nature and could apply in many domains of investigation beyond physics, as Bohr often noted. (He also made some suggestions for such applications, although they were mostly tentative and speculative in nature.) In quantum physics, the concept refers to certain particular physical phenomena, "quantum phenomena," the mutual exclusivity of which arises in view of the irreducible role of measuring instruments in the emergence of these phenomena. In Bohr's 1949 formulation, "evidence obtained under [all possible] different experimental conditions cannot be comprehended within a single picture [since parts of this evidence are mutually exclusive], but must be regarded as *complementary* in the sense that only the totality of the [observable] phenomena exhausts the possible information about the [quantum] objects [themselves]" ([2], v.2, p. 40; also [2], v. 3, p. 4).

The specific complementary configurations considered by Bohr (or by others) or a given definition (or interpretation) of the phenomena themselves involved may be different. Thus, certain complementary configurations considered at earlier stages of Bohr's thinking, specifically in the Como lecture, were abandoned by Bohr as his thinking developed, and he refined his understanding of quantum phenomena themselves and ultimately introduced a new concept of quantum phenomena in the wake of the EPR experiment. These revisions made the conception of complementarity in the narrow sense more experimentally grounded than in the Como lecture, which centered on certain more philosophically configured complementary entities. The *experimental grounding* of complementarity and the connections between the philosophical concepts in question, such as causality, and quantum phenomena was crucial to the Como argument as well. The difference between the Como and the subsequent versions of complementarity is defined by the relative balance of the physical and the philosophical aspects found in Bohr's corresponding arguments. In Bohr's post-EPR version of complementarity, the "information about the objects," invoked in Bohr's formulation cited above, only pertains to the *effects* of the interaction between these objects and the measuring instruments (as manifest in the latter) rather than anything concerning quantum objects themselves. Quantum phenomena themselves are now defined by Bohr in terms of such effects appearing under rigorously specified experimental conditions. Quantum objects are responsible for the emergence of these effects or phenomena, but only through their interactions with measuring instruments or other macro objects, without, it appears, allowing us to gain any knowledge about them or their behavior. Bohr's formulation also tells us that the complementarity of certain quantum phenomena is unavoidable and, hence, is essential for a comprehensive account of the totality of these phenomena. It follows that the concept of complementarity in the narrow sense captures a certain fundamental feature of nature, regardless of any theory or any interpretation of a given theory. It must, accordingly, be taken into account by any interpretation of quantum

[1] I shall not discuss in detail Bohr's post-EPR version of complementarity, which I have considered extensively elsewhere [3]. I shall, however, explain some of the key features of this interpretation insofar as they connect and help us to appreciate Bohr's earlier thinking and his trajectory toward his ultimate understanding of quantum phenomena and quantum theory.

mechanics, which must also give an interpretation to these phenomena themselves, in view of the apparently irreducible role of measuring instruments in forming them.

This interpretation becomes part of the second meaning of 'complementarity,' which may be designated as *broad*. It refers to Bohr's overall interpretation of, jointly, quantum phenomena and quantum mechanics, in part based on the narrow sense of complementarity just explained, but only in part. For, equally crucial is that Bohr sees quantum phenomena as defined by the irreducible role of measuring instruments in their constitution throughout his work. His understanding of this role and of the epistemology of quantum phenomena evolved, however. Ultimately, he is led to a very radical epistemology, according to which any description or even conception concerning quantum objects themselves or their behavior is no longer available and is even prohibited, or in Bohr's language "*in principle* excluded" ([2], v. 2, p. 62). The only phenomena available to quantum theory *are* the effects of the interaction between quantum objects and measuring instruments manifest in these instruments and describable by means of classical physics. The mathematical formalism of quantum mechanics is no longer seen as describing any physical processes (quantum or other) in the way classical mechanics describes the behavior of classical objects, but only provides the set of algorithms for predictions of the numerical data associated with the effects observed in measuring instruments and described by means of classical physics. These predictions are, in general, statistical, which, however, corresponds to the nature of quantum phenomena, since the identically prepared experiments generally lead to different recordings of their outcomes ([2], v. 2, p. 73).

2. COMPLEMENTARITY 1: THE SPIRIT OF COMO

As its title, "The Quantum Postulate and the Recent Development of Atomic Theory," suggests, Bohr's argument in the Como lecture is grounded in his concept of the quantum postulate, due to Planck's discovery of, as Planck called it, "the quantum of action," h. The concept never lost its significance for Bohr, even as other concepts, beginning with complementarity, moved to center stage. Indeed, Bohr's work on complementarity may be seen as a life-long effort to give the proper meaning to "the quantum postulate" and, thus, to "Planck's discovery of the *elementary quantum of action*" ([2], v. 1, p. 92; emphasis added). Beginning with Planck's discovery of his black body radiation law, quantum discontinuity (reflected in the quantum postulate) becomes a defining part of our understanding of the ultimate constitution of nature. The quantum postulate itself may formulated as follows:

> *Any observable phenomenon of quantum physics is either individual, discrete (discontinuous) or is a discrete sum, indeed a finite (if possibly very large) sum, of such individual phenomena or events; and any possible record of quantum events is either that of an individual phenomenon or the sum of such individual records.*

The summing-up pertains to records of such events occurring over certain periods of time, as in the case of particle-like quantum objects' collisions with the screen in the double-slit experiment. It follows that there are no continuous individual quantum phenomena. Certain collective observable phenomena, defined by the effects of the interaction between quantum objects and measuring instruments manifest in the latter, are discrete analogues of wave-like phenomena, insofar as we can, with due qualifications, speak of such wave-like features as "diffraction" or "interference." Such phenomena appear under certain experimental conditions, once a sufficiently large number of events are accumulated, as in the double-slit experiment, when there are no devices which allow us to know, even in principle, through which slit each quantum object involved passes. In other words, the individual phenomena comprising such composite wave-like phenomena are discrete or particle-like in the sense that they have features (such as punctual traces registered in silver screens) analogous to those found in the interactions between classical particle-like objects and measuring instruments.

The elaboration of the physical content of the quantum postulate just given follows Bohr's later views in presenting the postulate in terms of the effects of the interaction between quantum objects and measuring instruments manifest in those instruments, rather than in terms of the properties of quantum objects or their behavior, as my formulation of the postulate would in principle permit. The appeal to such properties precludes an adequate account of the situation according to the view developed in Bohr's post-EPR works, although even the Como lecture and, especially, "The Quantum of Action and the Description of Nature" make significant steps toward this view. In this view, all available phenomena are defined only in terms of certain, sometimes correlated, recorded effects, "practically irreversible amplification effects," such as the click of a photo-detector or the blackening of a grain of a photographic emulsion, rather than in terms of properties of quantum objects themselves ([2], v. 2, p. 51). The assignment of such properties is prevented by "the *impossibility of any sharp separation between the behavior of atomic objects and the interaction with the measuring instruments which serve to define the conditions under which the [observable] phenomena [in question in quantum physics] appear* ([2], v. 2, pp. 39-40). While grounding the indivisibility or wholeness of quantum phenomena in Bohr's sense, this impossibility also entails an epistemological discontinuity as regards quantum objects and processes (seen as being beyond any possible description or even conception), which discontinuity accompanies the discrete and hence in turn discontinuous nature of each observed phenomenon. Bohr's statement itself, technically, allows one to think in terms of certain properties of quantum objects at the time of their interaction with the measuring instruments, a view that Bohr entertains at earlier stages of his thinking. In Bohr's ultimate understanding, no reference to such properties, again, even single such properties (as against only, jointly, conjugate properties, due to the uncertainty relations) can be made unambiguously.

By the same token, all key quantum features—discreteness, discontinuity, individuality, indivisibility, atomicity, and, in a somewhat different way, wave-like continuity and probability—are transferred to the level of phenomena or effects in this sense. This transfer defines Bohr's new concept of 'atomicity' vis-à-vis the classical, Democritean idea of atomicity as "the limited divisibility of matter" ([2], v. 2, p. 33). This conception requires an adjustment of our usual understanding of atomicity, since the terms of atomicity now apply to certain physically composite entities (each consisting of an immense number of "atoms" in the usual sense), rather than to elementary physical entities, whether quantum objects themselves or even point-like traces of physical events. For, each such "atomic" entity involves the whole experimental arrangement, defined, again, by certain effects of the processes taking place by virtue of this arrangement. In other words, Bohr's concept of phenomenon or "atomicity" applies to individual phenomena as outcomes of single experiments. Collective configurations must be seen as, in turn, disconnected, discrete collectivities of individual phenomena. To quantum objects themselves one can no longer ascribe any conceivable physical or other properties, including "quantum" or "objects."

These concepts emerged as the result of a protracted struggle (over two decades) on Bohr's part, shaped by his debate with Einstein. The Como lecture is Bohr's only article on complementarity that was not affected by this debate, and it is tempting, even if meaningless, to contemplate how Bohr's thinking would have developed had Einstein not intervened critically into the debate concerning quantum theory. The lecture, however, helped Bohr to establish a platform for his subsequent thinking and his debate with Einstein, via the concept of the quantum postulate, seen as expressing the very essence of quantum theory, "characterized by the acknowledgment of a fundamental limitation in the classical physical ideas when applied to atomic phenomena." "The situation thus created," Bohr added, "is of a peculiar nature, since our interpretation of the experimental material rests essentially upon the classical concepts" ([2], v. 1, p. 53). These statements indicate the complexity (often missed) of Bohr's view of the role of the classical physical concepts as essential and yet limited, and specifically as inapplicable to quantum objects themselves and their behavior. The quantum postulate itself reflects or even *symbolizes* this situation. According to Bohr: "Notwithstanding the difficulties which, hence, are involved in the formulation of the quantum theory, it seems, as we shall see, that its essence may be expressed in the so-called quantum postulate, which attributed to any atomic processes an essential discontinuity, or rather individuality, completely foreign to the classical theories and symbolized by Planck's quantum of action" ([2], v. 1, p. 53).

Bohr's hesitation is worth noting, as is his appeal to the "essential" nature of these features or to the role of the classical concepts, used by Bohr in its direct sense of fundamental and irreducible, the same as in his appeal to "a *fundamental* limitation in the classical physical ideas." The essential discontinuity and the essential individuality were eventually reinterpreted in terms of Bohr's later concepts of phenomenon and atomicity, as just discussed, and connected by making this discontinuity or discreteness refer to that between individual phenomena in Bohr's sense, which defines all quantum phenomena. Bohr's appeal to the symbolic nature of Planck's quantum of action, h, is in part due to his symbolic view of all quantum-mechanical formalism. This formalism is seen as symbolic because, formally analogous to classical mechanics, it does not describe the physical behavior of quantum objects or any physical processes in space and time, including, it is worth stressing, between measurements. That latter is a common and persistent view (also expressed in terms of a "collapse" of the wave function during the act of measurement) which is sometimes attributed to Bohr as well, I would argue, mistakenly, at least as concerns his views subsequent to the Como lecture. As I shall discuss presently, the Como argument might suggest that Bohr had briefly contemplated a similar type of view at some point. According, however, to his view from 1929 on, the wave function only serves to predict the outcomes of certain specified experiments on the basis of the data obtained in certain previously performed experiments. Correlatively, Planck's quantum of action is also symbolic insofar as it should be seen as merely implying the strictly discrete, particle-like, nature of quantum objects, as opposed to their wave-like character, since both of these aspects still need to be considered, ultimately as pertaining to certain mutually exclusive effects registered in measuring instruments. Bohr elaborates as follows:

> This postulate implies a renunciation as regards the causal space-time co-ordination of atomic processes. Indeed our usual description of physical phenomena is based entirely on the idea that the phenomena concerned may be observed without disturbing them appreciably. This appears, for example, clearly in the theory of relativity, which has been so fruitful for the elucidation of the classical theories. As emphasized by Einstein, every observation or measurement ultimately rests on the coincidence of two independent events at the same space-time point. Just these coincidences will not be affected by any differences which the space-time co-ordination of different observers otherwise may exhibit. Now, the quantum postulate implies that any observation of atomic phenomena will involve an interaction with [between a quantum object and?] the agency of observation not to be neglected. Accordingly, an independent reality in the ordinary physical sense can neither be ascribed to the phenomenon [object?] nor to the agencies of observation. After all, the concept of observation is in so far arbitrary as it depends upon which objects are included in the system to be observed. Ultimately, every observation can, of course, be reduced to our sense perceptions. The circumstance, however, that in interpreting observations use has always to be made of theoretical notions entails that for every particular case it is a question of convenience at which point the concept of observation involving the quantum postulate with its inherent "irrationality" is brought in. ([2], v. 1, pp. 53-54)

The passage contains most of the ingredients of Bohr's subsequent conceptions and of his ultimate, post-EPR epistemology of quantum mechanics, although the key distinction between (unobservable and inconceivable) quantum objects and (observable) phenomena manifest in measuring instruments is not yet developed here. As I indicated in the text by my question marks, the passage appears to indicate some confusion or at least imprecision in expression in this respect. The same type of tension is reflected in the peculiar concept of "the individual," used as a noun in the article. This concept foreshadows the eventual move from applying any concepts to individual quantum objects to individual phenomena, defined in terms of measuring instruments impacted by quantum objects, to which, again, no description or conception, including that of individuality, could apply. An emphasis on and the very notion of "disturbance" also contribute to the problems of Bohr's argument in the Como lecture. At one level, this emphasis is useful in indicating that in the case of quantum phenomena the interference of measurement cannot be avoided and, hence, no independent behavior of quantum objects can be observed and, ultimately, theoretically represented. On the other hand, the concept of disturbance is problematic since it may imply the possibility of the, in principle, specifiable undisturbed behavior of quantum objects. For this reason the concept of "disturbance" is abandoned by Bohr in his later work, along with that of "creation" of the attributes of atomic objects by measurements ([2], v. 4, p. 104; [2], v. 2, pp. 64, 73). These concepts are replaced by a more rigorous concept of the irreducible role of measuring instruments in the constitution of quantum phenomena. For the moment, complementarity (in the narrow sense) is introduced in the Como lecture as the immediate consequence of the quantum postulate:

> This situation has far-reaching consequences. On one hand, the definition of the state of a physical system, as ordinarily understood [i.e. in classical physics], claims the elimination of all external disturbances. But in that case, according to the quantum postulate, any observation will be impossible, and, above all, the concepts of space and time lose their immediate sense. On the other hand, if in order to make observation possible we permit certain interactions with suitable agencies of measurement, not belonging to the system, an unambiguous definition of the state of the system is naturally no longer possible, and there could be no question of causality in the ordinary sense of the word. The very nature of the quantum theory thus forces us to regard the space-time co-ordination and the claim of causality, the union of which characterizes the classical theories, as complementary but exclusive features of the description, symbolizing the idealization of observation and definition respectively. Just as the relativity theory taught us that the convenience of distinguishing sharply between space and time rests solely on the smallness of the velocities ordinarily met with compared to the velocity of light, we learn from the quantum theory that the appropriateness of our usual space-time description depends entirely on the small value of the quantum of action as compared to the actions involved in ordinary sense perceptions. Indeed, in the description of atomic phenomena, the quantum postulate presented us with the task of developing a "complementarity" theory the consistency of which can be judged only by weighing the possibility of definition and observation. ([2], v. 1, pp. 54-55)

This task will take Bohr a while to complete, and it is certainly not yet accomplished in the Como lecture, as Bohr realized. The argument contains a number of problems, some of which had lasting consequences in quantum theory and its interpretation, although not in Bohr's own thinking. These difficulties are understandable and hardly surprising given the difficulty of the task. It took Bohr close to ten years and the debate with Einstein, leading to the EPR argument, and even longer, about twenty years, to finalize his "complementarity" theory. First of all, the complementarity "of the space-time co-ordination and causality" was to disappear quickly from Bohr's writings—and for good reasons, to be explained presently. The concept itself of complementarity is not sufficiently delineated either. Bohr speaks of "complementary but exclusive features of the description," without properly defining the meaning of "complementary." Ultimately, as explained above, Bohr came to define the complementary features more rigorously as mutually exclusive and, hence, never applicable simultaneously, and yet each possible to ascertain separately at a given point (depending on the alternative experimental set-ups we choose) and all of them necessary for a comprehensive theoretical framework accounting for the totality of quantum phenomena. In describing classical phenomena, such features could be ascertained simultaneously, which fact indeed defines classical physics; in the case of quantum phenomena, they cannot.

The particular claim in question concerning the complementarity of the space-time coordination and the claim of causality is problematic, even vis-à-vis Bohr's own argumentation in the lecture, although Bohr is of course right to say that the union of both is found and characterizes the classical theories in physics and even relativity. There could not be such a complementarity in quantum mechanics because there is *never* causality or determinism there, even at the level of effects and phenomena, manifest in measuring instruments, of the latter version, where this absence of causality becomes clear, let alone at the level of quantum processes themselves.[2] There is none in Bohr's later writings, and, as will be seen presently, this absence is in fact implicit or becomes required to make the overall argument consistent, given other propositions advanced by Bohr. In Bohr's later view, there could be, in the first place, no definition of the state of the system in question, and it is this lack of the definition of the state of the system that makes this claim impossible. In his reply to EPR, Bohr will speak of "a *final renunciation* of the classical ideal of causality" ([2], v. 4, p. 75; emphasis added) and of complementarity itself as "a rational *generalization* of the ... ideal of causality," which is, again, quite different from seeing causality as complementary to something else ([2], v. 2, p. 41; emphasis added). The Como lecture also speaks of "a consistent theory of atomic phenomena, which may be considered as a rational generalization of the causal space-time description of classical physics" ([2], v. 1, p. 87). Besides a significant nuance in the formulation itself, however, it still refers to the complementarity of coordination and causality, which is, again, manifestly absent in Bohr's later writing. The complementarity of observation and definition, and Bohr's appeal to these concepts themselves are gone as well.

The problem just discussed is far from inconsequential, and its troublesome implications are often found in the literature, for example, in the classic books of Dirac ([4]) and von Neumann ([5]), although Bohr's Como argument is not always the source or the only source of these difficulties. In particular, Bohr's elaborations just cited may suggest that there is somehow a causal evolution of the quantum system itself under investigation, governed by Schrödinger's equation, which evolution is discontinuously disrupted, or terminated altogether, by an act of observation or measurement. This type of view, again, often expressed in terms of the "collapse" of the wave function by an act of measurement came to represent many approaches to quantum mechanics and is persistent in the physics community. It is sometimes associated with the Copenhagen view and some attribute it to Bohr as well. No statement expressing or significantly supporting this view appears to be found in Bohr's writings; and even though the Como lecture is less definitive in this regard, it would by difficult to argue that Bohr quite held such a view even there, especially given his understanding of Schrödinger's wave mechanics as *symbolic*. As he says: "In fact, wave mechanics, just as the matrix theory, on this view represents a *symbolic transcription* of the problem of motion of classical mechanics adapted to the requirements of quantum theory" ([2], v. 1, p. 75; emphasis added). Bohr never uses the language of "collapse" either, and it would indeed be meaningless in Bohr's interpretation, at least, again, from 1929 on.

In Bohr's view, Schrödinger's equation may only be seen as charting exactly and, in this sense, *determining* the "course" of *probabilities* related to the outcomes of future measurements, *but only such probabilities*. Hence, the application of the term "deterministic" to Schrödinger's equation is at best misleading, although it is often found in

[2] It is true that one can conceive of or even theorize quantum processes as (ontologically) *causal* without ever allowing us to make deterministic predictions concerning the outcome of quantum experiments, even ideally, thus making our predictions statistical or even exactly the same as those of the standard quantum mechanics. All currently available approaches, such as Bohmian theories, that yield the same predictions as the standard quantum mechanics (which predictions are, again, fully in accord with all experimental findings thus far), are nonlocal and appear bound to be in view of Bell's and related theorems. In any event, that quantum processes can ever be assigned causality, hidden or not, is not the view Bohr takes in his later thinking. Bohr's Como argument has great difficulties in consistently maintaining causality even as complementary to the space-time coordination, and ultimately fails to do so.

the literature without any qualification. For Bohr, Schrödinger's equation or any other part of the mathematical machinery of quantum mechanics cannot be seen as describing or even referring, however indirectly, to any causal or (and these are linked in turn) continuous evolution or, again, any other properties of quantum systems themselves or their behavior, considered independently. As a "mechanics," quantum mechanics, in either form, is only a *symbolic* mechanics. That is, in some respects, it assumes the formal structure of classical mechanics, say, in its Hamiltonian form, but only serves as a set of algorithms for the predictions concerning the outcomes of certain possible experiments on the basis of certain previously performed experiments. All these experiments irreducibly involve measuring instruments, insofar as quantum objects cannot be meaningfully considered apart from the interaction with these instruments and insofar as all the data that can be meaningfully considered in quantum theory is only manifest (as classical data) in these instruments. These predictions are, again, in general, statistical, even and in particular as concerns the outcome of individual events. As I said, however, this circumstance also corresponds to the nature of quantum phenomena, since the identically prepared experiments in general lead to different recording of their outcomes. Such an identical preparation itself is always possible, since we can control the instruments themselves classically.

Thus, Bohr's overall argument in the Como lecture persistently undermines the possibility of a causal description of the independent behavior of quantum systems and, specifically, the possibility that Schrödinger's equation provides such a description. Against the letter of some of his claims, Bohr's argument may, in its spirit, be closer to abolishing causality altogether than keeping it as complementary to space-time coordination even in the Como lecture. This argument, as I said, still allows one to associate certain properties (under the constraints of the uncertainty relations) with quantum objects at the time of measurement, as Bohr appears to do in the lecture and elsewhere in his writings prior to and even in his reply to EPR. As Bohr's elaborations in the Como lecture introduce new concepts and make major steps forward, they also reflect difficulties and even confusions, inevitable in this process. Ultimately, the contributions far outweigh the problems, beginning with the idea of complementarity. The lecture thus is both a major accomplishment in its own right and a major platform for the future development of Bohr's ideas and argumentation concerning quantum phenomena, quantum theory, and complementarity. Bohr makes crucial steps on this road towards what can be more properly seen as complementarity in the spirit of Copenhagen within just several months from the publication of the Como lecture. The key works completed in 1929, "The Quantum of Action and the Description of Nature" and "Introductory Survey" to *Atomic Theory and the Description of Nature* (Volume 1 of *The Philosophical Writings of Niels Bohr* [2]), manifest these changes in Bohr's understanding of quantum phenomena and quantum mechanics.

3. COMPLEMENTARITY 2: THE RISE OF THE SPIRIT OF COPENHAGEN

Arguably the most significant change from the Como argument is that the space-time coordination, or the application of the position and time variables, is no longer considered as complementary to the claim of causality but instead to the applications of the laws of conservation of momentum and energy, respectively. Rather than only the complementarity of the application of two different concepts, this view expressly entails the complementarity of two different *experimental arrangements*, where the corresponding measurements and the applications of physical concepts and laws become mutually exclusive, in accordance with, and as an interpretation of, Heisenberg's uncertainty relations. The irreducibly statistical nature, again, *on the experimental grounds*, of our predictions concerning quantum phenomena, even and in particular individual ones, and of quantum mechanics itself, as a theory properly responding to this situation, is related to this particular complementarity as well. Quantum processes themselves are not only seen as irreducibly acausal but also, along with quantum objects, as being beyond any possible description or even conception. Although fully rigorously worked out by Bohr only by the late 1930s, this radical suspension of any possible quantum-level ontology was transparent in the thinking of Bohr and his followers already at this point, as Schrödinger noted in his cat-paradox paper, albeit by way of a very different assessment of this argumentation, which he saw as "a doctrine born of distress." Indeed, as Schrödinger also noted, the lack of causality is an immediate consequence of the assumption that quantum objects and processes are beyond any possible description or conception, since causality would be a partial conception or description concerning quantum processes. He said: "if a classical state does not exist at any moment, it can hardly change causally" ([6], p. 154). Once, however, causality is no longer applicable to our description of quantum processes or phenomena, it cannot be part of any complementarity either, as against the conservation laws, which can be fully maintained by quantum theory apart from causality, to which they are usually linked in classical physics.

Bohr's 1929 article "The Quantum of Action and the Description of Nature" is the first article to exhibit this decisive transition in Bohr's thought. Bohr changes his terminology from 'complementarity' to 'reciprocity,' perhaps in view of the problems of the Como lecture and the less than unenthusiastic response it had received in the physics community. Bohr quickly returns to the language of complementarity, which is more effective and more precise, ironically, especially in view of the adjustments he made in his argument in the article. One can clearly see advantages of the language of complementarity in Bohr's "Introductory Survey" (also written in 1929) to his *Atomic Theory and the Description of Nature* ([2], v.1), first published (in German) in 1931, and his important unpublished lecture of 1931, "Space-Time Continuity and Atomic Physics" ([7]. v. 6, pp. 361-70). The "Introductory Survey" even partly recasts the Como lecture in these new terms by stressing those of its aspects that are conducive this recasting, and, notably, it never mentions the complementarity of the space-time co-ordination and the claim of causality. The survey invokes our being forced "step by step to forego a causal description of the behavior of individual atoms in space and time" and "the renunciation of causality in the quantum-mechanical description" ([2], v. 1, p. 4). From this vantage, the Como argument may be seen as one of these steps. From the outset, Bohr (this continues the Como argument) defines quantum mechanics as a "symbolic" theory. Although radically different from classical mechanics in its key physical, mathematical, and epistemological features, quantum mechanics is argued to be "a natural generalization of the classical mechanics with which in beauty and self-consistency it may well be compared" ([2], v. 1, p. 92-93). Bohr writes:

> This goal [of developing quantum mechanics] has not been attained, still, without a *renunciation* of the causal space-time mode of description that characterizes the classical physical theory theories which have experienced such a profound clarification through the theory of relativity. In this respect, the quantum theory may be said to be a disappointment, for the atomic theory arose just from the attempt to accomplish such a description also in the case of phenomena, which, in our immediate sense impressions, do not appear as motions of material bodies. From the very beginning, however, one was not unprepared in this domain to come upon a failure of the forms of perception adapted to our ordinary sense impressions. We know now, it is true, that the often expressed skepticism with regard to the reality of atoms was exaggerated; for, indeed, the wonderful development of the art of experimentation has enabled us to study the effects of individual atoms. Nevertheless, the very recognition of the limited divisibility of physical processes, symbolized by the quantum of action, has justified the old doubt as to the range of our ordinary forms of perception when applied to atomic phenomena. Since, in the observation of these phenomena, we cannot neglect the interaction between the object and the instrument of observation, the question of the possibility of observation again comes to the foreground. Thus, we meet here, in a new light, the problem of the objectivity of phenomena which has always attracted so much attention in philosophical discussion. ([2], v. 1, pp. 92-93; emphasis added)

The contrast with the Como argument is striking. The complementarity of the space-time coordination and the claim of causality, and its relation to the question of observation and definition are gone, along with, to begin with, causality itself. What replaces them as the grounding feature of quantum epistemology is the *distinction*, which shapes all Bohr's arguments concerning complementarity from this point on, between the unobservable quantum *objects* and observable *phenomena*, defined through the *effects* of the interactions in question upon the measuring instruments. It is worth noting that, under these conditions, it is still in principle possible to speak of certain properties of quantum objects, unobservable but ascertainable through these effects (at least at the time of measurement), which view Bohr, again, eventually abandons. One can especially note the appearance and analytical *effectiveness* of the language of *effects*, which in turn becomes important for Bohr. If the "the wonderful development of the art of experimentation has enabled us to study the effects of individual atoms," it has also limited us to the observation and *descriptive* account of these effects (*predicted* by means of the quantum theory), rather than of quantum objects and processes.

These considerations imply, correlatively, both the irreducibly statistical nature of quantum mechanics and the irreducibility of certain complementary features (and both are in turn correlative to the uncertainty relations), but now, as against the Como lecture, defined in terms of experimentally defined observable phenomena. According to Bohr:

> We know now that for material particles as well as for light different conceptual pictures are necessary to account completely for the phenomena and to furnish a unique formulation of the statistical laws that govern the nature of observation. The more clearly it appears that a uniform

formulation of the quantum theory in classical terms is impossible, the more we admire Planck's happy intuition in coining the term "quantum of action" which directly indicates a renunciation of the [continuous] action principle, the central position of which in the classical description of nature [e.g. Hamiltonian mechanics] he himself emphasized on more than one occasion. This principle symbolizes, as it were, the peculiar reciprocal symmetry [i.e. complementarity] relation between the space-time description and the laws of conservation of energy and momentum, the great fruitfulness of which, already in classical physics, depends upon the fact that one may extensively apply them without following the course of phenomena [of the processes resulting in the appearance of phenomena?] in space and time. It is this very reciprocity [complementarity] which has been made use of in a most pregnant way in the quantum-mechanical formalism. As a matter of fact, the quantum of action [h] appears here only in relations in which space-time co-ordinates and momentum energy components, which are canonically conjugate quantities in the Hamiltonian sense, enter in a symmetrical and reciprocal manner [linked to Heisenberg's uncertainty relations]. In addition, the analogy between optics and mechanics, which has proved to be so fruitful for the recent developments of the quantum theory, depends intimately upon this reciprocity. ([2], v. 1, p. 94)

The main point is, again, the shift to a new form of complementary configurations, those defined by kinematical variables, on the one hand, and conservation laws, on the other, as against the complementarity of the space-time co-ordination and causality, dominant in the Como lecture. As I said, the possibility of referring to the properties of quantum objects themselves is not yet finally renounced, but is only constrained by the uncertainty relations. As in the Como lecture or elsewhere in Bohr's writing, the generally statistical nature of our predictions concerning quantum phenomena is seen in "The Quantum of Action and the Description of Nature" as experimentally given. It is a feature of nature rather than only of quantum mechanics, since, as explained above, the same experimental arrangements (which we can control in the manner of classical physics) may yield different outcomes. Quantum mechanics, as a statistical theory, merely properly responds to this aspect of quantum phenomena, as it appear to do to all other experimental data within the proper scope of quantum mechanics, which of course makes it an adequate theory of quantum phenomena. The key contribution of the article in this regard is that this fact is now seen or *interpreted* as a *consequence*, an *unavoidable* consequence, of the irreducible role of measuring instruments in the formation of any phenomena that can be meaningfully considered in quantum physics.

Christopher A. Fuchs speaks, in the context of quantum information theory (in its Bayesian form), of this type of argument as "a Paulian idea." It would be difficult to deny the point a Paulian *flavor*, at least at a certain point of his thinking on the subject (and Fuchs indeed appears to see it more as implicit or implied in, rather than stated by, Pauli). On the other hand, it would be even more difficult to argue that the argument does not originate in Bohr's argumentation, beginning with the Como lecture, where it is, again, dominated by the complementarity of space-time coordination and the claim of causality. Indeed, it is of some interest that earlier, in November of 1925, Pauli states: "I definitely believe that in the fundamental laws of satisfactory theory the concept [of] 'probability' should not appear" (Letter to Bohr, 17 November 1925, cited in [9], v. 6, p. 173). The "Paulian idea" appears nearly directly in the form just stated in "The Quantum of Action and the Description of Nature."[3] In the passage cited above, Bohr argues that "the very recognition of the limited divisibility of physical processes, symbolized by the quantum of action, has justified the old doubt as to the range of our ordinary forms of perception when applied to atomic phenomena. Since, *in the observation of these phenomena, we cannot neglect the interaction between the object and the instrument of observation, the question of the possibility of observation again comes to the foreground.* Thus, we meet here, in a new light, the problem of the objectivity of phenomena which has always attracted so much attention in philosophical discussion ([2], v. 1, pp. 92-93; emphasis added). The main point at the

[3] The article has Paulian connections, as does Bohr's later article, "On the Notions of Causality and Complementarity" ([2], v. 4, pp. 141-148), in many ways preparatory to "Discussion with Einstein on Epistemological Problems in Atomic Physics," with which this program is more or less brought to its completion (cf., [2], v. 2, p. 34). Pauli was involved in Bohr's work on the article under discussion at the moment, and he was an editor of the issue of *Dialectica*, in which the second article was originally published (*Dialectica* **2** [1948)]). He especially praised "The Quantum of Action and the Description of Nature" for the philosophical nature of its argumentation, which, he thought, enabled Bohr to say something new. Pauli was right. Bohr did say something new there, including, as Pauli perhaps perceived as well, concerning the nature of quantum probability. On Pauli's exchanges with Bohr on this article, see ([7], v. 6, pp. 193-195). It is of some interest that Bohr worked on the statistical aspects of quantum mechanics around that time ([7], v. 6, pp. 191-192). As he states in his letter to Pauli: "The fact is during recent months I have worked rather diligently on an investigation of *the statistical problems* in quantum theory. I really feel that I can now present *the question of observation* substantially more clearly" ([7], v. 6, p. 193; emphasis added). In other words, Bohr links the statistical problems of quantum theory and the question of observation. He also notes that the change from "complementarity" to "reciprocity" was a mistake ([7], v. 6, p. 195).

moment is the *essential* (again, in the direct sense of the term) link between the irreducible role of the measuring instruments in forming any given phenomena and the irreducibly statistical character of the theory predicting such phenomena, and more specifically, the latter as an unavoidable consequence of the former.

"The Quantum of Action and the Description of Nature" helps to set Bohr's ultimate program into operation, even if not to enact it to the degree he did in the wake of EPR's argument, although Einstein's earlier arguments of the proto-EPR type are on Bohr's mind already at this stage. By and large absent in the Como argument, Einstein enters Bohr's thinking and the scene of the debate concerning quantum phenomena and quantum theory, never to depart from either. Indeed, along with Bohr, he continues to dominate this continuing and seemingly interminable debate even now, most especially through his thinking of the type manifest in EPR's argument. "The Quantum of Action and the Description of Nature" brings Bohr to the threshold of his ultimate version of his interpretation of quantum mechanics as complementarity and to his radical version of quantum epistemology. It took Bohr another twenty years, however, first, to the Warsaw lecture, "Causality and Complementarity," of 1938 ([2], v. 4, pp. 83-91], via the 1935 exchange with EPR, and then, to his 1949 "Discussion with Einstein" ([2], v. 2, pp. 32-66], to finally cross this threshold and arrive to his ultimate destination in his journey, in describing which he famously said: "It was, in a way, my life" ([10]).

This destination was perhaps reached in part because Einstein was no longer there to make Bohr continue his journey. No significantly new ideas and arguments are found in Bohr's articles written after Einstein's death in 1955. Bohr appears to have needed Einstein to reach where he arrived, even though and perhaps because Einstein did not see the road taken by Bohr as possible for himself (he admitted that it was possible in principle) and never stopped his search for, to him, epistemologically more palatable alternatives. Bohr did not appear to have needed Einstein (only Heisenberg and Schrödinger) to start on this road by introducing complementarity in Como. He appears, however, to have needed Einstein to continue to move forward. On the other hand, helped by Einstein, Bohr reached quite far, arguably as far as it appears possible to reach on this particular road, at least for now.

ACKNOWLEDGEMENTS

I would like to thank Giacomo Mauro D'Ariano, Christopher A. Fuchs, Gregg Jaeger, Andrei Khrennikov, and N. David Mermin for productive exchanges concerning quantum theory.

REFERENCES

1. W. Heisenberg, *The Physical Principles of the Quantum Theory*, tr. K. Eckhart and F. C. Hoyt, New York: Dover, 1930, rpt. 1949.
2. N. Bohr, The Philosophical Writings of Niels Bohr, 3 vols., Ox Bow Press, Woodbridge, CN, 1987; The Philosophical Writings of Niels Bohr, Volume IV: Causality and Complementarity, Supplementary Papers, ed. J. Faye and H. J. Folse, Ox Bow Press, Woodbridge, CN, 1998.
3. A. Plotnitsky, *Reading Bohr: Physics and Philosophy*, Springer, Berlin, 2006.
4. P. A. M. Dirac, *The Principles of Quantum Mechanics*, Clarendon, Oxford, 1930 rpt. 1995.
5. J. von Neumann, *Mathematical Foundations of Quantum Mechanics*, tr. R. T. Beyer, Princeton, NJ: Princeton University Press, 1932, rpt. 1983.
6. E. Schrödinger, "The Present Situation in Quantum Mechanics," in J. A. Wheeler and W. H. Zurek, ed., *Quantum Theory and Measurement*, Princeton, NJ: Princeton University Press, 1983, 152-167.
7. Niels Bohr*: Collected Works*, 10 vols., Amsterdam: Elsevier, 1972-1996.
8. C. A. Fuchs, "Quantum Foundation in the Light of Quantum Information," in A. Gonis and P. E. A. Turchi, ed., *Decoherence and Its Implications in Quantum Computation and Information Transfer*, IOS Press, Amsterdam, 2001, 38-82; "Quantum Mechanics as Quantum Information, Mostly," *Journal of Modern Optics* **50** (2003), 987-1003.
9. J. Mehra and H. Rechenberg, *The Historical Development of Quantum Theory*, 6 vols., Springer, Berlin, 2001.
10. Niels Bohr, interview by Thomas Kuhn, Aage Petersen, and Eric Rüdinger, 17 November 1962, *Niels Bohr Archive*, Copenhagen.

Shall We Climb on the Shoulders of the Giants to Extend the R*eality* Horizon of Physics?

Chandrasekhar Roychoudhuri

Photonics Lab., Physics Department, U. Connecticut, Storrs, CT, USA, 06269.
Femto Macro Continuum, 7 Fieldstone Drive, Storrs, CT, USA, 06268.

Abstract. After a very successful flurry of activities for a few decades to maximize the benefits of the formalism of Quantum Mechanics to connect the micro and macro universe, the applied physics community has successfully engineered sustained technological innovations for human social advancements. However, a significant segment of the theoretical physics community put their endeavors essentially in inventing realities that are esthetically pleasing to our human logics (epistemology) rather than staying focused on discovering the actual physical realities in nature driven by cosmic logics (ontology). The purpose of this paper is an attempt to formulate a Reality Epistemology that can leverage our enormous successes in science to re-focus our attention to discovering nature's realities by understanding the physical processes behind all natural interactions that collectively make the cosmic evolution progressing forward. We underscore the deviation from seeking reality to justify the key premise of the paper. We can "see" (measure) the universe only through the "eyes" of the various sensors (detectors). None of these sensors are completely known to us as yet. All sensors also have inherently limited capabilities to respond to input signals and limited capabilities to "report" only a part of all that it experiences. We are thus forced to develop our mathematical theories mixing our human logics and incomplete information and hence they are all provisional and incomplete since they are predicting only correctly measured but limited report by the detector. Thus, we should be careful not to jump into conclusion that we have captured all the necessary cosmic logics behind the interactions involved. We dissect the measurement process in a generic way along with well defined steps to apply Reality Epistemology, which will jointly allow us to develop a scientific methodology of iteratively refining our "successful" human logics that can evolve towards our goal of capturing the cosmic logics. The core content of this paper was first presented at the 2007 QTRF-4 conference at the Vaxjo University [1].

1. HAS PHYSICS REALLY DEVIATED FROM SEEKING REALITY OF NATURE?

Newton is still considered the giant of all physicists who was constantly seeking actual reality and explicitly admitted that he does not understand what light is, just as Einstein did almost three hundred years later while underscoring the necessity of seeking reality of nature's processes [1-7]. The title of the paper derives from one of Newton's sayings: "If I have seen further than others, it is by standing upon the shoulders of giants." Today we are both fortunate and confused because our guiding giants have been divided among themselves for many many decades, if not for a couple of centuries after developing more than half a dozen successful theories that we have been failing for many decades to synthesize into one system, the way the cosmic universe appears to be. So, we are forced to choose giants of our own choice and develop our own path. QTRF-4 is a welcome collective effort in the right direction that is formally questioning the direction of evolution of physics. The recently published books by, Smolin [8], Laughlin [9] and Penrose [10] give us more impetus to question the current epistemology that equates the finality of a theory based more on the mathematical elegance and correctness of predictions for measurements, and less and less whether it is capable of mapping the actual processes undergoing in nature. The quality "elegance" has been raised to such a level that even measurements have become secondary to various "successful" string theories [8]. If we carefully analyze the 50 years' of contributions of classical physics before the birth of QM, we can observe that a faithful adherence to discover the reality (physical processes behind interactions) and the culture of pushing the boundary of theories beyond their limits, have actually nurtured the birth of QM. However, QM over the last 82 years, by insisting on the completeness of the theory, has effectively stifled the cultural process of giving birth to the next higher level theory.

CP962, *Quantum Theory, Reconsideration of Foundations – 4,*
edited by G. Adenier, A. Yu. Khrennikov, P. Lahti, V. I. Man'ko, and Th. M. Nieuwenhuizen

Nature has yielded an enormous amount of its working rules through our organization of diverse observations into coherent systems using logical mathematical modeling (theories). This implies to us that nature works using causal rules. Our logical premise is that causality in a large system cannot emerge if the component sub-systems are essentially non-causal; inverse is logically more acceptable. The emergent causal and logical behavior of the evolving cosmic system is essentially built out of sum of diverse interactions between the stable elementary particles - protons, neutrons and electrons. All these elementary interactions cannot be non-causal, as QM philosophy claims to be. If "...nobody understands Quantum Mechanics", yet the formalism works, then we ought to tell the next generation that there is something fundamentally wrong with the interpretations of the formalism. After overriding Ptolemy's geocentric vision, we are back on imposing homocentric vision on all the enquiring minds by re-directing them to solely rely upon our human mathematical logic, instead of attempting to visualize the actual processes in nature. Being stifled by the quantum philosophy, we are emulating the 'invention culture' from our technological successes and imposing that on to physics. We are now deeply engaged in inventing realities rather than discovering the actual realities behind nature's evolutionary processes. Since QM formalism works so well, it must contain more realities than so far we have acknowledged. Our Reality Epistemology will help extract such realities and eventually facilitate the evolution of higher order theories by finding the limitations of QM.

But, what is reality? Whatever I can measure (record) reproducibly under controlled (similar) conditions, are my reality related to the interactions between the identified interactants (sensor-sensee or detector-detectee) however minute amount of reality may that be compared to the vast cosmic ocean of unknowns. Since I know not any of the interactants and all the applicable forces and their separate and joint, linear and non-linear influences completely, I am obliged to use extend my rational imagination to visualize the processes undergoing between the interactants and develop some "limited" hypotheses congruent with the immediate observations. The best choice I have in modeling the interaction and the follow-on transformations experienced by the interactants, is to assume natural processes to be causal and logical.

We believe that we have created impediments towards our scientific progress by ignoring our eternal ignorance forced on us by nature the way we can gather information about nature. We can "see" (measure) the universe only indirectly through the "eyes" of the various sensors (detectors). None of these sensors are completely known to us as yet. All sensors also have inherently limited capabilities to "see" (respond) to input signals and limited capabilities to "report" only a part of all that it experiences. They all "see" through vision-limiting "goggles" and "speak" to us through band-limited "channels" which are characteristically unique for each of them and not quite known to us. We are forced to develop our mathematical theories mixing our human logics (hypotheses) and incomplete information and hence they are all provisional and incomplete since they are predicting only correctly measured but limited reports by the detectors. When we succeed in constructing an equation to model and predict a set of observations using human logic and limited information, we are automatically limiting our progress to integrate new behaviors of nature that are not logically congruent with the "limited" equation! Mathematics being human invented pure logic, it has to be logically "complete" and "closed", based on the original set of hypotheses, for it to work. Therefore, for any successful equation (theory) to work for a small segment of the undivided universe, by definition, it must have ignored many other potential interactions due to other forces and/or under logically very different contexts. Thus, the only way to integrate multiple successful theories, or partially solved jig-saw-puzzles of the universe, is to break them apart and try to re-assemble them as one bigger jig-saw-puzzle by selectively rejecting and/or modifying some of the human logics towards infallible cosmic logics. Therefore, we should be careful not to jump into conclusion with any working theory that we have captured all the necessary cosmic logics behind the set of interactions represented by the theory. We need to dissect the measurement processes in a generic way along with well defined steps to apply Reality Epistemology (RE), which will jointly allow us to develop a scientific methodology of iteratively refining our "limited successful" human logics into evolving cosmic logics that we are really after. It may be an indefinitely iterative process for us to find the ultimate theory of everything.

A correct approach needs to incorporate some validating or anchoring steps. Could that be integrated with our desire and need to design our sustainable evolution? It is the technologies that we have been discovering and inventing by emulating nature's various processes since the beginning of our emergence as bi-peds with enormous number of neural nets that are far more than the necessity of our survival against competition with the other species. Therefore, nature is our best peer. She is a creative system engineer undergoing logically self consistent steady evolution through innumerable processes of interactions between all possible entities in all possible ways creating galactic and biological technological marvels by utilizing only a small set of (four) interaction forces (or, potential gradients) to the extent we understand so far. It is interesting to note that we build large jig-saw-puzzles for our children by using only a few set patterns, which easily leads them to make many un-integrate-able little "solved" puzzles. Scientific history is replete with multiple theories correctly predicting the same measurements, some of which have been shown to be mathematically equivalent to each other and some are not.

It is wise for us to take the roles of creative reverse-system-engineers and then emulate some of the selected processes of nature to create beneficial technologies that are congruent with our sustainable evolution, for we need a long long time to understand the purpose of the cosmic evolution and our roles in it. Our theory driven technologies can become the anchoring points with nature's actual working rules (or, realities). Thus, the Reality Epistemology that we are proposing could be capable of integrating materialistic science and spiritualistic philosophy into one working system to accelerate our intellectual evolution. Such a wholistic vision can guide us more robustly through our innumerable successes and failures. History tells us that even our great thinkers have been extracting nature's working rules only in piecemeal and we have been successfully integrating them through ages into deeper and higher levels of working knowledge, but with a lot of ideological straggles amongst ourselves, which is healthy. Science has always been a "work in progress". Declaring the tentative *working rules* that are evolving with the evolution of our knowledge as the *laws* of nature imposes serious psychological impediment on the growth of the naturally enquiring minds of the young students. Science is not like building a pyramid by discovering the right fitting stones for we do not yet know the ultimate working structure of the universe!

Manifest and emergent material universe is real and causal. It is subtle, complex and *elusive*, but it is neither mystical nor *a non-causal illusion*! We need to appreciate that there are real ongoing physical processes behind the emergence of both the irreducible and stable elementary particles as well as the most complex systems built out of these elementary particles. Let us consider an elegant yet simple example that we have all experienced – a colorful rainbow, say, visible inside a man made fountain in Sun light. We know that photons do not possess any colors; neither do the water droplets inside the fountain! If you enter inside the fountain, you will not find the rainbow. Yet, it is an emergent reality in nature and not an illusion. It can be observed only by color-interpreting frequency sensitive sensors and only at the focal plane of an imaging system looking at the fountain and only when the Sun is behind the imaging sensor (observer). There is no real rainbow inside the fountain. The emergence of the mystical beauty of a rainbow is elusive but causal, and definitely not an illusion! The specific dispersive refracting property of the water droplet, which is at the root of rainbow color separation, is itself an emergent property of the spherical water molecules in bulk. Understanding the processes behind its formation help us understand the local weather conditions, besides enjoying its esthetic beauty, while advancing the optical sciences by validating a good number of optical principles through a single emergent phenomenon!

Even the elementary particles may be emergent undulating entities in the cosmic field rather than some immutable "material" particles. After all, mass is inter-convertible to energy and Schrödinger's equation "maps" them as some form of harmonic undulations in the cosmic field [11, 12]. It is worth noting that unlike Maxwell's wave equation for EM "waves" that equates second derivative of the field strength with time and space coordinate, Schrödinger's equation equates first time-derivative of the "particle" field strength with that of the second space-derivative of the field strength. Seeking reality will be elusive!

2. DISSECTING OBSERVATIONS AS "SUPERPOSITION EFFECTS AS MEASURED TRANSFORMATIONS (SEMT)" AT THE CORE OF DOING PHYSICS

Creating science out of methodical observations generally require appreciation of the following universal steps in nature. (i) We can scientifically measure only re-producible quantitative *transformations* that are experienced by our interactants (or detector-detectee, or sensor-sensee). (ii) Any transformation in measurable physical parameter requires *energy exchange* between the interactants. (iii) The energy exchange must be guided by at least one *force of interaction* between the interactants and it must be strong enough to facilitate the exchange of energy, which are usually constrained by unique characteristic of each interactant. (iv) All force rules being distance dependent, energy exchange between the interactants requires that they must experience each other as locally present or *physically superposed* entities (experience each other within their sphere of influence).

Therefore, all observations constitute measurable superposition effects. Interactants must be physically *superposed* within the range of the *interacting force* that will allow some *energy exchanger* followed by some *transformations* that is measurable for us. Superposition effect is thus an *active causal and local process*, not a passive mathematical principle! Interpretations of successful mathematical formulation must recognize this *Reality Epistemology* (RE). The objective of physics is to map and visualize the physical interaction processes that facilitate the energy exchange followed by transformations in the interactants.

2.1. Re-visiting SEMT in view of the four force rules we understand, so far!

(i). Gravitational force (GF): All cosmic entities, from galaxies, stars, planets, atoms and elementary particles, the entire observable material universes is effectively superposed on each other as far as GF is concerned. GF is weak; its range is long.

(ii). Electromagnetic force (EMF): Stability of atoms, molecules and their all possible transformations, including their interactions with electromagnetic waves (light, etc.) are all dictated by this force. A dominant part of the biospheric evolution is driven by this force. EMF is relatively stronger than GF, but the range is shorter. The superposition effects due to the EMF from the molecules of two different human bodies are essentially negligible (un-entangled) on a first order analysis.

(iii). Weak Nuclear force (WNF): Radioactivity and related isotopic nuclear transmutations are a by product of this force. The range of WNF is of the order of the size of the atomic nuclei. The superposition effects due to two radioactive atomic nuclei within the same bound molecule are negligible within the first order analysis.

(iv). Strong nuclear force (SNF): Our slow physical evolution relies on the stability of an array of nuclei held together by this SNF, built into stable atoms and molecules by the EMF and held on the surface of the Earth under the atmosphere by the GF. Different atoms within the same molecule are superposed as far as electromagnetic force is concerned, but their nuclei are not superposed as far as SNF is concerned within the first order analysis.

While the observable material and light based universe may generically appear as *non-local*, a careful analysis of SEMT tells us that all measured phenomena are necessarily *local* since all forces of interactions have a finite physical range! Entanglement is operative within the operating range of the force.

3. PROPOSED METHODOLOGY FOR REFINING WORKING HUMAN LOGIC INTO OPERATING COSMIC LOGIC [CC-LC-(ER)^2 EPISTEMOLOGY]

3.1 Defining CC-LC-(ER)^2 epistemology.

For centuries Physics has been synonymous with seeking *Conceptual Continuity (CC)* among a group of diverse but related set of observed phenomena by iteratively and creatively constructing *Logical Congruence (LC)* among all of them to find a higher level of organization leading to a coherent map or theory for the selected set of observations. Our *belief* in this CC-LC epistemology and intuitive *faith* in one continuous and logically functioning universe have been paying off enormously. Our cumulative successes indicate that nature's evolutionary processes do consist of logical patterns & organizations. Thus far, the CC-LC-epistemology has helped us "solve" several separate little pieces out of the giant cosmic jig-saw-puzzle. But we are having trouble in integrating them into one coherent puzzle.

As articulated earlier, all of our "successful" theories are constructed based on limited information about the set of phenomena it is modeling. But however limited, the very success implies that it has captured some cosmic truth, but not all. Thus, we must systematically apply the $(ER)_1$ epistemology, which is maximization of *Extracting* and *Extrapolating Realities* out of a given theory (equation). While the practices of successful predictions by any new theory are examples of $(ER)_1$, we have been neglecting the power of this component of the proposed epistemology chain by not focusing on finding the limitations of a theory. The case in point, we are considering QM as complete, rather than explicitly looking for its limitations that would pave the way for discovering higher level theories.

We believe that the "trouble" is not with physics [8], but lies with our epistemology. Current physics has been developed based essentially on reductionism, matter into elementary particles and radiations into photons. We have neglected to develop a methodology of thinking that would help appreciate the emergence of new complex properties and rules when a complex system is formed out of very many simpler elements. We now need to add another iterative feed back loop of $(ER)_2$ - *E*mergentism *and R*eductionism on to CC-LC-$(ER)_1$ and create a higher level of methodology, CC-LC-(ER)^2 epistemology. This reality seeking epistemology will help us iteratively refine reject re-define some of the founding human logics behind our current "successful" puzzle pieces (theories) and let them evolve closer and closer to the actual operating cosmic logics. Thereby, make the various theories more congruent (amenable) to each other towards unification, CC-LC epistemology but at a higher level. As we have underscored earlier, logically closed equations mapping successfully different subsets of cosmic phenomena based on incomplete knowledge of the universe will necessarily require modifications on their original fundamental premises (hypotheses) before they can accommodate, or amalgamate into, different cosmic logics of interactions. We do not have options but to start with human logics (epistemology) to start making small maps of nature. Then we use CC-LC-(ER)^2 epistemology to iteratively refine the human logics to become closer and closer to the cosmic logics, evolving towards grand unification of different theories.

3.2 An elucidating example.

Some 2500 years ago Gautam Buddha of India gave the best allegorical story on how to apply CC-LC-(ER)^2 to visualize and understand the subtle and elusive "material" universe. How would a group of people, blind from birth, describe and visualize an elephant? It applies equally well to us today as we are trying to describe and visualize the cosmic elephant. First we need to recognize that as far as scientific vision is concerned, we are literally blind. We never see or sense the world directly. Even our human vision is essentially a set of interpretations created by our brains that is convenient for our evolution, not what the actual image is on the retina. We see vivid colors in bright light and we recognize the same colors even in faint light, even though the photons do not have any colors at all. The images we "see" are erect, even though the actual image on the retina is always inverted! In seeking reality about the elephant, the blind people have to search for conceptual continuity (CC) among all their individual sensory inputs by iteratively applying all possible logical congruence (LC) among them. Even then they will only get the outer shape of the elephant. A deeper level of understanding about how such a shape can be a conscious living being requires the blind people to iteratively refine the model of elephant by first applying $(ER)_1$ - extracting and extrapolating their perceived realities to become commensurate with models of other living species. Then they need to apply $(ER)_2$ – *emergentism and reductionism*, to delve deeper into understanding the emergence of elephant's living behavior out of many parts and organs. Today, we "scientifically" understand the *emergence* (E) of any living being out of molecules and DNA's that are highly *reduced* (R) constituents, but we still do not fully understand the emergence of consciousness. Buddha's story also underscores that the existence of the elephant is real, irrespective of whether the blind people sensed it at all or understood its existence in the strict sense. So, philosophies giving serious credence to questions like, "did the tree fall if nobody heard of it?" is a useless diversion if we want to seriously explore the realities of the emergent cosmic universe. The bacteria in the woods are fully cognizant of the availability of lots of food from the fallen tree! Human philosophy cannot hinder their evolutionary physical needs.

3.3. Objective vs. subjective.

We also need to be aware regarding how to distinguish between the objective presence of the cosmic universe that has been evolving for billions of years, well before the appearance of human specie, and the inherently subjective manner by which we are for ever challenged to gather information about this universe. Personally, I really am not totally objective about the outside universe. I simply cannot be! I "see" what is really functionally important for my survival in the real world with my limited number of sensors, which themselves are limited in both sensing and "reporting" to me. My thinking and interpretations are also colored by my brain, my 100 trillion symbiotic microbes and bacteria, my family and social training and my personally acquired knowledge, politics and ego driven by my own successes and failures! By being consciously aware of the methodology of CC-LC-(ER)^2 epistemology, I can iteratively refine my subjective sensor inputs and the organizing human logics into higher and higher levels of objective truth and cosmic logics behind the evolving universe.

3.4. Why elegant mathematics is not enough?

Although it is obvious from the prolonged stagnant state of physics that elegance and symmetry of mathematics is not sufficient guide to explore nature, we present two simple examples to underscore the necessity of constantly applying CC-LC-(ER)^2 epistemology. First, let us revisit why we have rejected Ptolemy's geocentric model. It required several free parameters to allow Ptolemy to construct "epicycles" for each planet separately to accommodate relative "wiggle" motion relative to our Earth. Keppler, based on Copernicus' suggestion, showed that Helio-centric model fits the observations more coherently and logically without many free parameters except a central force of attraction by the Sun. Over the following centuries, Newton formalized the "central force" as gravity, Einstein generalized it to "curvature of space" and we are still at dilemmas as to how to accommodate the measured velocities of the stars in the outer periphery of the galaxies. The point is we need continuous refinements in our modeling based on discovering actual realities rather than inventing mathematically elegant ones. But, if we take the example of today's "successes" of various String Theories using many dozens of free parameters, Ptolemy's geocentric model can be revived with much less numbers of free parameters than the String Theories.

Let us look at another example with elementary mathematics. Pythagoras' relation can be replaced by a pair of relations that I discovered in my 7th grade school from a particular example of a right angled triangle with sides 5, 4 and 3, as many other students must have:

$$\left[c^2 = a^2 + b^2\right] \quad \text{vs.} \quad \left[c = 2a - b \text{ where (b/a)} = (3/4)\right] \tag{1}$$

Even though Pythagoras' quadratic relation can be derived from the pair of linear relations suggested above, which makes the two relations mathematically equivalent, my teacher favored the visualizing power or the geometric value of Pythagoras' relation. Because, one can literally construct the unit squares on each side of a right angled triangle and see for himself why Pythagoras' relation makes sense, which is not so obvious from the other approach based on geometric ratio of the sides. Advanced physics is replete with many such examples like (i) the equivalency of Heisenberg's matrix formulation vs. Schrödinger's "wave" equation and (ii) equivalency of Feynman's "path-integral" vs. Tomonaga-Schwinger's "variational method". (iii) Sudarshan showed that Wolf's classical coherence formulation is equivalent to Glauber's QM representation. Can one of the mathematical constructs guide us better than the other in seeking and visualizing the actual interaction processes in nature? This is a relevant question from the stand point of the epistemology we are proposing. The key point is to recognize that not all "working" human logic has a unique one-to-one relation to the cosmic logic. Thus we must develop a methodology of rational iteration process that can keep on refining our working human logic towards the "nirvana", the cosmic logic.

3.5. Imposing realities on the symbols and mathematical operators of successful equations.

Even in pure mathematics, equations to be correct, the meaning and operation of all the symbols and the connecting operators must be clearly defined. Based on our discussion in Section-2, on dissecting observations as "superposition effects as measured transformation" or SEMT, we must acknowledge that any equation that successfully predicts the outcomes of a wide set of interactions between different interactants, all the symbols must represent the states of the interactants and the operators must represent actual interactions (operations) between them. This is an essential component of the reality epistemology we are promoting.

The simplest example comes from the photo detection [13-15] process that requires careful attention to match the mathematical symbols representing actual physical process of light matter interactions, but not underscored in most of the books. Well formed optical beams (beyond near field diffraction zones), light beams simply pass through each other without modifying each others spatial and temporal energy distribution unless we insert detecting material dipoles within the volume of superposition which are quantum mechanically responsive to the frequencies of all the beams [3-7, 16-31]. If this were not true, we could not have discovered the unique and characteristic Hubble Doppler shift of light beams coming from trillions of individual stars residing in billions of galaxies whose beams are necessarily crossing through (being superposed) as they travel to the Earth. This has enormous consequences both in the classical quantum optics that we have been neglecting at an enormous cost to progress in physics!

Superposed or not, we "see" visible and higher frequency light beams only through the "eyes" of detecting material dipoles each one of them wearing unique "quantum goggles" [20]:

$$\cdot\ \text{Field}: E(t) = a(t)e^{i2\pi\nu t}; \ \text{Stimulation:} \ \psi(t) = \chi_{(1)} a(t) e^{i2\pi\nu t}; \ \text{Transformation:} \ D(t) = <\psi^*\psi> \tag{2}$$

D(t), the measurable transformation due to real physical superposition of the EM field and the detecting dipoles is a QM prescription that "works" and has two built in steps, taking square modulus of the dipole stimulation and the ensemble average. The susceptibility to polarization of the dipole $\chi_{(1)}$ contains all the quantum response properties of the molecules. Note that while normally we use only the linear (first order susceptibility), in reality all EM fields induce all possible linear and non-linear susceptibilities. We "blissfully" neglect this higher order effects until we encounter molecules with strong nonlinear polarizability that are becoming more and more common. In reality, the dipole stimulation should be written as:

$$\text{Stimulation:} \ \psi_{(n)}(t) = \sum\nolimits_n \chi_{(n)} E^n(t); \ \text{Transformation:} \ D(t) = <\psi_{(n)}^* \psi_{(n)}> \tag{3}$$

While Eq.3 already looks complex for general situations, it is even more complex in reality, because both the susceptibility and the EM field are vectors and accommodating the angle between them is quite complex [33]. For multiple EM fields superposed on the detecting molecules, taking the simple linear case of Eq.2, we have:

$$\text{Stimulation:} \ \psi(t) = \sum\nolimits_{q=1,2} \vec{\chi}_{(1)} \cdot \vec{a}(t-\tau_q) e^{i2\pi\nu(t-\tau_q)}; \ \text{Transformation:} \ D(t) = <\psi^*\psi> \tag{4}$$

We have introduced the vector notation on the dipole susceptibility and the EM field to underscore the absence of interference fringes due to orthogonally polarized light is due to the fact that the same dipole cannot carry out dipolar undulations into two orthogonal directions at the same time; it will chose to undulate to one or the other EM vector [12, 20, 25].

This simple recognition of the physical process of "seeing" light through the "eyes" of dipoles has many remarkable implications. First, any time light passes through any material and/or reflected or scattered by some material surface, some of their intrinsic physical properties (frequency, phase, amplitude, polarization) very likely will change. This is built into Maxwell's wave equation when one applies the "boundary conditions". Thus, if we think in terms of propagating photons, most of the time the re-directed photon is no longer the same photon that originally impinged on the surface of the medium.

Accordingly, the Bell's theorem [33] to be relevant at all for interference experiments, it has to be re-derived for each interferometer in terms of physical dipole undulations of not only the detector molecules but also of those of dielectric or metal coating boundary molecules of beam splitters and mirrors that introduce differential π-phase delays for "internal" vs. "external" reflections and the states of polarizations [34], etc. Generally, this π-phase factor is "accommodated" within the physical path difference between the different arms of an interferometer, which is equivalent to ignoring the physical processes that we want to underscore as doing real physics. This approach is benign as long as the Poynting vectors of the two light beams are non-collinear and of different wave front curvatures. When the Poynting vectors of the two collimated beams of equal amplitudes are collinear on a beam splitter [parameters are not explicitly introduced in the above equations], the total energy of both the beams (for same frequency only) will be re-directed in one of the two directions depending upon the coordination of the optical path delay and the π-phase delay between the "external" and "internal" reflections introduced by the boundary dipole molecules of the beam splitter. These boundary molecules cannot collectively redirect the wave front energy in one preferential direction without experiencing the simultaneous but differential π-phase stimulations from both the sides [27, 28] Measurable superposition effects are always due to simultaneous stimulations of the dipoles and hence the phenomenon is necessarily *local* [20].

Another implication of "seeing" light only through the "eyes" of detectors is how we use time-frequency Fourier theorem [18]. The theorem implies the possibility of synthesizing a pulse by superposing many Fourier frequencies as if the EM fields by themselves re-group their energy. But, we know that mode-locked lasers cannot function without the presence of intra-cavity saturable absorber or non-linear Kerr material [22, 24]. The dispersion theory based on this theorem also implies that dipoles of linear material medium must respond to the Fourier decomposed frequencies whenever it experiences a pulse irrespective of the shape and duration of the incident pulse that must always take a finite time to enter into the medium itself [19, 18].

The fourth implication relates to the coherence theory that is normally presented as "field-field" correlation even though the physical "correlation" comes from simultaneous response of the detecting dipoles to both the stimulations. Fortunately, for the current theoretical presentation in the normalized version, the square of the susceptibility factor $\chi^2_{(n)}$ drops out and the "degree of coherence" appears as due to "field-field" correlation. This makes this elegant theory rather unphysical. We are in the process of re-writing the coherence theory in terms of detection processes, which considers the susceptibility of the detecting molecules to be inseparable [27, 18]. Eq.5 helps one appreciate our point. We loose access to visualizing the physical processes by eliminating the susceptibility term $\chi^2_{(1)}$:

$$\gamma_{dipole}(\tau) = \left\langle \psi^*(t)\psi(t-\tau)\right\rangle / \left\langle |\psi(t)|^2 \right\rangle = \left\langle a^*(t)a(t-\tau)\right\rangle / \left\langle |a(t)|^2 \right\rangle \equiv \gamma_{field}(\tau) \tag{5}$$

Obviously, if we can discover a detector whose molecules undergo transformation simultaneously utilizing its multiple (sum of) non-linear polarizability, such a simple equivalency as in Eq.5 will not be possible. Further implications will be published elsewhere.

3.6. Applying CC-LC-(ER)^2 to model what a photon is.

First, we need to justify whether an attempt to re-define the photon is called for considering the current state of very broad acceptance of photons as indivisible quanta which "interfere only with itself"[35 Dirac]. "Self interference" (or, photons knows how to arrive only at the bright fringe locations) naturally dictates the acceptance of unusually strong non-causal concepts like, "non-locality", "delayed superposition", "multiple universe", etc. A summary of these main stream views and related references can be found from these review articles [36, 37]. Our position is that the emergence of causal macro universe, being totally based on quantum interactions in the micro universe, the latter interaction must also be causal. So we should try to model a causal "photon" to bring back *reality* in physics (the title of the paper).

A Fourier monochromatic mode of the vacuum is not a starting causal model for a photon since Fourier modes are physically non-causal, existing over all time that violates conservation of energy. So, the "CC-LC" component of our epistemology demands a causal model for the photon and when we press to also apply $(ER)_1$ (*extract and extrapolate reality*) out of various classical and quantum optics theories and observations, we find the following model. Our proposed photon is a mode of the vacuum as QED claims, but with two caveats. First, it is a space and time finite packet of EM wave evolving and propagating out following the classical Huygens-Fresnel (HF) diffraction model from the moment the emitting molecule releases in the "vacuum" the quantum of energy $\Delta E=h\nu$, with a carrier frequency ν under the space and time finite 3D envelope [23]. This far, our model is congruent with the correct demand of QM, $\Delta E=h\nu$. The next issue is to reconcile with the measured natural line width of spontaneous emission to be a Lorentzian. Classical physics (Lorentz) has solved the problem by proposing the emission envelope to be exponential whose Fourier transform is Lorentzian. We have analytically shown that the *time integrated* fringe broadening observed in classical spectrometers due to time-finite pulses do correspond to Fourier spectral intensity (square modulus of the Fourier transform of the temporal envelope) [31, 29, 21].

Let us now apply again $(ER)_1$ along with $(ER)_2$ (*emergentism and reductionism*). The HF diffraction model is holding out as a remarkably accurate model for light propagation from all the macro to nano photonic devices. So, it must have captured some cosmic logic in it. Its key proposition is that every single point on the wave front behaves as a new source point. We are proposing to accept this point to be literally true. This implies that the cosmic "vacuum" holds a stationary and uniform electromagnetic tension field (EMTF) everywhere in a state of equilibrium [11, 12]. The light wave (photon) is simply a propagating wave group that is an undulation of the EMTF induced by the released energy ΔE by an excited molecule undulating at a frequency ν. The photon is an emergent phenomenon out of the stationary EMTF. The model is quite congruent with all classical material based undulations that inherently propagate out with diffraction. The wave on the water surface is simply an undulation of the surface against the surface tension when displaced by an external energy source out of its state of equilibrium. Same is true for sound wave where the tension in equilibrium is the air pressure due to Earth's gravitational attraction on the air molecules. The similarity between the Maxwell's wave equation and the material based wave equation is remarkable. The displaced point out of the state of equilibrium, whether EMTF or water surface under tension, wants to comeback to its original state of equilibrium and delivers its "displacement energy" to the next domain making it the next ("secondary") source of wave while generating propagating wave and also validating Huygens' hypothesis over Newton's "corpuscular" model, although a space and time finite wave packet (energy conservation) do imply the "corpuscular" existence of light! Propagating wave is an emergent and collective phenomenon. The root of the generation and propagation of all wave-like phenomenons lay with a "field" (EMTF) or a material medium (water, etc.) under uniform tension in a state of equilibrium. Thus the cosmic space (EMTF) holds enormous amount of "static" energy, only a tiny fraction of which is manifest as propagating photon wave packets at any time whirling in every direction of the universe carrying the messages from distant atoms and molecules. EMTF itself quite possibly is the "Dark Energy" the astrophysicists have been looking for. No cosmic or local communication waves would have been possible without the existence of this EMTF in a state of quiet equilibrium! The next question is how to construct the stable particles out of it and then we will have a unified field theory inspired by the highly neglected classical hypothesis, the Huygens-Fresnel principle, supported by the various wave equations!

SUMMARY AND CONCLUSIONS

We hope that we have demonstrated that QM formalism contains more physical realities than we have so far appreciated. Schrödinger's ψ is not just abstract probability amplitude of an entity under study, but it represents the real physical dipole undulation stimulated by another undulating entity while physically present within each others range of the operating force of interaction [1, 20]. The purpose of physics is to visualize and understand the physical processes behind all interactions that have been keeping the "cosmic elephant" evolving logically and causally. Our description of the steps behind any measurement, SEMT (superposition effects as measured transformation), should convince the readers that measured interactions are causal and local. The logic is simple. The measurable state change or *transformation* (classical or quantum mechanical) require *energy exchange* allowed by an operating *force* provided the interactants are *local* to each other, implying that they are within each others force range. We have also provided a methodology of thinking that allows an iterative refinement of human logic (epistemology) towards cosmic logic (ontology). While all "working" theories definitely have captured some cosmic logics, none of them can claim to have definitely captured the ultimate cosmic logic because we are forced to construct any and all theories supported by measurements that can never gather complete information regarding: (i) all possible interactions capability between the interactants under study in any single experiment; (ii) it is impossible to be sure that we can collect all the information related all the transformations experienced by the interactants; (iii) this is further complicated by the fact that we do not know any of the interactants completely even today!

Based on these observations, we have developed a reality exploring epistemology, CC-LC-(ER)^2, which can help us iteratively enhance our "working" human logics towards the cosmic logics that we are trying to map. We have given a powerful example by presenting a model for photons that while accommodates all the QED demands and classical causality, and at the same time, leads us to make a quite plausible hypothesis that the cosmic medium contains an enormous amount of reserved energy ("Dark Energy") in the form of electromagnetic tension field (EMTF) in static equilibrium whose undulation induced by material dipoles manifest the space and time finite wave packets of light.

We believe that nature is a creative system engineer. So, our best hope to discover and understand nature's reality is to approach her as reverse engineers. Our power to visualize the subtle and invisible processes behind all classical and quantum process will be significantly enhanced when we recognize that we are literally blind for everything we construct with our imaginations are based on subjective interpretations of the limited sensorial inputs by our brain, which has been necessary for our successful evolution. Thus framing the question with human logic dictates the answer we construct, which may or may not be conducive to iterative refinement we have proposed to advance towards cosmic logic. Thus, our epistemology providing guidance on how to eliminate hypotheses those are devoid of cosmic reality even though they may be "working" within a limited theory. If our starting premise is that light energy constitutes indivisible quanta, we will keep on imposing the quantum properties of the detectors on to EM fields for we can see them only through the eyes of the detectors. If alternate realities exist, they can be discovered only if we accept such possibilities, as some well known dissenters have tried to pave the way for [2, 38-41] but failed repeatedly. In the seventies, we unsuccessfully tried to present [42-51] to the community that in view of non-interaction (non-interference) between well formed light beams, the use of the time-frequency Fourier theorem as a principle of nature lies at the root of miss-representing many of the classical and quantum optical phenomena because Fourier "monochromatic" frequencies, existing over all time, are necessarily non-causal and the approach introduces non-causality in our physical theories.

ACKNOWLEDGEMENTS

The author is grateful to the organizers of the QTRF-4 conference for giving him the opportunity to reach out a new audience through this publication.

REFERENCES

1. C. Roychoudhuri, Invited talk, 4th International Conference on "Quantum Theory-Foundational Reconsiderations", at Vaxjo U., Sweden, Jun.11-16 (2007); "Quantum Theory - Foundational Reconsiderations"; "Shall we climb on the shoulders of the giants to extend the REALITY horizon of Physics?"
2. A. Einstein, B. Podolsky and N. Rosen, Phys. Rev. **47**, 777 (1935); "Can quantum mechanical description of physical reality be considered complete?"
3. C. Roychoudhuri, A. F. Kracklauer & Kathy Creath, Eds. *The Nature of Light: What Are Photons?* CRC Press (2008); in preparation.
4. C. Roychoudhuri, A. F. Kracklauer & Kathy Creath, Eds., SPIE Proc. Vol. **6664** (2007), *The Nature of Light: What Are Photons?*
5. C. Roychoudhuri, A. F. Kracklauer & Kathy Creath, Eds., SPIE Proc. Vol. **5866** (2005), *The Nature of Light: What is a Photon?*
6. D. Lee and C. Roychoudhuri, Optics Express **11**(8), 944-51, (2003), [http://www.opticsexpress.org/abstract.cfm?URI=OPEX-11-8-944]; "Measuring properties of superposed light beams carrying different frequencies."
7. C. Roychoudhuri, Invited talk, ETOP conference at Toronto, Jun.3-5, 2007; to be published in 2007 by SPIE; "Can classical optical superposition principle get us out of quantum mysticism of non-locality and bring back REALITY to modern physics?"
8. L. Smolin, *Trouble with Physics,* Houghton Mifflin (2006).
9. R. Laughlin, *A Different Universe: Reinventing Physics from the Bottom Down*, (2006), Basic Books.
10. R. Penrose, *Road to Reality*, Alfred Knopf (2005).
11. C. Roychoudhuri, SPIE Proc. **Vol. 6664**, paper #2 (2007); "Can a deeper understanding of the measured behavior of light remove wave-particle duality?"
12. C. Roychoudhuri and C. V. Seaver: Invited talk, Proc. SPIE Vol. **6285**-01 (2006); "Are dark fringe locations devoid of energy of superposed fields?"

13. R. Loudon, *The quantum theory of light*, Oxford University Press (2000).
14. M. Scully and M. S. Zubairy, *Quantum optics*, Cambridge University Press (1997).
15. L. Mandel and E. Wolf, Optical coherence and quantum optics, Cambridge University Press (1995).
16. C. Roychoudhuri and P. Poulos, SPIE Proc. **Vol. 6664**, paper #12 (2007); "Can we get any better information about the nature of light by comparing radio and light wave detection processes?"
17. C. Roychoudhuri, N. Prasad and Q. Peng, SPIE Proc. **Vol. 6664**, paper #24 (2007); "Can the hypothesis 'photon interferes only with itself' be reconciled with superposition of light from multiple beams or sources?"
18. C. Roychoudhuri; Invited paper, Proc. SPIE Vol. **6667**, paper #18 (2007); "Bi-centenary of successes of Fourier theorem! Its power and limitations in optical system designs."
19. C. Roychoudhuri, N. Tirfessa, C. Kelley & R. Crudo,; SPIE Proceedings, Vol. **6468**, paper #53 (2007); "If EM fields do not operate on each other, why do we need many modes and large gain bandwidth to generate short pulses?"
20. C. Roychoudhuri, Phys. Essays **19** (3), September 2006; "Locality of superposition principle is dictated by detection processes."
21. C. Roychoudhuri and M. Tayahi; Intern. J. of Microwave and Optics Tech., July 2006; manuscript ID# IJMOT-2006-5-46: http://www.ijmot.com/papers/papermain.asp; "Spectral Super-Resolution by Understanding Superposition Principle & Detection Processes."
22. C. Roychoudhuri and N. Prasad, Invited paper, Proceedings of IEEE-LEOS Annual Conference (October, 2006), "Various ambiguities in re-constructing laser pulse parameters."
23. C. Roychoudhuri and N. Tirfessa, Proc. SPIE Vol. **6372,** paper-29 (2006), "Do we count indivisible photons or discrete quantum events experienced by detectors?"
24. C. Roychoudhuri, D. Lee and P. Poulos, Proc. SPIE Vol. **6290**, paper-02 (2006), "If EM fields do not operate on each other, how do we generate and manipulate laser pulses?"
25. C. Roychoudhuri and N. Tirfessa, Invited, Proc. SPIE Vol. **6292,** paper-01 (2006), "A critical look at the source characteristics used for time varying fringe interferometry."
26. C. Roychoudhuri and V. Lakshminarayanan, Proc. SPIE Vol. **6285,** paper-08 (2006), "Role of the retinal detector array in perceiving the superposition effects of light."
27. C. Roychoudhuri, Proc. SPIE Vol. **6108,** paper-50 (2006), "Reality of superposition principle and autocorrelation function for short pulses."
28. C. Roychoudhuri, SPIE Conf. Proc. Vol. **5866**, pp.26-35 (2005), "If superposed light beams do not re-distribute each others energy in the absence of detectors (material dipoles), can an indivisible single photon interfere by/with itself?"
29. C. Roychoudhuri, Proc. SPIE Vol. **5531**, pp.450-461(2004), "Propagating Fourier frequencies vs. carrier frequency of a pulse through spectrometers and other media."
30. C. Roychoudhuri & R. Roy, Guest Eds. *The Nature of Light: What is a Photon?*"; Optics & Photonics News Trends; special issue of OPN, October 2003, [http://www.osa-opn.org/abstract.cfm?URI=OPN-14-10-49].
31. C. Roychoudhuri, D. Lee, Y. Jiang, S. Kittaka, M. Nara, V. Serikov and M. Oikawa, Invited paper, Proc. SPIE Vol. **5246**, pp.333-344 (2003), "Limits of DWDM with gratings and Fabry-Perots and alternate solutions."
32. D. L. Mills, *Nonlinear Optics*, Springer (1998).
33. J. S. Bell, *Speakable and unspeakable in quantum mechanics*, Cambridge U. Press, (1997).
34. E. Hecht, *Optics*, Addison-Wesley, 1998.
35. P. A. M. Dirac, *The Principles of Quantum Mechanics*, Oxford University Press (1974).
36. Alain Aspect, "To be or not to be local", Nature Vol. **446**(19), p.866, April 2007.
37. A. Zeilinger, et al., Nature **433**, pp.230-238 (2005), "Happy centenary, Photon".
38. E. T. Jaynes, "Is QED Necessary?" in *Proceedings of the Second Rochester Conference on Coherence and Quantum Optics,* L. Mandel and E. Wolf (eds.), Plenum, New York, 1966, p. 21. See also: Jaynes, E. T., and F. W. Cummings, Proc. IEEE. **51,** 89 (1063), "Comparison of Quantum and Semiclassical Radiation Theory with Application to the Beam Maser". http://bayes.wustl.edu/etj/node1.html#quantum.beats.
39. Willis E. Lamb, Jr. and Marlan O. Scully, "The Photoelectric Effect without Photons", pp363-369, in *Polarization, matter and radiation*; Jubilee volume in honor of Alfred Kastler, Presses Universitaires de France, Paris (1969).
40. W. E. Lamb, Appl. Phys. **B60**, p77-84 (1995); "Anti-photon."
41. E. Panarella, "Nonlinear behavior of light at very low intensities: the photon clump model", p.105 in *Quantum Uncertainties – recent and future experiments and interpretations*, Eds. W. M. Honig, D. W. Kraft & E. Panarella, Plenum Press (1987). For a summary, see pp.218-228 of Ref. C. Roychoudhuri, K. Creath and A. Kracklauer, eds., *The Nature of Light: What Is a Photon,* SPIE Proceeding, Vol.**5866** (2005).
42. C. Roychoudhuri; J. Opt. Soc. Am.; **65**(12), 1418 (1976); "Response of Fabry-Perot Interferometers to Light Pulses of Very Short Duration". (The analysis of this paper is followed and cited in two books: a. "Fabry-Perot Interferometers"; G. Hernandez, Cambridge U., 1986 and b. "The Fabry-Perot Interferometer"; J. M. Vaughan; Adam Hilger, 1989).
43. C. Roychoudhuri; Bol. Inst. Tonantzintla **1**(5), 259 (1975); "Two Beam Interference Experiments and Some Quantum Concepts".
44. C. Roychoudhuri, R. Machorro & M. Cervantes; Bol. Inst. Tonantzintla, **2**(1), 55 (1976); "Some Interference Experiments and Quantum Concepts, II".
45. C. Roychoudhuri; Bol. Inst. Tonantzintla **2**(2), 101 (1976); "Is Fourier Decomposition Interpretation Applicable to Interference Spectroscopy?"

46. C. Roychoudhuri & S. Calixto; Boletin. Inst. Tonantzintla, **2**(3), 187 (1977); "Spectroscopy of Short Pulses."
47. C. Roychoudhuri, Boletin. Inst. Tonantzintla, **2**(3), 165 (1977); "Causality and Classical Interference and Diffraction Phenomena".
48. C. Roychoudhuri & S. Calixto; Boletin. Inst. Tonantzintla, **2**(3), 187 (1977); "Spectroscopy of Short Pulses."
49. C. Roychoudhuri, Boletin. Inst. Tonantzintla, **2**(3), 165 (1977); "Causality and Classical Interference and Diffraction Phenomena."
50. C. Roychoudhuri; Foundations of Physics, **8**(11/12), 845 (1978); "Heisenberg's Microscope - A Misleading Illustration."
51. C. Roychoudhuri, J. Siqueiros & E. Landgrave; p.87-94, Proc. Conf. *Optics in Four Dimensions*, Eds. M. A. Machado Gama & L. M. Narducci, American Institute of Physics (1981); "Concepts of spectroscopy of pulsed light."

Acquisition of Information is achieved by the Measurement Process in Classical and Quantum Physics

Paolo Rocchi * and Orlando Panella**

* *IBM, via Shangai 53, 00144 Roma, Italy*
** *INFN, via Pascoli, 06123 Perugia, Italy*

Abstract. No consensus seems to exist as to what constitutes a measurement which is still considered somewhat mysterious in many respects in quantum mechanics. At successive stages *mathematical theory of measure*, *metrology* and *measurement theory* tried to systematize this field but significant questions remain open about the nature of measurement, about the characterization of the observer, about the reliability of measurement processes etc. The present paper attempts to talk about these questions through the information science. We start from the idea, rather common and intuitive, that the measurement process basically acquires information. Next we expand this idea through four formal definitions and infer some corollaries regarding the measurement process from those definitions. Relativity emerges as the basic property of measurement from the present logical framework and this rather surprising result collides with the feeling of physicists who take measurement as a myth. In the closing this paper shows how the measurement relativity wholly consists with some effects calculated in QM and in Einstein's theory.

Keywords: Information science; Measurement theory; Quantum mechanics.

PACS: 03.65.Ta, 03.67.-a, 06.90.+v, 89.70.+c

INTRODUCTION

Normally measurement is taken as an operational sector whose major scope is to support scientists in different territories, and also professionals in various areas: sellers, workers, engineers etc. Some attempts have been made to provide a comprehensive framework for measurement. The term *metrology* has been coined to define a novel science which should embrace both experiment and theoretical determinations at any level of uncertainty in any field of science and technology. Despite this pompous manifest, in substance metrology defines, realizes and disseminates units of measurement [1]. The *abstract theory of measure* involved eminent authors (Borel, Lebesgue, Lyapunov, Kolmogorov, Carathéodory etc.) but does not justify the inner nature of measure which remains unclear. Modern books on *measurement theory* show the profile of a pragmatic discipline and include statistical methods, construction of instruments and probes, operations for measurements, laws and legal procedures, and other empirical topics [2]. In conclusion, the clear understanding of what a measure consists appears not close.

Several authors agree that information strongly affects the nature of a measure and we assume that the measurement process basically consists of information acquisition. In this way we shall examine the most significant features of measure through the lens provided by theoretical works on information, and try to answer some questions risen by philosophers and by theorists in classical and quantum physics, such as: May systematic errors of measure be erased? Why and how does a measurement process influence the outcomes? What is the observer characterization? Does the observer lie outside or inside a physical test? Is the intellectual awareness a part of the quantum measurement process?

CP962, *Quantum Theory, Reconsideration of Foundations – 4*,
edited by G. Adenier, A. Yu. Khrennikov, P. Lahti, V. I. Man'ko, and Th. M. Nieuwenhuizen

BEYOND THE JUNGLE

Various scientific theories about information have been put forward and stand often without the least explicit relationship to each other. We see so many differing definitions that also the classification of theories appears a tricky question [3]. The ensuing incomplete list gives an idea about the variety of schools:

- Semantic theory of information (Bar-Hillel and Carnap)
- Algorithmic theory of information (Solomonoff, Kolmogorov, Chaitin),
- Qualitative theory of information (Mazur),
- Economical theory of information (Marschak),
- Social theory of information (Goguen),
- Autopoietic theory of information (Maturana and Varela),
- Statistical theory of information (Shannon).

No adequate concept as well as understanding of information has been produced by theoretical researchers over the last decade of the 20th century [4]. The cultural scenario is so much confusing that theoretical physicists, who aim at exploiting the notions of the information and communication technology (ICT), normally cannot go ahead. The lack of a solid reference appears like a huge rock amid the route that obstructs the progress unless one finds a way to circumvent this barrier.

The present paper suggests getting around this obstacle by means of multidisciplinary conceptualization.

MULTIDISCIPLINARY APPROACH

Several disciplines, which revolve around the notion of information, deal with a broad variety of problems, nonetheless we can identify two principal views shared by authors. A first group (e.g. computer hardware, networking, neurophysiology) experiences the physical nature of information, namely they investigate the material form of information and its properties. A second group of disciplines (e.g. computer software, semiotics, logic, linguistics, psychology) communalize the semantic view to information namely they focus on the contents of texts, of messages, etc.

Quantum theorists are increasingly aware of the influence of cognitive sciences on the measurement problem and we take the perspective which integrates Shannon's work - on which technicians and engineers prevalently ground their theoretical studies - with the humanities. We specify this multidisciplinary way to information and to measure by means of the following principles which sum up the perspectives of the above mentioned two groups:

Principle 1: *A piece of information is a physical distinguishable item.*
Principle 2: *Information finds its definitive significance with respect to the human knowledge.*

We explode the foregoing principles in the following way.

A piece of information is present from background events and the receptor R is capable of detecting this piece if it is neat. The generic stuff E, whether an artifact (i.e. writings, pixels) or natural element (i.e. a quantum particle), is capable of informing on condition it is distinguishable. The entity E is impossible to detect save as E contrasts with an adjacent entity E^*. Discriminability is the special feature of all the pieces of information which we define in the following way.

Definition 1: *The entity E is said to be an item of information if E is distinct from a given comparison-entity E* with respect to the reference* R

$$E_{\mathrm{R}} = \mathrm{NOT}\ E^* \tag{1}$$

Where E and E are entities of the algebraic space* $\mathcal{E}$.

The vast majority of authors in semiotics and in linguistics accept that the item of information E replaces the object, the event or the idea NE; to wit E stands for NE. The former is frequently named *signifier* and the latter *signified.* We write the *signification* between the signifier E and the signified NE, adopting the Lacan symbolism [5]

Definition 2: *The entity E refers to the entity NE*

$$\frac{E}{NE} \tag{2}$$

From Definitions 1 and 2 we infer that the system for the acquisition of information includes two blocks: the former detects E and operates without any intelligence, the latter assigns the significance to the readout.

Definition 3: *The system to acquire information is equipped with the detection process* R *and the semantic process* S.

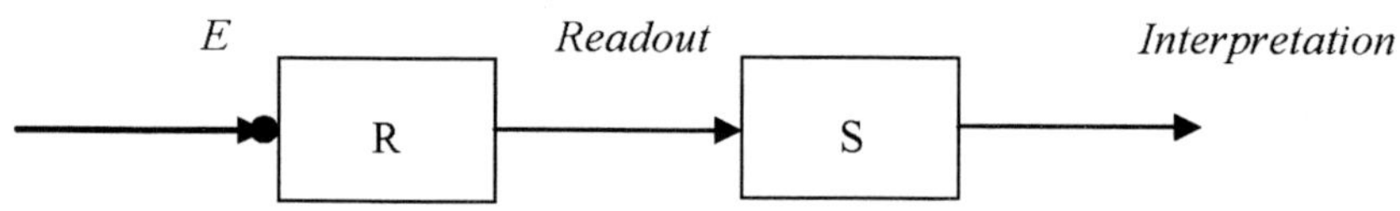

FIGURE 1. The R-S system

Information takes origin from sensation, the mechanical and initial act to acquire a sign, and then from the recognition stage S which determines the meaning of the acquired signal. In practice R-S is an artifact (i.e. measurement tool, detector), a biological system such as the human observer or a combination of both. Experience shows how the stages R and S may be generic, fuzzy and subjective due to human intervention, thus we define the most accurate conditions for R-S.

Constraint 1: *We assume E, E* are points in a metric mono-dimensional space, and from* (1) *we derive the condition for distinction*

$$E \neq E^* \tag{3}$$

Constraint 1: *We assume the signified NE is physically compatible with E and is a standard unit*

$$NE = EU \tag{4}$$

E and EU are exact and comparable entities due to (3) and (4), thus the semantic relation (2) becomes a ratio which provides a pure number

$$\frac{E}{EU} = Em \tag{5}$$

We conclude that $(Em{\cdot}EU)$ quantifies E to wit it provides the measure of E.

Definition 4 : *The following product is said the measure of E*

$$Em \cdot EU \tag{6}$$

The precision of the number *Em* gives the granularity of the measure and *EU* provides the meaning of the measure, thus (6) consists with Definitions 1 and 2.

The parallel Principles 1 and 2, the coupled Definitions 1 and 2, the operations R and S, the pair (6) show that measure has a double nature: physical and semantic. The dual origin of measure, in particular its reliance on the observer who establishes the interpretation of *E*, seemed a kind of scientific taboo [6]. Theorists frequently treats the human intervention in generic terms, because they fear the subjectivism and the generic behavior of the humans instead here the observer R-S finds a rigorous formalization.

The present paper does not develop the mathematics from the foregoing definitions; instead we shall discuss the major properties of measurement that is to say we go deep into the nature of measurement in order to improve our understanding of this topic.

RELATIVISM: THE BASIC PROPERTY OF MEASURE

Definitions 1 and 2 include the following set of arbitrary elements which experimentalists factually establish at their convenience

$$\{E^*, \mathrm{R}, NE, EU\} \tag{7}$$

In particular the reference is chosen so that R is close and physically compatible to *E* otherwise R cannot achieve the detection process. The comparison term E^* is selected so that *E* can contrasts. Sometimes the interpretation *NE* is a figment of human imagination and the semantic relationship does not follow any rule. Scientists determine the standard unit *EU* under the influence of sociological and political issues. In conclusion the elements (7) works out as essential parts for the acquisition of information but the present theory does not regulate them in terms of mathematics. The elements (7) are subjected to empirical criteria and not to the necessity of general reasons, thus we conclude that **measure is a relativistic quantity**. Relativism emerges as the fundamental property of a measure in the light of the present logical framework hence this seemingly paradoxical achievement is to be expanded.

The arbitrary items (7) bring about four genders of relativism which we call this way:
- *Couple relativity* due to E^*,
- *Reference relativity* due to R,
- *Semantic relativity* due to *NE*,
- *Unit relativity* due to *EU*.

Current literature amply discusses the doctrines of semantic arbitrariness as semantic relativity affects natural languages [7]. Physicists are aware that the selection of units influences the arrangement of systems of measurement. The criteria followed to fix sets of units which must be used to specify anything are illustrated in the domain of experimental sciences [8]. The couple relativity regards the comparison term which does not produce heavy effects[9]. Finally the detector R causes the most demanding consequences in physics which we discuss next.

The Reference Relativity Cannot be Developed within a Unified Theory

Eqn (1) establishes a functional relationship between *E* and E^*

$$E = f(E^*) \tag{8}$$

Thus the couple relativity may be treated within a unified mathematical frame [9], instead the reference relativity cannot be treated because of the generic relationship between *E* and R in (1)

$$E = E_{\mathrm{R}} \tag{9}$$

A broad variety of phenomena which inflict damages on measurements brings evidence of this corollary. The eclectic nature of the reference relativity emerges from the following cases randomly selected from classical physics.

1) The sensor R works within the minimum threshold mint_R and the maximum threshold maxt_R. If E does not fall in the available interval (mint_R, maxt_R), the measure of E does not exist.
2) The sensor may to some extent be sensitive for other properties than the property being measured. To exemplify, most instruments are influenced by the temperature of their environment.
3) If the output signal slowly changes independent of the measured property, this is defined as *drift*. Long term drift usually indicates a slow degradation of sensor properties over a long period of time.
4) Noise is a random deviation of the signal that varies in time. Noise can be reduced by signal processing, such as filtering, usually at the expense of the dynamic behavior of the sensor.
5) Hysteresis is an error caused by the fact that the sensor not instantly follows the change of the property being measured, and therefore involves the history of the measured property.

Errors, as from 1 to 5, make visible the reference relativity in classical physics. The efforts to compensate those errors by means of some kind of calibration strategy reduce the impact of those errors, but do not deny the reference relativity in principle.

In quantum field theoretical principles sustain the study of the measurement limits [10].

A) As first we quote the *uncertainty principle* by means of the Robertson-Schrödinger inequality. Given the Hermitian operators A and B and a system in the state ψ, there are probability distributions associated with the measurement of each A and B, giving rise to standard deviations ΔA_ψ and ΔB_ψ, then

$$\Delta A_\psi \Delta B_\psi \geq \frac{1}{2}\left|\langle [A,B] \rangle_\psi\right| \qquad (10)$$

Where $[A,B] = AB - BA$ is the commutator of A and B, and $\langle X \rangle_\psi$ denotes the expectation value. Due to this principle an uncertainty relation arises between any two observable quantities that can be defined by non-commuting operators e.g. position and momentum of a particle; angular position and angular momentum of a particle. The uncertain principle regards also the pair time-energy.

B) Quantum particles are actually waves with complex amplitude spread out in space and evolving according to indeterministic laws, except when a measurement occurs. During a measurement, the wave-function instantaneously changes into a state consistent with one of the possible measurement outcomes. The probability of any given outcome is dependent on the amplitude of the wave. In most cases, this means that the wave-function becomes suddenly localized in space, when before it was widely distributed. We express the state of a quantum system

$$|\psi\rangle = \sum_i |i\rangle \psi_i \qquad (11)$$

Where $|i\rangle$ form an orthonormal eigenvector basis. The $\psi_i = \langle i|\psi\rangle$ are the probability amplitude coefficients which we assume are normalized. When an observer measures the observable associated with the eigenbasis then the state of the wave function changes from $|\psi\rangle$ to just one of the $|i\rangle$s with probability $|\psi_i|^2$. All the other terms in the expansion of the wave function vanish into nothing due to the observer R-S. The act of measurement itself determines the state of the system as it is measured.

C) When a quantum state is entangled (i.e. two particles) it is possible to obtain the value of an observable of the particle E_2 with a measurement performed on particle E_1. The *EPR paradox* deals with the related violation of locality in ordinary QM in these types of experiments. In this case an observer performs measurement only on a particle but the state of the other is determined in a total indirect way. Let the observer R_{E2} is located much further from the birthplace of the particle E_1 and E_2 than the observer R_{E1}. Let the observer R_{E1} measures the projection of spin into the axis z for the particle E_1, and this value appears as $S_z(E_1)$. The particles E_1 and E_2 were in a singlet state at once after the decay of the object Q. Therefore the observer R_{E2} will find out a value

$$S_z(E_2) = S_z(E_1) \qquad (12)$$

For the particle E_2 with probability equal to the unity. Within the framework of standard quantum mechanics it means that the particle E_2 is in a state with definitive value of projection of spin $S_z(E_2)$ on the axis z.

Einstein forecasts some special effects on measurement which directly derive from the special relativity theory and the general relativity theory [11].

I) Take an event with space-time coordinates (t,x,y,z) in system R and (t',x',y',z') in R' which is moving at the speed v. The Lorentz transformation specifies that these coordinates are related in the following way

$$t' = \gamma\left(t - \frac{vx}{c^2}\right); \quad x' = \gamma(x - vt); \quad y' = y; \quad z' = z \tag{13}$$

where γ is the Lorentz factor

$$\gamma = \frac{1}{\sqrt{1 - \frac{v^2}{c^2}}} \tag{14}$$

The time lapse between two events is dependent on the relative speeds of the observers' reference frames. In fact we write the Lorentz transformation and its inverse in terms of coordinate differences and obtain

$$\Delta t' = \gamma\left(\Delta t - \frac{v\Delta x}{c^2}\right) \tag{15}$$

Take as an example two consecutive ticks of a clock characterized by Δx=0, the relation between the times of these ticks as measured in both systems is given by

$$\Delta t' = \gamma \Delta t \tag{16}$$

This equation shows the time dilatation, namely the time $\Delta t'$ between the two ticks as seen in the 'moving' frame R' is larger than the time Δt between these ticks as measured in the rest frame of the clock.

II) Suppose the length of an object Δx in the frame R, and we calculate the length of the same object in R'. Assuming that the end points are measured simultaneously in R', namely $\Delta t'$=0, we obtain using the Lorentz transform

$$\Delta x' = \frac{\Delta x}{\gamma} \tag{17}$$

This shows the length contraction to wit length $\Delta x'$ of the object measured in the moving frame R' is shorter than the length Δx in its own rest frame.

III) The mass of an object (also known as relativistic mass) increases with the speed according to the following expression

$$M = \gamma m \tag{18}$$

Where m is the invariant mass, and γ the Lorentz factor

IV) According to Einstein's general relativity the rotation of an object alters space and time, dragging a nearby object out of position compared to the predictions of Newtonian physics. In other words an observer who is distant from a rotating massive object and at rest with respect to its center of mass will find that the fastest clocks at a given distance from the object are not those which are at rest (as is the case for a non-rotating massive object). Instead, the fastest clocks will be found to have component of motion around the rotating object in the direction of the rotation. This event named *Lense-Thirring effect* or *frame dragging* occurs where a rotating massive object "drags" space-time along with its rotation. For example frame dragging causes the orientation of a gyroscope to change over time.

Empirical case studies justify effects 1-5, by contrast effects A-C derive from a theoretical frame not yet unified in QM. Einstein calculates effects I-IV which directly derive from the special relativity theory and from general relativity theory. This means that points 1-5, A-C and I-IV provide support, progressively stronger, to the reference relativity of measure. Three intellectual constructions – empirical methods, quantum theory, Einstein's theory – converge toward the same issue: **measurement is not an absolute process** due to the influence of arbitrary entities – in particular R. This conclusion seems to clash against the universal confidence on measurement as only measured results can validate a scientific theory. Physicists tend to refuse the relativism of measurement - evident in quantum mechanics - because they fear the scientific method may be taken unworthy. Instead this suspect is misleading because the measurement relativity derives from the precise set (7). We are capable of analyzing each arbitrary element and in consequence are capable of calculating its actions. If an experimentalist is unable to neutralize a distortion or a limitation caused by an arbitrary element in the practice, anyway he can forecast this negative effect with precision. E.g. the wave collapse restricts an experimental inquiry anyway the result obtained after the collapse is valid because we are aware of this effect. We can reasonably conclude that the measurement relativity – especially the reference relativity – does not entail that the scientific method deviates the human mind but simply holds that the scientific method is difficult to apply due to the relativistic effects.

Where the Observer Lies

The problems of inclusion/ exclusion of the observer from the physical experiment assume special significance because Principle 2 holds that the observer is necessary, moreover we have seen how the observer R-S causes additional effects. The reply to the question does not seem easy as it is insufficient to hold that the observer lies outside or inside the experiment.

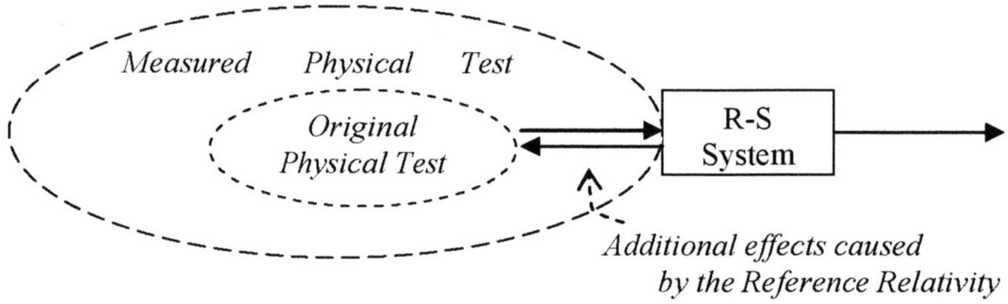

FIGURE 2.

We have exactly determined the operations of R-S thus we are able to divide *the original test* from *the measured test* that includes the interactions provoked by R-S. Therefore an observer lies outside the initial experiment and lies inside the measured experiment because of his interactions. This statement reinforces the notion that measurement is not neutral and the measurement relativity takes essential role.

The present corollary somehow agrees with the thought expressed by von Neumann, David Bohm, Henry Stapp, Freeman Dyson, and others who claim "consciousness causes collapse" [12]. In fact the consciousness may be seen as a part of the block S which makes an essential component of an observer.

Quantum Decoherence: an Aspect of the Measurement Relativity

The collapse constitutes the special and partially mysterious interference between a particle and the measurement system. There are two major schools toward the wave function collapse. The former rejects the collapse as a physical process and relates to it only as an illusion. Those who support this approach usually offer another interpretation of quantum mechanics, which avoids the wave function collapse. The latter schools supported by Niels Bohr and his Copenhagen interpretation accept the collapse as one of the elementary properties of nature. The present conceptualization matches with the last authors in particular it agrees with physicists who aim at identifying the fuzzy boundary between the quantum micro-world and the world where the classical intuition is applicable. The interaction between the microscopic level of the particle E and the macroscopic level of the system R-S appears as a special form of reference relativity which may be treated by means of the decoherence theory. Quantum decoherence was proposed in the context of the many-worlds interpretation, but it has also become an important part of some modern updates of the Copenhagen interpretation. Technically decoherence consists in detailing a model for the system-environment interactions and in deriving master equations for the reduced density matrix of the system.

Density Matrix Description of Decoherence.

When studying a point mass the inclusion of the unavoidable scattering processes with the particles of the environment gives the non-unitary part of the time evolution of the density matrix. In the coordinate representation the master equation for the density matrix reads

$$i\hbar\frac{\partial\rho(x,x')}{\partial t}=\left\langle x\middle|[H,\rho]\middle|x'\right\rangle-i\hbar\rho(x,x')F(x-x') \tag{19}$$

Where the quantity $F(x-x')$ can be connected to the scattering cross-section of the point mass. The unitary evolution based on the Schroedinger equation gives only the first term in the previous equation. Based on the assumptions that: 1) the recoil in the scattering is negligible (only the state of the scattered particle is changed during the interaction); 2) very many particles are scattered during a short time interval and 3) the incoming particles are distributed isotropically, one finds that $F(x-x')=\Lambda(x-x')^2$ with $\Lambda=k^2nv\sigma_{eff}$ (called the *localization rate*) where k is the momentum of the scattering particles, n their density and σ_{eff} their effective cross section with the point mass. Solving the master equation yields the suppression of off-diagonal terms of the reduced density matrix of decoherence

$$\rho(x,x',t)=\rho(x,x',0)\exp\left[-\Lambda t(x-x')^2\right] \tag{20}$$

See these references [13, 14, 15] for a review of explicit cases.

Wigner Function Description of Decoherence

A commonly used representation of both pure and mixed states, alternative to the density matrix, is the Wigner function which is related (for continuous systems) to the density matrix representation by

$$W(x,p)=\frac{1}{\pi\hbar}\int_{-\infty}^{+\infty}dye^{2ipy/\hbar}\rho(x-y,x+y) \tag{21}$$

It is often a useful tool to study correlations between position and momentum. If the system follows a Schroedinger time evolution with a potential $V(x)$, the time-dependence of the Wigner function is given by the so-called *Moyal bracket* $\{H,W\}_{MB}$

$$\frac{\partial}{\partial t}W(x,p,t)=\{H,W\}_{MB}=\{H,W\}_{PB}+\sum_{n\geq1}\frac{\hbar^{2n}(-1)^n}{2^{2n}(2n+1)!}\partial_x^{2n+1}V(x)\partial_p^{2n+1}W(x,p) \tag{22}$$

Where $\{H,W\}_{PB}$ is the usual Poisson bracket. As in the density matrix representation the general consequence of non unitary dynamics - due to the environmental interactions - is the damping of oscillations in the Wigner function. In this representation decoherence acts like diffusion in the momentum variable. In the general case the equation of motion of the Wigner function take the form of a Fokker-Planck equation

$$\frac{\partial}{\partial t}W(x,p,t)=\left(-\frac{p}{m}\frac{\partial}{\partial x}+m\omega^2x\frac{\partial}{\partial p}+\Lambda\frac{\partial^2}{\partial p^2}\right)W(x,p,t) \tag{23}$$

The last term is easily recognized as a diffusion term that will generate, as in the diffusion equation, the damping of oscillations in momentum space [16].

These topics are going to be further developed but the present paper do not assist theorists to complete the quantum decoherence calculus because the reference relativity cannot be developed within a unified mathematical theory due to (9).

CONCLUSIONS

The measurement problem collects the set of key questions that every interpretation of quantum mechanics must address. This contribution offers a new approach to the measurement problem using the concepts shared in the information domain.

The present theoretical study yields the following results:

- The dual nature of measure,
- The necessary intervention of the observer,
- The relativistic nature of measurement.

The last point appears significant since elaborated procedures show how the measurement process is the most accurate system for the acquisition of information. Units of measurement established by international treaties ensure a fair base to scientific knowledge. The present theory concords with these issues through (3), (4) and (5). Constraints 1 and 2 ensure that the measurement processes are the most reliable and the most objective processes with respect to other processes that acquire information, but this high position does not imply that measurement is a zero-defects process. The arbitrary elements (7) provoke difficulties in measuring. Physicists are aware of these obstacles but they are reluctant to acknowledge the systematic defects of the measurement systems due to the eminent role of tests in the scientific method. Thinkers fear that criticism on measurement shakes up the foundations of the scientific knowledge and the measure relativism could deny the validity of the scientific method. Instead this fear has not ground because (7) identifies the arbitrary entities of a measure. The present conclusions match with important phenomena devised by Einstein and by quantum theorists. Points 1-5, A-C and I-IV pertain to a unified scenario. We finally guess that the reference relativity can remove heavy prejudices upon the mythical conception of measurement.

ACKNOWLEDGMENTS

One of the authors (O.P.) wishes to acknowledge partial support from the Fondazione Cassa di Risparmio di Spoleto (Italy).

REFERENCES

1. R.H. Harlow, R.L. Thompson, C. Dotson - *Fundamentals of Dimensional Metrology* - Thomson Delmar Learning (2003).
2. I. Gertsbakh, E. Gertsbakh - *Measurement Theory for Engineers* - Springer (2003).
3. L. Floridi - Open Problems in the Philosophy of Information - *Metaphilosophy*, 35 (4), 554-582 (2004).
4. D. Israel, J. Perry - What is Information? - in P. Hanson (ed.) *Information, Language and Cognition.* University of British Columbia Press (1990).
5. J. Lacan - *Écrits* - Routledge (1977).
6. J. Bub – Quantum Mechanics is About Quantum Information - *Foundation of Physics*,34 ,541-560 (2005)
7. J.D. Culler - *Ferdinand De Saussure* - Cornell University Press (1986).
8. D.M. Anthony - *Engineering Metrology* - Pergamon Press (1986).
9. P.Rocchi, L. Gianfagna - An Introduction to the Problem of the Existence of Classical and Quantum Information - *Proc. Quantum Theory:Reconsideration of Foundations* (QTRF-3), 248-258 (2006).
10. R.A. Healey, G.Hellman - *Quantum Measurement: Beyond Paradox* - University of Minnesota Press (1998).
11. H.R. Brown - *Physical Relativity: Space-time Structure from a Dynamical Perspective* - Oxford University Press (2005).
12. D. Bohm - *Wholeness and the Implicate Order* - Routledge (1980).
13. E. Joos, H.D. Zeh, - Zeitschrift fur Physik B - *Condensed Matter*, 59, 223-243 (1985).
14. M.R. Gallis, G.N. Fleming, *Phys. Rev. A* 42, 38-48 (1990).
15. G. C. Ghirardi, A. Rimini, T. Weber, *Phys. Rev. D* 34, 470-491 (1986).
16. J.P. Paz, S. Habib, W.H. Zureck, *Phys. Rev. D.* 47 488-501 (1993).

Can the Quantum Measurement Problem be Resolved within the Framework of Schroedinger Dynamics and Quantum Probability?

Geoffrey Sewell

Department of Physics, Queen Mary, University of London, Mile End Road, London E1 4NS, UK.
e-mail: g.l.sewell@qmul.ac.uk

Abstract. We provide an affirmative answer to the question posed in the title. Our argument is based on a treatment of the Schroedinger dynamics of the composite, S_c, of a quantum microsystem, S, and a macroscopic measuring apparatus, $\mathscr{I}$, consisting of N interacting particles. The pointer positions of this apparatus are represented by subspaces of its representative Hilbert space that are simultaneous eigenspaces of coarse-grained intercommuting macroscopic observables. By taking explicit account of their macroscopicality, we prove that, for suitably designed apparatus $\mathscr{I}$, the evolution of the composite S_c leads both to the reduction of the wave-packet of S and to a one-to-one correspondence between the resultant state of this microsystem and the pointer position of $\mathscr{I}$, up to utterly neglible corrections that decrease exponentially with N.

Keywords: Schroedinger dynamics of microsystem-cum-measuring instrument, macroscopic phase cells as pointer positions, macroscopic decoherence, reduction of wave packet of microsystem, large deviations.
PACS: 03.65.Ta, 02.50.Cw

1. INTRODUCTORY DISCUSSION

According toVon Neumann's [1] phenomenological picture of the measurement process, the coupling of a microsystem, S, to a measuring instrument, $\mathscr{I}$, leads to the following two effects.

(I) It converts a pure state of S, as given by a linear combination $\sum_{r=1}^{n} c_r u_r$ of its orthonormal energy eigenstates u_r, into a statistical mixture of these states for which $|c_r|^2$ is the probability of finding this system in the state u_r. This effect is often termed the 'reduction of the wave packet'.

(II) It sends a certain set of classical, i.e. intercommuting, macroscopic variables M of $\mathscr{I}$ to values, indicated by pointers, that specify which of the states u_r of S is realised.

This leaves us with the following basic question of quantum measurement theory, which is clearly pertinent to that of the completeness of quantum mechanics.
$\mathscr{Q}$. *Can the standard quantum dynamics of the composite* $(S+\mathscr{I})$, *allied to a suitable choice of the macroscopic observables M, lead to the effects (I) and (II)?*

This is the contentious question that we address here and, as in [2,3], our response to it is affirmative, as is that of some other authors [4-6]. Our present objective is to describe the mathematical structure of our general argument that leads to this conclusion and that is supported by treatments of some concrete models. We remark that our affirmative response to $\mathscr{Q}$ conflicts with that of Von Neumann [1] and Wigner [7], who argued that the observation of the pointer of $\mathscr{I}$ requires another measuring instrument, $\mathscr{I}_2$, which in turn requires yet another instrument, and so on, in such a way that the whole process involves an infinite regression ending up in the observer's brain! [1] In our view, the essential flaw in the Von Neumann- Wigner theory is its failure to take explicit account of the macroscopicality of the observables corresponding to the pointer positions of the measuring apparatus. By contrast, we build this macroscopicality into our treatment, and thereby evince a qualitative difference between the characteristic properties of the microsystem S and the macrosystem $\mathscr{I}$ that removes the need for a Von Neumann-Wigner regression.

[1] Others have taken the view that the measurement problem cannot be resolved without modification of the Schroedinger dynamics of S, due to *either* its interaction with the 'rest of the Universe' [8-11] *or* a certain postulated nonlinearity that leads to a classical deterministic evolution of its macroscopic observables [12].

CP962, *Quantum Theory, Reconsideration of Foundations – 4*,
edited by G. Adenier, A. Yu. Khrennikov, P. Lahti, V. I. Man'ko, and Th. M. Nieuwenhuizen

As a preliminary to setting up a mathematical theory of the measurement process, we note that it is clear from the works of Bohr [13], Jauch [14] and Van Kampen [5] that such a theory demands both a characterisation of the macroscopicality of the observables M and an amplification property of the $S-\mathcal{I}$ coupling whereby different microstates of S give rise to macroscopically different states of $\mathcal{I}$. Evidently, this implies that the initial state in which $\mathcal{I}$ is prepared must be unstable against microscopic changes in the state of S. On the other hand, as emphasised by Whitten- Wolfe and Emch [15, 16], the correspondence between the microstate of S and the eventual observed macrostate of $\mathcal{I}$ must be stable against macroscopically small changes in the initial state of this instrument, of the kind that are inevitable in experimental procedures. Thus, the system $(S+\mathcal{I})$ must exhibit a striking combination of stability and instability properties.

Rigorous constructive treatments of the measurement process, which take account of the above considerations, have been provided by Hepp [17] and Whitten-Wolfe and Emch [15, 16] on the basis of models for which $\mathcal{I}$ is idealised as an infinitely extended system of finite density, for which the macroscopic observables M are intercommuting global intensive ones. This picture of $\mathcal{I}$ corresponds to that employed for the statistical mechanical description of large systems in the thermodynamic limit [18-20], and it has the merit of sharply distinguishing between macroscopically different states, since different values of M correspond to disjoint primary representations of the observables. Moreover, in the treatments of the measurement problem based on this picture, the models of Hepp [17] and Whitten-Wolfe and Emch [15, 16] do indeed exhibit the required reduction of the wave-packet and the one-to-one correspondence between the pointer position of $\mathcal{I}$ and the resultant state of S; and these results are stable against all localised perturbations of the initial state of $\mathcal{I}$. On the debit side, however, Hepp's model requires an infinite time for the measurement to be effected (cf. Bell [21]), while although that of Whitten-Wolfe and Emch achieves its measurements in finite times, it does so only by dint of a physically unnatural, globally extended $S-\mathcal{I}$ interaction.

In view of these observations, we base our treatment of the measuring process on the generic model of $S_c=(S+\mathcal{I})$ for which $\mathcal{I}$ is a large but finite N- particle system. Our main aim is to determine whether there are viable conditions under which this model yields the same essential results as the infinite one, but within a finite realistic observational time. In fact, we have achieved this aim in recent works [2, 3], in which we showed that the quantum mechanics of the finite model does indeed contain the structures required for an affirmative answer to the question $\mathcal{Q}$, as illustrated by an explicit treatment of the finite version of the Coleman-Hepp model [17]. This result provides rigorous mathematical substantiation of the arguments of Refs. [4-6] which led to the same main conclusion. A key feature of our treatment is the representation of the macro-observables M and the pointer positions of $\mathcal{I}$ within the framework proposed by Van Kampen [22] and Emch [23], whereby M comprises a set of coarse- grained intercommuting extensive observables and the pointer positions of $\mathcal{I}$ correspond to their simultaneous eigenspaces. These are mutually orthogonal subspaces of the Hilbert space of the pure states of $\mathcal{I}$ and are the natural analogues of classical phase cells. Most importantly, their dimensionalities are astronomically large, increasing exponentially with N, and their role in the present treatment is analogous to that of the inequivalent representation spaces in the infinite system models of Refs. [15- 17]. As a result, the finite system model yields the essential positive results of the infinite one, but within a finite, realistic observational time. To be precise, it exhibits the above properties (I) and (II), in a stable manner, up to corrections that decrease exponentially with N and that are therefore utterly negligible by any standards of experimental physics.

The present note is devoted to a description of our mathematical scheme that has led to these results. We start in Section 2 by constructing the generic model of S_c and formulating both the time-dependent expectation values of the observables of S and their conditional expectation values, given the values of the macro-observables M of $\mathcal{I}$, subject to the assumption that S and $\mathcal{I}$ are independently prepared and then coupled together at time $t=0$. In Section 3, we formulate the conditions on the dynamics of the model and the structure of the macro- observables M under which it exhibits the properties (I) and (II) of a measurement process in a stable manner. We then briefly describe two models that fulfill these conditions, referring the reader to other articles for their detailed treatments. In Section 4, we probe further into the probabilistic structure of the model in order to pin-point, in general terms, the source of the effects (I) and (II). There we find that this is provided by a *large deviation* principle, which represents a rather general collective property of many-particle systems [24] and constitutes a generalisation to certain nonequilibrium states of Einstein's formula, $P=\text{const.}\exp(S)$, which relates the equilibrium probability distribution, P, of chosen macroscopic variables to the entropy function, S, of these variables. We conclude, in Section 5, with a brief resume of the picture presented here.

2. THE GENERIC MODEL

We assume that the algebras of bounded observables of the microsystem S, the instrument $\mathcal{I}$ and their composite $S_c = (S + \mathcal{I})$ are those of the bounded operators in separable Hilbert spaces $\mathcal{H}, \mathcal{K}$ and $\mathcal{H} \otimes \mathcal{K}$, respectively. Correspondingly, the states of these systems are represented by the density matrices in the respective spaces. The density matrices for the pure states of any of these systems are then the projection operators $P(f)$ of their normalised vectors f. For simplicity we assume that $\mathcal{H}$ is of finite dimensionality n.

We base the macroscopic description of $\mathcal{I}$ pertinent to the measuring process on an abelian subalgebra $\mathcal{M}$ of $\mathcal{B}$, which is generated by coarse-grained macroscopic observables (cf. [22, 23]): these are typically extensive variables of parts or the whole of $\mathcal{I}$. The choice of $\mathcal{M}$ yields a partition of $\mathcal{K}$ into the simultaneous eigenspaces $\mathcal{K}_\alpha$ of its elements. These subspaces of $\mathcal{K}$ are termed 'phase cells' as they are the canonical analogues of classical phase cells. We take them to represent the macrostates of $\mathcal{I}$ and we assume that they are unequivocally indicated by the 'pointer positions' of this instrument. Most importantly, the dimensionality of each $\mathcal{K}_\alpha$ is astronomically large, since it increases exponentially with the corresponding entropy and thus with N.

Since $\mathcal{I}$ is designed so that the pointer readings are in one-to-one correspondence with the eigenstates $u_1, \ldots, u_n$ of S, we assume that the index α of its macrostates also runs from 1 to n. Hence, denoting the projection operator for $\mathcal{K}_\alpha$ by Π_α, it follows from the above specifications that

$$\Pi_\alpha \Pi_\beta = \Pi_\alpha \delta_{\alpha\beta}, \tag{1}$$

$$\sum_{\alpha=1}^{n} \Pi_\alpha = I_{\mathcal{K}} \tag{2}$$

and that each element M of $\mathcal{M}$ takes the form

$$M = \sum_{\alpha=1}^{n} M_\alpha \Pi_\alpha, \tag{3}$$

where the M_α's are scalars.

We assume that S_c is a conservative system, whose Hamiltonian operator H_c, in $\mathcal{H} \otimes \mathcal{K}$, takes the form

$$H_c = H \otimes I_{\mathcal{K}} + I_{\mathcal{H}} \otimes K + V, \tag{4}$$

where H and K are the Hamiltonians of S and $\mathcal{I}$, respectively, and V is the $S - \mathcal{I}$ interaction energy. Further, we assume that $\mathcal{I}$ is an instrument of the first kind [14], in that the interaction V induces no transitions between the eigenstates of H. Thus, since the latter comprise an orthogonal basis set $(u_1, \ldots, u_n)$ of $\mathcal{H}$, with energy levels $(\varepsilon_1, \ldots, \varepsilon_n)$, respectively, the operators H and V take the forms $\sum_{r=1}^{n} \varepsilon_r P(u_r)$ and $\sum_{r=1}^{n} P(u_r) \otimes V_r$, respectively, where the V_r's are self-adjoint operators in $\mathcal{K}$. Hence, by Eq. (4), H_c reduces to the form $H_c = \sum_{r=1}^{n} P(u_r) \otimes K_r$, where

$$K_r = K + V_r + \varepsilon_r I_{\mathcal{K}}. \tag{5}$$

Consequently the one-parameter unitary group $\{U_c(t) = \exp(iH_c t) | t \in \mathbf{R}\}$, which governs the dynamics of S_c, is given by the formula

$$U_c(t) = \sum_{r=1}^{n} P(u_r) \otimes U_r(t), \tag{6}$$

where

$$U_r(t) = \exp(iK_r t). \tag{7}$$

Note. Eqs (5)-(7) signify that different eigenstates u_r of S give rise to different evolutions of $\mathcal{I}$. This is crucial to the realisability of the effect (II) of Sec. 1, whereby the pointer of $\mathcal{I}$ is driven into a position determined by the eigenstate of S.

Initial Conditions. We assume that the the systems S and $\mathcal{I}$ are coupled together at time $t = 0$ following independent preparation of S in a pure state and $\mathcal{I}$ in a mixed one, as represented by a normalised vector ψ and a density matrix Ω, respectively. Thus the initial state of the composite S_c is given by the density matrix

$$\Phi = P(\psi) \otimes \Omega \tag{8}$$

and its evolute at time t (> 0) is

$$U_c^\star(t) \Phi U_c(t) := \Phi(t). \tag{9}$$

Further, since ψ is a normalised vector, it is a linear combination of the basis vectors $(u_1,..,u_n)$ and hence takes the form

$$\psi = \sum_{r=1}^{n} c_r u_r, \tag{10}$$

where $\sum_{r=1}^{n} |c_r|^2 = 1$. Hence, by Eqs. (8)-(10),

$$\Phi(t) = \sum_{r,s=1}^{n} \bar{c}_r c_s P_{r,s} \otimes \Omega_{r,s}(t), \tag{11}$$

where $P_{r,s}$ is the operator in $\mathscr{H}$ defined by the equation

$$P_{r,s} f = (u_s, f) u_r \ \forall \ f \in \mathscr{H} \tag{12}$$

and

$$\Omega_{r,s}(t) = U_r^{\star}(t) \Omega U_s(t). \tag{13}$$

We note that $\Omega_{r,r}(t)$ is just the evolute of Ω corresponding to the eigenstate u_r of S.

Expectation and Conditional Expectation Values of Observables. The observables of S_c with which we shall be concerned are just the self-adjoint elements of $\mathscr{A} \otimes \mathscr{M}$. Their expectation values for the time-dependent state $\Phi(t)$ are given by the formula

$$E(A \otimes M) = \mathrm{Tr}\big(\Phi(t)[A \otimes M]\big) \ \forall \ A \in \mathscr{A}, \ M \in \mathscr{M}. \tag{14}$$

In particular, the expectation values of the observables of S are given by the equation

$$E(A) = E(A \otimes I_{\mathscr{K}}) \ \forall A \in \mathscr{A}, \tag{15}$$

while the probability that the macrostate of $\mathscr{I}$ corresponds to the cell $\mathscr{K}_\alpha$ is

$$w_\alpha = E(I_{\mathscr{H}} \otimes \Pi_\alpha). \tag{16}$$

Further, in view of the abelian property of the algebra $\mathscr{M}$, the functional E induces a conditional expectation of $A(\in \mathscr{A})$ with respect to $\mathscr{M}$, namely a linear map, $E(.|\mathscr{M})$, of $\mathscr{A}$ into $\mathscr{M}$ that preserves positivity and normalisation and that satisfies the compatibility condition

$$E\big(E(A|\mathscr{M})M\big) = E(A \otimes M) \ \forall \ A \in \mathscr{A}, \ M \in \mathscr{M}. \tag{17}$$

Moreover, since $E(A|\mathscr{M})$ is an element of $\mathscr{M}$ and therefore takes the form $\sum_{\alpha=1}^{n} f_\alpha(A) \Pi_\alpha$, where the f_α's are linear, positive, normalised functionals on $\mathscr{A}$, it follows from this observation and Eqs. (1)-(3), (16) and (17) that $f_\alpha(A) = E(A \otimes \Pi_\alpha)/w_\alpha$ and hence that

$$E(A|\mathscr{M}) = \sum_{\alpha=1}^{n} E(A \otimes \Pi_\alpha) \Pi_\alpha / w_\alpha.$$

Since Π_α is the projector for the cell $\mathscr{K}_\alpha$, its coefficient is the conditional expectation value $E(A|\mathscr{K}_\alpha)$ of A, given the macrostate $\mathscr{K}_\alpha$ of $\mathscr{I}$. Thus

$$E(A|\mathscr{K}_\alpha) = E(A \otimes \Pi_\alpha)/w_\alpha \ \forall \ A \in \mathscr{A}, \ w_\alpha \neq 0. \tag{18}$$

We now relate $E(A)$, $E(A|\mathscr{K}_\alpha)$ and w_α to the key dynamical quantity

$$F_{r,s:\alpha} = Tr\big(\Omega_{r,s}(t) \Pi_\alpha\big) \ \forall \ r, s, \alpha \in \{1,..,n\}, \tag{19}$$

noting in particular that $F_{r,r:\alpha}$ is the probability that $\mathscr{K}_\alpha$ is the macrostate of $\mathscr{I}$ at time t when u_r is the state of S. By Eqs. (2), (11)-(13), (16), (18) and (19), $E(A)$, $E(A|\mathscr{K}_\alpha)$ and w_α are related to F by the formulae

$$E(A) = \sum_{r=1}^{n} |c_r|^2 (u_r, A u_r) + \sum_{r \neq s; r,s=1}^{n} \sum_{\alpha=1}^{n} F_{r,s:\alpha} \bar{c}_r c_s (u_r, A u_s) \ \forall \ A \in \mathscr{A}, \tag{20}$$

$$E(A|\mathscr{K}_\alpha) = \sum_{r,s=1}^{n} F_{r,s;\alpha} \bar{c}_r c_s (u_r, A u_s)/w_\alpha \ \forall \ A \in \mathscr{A}, \ w_\alpha \neq 0. \tag{21}$$

and

$$w_\alpha = \sum_{r,s=1}^{n} \bar{c}_r c_s F_{r,s;\alpha}. \tag{22}$$

The following key properties of F ensue from Eqns. (2), (13) and (19).

$$\sum_{\alpha=1}^{n} F_{r,r:\alpha} = 1, \tag{23}$$

$$1 \geq F_{r,r:\alpha} \geq 0, \tag{24}$$

$$F_{r,s:\alpha} = \overline{F}_{s,r:\alpha} \tag{25}$$

and, for $z_1,..,z_n \in \mathbf{C}$, the sesquilinear form $\sum_{r,s=1}^{n} \bar{z}_r z_s F_{r,s;\alpha}$ is positive. Hence

$$F_{r,r;\alpha} F_{s,s;\alpha} \geq |F_{r,s;\alpha}|^2. \tag{26}$$

3. THE MEASUREMENT PROCESS

Suppose now that a reading of the pointer of $\mathscr{I}$ is made at time t. Then, according to the standard probabilistic interpretation of quantum mechanics, it follows from the above specifications that

(i) $E(A)$ is the expectation value of the observable A of S immediately before the reading;

(ii) w_α is the probability that the reading yields the result that the macrostate of $\mathscr{I}$ corresponds to the cell $\mathscr{K}_\alpha$; and,

(iii) in that case, $E(A|\mathscr{K}_\alpha)$ is the expectation value of A immediately after the measurement.

Hence the conditions for the realisation of the demands (I) and (II) of Von Neumann's phenomenological picture, described in Sec. 1, are that, for t greater than some critical, realistic value, τ, and less, in order of magnitude, than the Poincaré recurrence times,

(a)

$$E(A)=\sum_{r=1}^{n}|c_r|^2(u_r,Au_r)\ \forall\ A\in\mathscr{A} \tag{27}$$

and

(b) there is a unique invertible transformation ϕ of the set $(1,2,..,n)$ such that

$$E(A|\mathscr{K}_\alpha)=(u_{\phi(\alpha)},Au_{\phi(\alpha)})\ \forall\ A\in\mathscr{A}. \tag{28}$$

In other words, the pointer reading α signifies that the resultant state of S is $u_{\phi(\alpha)}$.

Proposition. *The combination of conditions (a) and (b) is equivalent to the formula*

$$F_{r,r;\phi^{-1}(r)}=1\ \forall\ r=1,..,n. \tag{29}$$

Hence this formula is equivalent to the Von Neumann conditions (I) and (II).

Proof. By Eqs. (24) and (26), Eq. (29) implies that $F_{r,s;\alpha}$ vanishes unless $r=s=\phi(\alpha)$ and is therefore equivalent to the formula

$$F_{r,s;\alpha}=\delta_{r,\phi(\alpha)}\delta_{s,\phi(\alpha)}\ \forall\ r,s,\alpha=1,..,n. \tag{30}$$

Assuming this formula , it follows immediately from Eqs. (20)-(22) that conditions (a) and (b) are fulfilled. Conversely, assuming condition (b), it follows from a comparison of the form of $w_\alpha E(A|\mathscr{K}_\alpha)$ obtained from Eqs. (22) and (28) with that given by Eq. (21) that

$$\sum_{r,s=1}^{n}\bar{c}_r c_s F_{r,s;\alpha}(u_{\phi(\alpha},Au_{\phi(\alpha)})=\sum_{r,s=1}^{n}\bar{c}_r c_s F_{r,s;\alpha}(u_r,Au_s)\ \forall\ A\in\mathscr{A},\ c_1,\ ,c_n\in\mathbf{C}.$$

On equating coefficients of $\bar{c}_r c_s$ in this formula, we see that

$$F_{r,s;\alpha}\left[(u_{\phi(\alpha)},Au_{\phi(\alpha)})-(u_r,Au_s)\right]=0\ \forall\ A\in\mathscr{A},\ r,s=1,2..,n.$$

In view of the orthonormality of the u_r's and the invertibility of ϕ, this last formula is equivalent to Eq. (30) and thus to Eq. (29).

Weakening of the Condition (29). In fact, that condition is extremely stringent and one sees both from the study of tractable models [2-4] and from a general probabilistic argument, presented in Sec. 4, that it should be weakened by a corrective term, due to macroscopic fluctuations, that decreases exponentially with N for large N. In other words, the sharp condition given by Eq. (29) should be weakened to the form wherein the difference between the two sides of that equation does not exceed $\exp(-cN)$, where c a positive constant of the order of unity. Hence, in view of Eqs. (23) and (24), this weaker condition is that

$$0\leq 1-F_{r,r;\phi^{-1}(r)}\leq\exp(-cN). \tag{31}$$

Correspondingly, it follows [2, 3] from arguments parallel to those employed to pass from Eqs. (27) and (28) to Eq. (30) in the proof of the above Proposition that the weakening of condition (29) to Eq. (31) implies corrections of order $\exp(-cN/2)$ to Eqs. (27) and (28), i.e. to the Von Neumann conditions (I) and (II). These corrections are utterly negligible from a physical standpoint, since N is typically of the order of 10^{8d}, where d is the dimensionality of the instrument $\mathscr{I}$.

A further essential property of an efficacious measuring instrument is that it should be stable against local perturbations of its initial state Ω (cf. [15, 16]). We express this condition in the following form.
$\mathscr{S}$. The formula (31) is stable under perturbations of the initial state Ω of $\mathscr{I}$ that are localised in the sense of leaving this state unchanged outside a ball whose volume is $O(1)$ with respect to N.

Thus, we characterise a quantum measurement process by the conditions given by Eq. (31) and $\mathscr{S}$. These conditions are viable, since they have been shown to be fulfilled by concrete models that may be described as follows: we refer the reader to the cited articles for full treatments of them.

Model 1 [2, 3]. This is a finite version of the Coleman-Hepp model [17], which is designed to measure the z-component, say, of the spin of an electron. The model consists of an electron, $\mathscr{E}$, and a chain, $\mathscr{C}$, of N Pauli spins. We regard the electron $\mathscr{E}$ as a composite of its own Pauli spin, S, and a spinless particle, $\mathscr{P}$, that carries its orbital motion. We then take the spin S to be the microsystem under observation and the composite of $\mathscr{P}$ and $\mathscr{C}$ to be the measuring instrument $\mathscr{I}$. We take the phase cells of $\mathscr{I}$ to be the subspaces $\mathscr{K}_{\pm}$ of its canonically defined Hilbert space $\mathscr{K}$ that correspond to positive and negative polarisations, respectively. We assume that $\mathscr{E}$ and $\mathscr{C}$ are coupled together at $t=0$ following independent preparations of S and $\mathscr{P}$ in pure states and $\mathscr{C}$ in a state of equilibrium [2] subject to a constraint, subsequently removed, that fixes its polarisation to a value directed along Oz: the initial state of $\mathscr{P}$ is assumed to be a wave packet localised at one end of $\mathscr{C}$ and moving towards the other end of that chain. The $\mathscr{E}-\mathscr{C}$ coupling, following this preparation, is assumed to be of the form $P_{-}\otimes V$, where P_{-} is the projection operator for the eigenstate of S for which the z-component, s_z, of its spin is negative and V is an interaction between $\mathscr{P}$ and the spins comprising $\mathscr{C}$ that reverses the latter ones in turn as the electron passes by them. Thus, up to corrections due to fluctuation effects, the resultant macrostate of $\mathscr{I}$ corresponds to the cell $\mathscr{K}_{+}$ or $\mathscr{K}_{-}$ according to whether s_z is in its eigenstate with positive or negative eigenvalue. To be precise [2, 3], taking fluctuations properly into account, the system $\mathscr{I}=(\mathscr{P}+\mathscr{C})$ satisfies the conditions given by Eq. (31) and $\mathscr{S}$ and is therefore an effective measuring instrument for the spin S; while the time taken to effect the measurement is essentially that required for the electron to run the length of the chain.

Model 2. This is the Allahverdyan-Balian-Nieuwenhuizen (ABN) model [4]. It consists of a localised Pauli spin, S, an Ising-Weiss ferromagnet, $\mathscr{F}$, and a heat bath $\mathscr{B}$. As described in the terms of the present article, S is the microsystem under observation and the composite $(\mathscr{F}+\mathscr{B})$ is the instrument $\mathscr{I}$ that measures the z-component, s_z, of its spin. The phase cells of $\mathscr{I}$ are the subspaces $\mathscr{K}_{\pm}$ of its canonically defined Hilbert space $\mathscr{K}$ that carry positive and negative polarisations, respectively, along Oz. It is assumed that the initial states of S, $\mathscr{F}$ and $\mathscr{B}$ are uncorrelated with $\mathscr{F}$ in a polarisation-free metastable state below its transition temperature and $\mathscr{B}$ in a thermal equilibrium state at that temperature. It is assumed that the subsequent couplings of the components of the model comprise (a) long range interactions between the z- components of the spin of S and those of $\mathscr{F}$ and (b) interactions between $\mathscr{B}$ and $\mathscr{F}$ that drive the latter into an equilibrium state with positive or negative polarisation along Oz according to whether s_z is positive or negative. Under these conditions, $\mathscr{I}$ serves as an effective measuring apparatus for S [4], i.e. it satisfies Eq. (31) and the stability condition[3]. Again, as in Model 1, the cells $\mathscr{K}_{\pm}$ represent the pointer positions for the eigenstates of the s_z with positive and negative eigenvalues, respectively.

4. ROLE OF THE LARGE DEVIATION PRINCIPLE

We now aim to probe more deeply into the quantum statistics of the measurement process in order to excavate the source of the properties of $\mathscr{I}$ represented by Eq. (31) and the stability condition $\mathscr{S}$. For simplicity, we confine our attention here to the case where the pointer reading corresponds to the value of a single coarse grained macroscopic observable[4] M of $\mathscr{I}$ and thus where the algebra $\mathscr{M}$ is generated by bounded functions of M. Further, we assume that M is a coarsened version of a fine-grained extensive observable M_f in the following sense (cf. [22, 23]).

Defining the intensive variable

$$m := N^{-1}M_f, \tag{32}$$

[2] This is the state that maximises the entropy of $\mathscr{C}$ subject to the specified constraint, and by the subadditivity of entropy [25] it takes the form $\otimes_{n=1}^{N}\frac{1}{2}(I+m\sigma_{n,z})$, where m is its polarisation and $\sigma_{n,x}$ is the z- component of the n'th spin of $\mathscr{C}$.

[3] This condition was not treated in [4], but a simple application of the method of Refs. [2, 3] demonstrates that the ABN model does satisfy it

[4] Generalisation to the case of several intercommuting ones is straightforward.

we assume that m is a bounded operator in $\mathscr{K}$ with pure point spectrum, whose highest and lowest points are the extremals of a closed interval Δ of $\mathbf{R}$. Further, we assume that this spectrum simulates a continuum for large N, in that the maximum spacing between the eigenvalues of m tends to zero as N tends to infinity.

We construct the coarse-grained version M of M_f by dividing Δ into a set of disjoint intervals of equal length $\{\mathscr{J}_\alpha|\alpha = 1, .,n\}$ and defining $\mathscr{K}_\alpha$ to be the subspace of $\mathscr{K}$ spanned by the eigenvectors of m whose eigenvalues lie in $\mathscr{J}_\alpha$. We then define m_α to be the arithmetic mean of those eigenvalues and define the coarse- grained observable M by the formula

$$M = \sum_{\alpha=1}^{n} N m_\alpha \Pi_\alpha, \tag{33}$$

where, as previously, Π_α is the projection operator for $\mathscr{K}_\alpha$.

We recall that, as noted following Eq. (19), $F_{r,r;\alpha}$ is the probability that $\mathscr{K}_\alpha$ is the macrostate of $\mathscr{I}$ at time t that corresponds to the eigenstate u_r of S. Furthermore, the density matrix $\Omega_{r,r}(t)$, defined by Eq. (13), represents the state of the *full system* $\mathscr{I}$ at time t corresponding to this eigenstate of S. Hence it follows from the above construction of M in terms of the fine-grained intensive observable m, which we may treat as a classical variable, that

$$F_{r,r;\alpha} = P_r(m \in \mathscr{J}_\alpha), \tag{34}$$

where P_r is the probability measure on Δ induced by the state $\Omega_{r;r}(t)$.

We now assume that this probability satisfies a *large deviation principle* [24], which is widely applicable to intensive macroscopic variables of many-particle systems. As noted in Sec. 1, this is a natural generalisation to nonequilibrium situations of Einstein's formula $P = \text{const.}\exp(S)$ for the relationship at thermal equilibrium between the entropy S, expressed as function of macroscopic variables and their probability distribution P. Thus, in the present situation, the large deviation principle asserts that the density of the probability measure P_r on the variable m takes the form $f_{r,N}(m)\exp\big(N\sigma_r(m)\big)$, where the function σ_r is N-independent and plays the role of a generalised specific entropy and $N^{-1}\log\big(f_{N,r}(m)\big)$ tends to zero as N tends to infinity. Thus, for large N, the density of P_r is effectively governed by the exponential term $\exp(N\sigma_r)$.

In order to demonstrate that, under very mild conditions on σ_r, both Eq. (31) and the stability condition $\mathscr{S}$ are satisfied, we suppose that

(a) for each $r \in (1,2, .,n)$, σ_r attains its maximum at precisely one value, m_r, of m;
(b) m_r lies in the interior of one of the intervals $\mathscr{J}_\alpha$, namely $\mathscr{J}_{\phi^{-1}(r)}$, and $\phi^{-1}(r) \neq \phi^{-1}(s)$ if $r \neq s$;
(c) $\sigma_r(m_r) - \sigma_r(m)$ is greater than some positive constant c for all $r \in (1,2, .,n)$ and $m \notin \mathscr{J}_{\phi^{-1}(r)}$; and
(d) the function σ_r is unaffected by localised perturbations, as specified in condition $\mathscr{S}$, of the initial state Ω of $\mathscr{I}$.

These conditions are certainly viable and, indeed, they are fulfilled by the models of Refs. [2-4]. Further, when taken in conjunction with the above specified large deviation principle for P_r, they imply that a is an invertible transformation of the set $(1,2, .,n)$ and that

$$P_r(m \in \mathscr{J}_\alpha) < \exp(-cN) \text{ for } \alpha \neq \phi^{-1}(r),$$

where c is a positive constant of the order of unity. In view of Eq. (34), this signifies that the instrument $\mathscr{I}$ satisfies the condition (31); and property (d) ensures that it fulfills the stability condition $\mathscr{S}$.

5. CONCLUDING REMARKS

We have provided a general mathematical scheme for the description of quantum measurement theory within the framework of Schroedinger dynamics and quantum probability. This scheme is manifestly realisable and involves nothing more than the dynamics of the composite of a quantum microsystem S and a macroscopic measuring instrument $\mathscr{I}$. Thus, it does not require any appeal either to cosmological actions or to Von Neumann-Wigner psycho-physical parallelism, with its infinite hierarchy of measuring instruments. Indeed the only participation of human intelligence in the measurement process is in the design of the instrument $\mathscr{I}$ and the interpretation of readings of its pointer positions.

REFERENCES

1. J. Von Neumann: *Mathematical Foundations of Quantum Mechanics*, Princeton University Press, Princeton, NJ, 195

2. G. L.Sewell: Rep. Math. Phys. **56**, 271, 2005
3. G. L. Sewell: Markov Processes and Rel. Fields **13**, 425, 2007
4. A. E. Allahverdyan, R. Balian and Th. M. Nieuwenhuizen. Eur. Phys. Lett. **61**, 452, 2003
5. N. G. Van Kampen: *Physica* A **153**, 97, 1988.
6. A. Peres: Am. J. Phys. **54**, 688 , 1986
7. E. P.Wigner: Pp. 171-84 of *Symmetries and Reflections*, Indiana University Press, Bloomington, 1967.
8. N. Gisin: *Phys. Rev. Lett.* **52**, 1657, 1984
9. E. Joos and H. D. Zeh: *Z. Phys.* B **59**, 223, 1985
10. L. Diosi: *J. Phys.* A **21**, 2885, 1988
11. I. Percival: *Quantum State Diffusion*, Cambridge Univ. Press, Cambridge, 1998.
12. G. C. Ghirardi, A. Rimini and T. Weber: *Phys. Rev.* D **34**, 470, 1986
13. N. Bohr: *Discussion with Einstein on epistomological problems in atomic physics*, Pp. 200-241 of *Albert Einstein: Philosopher-Scientist*, Ed. P. A. Schilp, The Library of Living Philosophers, Evanston, IL, 1949.
14. J, M. Jauch: *Foundations of Quantum Mechanics*, Addison Wesley, Reading, MA, 1968.
15. B. Whitten-Wolfe and G. G. Emch: *Helv. Phys. Acta* **49**, 45, 1976
16. G. G. Emch: Pp. 255-264 of *Quantum Information and Communication*, E. Donkor, A. R. Pirich and H. E. Brandt, Eds., Intern. Soc. Opt. Eng. (SPIE) Proceedings 5105, 2003
17. K.Hepp: *Helv. Phys. Acta* **45**, 237, 1972
18. D. Ruelle: *Statistical Mechanics*, W. A. Benjamin, New York, 1969.
19. G. G. Emch: *Algebraic methods in Statistical Mechanics and Quantum Field Theory*, Wiley, New York, 1972.
20. G. L. Sewell: *Quantum Mechanics and its Emergent Macrophysics*, Princeton University Press, Princeton, 2002.
21. J. S. Bell: *Helv. Phys. Acta* **48**, 93, 1975
22. N. G. Van Kampen: *Physica* **20**, 603, 1954
23. G. G. Emch: *Helv. Phys. Acta* **37**, 532, 1964
24. R. S. Ellis: *Entropy, Large Deviations and Statistical Mechanics*, Springer, New York, 1985
25. E. H. Lieb and M. B. Ruskai: J. Math. Phys. **14**, 1938, 1973

Derivation of the Radiative Transfer Equation for Polarized Light

Wilhelm von Waldenfels

Institut für Angewandte Mathematik der Universität Heidelberg, Im Neuenheimer Feld 294 69120 Heidelberg Germany

Abstract. The transfer equation for polarized light is related to a non-commutative Markov process. Using scattering by two level atoms, where the lower state is an s-state and the upper state is a p-state, and the limits of geometrical optics and singular coupling and assuming, that the scatterers form a Poisson point process, the infinitesimal law of the Markov process can be determined. That is equivalent to the transfer equation.

Keywords: Radiative Transfer, Scattering, Non-commutative Markov process.
PACS: 95.30 Jx,02.50 Ga

INTRODUCTION

The intensity of monochromatic polarized light $\mathfrak{I}_t(z,n)$ is a function of position $z \in \mathbb{R}^3$ and of direction $n \in \mathbb{S}^2$,i.e. the unit sphere in $\mathbb{R}^3$. Its value at (z,n) is a three dimensional, positive definite matrix orthogonal to n, i.e.$\mathfrak{I}_t(z,n)n = 0$. Its trace describes the local energy. The quotient $\mathfrak{I}_t(z,n)/(\mathrm{Tr}\mathfrak{I}_t(z,n))$ is a positive definite matrix of trace 1, describing the polarization and is called *polarization matrix* We prefer the description by positive definite matrices to the use of Stokes' vector.

If there are N photons in the considered area, then

$$\mathfrak{I}_t(z,n) = N\mathfrak{P}_t(z,n) \tag{1}$$

and $\mathfrak{P}_t$ is a kind of probability density generalized in a noncommutative way. We have

$$\int_{\mathbb{R}^3}\int_{\mathbb{S}^2} \mathrm{Tr}\mathfrak{P}_t(z,n)\mathrm{d}z\mathrm{d}n = 1. \tag{2}$$

Assume that the photon is at time t at the place (z,n) and has the polarization matrix ρ. Having simultaneously a fixed place and a fixed direction is not compatible with the wave character of light. But it is compatible in the limit of geometrical optics . We consider scattering by two level atoms, where the lower level is a non-degenerate s-state and the upper state is a three fold degenerate p- state. We assume the atoms on random positions forming a Poisson point process. We use the weak coupling limit and shrink the extension of a photon to a point.These approximations induce a kind of non-commutative Markovianity. A single photon at (z,n) is given by $\rho\delta_{z,n}$. Due to free evolution and scattering in an infinitesimal time $\mathrm{d}t$

$$\rho\delta_{z,n} \mapsto (1-\sigma\mathrm{d}t)\rho\delta_{z+n\mathrm{d}t,n} + \frac{3\sigma\mathrm{d}t}{8\pi}\int_{\mathbb{S}^2}\mathrm{d}n' P(n')\rho P(n')\delta_{z,n'} \tag{3}$$

where $P(n)$ is the projector in $\mathbb{R}^3$ onto the plane orthogonal to n, so

$$P(n)_{ik} = \delta_{ik} - n_i n_k. \tag{4}$$

One obtains from there an equation for $\mathrm{d}\mathfrak{P}_t/\mathrm{d}t$ and finally the transfer equation

$$\frac{\mathrm{d}}{\mathrm{d}t}\mathfrak{I}_t(z,n) = -(n\nabla)\mathfrak{I}_t(z,n) - (\sigma+\alpha)\mathfrak{I}_t(z,n) + \frac{3\sigma}{8\pi}\int_{\mathbb{S}^2}\mathrm{d}n' P(n)\mathfrak{I}_t(z,n')P(n). \tag{5}$$

The absorption coefficient α has been put in by hand, as Chandrasekhar [1] does it. Remark, that the speed of light is 1 .Equation (5) has been derived in section 2. ,equation (3) in section 3.

The transfer equation for polarized light has been established by Chandrasekhar [1]. An other derivation is due to E.Landi degl' Innocenti [3]. The methods we are using are a generalization of those used in [4], [5].

CP962, *Quantum Theory, Reconsideration of Foundations – 4*,
edited by G. Adenier, A. Yu. Khrennikov, P. Lahti, V. I. Man'ko, and Th. M. Nieuwenhuizen

TRANSFER OF POLARIZED LIGHT AS A NONCOMMUTATIVE MARKOV PROCESS

We recall at first the transfer equation for unpolarized light. Its intensity $I_t(z,n)$ is a real number ≥ 0 and if there are N photons, $I_t(z,n) = NP_t(z,n)$, where $P_t(z,n)$ is a classical probability density. As it has been shown in [4], P_t is the probability density of a classical Markov process Ξ_t on $\mathbb{R}^3 \times \mathbb{S}^2$, the infinitesimal law of which (if we do not regard absorption)is given in the following way. If $\Xi_t = (z,n)$, then

$$\Xi_{t+\mathrm{d}t} = \begin{cases} z+n\mathrm{d}t & \text{with probability} \quad 1-\sigma\mathrm{d}t \\ (z,n') & \text{with probability} \quad \sigma\mathrm{d}t\mathrm{d}n'/(4\pi) \end{cases},$$

where $\mathrm{d}n$ is the surface elenet on the unit sphere $\mathbb{S}^2$. A Markov process defines a Markov semigroup $T_t, t \geq 0$ describing the time development with $T_{t+s} = T_tT_s$ and $T_0 = \mathbf{1} =$ identity . One has $T_tP_s = P_{t+s}$. The infinitesimal law can be formulated in the following way

$$T_{\mathrm{d}t}\delta_{z,n} = (1-\sigma\mathrm{d}t)\delta_{z+n\mathrm{d}t,n} + \frac{\sigma\mathrm{d}t}{4\pi}\int_{\mathbb{S}^2}\delta_{z,n'}\mathrm{d}n',$$

where $\delta_{z,n}$ is the point mass at (z,n). As will be shown below for polarized light, it is easy to derive the transport equation from the last equation.

In the case of polarized light, one has to consider a matrix valued noncommutative probability density $\mathfrak{P}_t$ and the straight forward definition of a classical Markov process cannot be used any more. Nevertheless considering the physical situation a semigroup $T_t, t \geq 0$ should exist,due to the approximations explained in the next section : the changement of the state of the photon depends only on the present state.The operators T_t are linear and positivity preserving in the sense, that a function or measure with values in the positive definite matrices is mapped into an object with the same properties.

The state of a photon is given by the position z,the direction n and by the polarization ρ. As it will be shown in the next section

$$T_{\mathrm{d}t}\rho\,\delta_{z,n} = (1-\sigma\mathrm{d}t)\rho\,\delta_{z+n\mathrm{d}t,n} + \frac{3\sigma\mathrm{d}t}{8\pi}\int_{\mathbb{S}^2}\mathrm{d}n'P(n')\rho P(n')\delta_{z,n'} \tag{6}$$

Due to linearity

$$\begin{aligned}\mathfrak{P}_{t+\mathrm{d}t} = T_{\mathrm{d}t}\mathfrak{P}_t &= T_{\mathrm{d}t}\iint \mathrm{d}z\mathrm{d}n\mathrm{Tr}\mathfrak{P}_t(z,n)\rho_t(z,n)\delta_{z,n} \\ &= \iint \mathrm{d}z\mathrm{d}n\mathrm{Tr}\mathfrak{P}_t(z,n)T_{\mathrm{d}t}(\rho_t(z,n)\delta_{z,n})\end{aligned}$$

with

$$\rho_t(z,n) = \frac{\mathfrak{P}_t(z,n)}{\mathrm{Tr}\mathfrak{P}_t(z,n)}.$$

Inserting (6) we obtain

$$\begin{aligned}\mathfrak{P}_{t+\mathrm{d}t} = \iint \mathrm{d}z\mathrm{d}n\mathfrak{P}_{t+\mathrm{d}t}(z,n)\delta_{z,n} &= (1-\sigma\mathrm{d}t)\iint \mathrm{d}z\mathrm{d}n\mathfrak{P}_t(z,n)\delta_{z+n\mathrm{d}t,n} \\ &+\frac{3\sigma\mathrm{d}t}{8\pi}\iiint \mathrm{d}z\mathrm{d}n\mathrm{d}n'P(n')\mathfrak{P}_t(z,n)P(n')\delta_{z,n}.\end{aligned}$$

In the first term intoduce $z - n\mathrm{d}t$ as new variable,in the second term interchange the role of the variables n and n', assume that $\mathfrak{P}_t$ is differentiable, expand to the first order and abtain

$$\frac{\mathrm{d}}{\mathrm{d}t}\mathfrak{P}_t(z,n) = -(n\nabla)\mathfrak{P}_t(z,n) - \sigma\mathfrak{P}_t(z,n) + \frac{3\sigma}{8\pi}\int_{\mathbb{S}^2}\mathrm{d}n'P(n)\mathfrak{P}_t(z,n')P(n). \tag{7}$$

Multply by the number of photons N and introduce an absorption coefficient α by hand and arrive at the transfer equation (5).

DERIVATION OF THE TRANSITION PROBABILITY

We use the formalism of quantum mechanics and employ Dirac's notation. We consider the Hilbert space $\mathfrak{H}_0 = L^2(\mathbb{R}^3,\mathbb{C}^3) = L^2(\mathbb{R}^3,\mathbb{C}) \otimes \mathbb{C}^3$ of $\mathbb{C}^3$- valued square integrable functions. We regard the its elements as functions of the wave number k and its Fourier transforms $\hat{f}$ as functions of the position x

$$\hat{f}(x) = (2\pi)^{-3/2} \int \exp^{i(kx)} f(k) dk, \qquad f(k) = (2\pi)^{-3/2} \int \exp^{-i(kx)} \hat{f}(x) dk. \tag{8}$$

We consider three dimensional vectors as column vectors $x = \begin{pmatrix} x_1 \\ x_2 \\ x_3 \end{pmatrix}$ and denote by $x^T = (x_1,x_2,x_3)$ the transposed and by $x^* = (\bar{x}_1,\bar{x}_2,\bar{x}_3)$ the adjoint vector. The scalar product is denoted by $(xy) = x^*y$ or simply by xy, if no misunderstanding is possible. The scalar product in $\mathfrak{H}_0$ is defined by

$$\langle f|g\rangle = \int \mathrm{d}k f^*(k) g(k) = \int \mathrm{d}k \sum_j \bar{f}_j(k) g_j(k)$$

Define the operators K_i and X_i for $i = 1,2,3$ by

$$(K_i f)(k) = k_i f(k), \qquad (X_i f)(k) = i\partial/(\partial k_i) f(k). \tag{9}$$

One has the usual commutation relations $[K_j, X_l] = -i\delta_{j,l}$. Denote by δ_k the generalized function $\delta_k(k') = \delta(k-k')$. The functions $\delta_k, k \in \mathbb{R}^3$ form a generalized basis of $L^2(\mathbb{R}^3)$.If $e_i, i = 1,2,3$ is an orthonormal basis of $\mathbb{R}^3$,the functions $\{e_i\delta_k;\, i = 1,2,3,\ k \in \mathbb{R}^3\}$ form a generalized orthonormal basis of $\mathfrak{H}_0$.One has

$$K_i e_j \delta_k = k_i e_j \delta_k$$

Assume $\alpha, \beta \in \mathbb{R}^3$. Then

$$(\exp i(\alpha X) f)(k) = f(k-\alpha), \qquad (\exp i(\beta K) f)(k) = \exp i(\beta k) f(k). \tag{10}$$

Define the Weyl operator

$$\exp\left(i(\alpha X) + i(\beta K)\right) = \exp^{-i\alpha\beta/2} \exp i(\beta K) \exp i(\alpha X). \tag{11}$$

If $f,g \in \mathfrak{H}_0$ and M is a 3×3 matrix we define the Fourier Weyl transform for $|g\rangle\langle f|$

$$\begin{aligned} W(\alpha,\beta;M;f,g) &= \langle f|M \exp(i(\alpha X) + i(\beta K))|g\rangle & (12)\\ &= \int \mathrm{d}k \sum_{j,l} \bar{f}_j(k) M_{jl} \exp(-i\alpha\beta/2 + i\beta k) g_l(k-\alpha). & (13) \end{aligned}$$

This notion can be extended to density matrices easily. A density matrix is determined by its Fourier Weyl transform in a unique way.

A polarized lightwave ψ is given by a function in $\mathfrak{H}_0$ such that $(k\psi(k)) = \sum k_i \psi_i(k) = 0$ for all k. We denote by $\mathfrak{H} = \{\psi \in \mathfrak{H}_0 : (k\psi(k)) = 0\}$ the Hilbert space of all those functions. Define by $k^0 = k/|k|$ the unit vector in the direction of k and choose for any k two vectors $e^1(k), e^2(k)$, such that $\{k^0, e^1(k), e^2(k)\}$ form a threedimensional orthonormal basis. The $e^i(k)$ can be chosen in a piecewise continuous way. We have

$$\psi(k) = \psi^1(k) e^1(k) + \psi^2(k) e^2(k),$$

where the ψ^i are scalar functions.

A photon of approximative frequency ω_0 is a *quasimonochromatic* light wave of the form

$$\begin{aligned} \hat{\psi}(x) &= \exp(i\omega_0 n x)\hat{\varphi}(x) \\ \psi(k) &= \varphi(k - n\omega_0) = e^1(k)\varphi^1(k - n\omega_0) + e^2(k)\varphi^2(k - n\omega_0) \\ \|\psi\|^2 &= \int \mathrm{d}k |\psi(k)|^2 = 1, \end{aligned} \tag{14}$$

where n is a unit vector and ω_0 is considered to be big. We may assume, that the $e^i(k)$ are continuous in the neighborhood of n.

We want to perform the *limit of gemetrical optics* and calculate

$$W_{\omega_0}(\alpha,\beta;M;\psi,\psi) = \langle\psi|M\exp i(\alpha X) + i(\beta K/\omega_0)|\psi\rangle$$

and obtain

$$\int \mathrm{d}k \sum_{jl} e^j(k)^T M e^l(k)\bar{\varphi}^j(k-n\omega_0)\exp(-i\alpha\beta/(2\omega_0)+i\beta k/\omega_0)\varphi^l(k-n\omega_0-\alpha)$$

Introduce $k-n\omega_0$ as new variable and observe that for $\omega_0 \to \infty$ one has $e^i(k+n\omega_0) \to e^i(n)$ and get

$$\int \mathrm{d}k \sum_{jl} e^j(n)^T M e^l(n)\exp(i\beta n)\bar{\varphi}^j(k)\varphi^l(k-\alpha).$$

So

$$W_{\omega_0}(\alpha,\beta;M;\psi,\psi) \to \iint_{\mathbb{R}^3\times\mathbb{S}^2} \mathrm{d}x\mathrm{d}u \exp(i\alpha x + i\beta u)\ \mathrm{Tr}(\rho(x)M)\delta_n(u) \tag{15}$$

with

$$\rho(x) = \sum_{jl}\hat{\varphi}^j(x)\bar{\hat{\varphi}}^l(x)e^j(n)e^l(n)^T.$$

Here $\mathbb{S}^2$ is the unit sphere in the threedimensional space and δ_n is the point measure on $n \in \mathbb{S}^2$.The quantity

$$\mathfrak{p}(x,u) = \rho(x)\delta_n(u) \tag{16}$$

is a measure on $\mathbb{R}^3 \times \mathbb{S}^2$ with values in the set of positiv definite three dimensional matrices A such that $An = 0$. As $\iint \mathrm{Tr}\,\mathfrak{p}(x,u)\mathrm{d}x\mathrm{d}u = 1$, the measure $\mathfrak{p}$ is the natural generalization of a probability measure.

The free Hamiltonian of a photon is

$$H_0 = |K| = \sqrt{K_1^2+K_2^2+K_3^2}. \tag{17}$$

The free evolution of the photon is given by

$$\psi^{(0)}(t) = \exp(-iH_0t)\psi. \tag{18}$$

Observe $|\omega_0 n - k| - \omega_0 \to (kn)$ and obtain for $\omega_0 \to \infty$ with the help of similar calculations as in [5]

$$W_{\omega_0}(\alpha,\beta;M;\psi^{(0)}(t),\psi^{(0)}(t)) \to \int \mathrm{d}k\mathrm{d}u \exp(i\alpha x + i\beta u)\mathfrak{p}_t^{(0)}(x,u) \tag{19}$$

with

$$\mathfrak{p}_t^{(0)}(x,u) = \rho(x-nt)\delta_n(u). \tag{20}$$

. We assume,that the photon is scattered by a two level atom situated at the point x_0. The lower level is a non degenerate s state, the upper level is threefold degenerate p state. Following the usual scattering theory [2],we may introduce an interaction Hamiltonian

$$H_I = \exp(-iKx_0)VV^*\exp(+iKx_0) \tag{21}$$

Where

$$\begin{aligned} V:\mathbb{C}^3 &\to \mathfrak{H}: V = \int \mathrm{d}k\, c(k)P(k)|\delta_k\rangle \\ V^*:\mathfrak{H} &\to \mathbb{C}^3: V^* = \int \mathrm{d}k\, \bar{c}(k)P(k)\langle\delta_k| \end{aligned} \tag{22}$$

and recall (4)that $P(k) = \sum_{j=1,2}e^j(e^j)^T = \mathbf{1} - k^0(k^0)^T$ is the projector onto the orthogonal complement of k. So for $x \in \mathbb{C}^3, f \in \mathfrak{H}, f(k) = \sum f^j(k)e^j(k)$ we have

$$\begin{aligned} (Vx)(k) &= c(k)P(k)x \\ V^*f &= \int \sum \bar{c}(k)f^j(k)e^j(k)\mathrm{d}k = \int \bar{c}(k)f(k)\mathrm{d}k \end{aligned}$$

We assume

$$c(k) = \omega_0^{-1/2} g(|k| - \omega_0) \tag{23}$$

where g is real for simplicity and there exist two constants $\Delta\omega_1, \Delta\omega_2$

$$0 < \Delta\omega_1 << \Delta\omega_2 << \omega_0$$

such that $g(\omega) = 0$ for $\omega > \Delta\omega_2$ and $g(\omega) = g$ for $\omega < \Delta\omega_1$ and furthermore $g(\omega)^2 = g(-\omega)^2$. We take care of these conditions in letting at first $\omega_0 \to \infty$ *(limit of geometrical optics)* and then going with $g(\omega)$ to g such that $g(\omega)^2 = g(-\omega)^2$ *(singular coupling limit)*.

We expand to the second order

$$\psi(t) = e^{-iHt}\psi = \psi^{(0)}(t) - i\psi^{(1)}(t) - \psi^{(2)}(t) \tag{24}$$

with

$$\psi^{(0)}(t) = e^{-iH_0 t}\psi \tag{25}$$

$$\psi^{(1)}(t) = \int_0^t e^{-i(t-t_1)H_0} H_I e^{-iH_0 t_1} \mathrm{d}t_1 \tag{26}$$

$$\psi^{(2)}(t) = \int\int_{0<t_1<t_2<t} e^{-i(t-t_2)H_0} H_I e^{-i(t_2-t_1)H_0} H_I e^{-it_1 H_0} \mathrm{d}t_1 \mathrm{d}t_2. \tag{27}$$

In order to continue we have to perform some auxiliary calculations similar to those in [5].

$$V^* \exp^{iKx_0 + iH_0 t_1} M \exp^{i\alpha X + i\beta K/\omega_0} \exp^{-iH_0 t} |\psi\rangle =$$
$$\omega_0^{-1/2} \int g(|k| - \omega_0) \exp i(kx_0 + |k|t_1 - \alpha\beta/2\omega_0 + \beta k/\omega_0 - |k-\alpha|t) P(k) M \sum_j \varphi^j(k - \alpha - n\omega_0) e^j(k-\alpha) \mathrm{d}k$$
$$= \omega_0^{-1/2} \exp(i\omega_0(nx_0 - (t - t_1)) C_1(t_1, t; \omega_0)$$

Proceeding like for the derivation of equation (20) we obtain that for $\omega_0 \to \infty$ the quantity $C_1(t_1, t; \omega_0)$ converges to

$$C_1(t_1, t) = \int g(kn) \exp i(kx_0 - kn(t - t_1) + \beta n + \alpha nt) \sum_j P(n) M e^j(n) \varphi^j(k - \alpha) \mathrm{d}k.$$

Perform the singular coupling limit

$$C_1(t_1, t) \to (2\pi)^{3/2} g \exp i(\alpha x_0 + \alpha n t_1 + \beta n) \sum_j P(n) M e^j(n) \hat{\varphi}^j(k)(x_0 - n(t - t_1))$$

By putting $\alpha = \beta = t_1 = 0, M = 1$ we obtain

$$V^* \exp(iKx_0 - iH_0 t)|\psi\rangle = \omega_0^{-1/2} \exp(i\omega_0(nx_0 - t)) C_2(t; \omega_0)$$

and

$$C_2(t; \omega_0) \to C_2(t) = \int g(kn) \exp i(kx_0 - knt) \sum_j e^j(n) \varphi^j(k) \mathrm{d}k$$

In the singular coupling limit

$$C_2(t) \to (2\pi)^{3/2} g \sum_j e^j(n) \varphi^j(x_0 - nt)$$

We have

$$\begin{aligned} V^* & \exp(iKx_0 + iH_0 t_2) M \exp(i\alpha X + i\beta K/\omega_0) \exp(-iH_0 t_1 - iKx_0) V \\ = & \omega_0^{-1} \int P(k) M g(|k| - \omega_0) \mathrm{d}k \\ \times & \exp(i(kx_0 + |k|t_2 + \beta k/\omega_0 - \alpha\beta/2\omega_0 - (k-\alpha)x_0 - |k-\alpha|t_1)) \\ \times & g(|k-\alpha| - \omega_0) P(k) = \omega_0 \exp(i\omega_0(t_2 - t_1)) C_3(t_2, t_1; \omega_0) \end{aligned}$$

Using polar coordinates and the same procedures as above one obtains

$$C_3(t_2,t_1;\omega_0) \;\rightarrow\; C_3(t_2,t_1) \;=\; \int_{\mathbb{S}^2} \mathrm{d}e P(e) M P(e) \int \mathrm{d}\omega g(\omega) \exp(i(\omega(t_2 - t_1) + \beta e + \alpha x_0 + \alpha e t_1) g(\omega - \alpha e)$$

In the singular coupling limit

$$C_3(t_2,t_1) \rightarrow 2\pi g^2 \delta(t_2 - t_1) \int_{\mathbb{S}^2} P(e) M P(e) \exp i(\beta e + \alpha x_0 + \alpha e t_1)$$

One has

$$V^* \exp i(K x_0 + H_0 t - K x_0) V = \omega_0 e^{\omega_0 t} C_4(t;\omega_0)$$

$$C_4(t,\omega_0) \rightarrow C_4(t) = \frac{8\pi}{3} \int \mathrm{d}\omega e^{i\omega t} g(\omega)^2$$

Due to the symmetry $g(\omega)^2 = g(-\omega)^2$ and we have $\int e^{i\omega s} g(\omega)^2 1_{s>0} \mathrm{d}\omega \rightarrow \pi g^2 \delta(s)$. and in the singular coupling limit

$$1_{t<0} C_4(t) \rightarrow \frac{8\pi^2}{3} \delta(t)$$

We obtain

$$W_{\omega_0}(\alpha,\beta;M;\psi^{(1)}(t),\psi^{(0)}(t)) = O(\omega_0^{-1}) \rightarrow 0$$

and

$$\begin{aligned} W_{\omega_0}(\alpha,\beta;M;\psi^{(1)}(t),\psi^{(1)}(t)) &= \int_0^t \int_0^t \mathrm{d}t_1 \mathrm{d}t_2 \langle \psi | \exp i(H_0(t-t_2) - K x_0) V \\ & V^* \exp i(K x_0 + H_0 t_2) M \exp i(\alpha X + \beta K/\omega_0) \exp i(-H_0 t_1 - K x_0) V V^* \exp i(K x_0 - H_0(t-t_1)) | \psi \rangle \\ & \rightarrow \int_0^t \int_0^t \mathrm{d}t_1 \mathrm{d}t_2 \overline{C_2(t-t_2)} C_3(t_2,t_1) C_2(t-t_1) \end{aligned}$$

Perform the singular coupling limit and obtain

$$\begin{aligned} & g^4 (2\pi)^4 \int_0^t \mathrm{d}t_1 \int \mathrm{d}e \exp i(\beta e + \alpha x_0 + \alpha e t_1) \sum_{jl} e^j(n) \bar{\varphi}^j(x_0 - n(t-t_1) P(e) M P(e) e^l(n) \varphi^l(x_0 - n(t-t_1)) \\ = \;& g^4 (2\pi)^4 \int_0^t \mathrm{d}t_1 \int \mathrm{d}e \exp i(\beta e + \alpha x_0 + \alpha t_1) \mathrm{Tr}(M P(e) \rho(x_0 - n(t-t_1) P(e)) \end{aligned}$$

Calculate

$$\begin{aligned} & W_{\omega_0}(\alpha,\beta;\psi_2(t),\psi_0 t)) \\ = \;& \int\int_{0<t_1<t_2<t} \mathrm{d}t_1 \mathrm{d}t_2 \langle \psi | e^{iH_0(t-t_2) - iKx_0} V V^* e^{iKx_0 + iH_0(t_2-t_1) - iKx_0} V V^* e^{iKx_0 + H_0 t_1} e^{i\alpha X + i\beta K/\omega_0} e^{-iH_0 t} | \psi \rangle \\ & \rightarrow \int\int_{0<t_1<t_2<t} \mathrm{d}t_1 \mathrm{d}t_2 \overline{C_2(t_2-t_1)} C_4(t_2-t_1) C_1(t_1,t) \end{aligned}$$

In the singular coupling limit this expression converges to

$$\begin{aligned} & 2(2\pi)^5 g^4/3 \int_0^t \mathrm{d}t_1 \exp(i(\alpha x_0 + n\alpha t_1 + \beta n)) \sum_{jl} e^j(n) \bar{\varphi}^j(x_0 - n(t-t_1)) M e^l(n) \varphi^j(x_0 - n(t-t_1)) \\ = \;& 2(2\pi)^5 g^4/3 \int_0^t \mathrm{d}t_1 \exp(i(\alpha x_0 + n\alpha t_1 + \beta n)) \mathrm{Tr}(M \rho(x_0 - n(t-t_1)) \end{aligned}$$

Write for short $W_{\omega_0}(\alpha,\beta;\psi^{(j)},\psi^{(l)}) = W(j,l)$,then up to the second order in time t

$$W_{\omega_0}(\alpha,\beta;M;\psi(t),\psi(t)) = W(0,0) - iW(0,1) + iW(1,0) - W(2,0) + W(1,1) - W(0,2)$$

Now $W(0,0)$ has been calculated in (20), $W(1,1)$ and $W(2,0)$ above. One has $W(0,2) = W(2,0)$ and $W(1,0)$ differs from $W(2;0)$ by a real factor $\sim \omega_0^{-1}$, so the contribution $W(1,0) - W(0,1)$ vanishes and a more detailed calculation shows, that it vanishes even in the next order in ω_0^{-1}, too.

Proposition . *Expand in the Fourier Weyl transform*

$$W_{\omega_0}(\alpha,\beta;M;\psi(t),\psi(t))$$

to the second order and perform at first the limit of geometrical optics, and then the singular coupling limit. Then the Fourier Weyl transform converges to $\int e^{i\alpha x+i\beta u}\mathrm{Tr}(M\mathfrak{p}_t(x,u))dxdu$ *with*

$$\begin{aligned}\mathfrak{p}_t(x,u) &= \rho(x-nt)\delta_n(u) - \sigma_0\int_0^t \delta(x_0+nt_1-x)\rho(x-n(t-t_1))\mathrm{d}t_1\delta_n(u)\\ &+\frac{3\sigma_0}{8\pi}\int_0^t \mathrm{d}t_1\int_{\mathbb{S}^2}\mathrm{d}e\,\delta(x-x_0-et_1)P(e)\rho(x_0-n(t-t_1))P(e)\delta_e(u)\end{aligned} \tag{28}$$

where $\sigma_0 = 4(2\pi)^5 g^4/3$.
Use the properties of the δ function and rewrite (28)

$$\begin{aligned}\mathfrak{p}_t(x,u) &= \rho(x-nt)\delta_n(u) - \sigma_0 t\,\delta(x_0+nt_1-x)\rho(x-nt))\delta_n(u)\\ &+\frac{3\sigma_0}{8\pi}\int_0^t \mathrm{d}t_1\int_{\mathbb{S}^2}\mathrm{d}e\,\delta(x-x_0-et_1)P(e)\rho(x-et_1-n(t-t_1))P(e)\delta_e(u)\end{aligned}$$

We assume that the positions x_0 of the scattering atoms form a Poisson point process with density $\mathcal{N}$, then the average is

$$\begin{aligned}\mathbb{E}\mathfrak{p}_t(x,u) &= \rho(x-nt)\delta_n(u) - \mathcal{N}\sigma_0 t\rho(x-nt)\delta_n(u)\\ &+\frac{3\mathcal{N}\sigma_0}{8\pi}\int_{\mathbb{S}^2}\mathrm{d}e\int_0^t \mathrm{d}t_1 P(e)\rho(x-et_1-n(t-t_1))P(e)\delta_e(u).\end{aligned} \tag{29}$$

At a mesosopical point of view we may replace $\rho(x) \Rightarrow \rho\delta(x-z)$, where ρ is a 3×3 density matrix with $\rho n = 0$. Put

$$\sigma = \mathcal{N}\sigma_0 \tag{30}$$

Then the last equation reads without mentioning the average

$$\begin{aligned}\mathfrak{p}_t(x,u) &= (1-\sigma t)\rho\delta(x-z-nt)\delta_n(u)\\ &+\frac{3\sigma}{8\pi}\int_{\mathbb{S}^2}\mathrm{d}e\int_0^t \mathrm{d}t_1 P(e)\rho P(e)\delta(x-et_1-n(t-t_1)-z)\delta_e(u).\end{aligned} \tag{31}$$

Write $\delta_{z,n}(x,u) = \delta(x-z)\delta_n(u)$, introduce a time development operator T_t and rewrite (31)

$$T_t\rho\delta_{z,n} = (1-\sigma t)\rho\delta_{z+nt,n} + \frac{3\sigma}{8\pi}\int_{\mathbb{S}^2}\mathrm{d}e\int_0^t \mathrm{d}t_1 P(e)\rho P(e)\delta_{z+et_1+n(t-t_1),e}. \tag{32}$$

Expand to the the first order, replace t by $\mathrm{d}t$ and obtain (6).

REFERENCES

1. S.Chandrasekhar: *Radiative Transfer* Dover 1960
2. L.Landau,E.M.Lifschitz. *Quantenelektrodynamik*. Akademie-Verlag Berlin 1961.
3. E.Landi degl'Innocenti.Polarization in spectral lines. Solar Physics 85 (1983) 3-31.
4. W. von Waldenfels. The radiative transfer equation is the potential equation of a Markov process. To appear in :Numerical 3D radiative transfer. Springer Verlag.
5. W. von Waldenfels. Derivation of the radiative transfer equation.

CONTRIBUTED PRESENTATIONS

Nonlinear Dynamical Equation for Irreversible, Steepest-Entropy-Ascent Relaxation to Stable Equilibrium

Gian Paolo Beretta

Dipartimento di Ingegneria Meccanica, Università degli Studi di Brescia, via Branze 38, 25123 Brescia, Italy

Abstract. We discuss the structure and main features of the nonlinear evolution equation proposed by this author as the fundamental dynamical law within the framework of Quantum Thermodynamics. The nonlinear equation generates a dynamical group providing a unique deterministic description of irreversible, conservative relaxation towards equilibrium from any non-equilibrium state, and satisfies a very restrictive stability requirement equivalent to Hatsopoulos-Keenan statement of the second law of thermodynamics. Here, we emphasize its mathematical structure and its applicability also within other contexts, such as Classical and Quantum Statistical Mechanics, and Information Theory.

Keywords: Nonlinear Dynamics Dynamics; Irreversibility; Entropy; Information Theory; Quantum Thermodynamics
PACS: 03.65.Ta, 11.10.Lm, 04.60.-m, 05.45.-a

INTRODUCTION

The problem of understanding entropy and irreversibility has been tackled by a large number of preeminent scientists during the past century. Schools of thought have formed and flourished around different perspectives of the problem. But a definitive solution has yet to be found.

We address a mathematical problem very relevant to the question of nonequilibrium and irreversibility, namely, that of "designing" a general evolution equation capable of describing irreversible but conservative relaxation towards equilibrium. Our objective is to present an interesting mathematical solution to this "design" problem, namely, a new nonlinear evolution equation that satisfies a set of very stringent, relevant requirements [1].

We first define four essentially different contexts within which the new equation is clearly relevant, with entirely different interpretations: Classical Statistical Mechanics (CSM), Classical Information Theory (CIT), Quantum Statistical Mechanics (QSM), Quantum Thermodynamics (QT). Then, we list the "design specifications" that we intend to impose on the desired evolution equation. We review some useful well-known mathematics involving Gram determinants and, finally, present our nonlinear evolution equation which meets the stringent design specifications.

The discussion here is purposely devoided of our school's quite unorthodox perspective, views and hypotheses on the physical meaning of entropy and irreversibility. This is not only due to obvious space limitations, but mainly because we feel that the proposed nonlinear equation constitutes an important mathematical-physics advance in itself not only in the Quantum Thermodynamics contexts for which it was originally designed and developed [2–7], but also in other contexts, such as CSM, CIT, QSM, as well as possibly in Quantum Information, Biology, Sociology and Economics.

FRAMEWORK A: CLASSICAL STATISTICAL MECHANICS

Let Ω be a phase space, and $\mathscr{L}$ the set of real, square-integrable functions $A,B,\dots$ on Ω, equipped with the inner product $(\cdot|\cdot)$ defined by

$$(A|B) = \mathrm{Tr}(AB) = \int_\Omega AB\, d\Omega \tag{1}$$

where $\mathrm{Tr}(\cdot)$ in this framework denotes $\int_\Omega \cdot\, d\Omega$. We denote by $\mathscr{P}$ the subset of all nonnegative-definite, normalized functions (distributions) ρ in $\mathscr{L}$, i.e.,

$$\mathscr{P} = \{\rho \text{ in } \mathscr{L} \,|\, \rho \geq 0, \mathrm{Tr}(\rho) = \int_\Omega \rho\, d\Omega = 1\} \tag{2}$$

We will then consider a set $\{H, N_1, \dots, N_r\}$ of functions in $\mathscr{L}$.

CP962, *Quantum Theory, Reconsideration of Foundations – 4*,
edited by G. Adenier, A. Yu. Khrennikov, P. Lahti, V. I. Man'ko, and Th. M. Nieuwenhuizen

In Classical Statistical Mechanics, ρ is the Gibbs density-of-phase distribution function which represents the index of statistics from a generally heterogeneous ensemble of identical systems (with associated phase space Ω) distributed over a range of possible classical mechanical states (the support of ρ). H is the Hamiltonian function, and N_i the number-of-particle function for particles of type i.

FRAMEWORK B: CLASSICAL INFORMATION THEORY

Let $\mathscr{L}$ be the set of all $n \times n$ real, diagonal matrixes $A = \mathrm{diag}(a_j)$, $B = \mathrm{diag}(b_j)$, $\ldots$ ($n \leq \infty$), equipped with the inner product $(\cdot|\cdot)$ defined by

$$(A|B) = \mathrm{Tr}(AB) = \Sigma_{j=1}^{n} a_j b_j \tag{3}$$

We denote by $\mathscr{P}$ the subset of all nonnegative-definite, unit-trace matrixes ρ in $\mathscr{L}$, i.e.,

$$\mathscr{P} = \{\rho = \mathrm{diag}(p_j) \,|\, p_j \geq 0,\ \mathrm{Tr}(\rho) = \Sigma_{j=1}^{n} p_j = 1\} \tag{4}$$

Later we consider a set $\{H, N_1, \ldots, N_r\}$ of diagonal matrixes $H = \mathrm{diag}(e_j)$, $N_1 = \mathrm{diag}(n_{1j})$, $\ldots$, $N_r = \mathrm{diag}(n_{rj})$ in $\mathscr{L}$.

In Information Theory [8], $\rho = \mathrm{diag}(p_j)$ represents the probability assignment to a set of n events, p_j being the probability of occurrence of the j-th event. $H, N_1, \ldots, N_r$ are characteristic features of the events in the set, taking on the values $e_j, n_{1j}, \ldots, n_{rj}$, respectively, for the j-th event.

FRAMEWORK C: QUANTUM STATISTICAL MECHANICS

Let $\mathscr{H}$ be a Hilbert space ($\dim \mathscr{H} \leq \infty$), and $\mathscr{L}$ the set of all linear operators A, B, $\ldots$ on $\mathscr{H}$, equipped with the real inner product $(\cdot|\cdot)$ defined by

$$(A|B) = \mathrm{Tr}\left(A^\dagger B + B^\dagger A\right)/2 \tag{5}$$

where $A^\dagger$ denotes the adjoint of operator A and $\mathrm{Tr}(\cdot)$ the trace functional. We denote by $\mathscr{P}$ the set of all self-adjoint, nonnegative-definite, unit-trace operators ρ in $\mathscr{L}$, i.e.,

$$\mathscr{P} = \{\rho \text{ in } \mathscr{L} | \rho^\dagger = \rho, \rho \geq 0, \mathrm{Tr}\rho = 1\} \tag{6}$$

We will then consider a set $\{H, N_1, \ldots, N_r\}$ of self-adjoint operators in $\mathscr{L}$, where each N_i commutes with H, i.e., is such that $HN_i = N_i H$, for $i = 1, \ldots, r$.

In Quantum Statistical Mechanics, ρ is the von Neumann statistical or density operator which represents the index of statistics from a generally heterogeneous ensemble of identical systems (same Hilbert space $\mathscr{H}$ and operators $\{H, N_1, \ldots, N_r\}$) distributed over a range of generally different quantum mechanical states. If each individual member of the ensemble is isolated and uncorrelated from the rest of the universe, its state is described according to Quantum Mechanics by an idempotent density operator ($\rho^2 = \rho = P_{|\psi\rangle} = \frac{|\psi\rangle\langle\psi|}{\langle\psi|\psi\rangle}$), i.e., a projection operator onto the span of some vector $|\psi\rangle$ in $\mathscr{H}$. If the ensemble is heterogeneous, its individual member systems may be in different states, $P_{|\psi_1\rangle}$, $P_{|\psi_2\rangle}$, and so on.

H is the Hamiltonian operator, and operator N_i, for $i = 1, \ldots, r$, is the number operator for particles of type i in the system (if the system has a fixed number n_i of particles of type i, then $N_i = n_i I$, where I is the identity operator on $\mathscr{H}$).

FRAMEWORK D: QUANTUM THERMODYNAMICS

In our formulation of Quantum Thermodynamics [2–6], the mathematical framework is the same as that just summarized for QSM, but the fundamental difference is in the physical meaning of the density operator. Indeed, QT assumes that the true individual quantum state of a system isolated and uncorrelated from the rest of the universe is represented by a density operator ρ which are not necessarily idempotent. Over the set of idempotent ρ's, QT coincides with Quantum Mechanics, but it differs fundamentally from it because it assumes a broader set of possible states, corresponding to the set of non-idempotent ρ's. This way, the functional $S(\rho)$ (defined in the next section) describes in QT an intrinsic (non-statistical) fundamental state property of the individual system. This is very different from the meaning that the von Neumann functional $S(\rho)$ has in QSM, where it measures the degree of heterogeneity of the ensemble whose statistics are represented by ρ.

MEAN VALUE FUNCTIONALS AND S-FUNCTIONAL

From here on, our notation allows us to treat at once the four contexts just defined. For reasons to become apparent below, the elements H, $N_1,\ldots,N_r$ introduced in either context, will be called the generators of the motion (MG). We assume that such sets always contain at least element H, that we call the Hamiltonian MG.

For each MG, we then define a mean value functional on $\mathscr{P}$ as follows

$$m(\rho;H) = \mathrm{Tr}\rho H = (\sqrt{\rho}|\sqrt{\rho}H) \quad , \qquad m(\rho;N_i) = \mathrm{Tr}\rho N_i = (\sqrt{\rho}|\sqrt{\rho}N_i) \tag{7}$$

Moreover, we define the S-functional [9] on $\mathscr{P}$ as

$$S(\rho) = -k\mathrm{Tr}(\rho\ln\rho) = -k(\sqrt{\rho}|\sqrt{\rho}\ln\rho) \tag{8}$$

Depending on the context, the S-functional represents the statistical uncertainty as to the actual classical or quantum state of a system, the information carried by the occurrence of one of the possible events (or the degree of uncertainty as to which will be the next event), or the thermodynamic entropy.

For each given set of values $\langle H\rangle$, $\langle N_1\rangle$, ..., $\langle N_r\rangle$, in the range of the mean value functionals (Eqs. 7) corresponding to the GM's, we consider the subset of all elements ρ in $\mathscr{P}$ that share the given mean values, i.e.,

$$\mathscr{P}_{\{\langle H\rangle,\langle N_1\rangle,\ldots,\langle N_r\rangle\}} = \{\rho \text{ in } \mathscr{P}\,|\, m(\rho;H) = \langle H\rangle, m(\rho;N_i) = \langle N_i\rangle \text{ for } i=1,\ldots,r\} \tag{9}$$

On each such subset, i.e., for fixed mean values $\langle H\rangle$, $\langle N_1\rangle$, ..., $\langle N_r\rangle$ of the generators of the motion, the S-functional (Eq. 8) achieves a unique maximum at the point

$$\rho = e^{-\alpha}\exp\Big(-\beta H + \Sigma_{i=1}^r \nu_i N_i\Big) \quad \text{where} \quad \alpha = \ln\mathrm{Tr}\Big[\exp\Big(-\beta H + \Sigma_{i=1}^r \nu_i N_i\Big)\Big] \tag{10}$$

and, of course, $\beta = \beta(\langle H\rangle,\langle N_1\rangle,\ldots,\langle N_r\rangle)$ and $\nu_i = \nu_i(\langle H\rangle,\langle N_1\rangle,\ldots,\langle N_r\rangle)$. It is noteworthy that the maximum-S points satisfy the condition

$$\sqrt{\rho}\ln\rho = -a\sqrt{\rho} - b\sqrt{\rho}H + \Sigma_{i=1}^r c_i\sqrt{\rho}N_i \tag{11}$$

for some real numbers a, b and c_i, $i=1,\ldots,r$. In words, the maximum-S element ρ is such that $\sqrt{\rho}\ln\rho$ lies in the linear manifold generated by elements $\sqrt{\rho}$, $\sqrt{\rho}H$, $\sqrt{\rho}N_1,\ldots,N_r$. Condition 11 is satisfied not only by the maximum-S elements given by Eq. 10, but also by the elements given by

$$\rho = e^{-a}B\exp\Big(-bH + \Sigma_{i=1}^r c_i N_i\Big)B \quad \text{where} \quad a = \ln\mathrm{Tr}\Big[B\exp\Big(-bH + \Sigma_{i=1}^r c_i N_i\Big)B\Big] \tag{12}$$

where $b = b(B;\langle H\rangle,\langle N_1\rangle,\ldots,\langle N_r\rangle)$, $c_i = c_i(B;\langle H\rangle,\langle N_1\rangle,\ldots,\langle N_r\rangle)$ and B is any idempotent element in $\mathscr{L}$ (i.e., $B^2 = B$). Clearly, Eq. 12 reduces to Eq. 10 iff $B = I$ (I = constant function equal to 1 on the whole Ω in CSM; $I = \mathrm{diag}(1)$ in CIT; I = identity operator on $\mathscr{H}$ in QSM and QT).

DYNAMICAL LAW DESIGN SPECIFICATIONS

Our scope is to design a function $F(\cdot)$ such that every solution $\rho(t)$ of the autonomous differential equation

$$\frac{d}{dt}\rho(t) = F(\rho(t)) \tag{13}$$

with $\rho(0)$ anywhere in $\mathscr{P}$ satisfies the following conditions for all t's, $-\infty < t < \infty$:

(i) $\rho(t)$ lies entirely in $\mathscr{P}$ (no forward nor backward escape times);

(ii) $m(\rho(t);H) = m(\rho(0);H)$, and $m(\rho(t);N_1) = m(\rho(0);N_i)$ for $i=1,\ldots,r$;

(iii) $S(\rho(t+u)) \geq S(\rho(t))$ for all $u > 0$;

(iv) within each subset $\mathscr{P}_{\{\langle H\rangle,\langle N_1\rangle,\ldots,\langle N_r\rangle\}}$, the maximum-$S$ element given by Eq. 10) is the only equilibrium solution that is stable according to Lyapunov [9]; all other equilibrium elements must not be stable.

Notice that requirement (iv) is most restrictive. For example, within QSM, it rules out the von Neumann evolution equation ($F(\rho) = -i(H\rho - \rho H)/\hbar$) because all the stationary density matrices (ρ such that $H\rho = \rho H$) are stable according to Lyapunov and, in general, there are many more than a single one within each set $\mathscr{P}_{\{\langle H\rangle,\langle N_1\rangle,\ldots,\langle N_r\rangle\}}$.

SOME NECESSARY MATHEMATICAL BACKGROUND

Given a subset of elements $A, B, \ldots, Z$ in $\mathscr{L}$, we denote by $M(A, B, \ldots, Z)$ the Gram matrix

$$\begin{bmatrix} (A|A) & (A|B) & \ldots & (A|Z) \\ (B|A) & (B|B) & \ldots & (B|Z) \\ \vdots & \vdots & \ddots & \vdots \\ (Z|A) & (Z|B) & \ldots & (Z|Z) \end{bmatrix} \tag{14}$$

where $(\cdot|\cdot)$ is the real symmetric inner product defined on $\mathscr{L}$. We denote by $G(A, B, \ldots, Z)$ the Gram determinant of $A, B, \ldots, Z$ with respect to inner product $(\cdot|\cdot)$, i.e., $G(A, B, \ldots, Z) = \det[M(A, B, \ldots, Z)]$. Matrix $M(A, B, \ldots, Z)$ is nonnegative definite and $G(A, B, \ldots, Z)$ is nonnegative. Elements $A, B, \ldots, Z$ are linearly independent (LI) iff their Gram determinant $G(A, B, \ldots, Z)$ is nonzero and, hence, strictly positive.

Given a subset of elements $A, B, \ldots, Z$ in $\mathscr{L}$, we denote by $L(A, B, \ldots, Z)$ the linear manifold spanned by all linear combinations with real coefficients of the elements $A, B, \ldots, Z$. With respect to the inner product $(\cdot|\cdot)$ defined on $\mathscr{L}$, we denote the projection of a given element V in $\mathscr{L}$ onto a linear manifold L by the symbol $(V)_L$. $(V)_L$ is the unique element in L such that $((V)_L|X) = (V|X)$ for all X in L.

The theory of Gram determinants, very seldom used in the physics literature, offers a useful explicit way of writing the projection $(V)_L$ of V onto a given linear manifold L. Let the given linear manifold be $L = L(A, B, \ldots, Z)$, where elements $A, B, \ldots, Z$ need not be LI. Select any subset of LI elements $E_1, E_2, \ldots, E_m$ spanning L, i.e., such that $G(E_1, E_2, \ldots, E_m) > 0$ and $L(E_1, E_2, \ldots, E_m) = L$. By the definition of $(V)_L$, $((V)_L|E_j) = (V|E_j)$ for every $j = 1, 2, \ldots, m$, and $(V)_L = \sum_{i=1}^m v_i E_i$, where v_i are real scalars. Thus,

$$\sum_{i=1}^m v_i (E_i|E_j) = (V|E_j) \text{ for } j = 1, 2, \ldots, m \tag{15}$$

Because $(E_i|E_j) = [M(E_1, E_2, \ldots, E_m)]_{ij}$ and the elements $E_1, E_2, \ldots, E_m$ are LI, Eqs. 15 are LI and can be solved for the v_i's to yield

$$v_i = \sum_{i=1}^m (V|E_j) [M(E_1, E_2, \ldots, E_m)^{-1}]_{ji} \quad \text{for } i = i = 1, 2, \ldots, m \tag{16}$$

and, therefore,

$$(V)_L = \sum_{i=1}^m \sum_{j=1}^m (V|E_j) [M(E_1, E_2, \ldots, E_m)^{-1}]_{ji} E_i \tag{17}$$

Alternatively, Cramer's rule yields the equivalent, more elegant expression

$$(V)_L = -\frac{1}{G(E_1, E_2, \ldots, E_m)} \det \begin{bmatrix} 0 & E_1 & E_2 & \ldots & E_m \\ (E_1|V) & (E_1|E_1) & (E_1|E_2) & \ldots & (E_1|E_m) \\ (E_2|V) & (E_2|E_1) & (E_2|E_2) & \ldots & (E_2|E_m) \\ \vdots & \vdots & \vdots & \ddots & \vdots \\ (E_m|V) & (E_m|E_1) & (E_m|E_2) & \ldots & (E_m|E_m) \end{bmatrix} \tag{18}$$

Below, for any given element ρ in the set $\mathscr{P}$, we will need to consider the projection of $\sqrt{\rho}\ln\rho$ onto the linear manifold $L(\sqrt{\rho}, \sqrt{\rho}H, \sqrt{\rho}N_1, \ldots, \sqrt{\rho}N_r)$ where $\sqrt{\rho}, \sqrt{\rho}H, \sqrt{\rho}N_1, \ldots, \sqrt{\rho}N_r$ are not necessarily LI. Using Eq. 18 and Definitions 1, 3 and 5 of the inner product $(\cdot|\cdot)$, we find

$$(\sqrt{\rho}\ln\rho)_{L(\sqrt{\rho}, \sqrt{\rho}H, \sqrt{\rho}N_1, \ldots, \sqrt{\rho}N_r)} = -\frac{1}{G(\sqrt{\rho}R_0, \sqrt{\rho}R_1, \ldots \sqrt{\rho}R_z)} \times \tag{19}$$

$$\times \det \begin{bmatrix} 0 & \sqrt{\rho}R_0 & \sqrt{\rho}R_1 & \ldots & \sqrt{\rho}R_z \\ \mathrm{Tr}\rho\ln\rho & \mathrm{Tr}\rho R_0^2 & \frac{1}{2}\mathrm{Tr}\rho\{R_0, R_1\} & \ldots & \frac{1}{2}\mathrm{Tr}\rho\{R_0, R_z\} \\ \mathrm{Tr}\rho R_1\ln\rho & \frac{1}{2}\mathrm{Tr}\rho\{R_1, R_0\} & \mathrm{Tr}\rho R_1 & \ldots & \frac{1}{2}\mathrm{Tr}\rho\{R_1, R_z\} \\ \vdots & \vdots & \vdots & \ddots & \vdots \\ \mathrm{Tr}\rho R_z\ln\rho & \frac{1}{2}\mathrm{Tr}\rho\{R_z, R_0\} & \frac{1}{2}\mathrm{Tr}\rho\{R_z, R_1\} & \ldots & \mathrm{Tr}\rho R_z^2 \end{bmatrix} \tag{20}$$

where $\{A, B\} = AB + BA$, and $R_0, R_1, \ldots, R_z$ are a subset of elements in $\mathscr{L}$ such that $L(\sqrt{\rho}R_0, \sqrt{\rho}R_1, \ldots, \sqrt{\rho}R_z) = L(\sqrt{\rho}, \sqrt{\rho}H, \sqrt{\rho}N_1, \ldots, \sqrt{\rho}N_r)$ and $G(\sqrt{\rho}R_0, \sqrt{\rho}R_1, \ldots, \sqrt{\rho}R_z) > 0$.

STEPEEST-S-ASCENT NONLINEAR EVOLUTION EQUATION

With the above background, the nonlinear evolution equation proposed by the author [3–6, 10] to meet our design specifications takes the compact form:

$$\frac{d}{dt}\rho(t) = -\frac{i}{\hbar}[H,\rho] - \frac{1}{2\tau}\left(\sqrt{\rho}D(\rho) + D^{\dagger}(\rho)\sqrt{\rho}\right) \tag{21}$$

$$D(\rho) = \sqrt{\rho}\ln\rho - (\sqrt{\rho}\ln\rho)_{L(\sqrt{\rho},\sqrt{\rho}H,\sqrt{\rho}N_1,\ldots,\sqrt{\rho}N_r)} \tag{22}$$

where $[H,\rho] = H\rho - \rho H$ (= 0 within CSM and CIT), $\hbar$ is the reduced Planck constant (playing a role only within QSM and QT), τ is a characteristic time constant, $H, N_1, \ldots, N_r$ are fixed GM's.

Notice that $D(\rho(t))$ is orthogonal to the linear manifold spanned by $\sqrt{\rho}(t), \sqrt{\rho}(t)H, \sqrt{\rho}(t)N_1, \ldots, \sqrt{\rho}(t)N_r$ and the term $i[H,\rho]$ is orthogonal to $\sqrt{\rho}D(\rho) + D^{\dagger}(\rho)\sqrt{\rho}$.

Within QT, if the isolated and uncorrelated system AB is composed of subsystems A (Alice) and B (Bob) that are either interacting ($H = H_A \otimes I_B + I_A \otimes H_B + V_{AB}$) or correlated ($S \neq S_A \otimes I_B + I_A \otimes S_B$, where $S = -k_B(P_{\mathrm{Ran}\rho})\ln\rho$), or both, the proposed equation takes the form [6, 10] (we assume for simplicity that A and B have no non-Hamiltonian GM's):

$$\frac{d\rho}{dt} = -\frac{i}{\hbar}[H,\rho] + \frac{1}{2k_B\tau_A}\left(\sqrt{\rho_A}D_A + D_A^{\dagger}\sqrt{\rho_A}\right)\otimes\rho_B + \frac{1}{2k_B\tau_B}\rho_B\otimes\left(\sqrt{\rho_B}D_B + D_B^{\dagger}\sqrt{\rho_B}\right) \tag{23}$$

$$\begin{aligned} D_A &= \sqrt{\rho_A}(S)^A - [\sqrt{\rho_A}(S)^A]_{L\{\sqrt{\rho_A},\sqrt{\rho_A}(H)^A\}} & \quad D_B &= \sqrt{\rho_B}(S)^B - [\sqrt{\rho_B}(S)^B]_{L\{\sqrt{\rho_B},\sqrt{\rho_B}(H)^B\}} \\ (H)^A &= \mathrm{Tr}_B[(I_A\otimes\rho_B)H] & (H)^B &= \mathrm{Tr}_A[(\rho_A\otimes I_B)H] \\ (S)^A &= \mathrm{Tr}_B[(I_A\otimes\rho_B)S] & (S)^B &= \mathrm{Tr}_A[(\rho_A\otimes I_B)S] \end{aligned} \tag{24}$$

Despite the nonlinearity, the structure of the non-Hamiltonian terms in the equation prevents "no-signaling" violations.

All zero entropy states ($\rho^2 = \rho$), even if A and B are entangled, obey the Schroedinger equation $\frac{d\rho}{dt} = -\frac{i}{\hbar}[H,\rho]$, thus not contradicting any of the results of QM, although within QT, these solutions, including the stationary states of QM ($\rho^2 = \rho$, $\rho H = H\rho$) are very weakly unstable limit cycles or equilibrium states [11].

The proofs that this equations satisfy our design specification are in Refs. [5–7, 10, 12]. Several other intriguing features including Onsager's reciprocal relations are discussed in Refs. [11–13]. Eq. 23 generalizes Eq. 21 to composite systems consistently with the additional design specs required to avoid non-locality paradoxes [6, 10].

We believe that in view of its intriguing properties, this evolution equation constitutes an important mathematical "tool" for a variety of non-equilibrium relaxation problems, not only within our QT, but also within different contexts, such as CSM, CIT, and QSM, as discussed here, as well as Quantum Information, Biology, Sociology and Economics. Even in the Thermodynamics context, where different schools of thought notoriously have contrasting perspectives on the physical meaning of entropy and irreversibility, many important insights can nevertheless be extracted from the richness of structure and the well-behaved and self-consistent features of this relevant nonlinear dynamical equation.

REFERENCES

1. G.P. Beretta, Mod. Phys. Lett. A **20**, 977 (2005).
2. G.N. Hatsopoulos and E.P. Gyftopoulos, Found. Phys. **6**, 15, 127, 439, 561 (1976).
3. G.P. Beretta, Sc.D. thesis, M.I.T., 1981, unpublished, e-print quant-ph/0509116.
4. G.P. Beretta, in *Frontiers of Nonequilibrium Statistical Physics,* NATO Advanced Study Institute, Santa Fe, 1984, Eds. G.T. Moore and M.O. Scully (NATO ASI Series B: Physics **135**, Plenum Press, NY, 1986), p. 193 and p. 205. G.P. Beretta, in *The Physics of Phase Space,* Eds. Y.S. Kim and W.W. Zachary (Lecture Notes in Physics **278**, Springer-Verlag, NY, 1986), p. 441.
5. G.P. Beretta, E.P. Gyftopoulos, J.L. Park, and G.N. Hatsopoulos, Nuovo Cimento B **82**, 169 (1984).
6. G.P. Beretta, E.P. Gyftopoulos, and J.L. Park, Nuovo Cimento B **87**, 77 (1985). J. Maddox, Nature **316**, 11 (1985).
7. G.P. Beretta, Int. J. Theor. Phys. **24**, 119 (1985).
8. E.T. Jaynes,Phys. Rev., Vol. 106, 620 (1957); Vol. 108, 171 (1957).
9. G.P. Beretta, J. Math. Phys. **27**, 305 (1986).
10. G.P. Beretta, e-print quant-ph/0112046.
11. G.P. Beretta, Phys. Rev. E **73**, 026113 (2006).
12. S. Gheorghiu-Svirschevski, Phys. Rev. A **63**, 022105, 054102 (2001).
13. G.P. Beretta, Found. Phys. **17**, 365 (1987).
14. References [4–7, 9, 13] are available online at www.quantumthermodynamics.org.

The Continuity Equation as a Bridge between Quantum and Classical Probabilities

N. L. Chuprikov

Department of Theoretical Physics, Tomsk State Pedagogical University, 634041, Tomsk, Russia

Abstract. We argue that quantum mechanics, in reality, respects the principles of a macroscopic realism (PMRs). The current quantum theory to tell us that "...the state of a system can be instantaneously changed by a distant measurement..." is not a physical one. Its key statements - that *the violation of Bell's inequality in the EPR-Bell experiments verifies nonlocality*, and *nonlocal correlations respect special relativity* - are false. Both the EPR-Bell experiments and theorems to support the "non-signalling principle" resort to the quantum mechanical averaging rule which is inapplicable to Cat states because such averaging violates the correspondence principle (CP). The main idea of the paper is exemplified by our model of a one-dimensional completed scattering.

Keywords: nonlocality; entanglement; wave-particle duality; superposition principle
PACS: 03.65.Ca, 03.65.Xp

INTRODUCTION

At present, due to the EPR-Bell and Schrödinger's cat (or, simply, Cat) paradoxes to appear for Cat states - coherent superpositions of macroscopically distinct states (CSMDSs) - there is a widespread opinion that the linear formalism of QM is inapplicable to the macro-world, that the superposition principle contradicts the PMRs [1]. However, this is not the case. We begin our analysis with the current vision of the paradoxes, which is based eventually on the orthodox interpretation (OI) of the wave-particle duality. Its main statements are divided here into three lessons.

THREE LESSONS OF THE ORTHODOX INTERPRETATION

'Nonlocality' and 'observer-dependence' of Cat states is, perhaps, the main mystery of modern QM. Both features are usually associated with compound systems. But the notion of Cat states is applicable to a single electron, too. In fact, the OI's attitude toward the one-particle wave function lies in the bottom of the mystery of Cat states.

Lesson 1: A 1D completed scattering. The one-electron wave function describes an electron in a single experiment. An electron cannot be treated as corpuscle or wave. It is *both* corpuscle *and* wave simultaneously (such an electron will be called here "the OI's electron"). Being scattered on a 1D potential barrier, it is literally in the superposition of the states 'transmitted electron' and 'reflected electron' to occupy macroscopically distinct spatial regions. To say that a scattered OI's electron is *either* transmitted *or* reflected by the barrier is erroneous in principle. One cannot conceive the OI's electron as a part of the universe (= external physical world). The time spent by the OI's electron in the barrier region has physical sense only in the context of a particular experimental situation (observer-dependence) and can be anomalously short or even negative by value (nonlocality) (see, e.g., [2,3]).

Lesson 2: The EPR-Bell paradox. Bell's assumption about the existence of local hidden variables leads to an inequality which is violated by QM and experiment. So that Bell's assumption is wrong and two electrons, in the Cat state, have no pre-existing properties and exhibit nonlocal features: inspecting the state of one of these electrons instantly changes that of its partner to be well far from the former. But nonlocal correlations obey the "no-signaling principle" (see, e.g., [4]).

Lesson 3: The Cat paradox. The 'observer-dependence' and 'nonlocality' of the OI's universe, at the micro-level, pass inevitably to its macro-level and hence the cat, in the Cat state, is both alive and died simultaneously. The

CP962, *Quantum Theory, Reconsideration of Foundations – 4*,
edited by G. Adenier, A. Yu. Khrennikov, P. Lahti, V. I. Man'ko, and Th. M. Nieuwenhuizen

superposition principle contradicts the PMRs, because in our perception the cat always appears in a definite state, i.e., it is *either* alive *or* died. This means that the Cat state is needed in a macro-objectification and the Cat paradox should be considered as a macro-objectification problem. QM must be replaced by another theory which would imply the existence of some physical process to suppress, at the macro-level, the action of the superposition principle and thereby to convert a pure Cat state into a statistical mixture.

The grand total of the lessons. Neither the superposition principle nor the PMRs govern the whole OI's universe; it is not a (local observer-independent) physical world and needed in a macro-objectification. In fact, there are neither first-principles to govern the external physical world nor the world itself (see also [5]).

WHAT IS WRONG IN THESE LESSONS?

Our next step is to show that, in reality, the superposition principle and PMRs govern together Cat states, respecting each other. To show this, we have to reconsider the above lessons in the reverse order.

The preparation problem for OI's Cat states. Let us consider the Schrödinger's thought experiment with an electron scattered on a 1D potential barrier (instead of a decaying nucleus), the vial with a poison and long-suffering cat. All are in a closed box. The relationship between them is assumed to be causal. The electron's source and the vial are supposed to be far enough from the potential barrier, the vial being in the transmission spatial region. Thus, we deal here with a 1D *completed* scattering.

We consider the electron and cat as parts of the compound system 'electron+cat' and suppose that this system is in a pure state

$$|\Psi\rangle_{e+c} = |\psi_{tr}^{end}\rangle \cdot |0\rangle_c + |\psi_{ref}^{end}\rangle \cdot |1\rangle_c; \qquad (1)$$

$|\psi_{tr}^{end}\rangle$ and $|\psi_{ref}^{end}\rangle$ are pure states of transmitted and reflected electrons; $|0\rangle_c$ and $|1\rangle_c$ are normalized pure states of a died and alive cat; $\langle\psi_{tr}^{end}|\psi_{tr}^{end}\rangle = \langle\psi_{ref}^{end}|\psi_{ref}^{end}\rangle = 1/2$.

As was said above, by the current vision of the Cat paradox the cat, in this state, is both alive and died simultaneously. In our perception it appears in a definite state due to an unavoidable localization process (see, e.g., [6]) to "macro-objectificate" the Cat state. It is *irreversible* and effective for macro-systems.

However, it is legitimate to ask: "How to prepare the state which is needed in a macro-objectification?" It is evident that the cat is definitely alive before preparing the state (1). Thus, its preparation is a "macro-disobjectification" process.

We could imagine this process as follows. At the first stage the electron's source (to emit only one electron, in each experiment), the vial with a poison and the cat are in a closed box, but the source of electrons has not yet been switched on. Thus, the cat is definitely alive at this stage. At the second stage, an observer to be outside of the box launches the electron's source. As a result, the state of an emitted electron begins evolving from the initial one to the superposition $|\psi_{tr}^{end}\rangle + |\psi_{ref}^{end}\rangle$.

We might expect that namely at this stage the cat becomes neither alive nor died. However, the question of how the scattered OI's electron "macro-disobjectificates" (delocalizes) the cat's state $|1\rangle_c$ arises. Indeed, any direct influence of the electron on the cat is negligible; this process implies the existence in Nature of some communicator (amplifier) to transfer the 'nonlocality' of the scattered OI's electron to the yet alive cat. However, such transfer looks quite unrealistic. On the road from the micro- to macro-level of the OI's universe, all unavoidable stochastic physical processes could facilitate anything but a "macro-disobjectification process".

So that "the OI's Cat state", where the cat is both alive and died simultaneously, cannot be prepared in principle: the current interpretation of the Cat paradox simply has no physical sense. To keep the universe as a knowable real world governed both by the superposition principle and the PMRs, we must solve it at the micro-level. The Cat state (1) to imply a *causal* relationship between an electron and cat is needed in extending the application of the PMRs to a single electron. Then the cat is *either* alive *or* died because an electron is *either* reflected *or* transmitted by a potential barrier.

One could say that we have arrived at a priori deadlock conclusion, because the PMRs discard nonlocality to be an experimental fact. However, the latter is wrong.

What do the EPR-Bell experiments falsify, in reality? One has to take into account that all statistical experiments, including EPR-Bell ones, consist at least of three stages: 1) obtaining experimental data; 2) their sampling; and 3) their averaging.

Of course, the first stage of the EPR-Bell experiments is beyond doubts. However, already the second one raises questions. Indeed, as it has been shown in [7], raw data obtained in the optical EPR-Bell experiments [8], examined under the fair sampling assumption, impugn the "non-signaling principle". By [7], either the fair sampling assumption or the "non-signaling principle" is wrong.

However, the fair sampling assumption is a *requirement* for statistical experiments, while the "non-signaling principle" is a *theoretical prediction* whose validity is needed in verification. Thus, in fact, the study [7] shows that "nonlocal correlations" do not obey the "non-signaling principle" and, thus, the experiments [8] falsify the theorems like [4] and assumptions to underlie them.

The crucial stage of such experiments is just the third one. Based (like [4]) on the assumption that all postulates of QM are applicable to Cat states, these experiments resort to the averaging over the whole two-electron Cat state. But the PMRs forbid such averaging. *The EPR-Bell experiments based on the fair sampling assumption falsify the current practice of introducing observables (e.g., correlations) for Cat states, rather than Bell's assumption on the existence of local hidden variables.*

We have to note that our criticism of the current practice of application of Born's averaging rule to Cat states complements and develops the approaches (see [9] and references therein) focused on the analysis of Bell's theorem itself. They point to Vorob'ev's theorem in probability theory to forbid averaging over hidden variables ascribed to different sets of EPR-Bell experiments with differently oriented detectors.

The Cat paradox as a correspondence problem. In solving the Cat paradox, we must discard the OI's attitude toward the one-electron wave function: the wave function cannot be in principle associated with a single electron in a single run, because the former evolves deterministically while the latter behaves stochastically.

Only the statistical interpretation (SI) of QM (see, e.g., [10]) reflects properly the nature of the wave function: in the x- (p-)representation, the wave function provides the x- (p-)distribution for *point-like* electrons in a quantum ensemble. The SI implies that, in a single one-particle experiment, an observer can measure (directly or no) a *true* random value of the particle's position (or momentum). So that, *the Cat paradox, within the SI, is neither macro-objectification nor measurement problem.*

The analysis of the state $\left|\psi_{tr}^{end}\right\rangle + \left|\psi_{ref}^{end}\right\rangle$ suggests that this paradox should be treated as *a correspondence problem*. A point-like electron cannot occupy two or more macroscopically distinct spatial regions, both in classical and quantum mechanics. Thus, putting a ban on the "either-or" scenario, in the current interpretation of this (micro-)Cat state, is inconsistent with the correspondence principle (CP).

CONCEPTS OF COMBINED AND ELEMENTARY STATES

Note that, for this superposition, standard QM gives not only a false interpretation, but also a priori false predictions. Indeed, by the current understanding of the CP, *any* one-particle time-dependent wave function is a counterpart to a single classical one-particle trajectory. Hence, averaging over *any* one-particle wave function gives the *expectation* (i.e., the most probable) values of one-particle observables. This correspondence rule should also be valid for $\left|\psi_{tr}^{end}\right\rangle + \left|\psi_{ref}^{end}\right\rangle$. However, averaging x or p over this state does not give *expectation* values of these one-particle observables.

The above correspondence rule is evident must be revised. QM must distinguish a pure one-particle (sub-)ensemble which has available to it two or more macroscopically distinct states but occupies at any given time a definite one of those states (we rephrase one of the PMRs [1]). By our terminology, such ensembles, their states and corresponding processes are elementary. Otherwise, they are *combined*.

Only an *elementary* time-dependent one-particle pure state can be considered as a quantum counterpart to a single classical one-particle trajectory. A *combined* time-dependent one-particle pure state represents a coherent superposition of N ($N > 1$) macroscopically distinct elementary sub-states. It is a counterpart to N trajectories.

The CP allows introducing observables only for elementary states. It implies the availability of experimental schemes to allow measuring observables for each elementary sub-state, without destroying interference between them.

In QM based on the CP, Cat states are combined ones and, in particular, a 1D completed scattering is a combined process to consist from two elementary sub-processes - transmission and reflection. Thus, all one-particle observables, including characteristic times, can be introduced only for either sub-process, individually.

So, quantum description of combined states must provide (i) rules of decomposing such states into elementary sub-states and also (ii) experimental schemes to allow measuring observables of elementary sub-states, without destroying the interference pattern. However, at present QM does not obey this requirement. The standard model of a 1D completed scattering does not suggest decomposing this process into sub-processes. Moreover, it is a commonplace that the linear formalism of QM does not allow, in principle, such decomposition. However, a macrorealistic model [11,12] of this process shows that this is not the case.

In the case of a symmetric potential barrier, it shows explicitly that the wave function $\Psi_{full}(x,t)$ to describe a 1D completed scattering can be uniquely presented as a CSMDS, at all stages of scattering: $\Psi_{full}(x,t) = \psi_{tr}(x,t) + \psi_{ref}(x,t)$ where $\psi_{tr}(x,t)$ and $\psi_{ref}(x,t)$ describe transmission and reflection, respectively. Either sub-process is characterized (see [12]) by the Larmor time to give the time spent, on the average, by an electron in the barrier region. This characteristic time can be measured with the help of the Larmor-clock procedure, without destroying interference between the sub-processes. Of great importance is that the Larmor time for transmission does not imply the Hartman effect predicted by the standard model of the process (see, e.g., [2,13]), whose interpretation and experimental verification are extremely controversial [3,14].

Note that this procedure does not allow measuring the so called group and signal times. This fact supports our belief that, in a completed scattering, the passage of an electron through the barrier region is an observer-independent process and hence QM must give a unique definition of the time spent by a particle in this region.

Of importance is also to stress that only the superposition of $\psi_{tr}(x,t)$ and $\psi_{ref}(x,t)$ obeys the Schrödinger equation, while each of them obeys the continuity one (i.e., the sub-processes interact with each other). This fact says about the crucial role played by the continuity equation in developing a macrorealistic description of the process.

REFERENCES

1. A.J. Leggett,. *J. Phys.: Condens. Matter,* **14**, R415–R451 (2002)
2. J.G. Muga, C.R. Leavens, *Physics Reports,*. **338**, 353-438 (2000)
3. H.G. Winful,. *Physics Reports,* **436**, 1–69 (2006)
4. G.C. Ghirardi, R. Grassi, A. Rimini and T. Weber, *Europhys. Lett.* **6**, 95-100 (1988)
5. J.B. Hartle, *J. Phys. A: Math. Gen.*, **40**, 3101–3121 (2007)
6. G.C. Ghirardi, *J. Phys. A: Math. Gen.*, **40**, 2891–2917 (2007)
7. G. Adenier, A.Yu. Khrennikov, *J. Phys. B: At. Mol. Phys.*, **40**, 131-141 (2007)
8. G. Weihs *et al*, *Phys. Rev. Lett.*, **81**, 5039-5043 (1998)
9. A. Khrennikov, In *Foundations of Probability and Physics - 4*, edited by Adenier G., Fuchs C.A. and Khrennikov A.Yu., AIP Conference Proceedings, **889**,. Melville, New York 2007, pp. 7-17
10. L.E. Ballentine, Quantum mechanics. Prentice Hall, Englewood Cliffs, NJ (1990)
11. N.L. Chuprikov, *Russian Physics,* **49**, 119-126 (2006)
12. N.L. Chuprikov, *Russian Physics,* **49**, 314-325 (2006)
13. M. Buttiker,. *Phys. Rev.*, **B27**, 6178-6188 (1983)
14. A. Ranfagni *et al,:Phys. Lett.* A, **352**, 473-477 (2006)

Kaonic Quantitative Complementarity and Quantum Erasers

A. Bramon*, G. Garbarino† and B. C. Hiesmayr**

*Grup de Física Teòrica, Universitat Autònoma de Barcelona, E–08193 Bellaterra, Spain
†Dipartimento di Fisica Teorica, Università di Torino and INFN, Sezione di Torino, I–10125 Torino, Italy
**Faculty of Physics, University of Vienna, A–1090 Vienna, Austria

Abstract. We illustrate two basic aspects of quantum mechanics applied to the neutral kaon system. A quantitative formulation of Bohr's complementarity principle is first described for the free space evolution of single kaons and entangled kaon pairs. We then show that the neutral kaon system is also suitable for an optimal demonstration of the "quantum eraser", including its operation in the "delayed choice" mode.

Keywords: Double–slit Experiment, Quantitative Complementarity, Quantum Eraser, Delayed Choice
PACS: 03.65.-w, 03.65.Ud,14.40.Aq

INTRODUCTION

Fundamental issues in quantum mechanics can be conveniently illustrated by the basic double–slit experiment. One has to explore the consequences of the complementarity relation existing between the observation of interference fringes (wave–like behaviour) and the acquisition of "which way" information (particle–like behaviour).

In a double–slit–like experiment, interference patterns are observed if and only if it is impossible to know, *even in principle*, which path the quantum took. Interference disappears if there is a way to know —e.g. through a *quantum marking* procedure— which path the quantum took; it is completely irrelevant whether the outcome of the eventual "which way" observation is read out or not: interference is in any way lost. But, if that "which way" mark is erased by a suitable measurement —*quantum erasure* [1]—, interferences can be restored.

The acquisition of "which way" information in an interferometer is always accompanied by a reduction of the interference fringe visibility. This well known, *qualitative* observation has been recently translated into modern and *quantitative* expressions of Bohr's complementarity [2–5]. These complementarity relations, as well as the quantum eraser effect, have been investigated and confirmed in a variety of recent interferometric experiments with atoms [5] or photon pairs [6].

In this contribution we extend these considerations to neutral kaons by exploiting a clear analogy between the neutral kaon propagation in free space and the propagation of an interfering object in a double–slit device. Indeed, neutral kaons present the phenomenon of K^0–$\bar{K}^0$ oscillations, which play the role of the interference fringes in a two–way interferometer. The short– and long–lived kaon states, K_S and K_L, being characterized by remarkably different decay widths in free space are the analogs of the two separated trajectories in interferometric devices. Therefore, the better one can know if the propagation proceeds through the K_S or K_L component —i.e., the more "*which width*" information one has—, the less pronounced will be the K_S–K_L interference and the visibility of the K^0–$\bar{K}^0$ oscillations.

QUANTUM MECHANICS OF NEUTRAL KAONS

Two alternative bases, each one associated to a possible measurement, are accessible to neutral kaons [7].

The strangeness basis, $\{K^0, \bar{K}^0\}$ with $\langle K^0|\bar{K}^0\rangle = 0$, is the appropriate one to discuss strong production and reactions of kaons, as well as kaon strangeness measurements. The latter are performed by inserting along the kaon trajectory a piece of ordinary matter which can induce strangeness conserving kaon–nucleon strong interactions. The incident neutral kaon is then identified either as a K^0 or a $\bar{K}^0$ in a projective measurement.

CP962, *Quantum Theory, Reconsideration of Foundations – 4*,
edited by G. Adenier, A. Yu. Khrennikov, P. Lahti, V. I. Man'ko, and Th. M. Nieuwenhuizen

The second basis consists of the K_S and K_L states having well defined masses $m_{S(L)}$ and decay widths $\Gamma_{S(L)} = 1/\tau_{S(L)}$; it is the appropriate one to discuss neutral kaon propagation in free space, with:

$$|K_{S(L)}(\tau)\rangle = e^{-i\lambda_{S(L)}\tau}|K_{S(L)}\rangle, \quad \lambda_{S(L)} = m_{S(L)} - i\Gamma_{S(L)}/2. \tag{1}$$

The K_S and K_L eigenstates do not oscillate into each other in time, but, since $\Gamma_S \simeq 579\,\Gamma_L$, the short–lived component of a given neutral kaon extincts much faster than the long–lived one. Because of this lifetime "mark", knowing if this kaon has propagated in free space *either* as K_S *or* K_L is thus possible by detecting at which time it decays. If kaons decaying before (after) 4.8 τ_S after production are identified as K_S's (K_L's), misidentifications amount to only a few parts in 10^{-3} [7]. Such a discrimination, linked to the acquisition of "which width" information, is thus in analogy with usual "which way" discussions in two–way interferometers.

By neglecting small CP violating effects, which are of the same order as the mentioned K_S vs K_L misidentifications, the following relationship exist between the two kaon bases:

$$|K^0\rangle = \frac{1}{\sqrt{2}}\left[|K_S\rangle + |K_L\rangle\right], \quad |\bar{K}^0\rangle = \frac{1}{\sqrt{2}}\left[|K_S\rangle - |K_L\rangle\right]. \tag{2}$$

Note that the strangeness and lifetime measurements completely exclude each other: the former require the insertion of nucleonic matter, the latter propagation in free space. Bohr's complementarity principle is thus at work: if strangeness (lifetime) is known, both outcomes for lifetime (strangeness) are equally probable.

QUANTITATIVE COMPLEMENTARITY

Single kaons

Eqs. (1) and (2) imply that the (proper) time evolution of a kaon state produced as a K^0 at time $\tau = 0$ is given by

$$|K^0(\tau)\rangle = \frac{1}{\sqrt{2}}\left[e^{-i\lambda_S\tau}|K_S\rangle + e^{-i\lambda_L\tau}|K_L\rangle\right]. \tag{3}$$

By normalizing to kaons surviving up to time τ, the previous state can be more conveniently rewritten as $|K^0(\tau)\rangle = (|K_S\rangle + e^{-\frac{1}{2}\Delta\Gamma\tau}e^{-i\Delta m\tau}|K_L\rangle)/(\sqrt{1+e^{-\Delta\Gamma\tau}})$, where $\Delta\Gamma \equiv \Gamma_L - \Gamma_S$ and $\Delta m \equiv m_L - m_S$. By using the $\{K^0, \bar{K}^0\}$ basis one equivalently obtains

$$|K^0(\tau)\rangle = \frac{1+e^{-\frac{1}{2}\Delta\Gamma\tau}e^{-i\Delta m\tau}}{\sqrt{2(1+e^{-\Delta\Gamma\tau})}}|K^0\rangle + \frac{1-e^{-\frac{1}{2}\Delta\Gamma\tau}e^{-i\Delta m\tau}}{\sqrt{2(1+e^{-\Delta\Gamma\tau})}}|\bar{K}^0\rangle. \tag{4}$$

One thus easily obtains the following τ–dependent transition probabilities:

$$\left|\langle K^0|K^0(\tau)\rangle\right|^2 = \frac{1}{2}\{1+\mathcal{V}_0(\tau)\cos(\Delta m\,\tau)\}, \quad \left|\langle \bar{K}^0|K^0(\tau)\rangle\right|^2 = \frac{1}{2}\{1-\mathcal{V}_0(\tau)\cos(\Delta m\,\tau)\}, \tag{5}$$

where

$$\mathcal{V}_0(\tau) = 1/\cosh(\Delta\Gamma\,\tau/2) \tag{6}$$

is the time–dependent fringe visibility of the $\bar{K}^0$–K^0 oscillations. On the contrary, no K_S–K_L oscillations are expected:

$$w_L(\tau) \equiv \left|\langle K_L|K^0(\tau)\rangle\right|^2 = \frac{e^{-\Delta\Gamma\tau}}{1+e^{-\Delta\Gamma\tau}}, \quad w_S(\tau) \equiv \left|\langle K_S|K^0(\tau)\rangle\right|^2 = \frac{1}{1+e^{-\Delta\Gamma\tau}}. \tag{7}$$

These observations admit the following interpretation [8–10]. As soon as a K^0 is produced, it starts propagating in free space in the coherent superposition of K_S and K_L [see Eq. (3)] and mimics the two–way propagation of any system beyond a symmetrical double–slit. In the familiar double–slit case, the system follows the two paths without "jumping" from one to the other in the same way as K_S–K_L oscillations are "forbidden". In the kaon case, however, there are not two separated trajectories but a single path comprising *automatically* (i.e., with no need of any double–slit–like apparatus) the two components K_S and K_L.

For initial K^0's surviving up to time τ, the probabilities for K_S and K_L propagation are given by the w_S and w_L of Eq. (7). One thus obtains the "width predictability" (corresponding to the "path predictability" defined in Ref. [2]):

$$\mathcal{P}(\tau) \equiv \left|w_S(\tau) - w_L(\tau)\right| = \left|\tanh(\Delta\Gamma\tau/2)\right|. \tag{8}$$

$\mathscr{P}(\tau)$ quantifies the a priori (i.e., before any measurement is performed) "which width" knowledge we have from the fact that a kaon which was created as a K^0 at time $\tau = 0$ has survived up to time τ. At $\tau = 0$, both components start propagating with the same probability, $w_S(0) = w_L(0) = 1/2$, and the "width predictability" vanishes, $\mathscr{P}(0) = 0$. In other words, there is no information on which component actually propagates and the visibility of strangeness oscillations is maximal, $\mathscr{V}_0(0) = 1$. However, since the K_S and K_L components are intrinsically "marked" by their different lifetimes, "which width" information is obtained for initial K^0's surviving up to time τ and the corresponding oscillation visibility is thus reduced. Fringe visibility and "width predictability" fulfill the equation:

$$\mathscr{V}_0^2(\tau) + \mathscr{P}^2(\tau) = 1, \tag{9}$$

which is a generalization of the quantitative duality relation of Ref. [2] to situations where $\mathscr{P}$ and $\mathscr{V}_0$ are not constants.

Note that the same expressions of Eqs. (6) and (8) for $\mathscr{V}_0(\tau)$ and $\mathscr{P}(\tau)$ are valid when the kaon state produced at $\tau = 0$ is a $\bar{K}^0$. Also in this case the state remains quantum mechanically pure. The two terms in Eq. (9) add up to one rather than fulfilling the less stringent relation $\mathscr{V}_0^2(\tau) + \mathscr{P}^2(\tau) < 1$ for mixed states.

Recent measurements by the CPLEAR Collaboration at CERN [11] have been interpreted [9] in terms of the above quantitative duality. The proton–antiproton annihilation processes, $p\bar{p} \to K^-\pi^+K^0$ or $p\bar{p} \to K^+\pi^-\bar{K}^0$, were used to produce either a K^0 or a $\bar{K}^0$ initial state, which were then allowed to propagate in free space. In a first experiment, strangeness oscillations were observed by detecting semileptonic neutral kaon decays. In another experiment, K^0–$\bar{K}^0$ oscillations were observed via strangeness measurements monitored by kaon–nucleon strong interactions. The experimental data collected in both CPLEAR experiments are seen to fulfill the statement for quantitative duality of Eq. (9). Further experiments are planned for the near future [12] at the ϕ–factory Daphne [13], which copiously produces neutral kaons via strong $\phi \to K^0\bar{K}^0$ decays.

Entangled kaon pairs

The a priori knowledge on the kaon lifetime expressed by the predictability $\mathscr{P}(\tau)$ comes exclusively from knowing that the kaon is undecayed at time τ. This knowledge can be improved if a measurement is performed on the kaon state [3–5]. Since kaon measurements are completely destructive, two–particle entanglement is necessary to this end. Working with entangled kaons, one can perform a measurement on one member (the *meter*) in order to increase the information on the propagation mode of its partner (the *object*) without the latter being annihilated.

To this aim, consider the decay of the ϕ–meson [13] (or S–wave $p\bar{p}$ annihilation [11]) into $K^0\bar{K}^0$ pairs. Just after the decay, i.e., at time $\tau = 0$, one starts with the following maximally entangled state:

$$|\phi(0)\rangle = \frac{1}{\sqrt{2}}\left[|K^0\rangle_l|\bar{K}^0\rangle_r - |\bar{K}^0\rangle_l|K^0\rangle_r\right] = \frac{1}{\sqrt{2}}\left[|K_L\rangle_l|K_S\rangle_r - |K_S\rangle_l|K_L\rangle_r\right], \tag{10}$$

where l and r denote the "left" and "right" kaon directions of motion. The state evolution up to time τ_l (τ_r) along the left (right) is easily obtained by using the lifetime basis. By considering kaon pairs with both left and right members surviving up to these times, and introducing $\Delta\tau \equiv \tau_l - \tau_r$, one thus obtains the normalized state:

$$|\phi(\Delta\tau)\rangle = \frac{1}{\sqrt{1+e^{\Delta\Gamma\Delta\tau}}}\left[|K_L\rangle_l|K_S\rangle_r - e^{i\Delta m\Delta\tau}e^{\frac{1}{2}\Delta\Gamma\Delta\tau}|K_S\rangle_l|K_L\rangle_r\right]. \tag{11}$$

Note the analogy between this entangled kaon state and the polarization–entangled two–photon state $|\Psi\rangle = [|H\rangle_l|V\rangle_r - e^{i\Delta\phi}|V\rangle_l|H\rangle_r]/\sqrt{2}$ widely used in optical experiments [8]. Eqs. (2) and (11) supply the various joint probabilities $P(K_l, K_r)$ for detecting a K_l (K_r) on the left (right) at time τ_l (τ_r) with $K_{l,r} = K^0, \bar{K}^0, K_S$ or K_L.

Let us now consider the following two experimental arrangements [8]. In both of them one measures the strangeness of the left moving kaon by inserting a dense slab of (nucleonic) matter at different distances along its trajectory; since each distance corresponds to a given time–of–flight τ_l, one can measure at different values of τ_l and look for strangeness oscillations of this (object) kaon. On the other hand, the measurement on the right moving (meter) kaon is always performed at a *fixed* time τ_r^0; but on this kaon, one of two alternative measurements, $\mathscr{M} = \mathscr{S}$ or $\mathscr{L}$, is performed by either inserting or not a strangeness detector at time τ_r^0. In the first case, $\mathscr{M} = \mathscr{S}$, strangeness is measured on both kaons and four τ_l–dependent joint probabilities $P(K_l, K_r)$ with $K_{l,r} = K^0$ or $\bar{K}^0$ can be recorded. In the second case, by removing the strangeness detector one observes the lifetime of the right moving kaon, $\mathscr{M} = \mathscr{S}$, and measures the four joint probabilities $P(K_l, K_r)$ with $K_l = K^0$ or $\bar{K}^0$ and $K_r = K_S$ or K_L.

Following Englert [3], we can then define the "which width knowledge" for the object kaon, $\mathscr{K}_{\mathscr{M}}(\tau_l)$, which depends on the measurement $\mathscr{M}$ performed on the meter kaon. Note that an experiment can never decrease the information provided by the formerly discussed predictability $\mathscr{P}(\tau_l)$, and one necessarily has $\mathscr{K}_{\mathscr{M}}(\tau_l) \geq \mathscr{P}(\tau_l) \equiv |\tanh(\Delta\Gamma\tau_l/2)|$.

One can similarly consider the corresponding visibility of the object kaon strangeness oscillations, $\mathcal{V}_{\mathcal{M}}(\tau_l)$, which now satisfies $\mathcal{V}_{\mathcal{M}}(\tau_l) \leq \mathcal{V}_0(\tau_l) \equiv 1/\cosh(\Delta\Gamma\,\tau_l/2)$. It is easy to see that $\mathcal{K}_{\mathcal{M}}(\tau_l)$ and $\mathcal{V}_{\mathcal{M}}(\tau_l)$ are linked by a new quantitative duality relation [4, 5]:

$$\mathcal{V}^2_{\mathcal{M}}(\tau_l) + \mathcal{K}^2_{\mathcal{M}}(\tau_l) = 1. \tag{12}$$

Without the strangeness detector on the right beam, $\mathcal{M} = \mathcal{L}$, one acquires "which width" information for both the meter and object kaons and therefore no strangeness oscillations (in τ_l) are predicted:

$$\mathcal{V}_{\mathcal{L}}(\tau_l) = 0 \;\forall\, \tau_l, \tag{13}$$

for any of the four possible joint detection probabilities $P(K_l, K_r)$ with $K_l = K^0$ or $\bar{K}^0$ and $K_r = K_S$ or K_L. Correspondingly, one obtains the maximal value for the object kaon "which width knowledge" [3]:

$$\mathcal{K}_{\mathcal{L}}(\tau_l) \equiv \left|P(K_S(\tau_l), K_S(\tau_r^0)) - P(K_L(\tau_l), K_S(\tau_r^0))\right| + \left|P(K_S(\tau_l), K_L(\tau_r^0)) - P(K_L(\tau_l), K_L(\tau_r^0))\right| = 1 \;\forall\, \tau_l, \tag{14}$$

and the quantitative duality relation of Eq. (12) is trivially satisfied.

Considering instead a strangeness measurement along the right beam, $\mathcal{M} = \mathcal{S}$, one obtains the four joint probabilities $P(K_l, K_r)$ with $K_{l,r} = K^0$ or $\bar{K}^0$ which show τ_l–dependent strangeness oscillations and anti–oscillations with visibility:

$$\mathcal{V}_{\mathcal{S}}(\tau_l) = 1/\cosh\left(\Delta\Gamma\,(\tau_l - \tau_r^0)/2\right). \tag{15}$$

This visibility is maximal ($\mathcal{V}_{\mathcal{S}} = 1$) for $\tau_l = \tau_r^0$ (no "which width" information available), but the strangeness oscillations disappear ($\mathcal{V}_{\mathcal{S}} \to 0$) for $\tau_l \to \infty$ (full "which width" information). The "which width" information on the object kaon, which is available as a result of the joint strangeness measurement, can be expressed in terms of the "which width knowledge" [3]:

$$\mathcal{K}_{\mathcal{S}}(\tau_l) \equiv \left|P[K_S(\tau_l), K^0(\tau_r^0)] - P[K_L(\tau_l), K^0(\tau_r^0)]\right| + \left|P[K_S(\tau_l), \bar{K}^0(\tau_r^0)] - P[K_L(\tau_l), \bar{K}^0(\tau_r^0)]\right| = \left|\tanh\left(\Delta\Gamma(\tau_l - \tau_r^0)/2\right)\right|,$$

which, together with the visibility (15), satisfies again the quantitative duality requirement (12).

Note that the two types of alternative measurements performed on the right moving kaon are clearly different. If strangeness is measured, $\mathcal{M} = \mathcal{S}$, no "which width" information is obtained in addition to the information that was already known a priori, thus $\mathcal{K}_{\mathcal{S}}(\tau_l) \equiv \mathcal{P}(\tau_l)$. In the other case, $\mathcal{M} = \mathcal{L}$, the opposite situation is achieved and the "which width" knowledge is maximally increased, $\mathcal{K}_{\mathcal{L}}(\tau_l) \equiv 1 > \mathcal{P}(\tau_l)$. According to the definition of Ref. [3], this corresponds to a "width distinguishability" $\mathcal{D}(\tau_l) \equiv \mathrm{Max}_{\mathcal{M}}\{\mathcal{K}_{\mathcal{M}}(\tau_l)\} = 1$.

QUANTUM ERASER

The discussion of previous Section can be reinterpreted in terms of the *quantum eraser* [1], whose protocol consists of three consecutive steps [14].

In a first step one starts with a single (object) kaon which is a coherent superposition of K_S and K_L states such as the one in Eq. (3), which provides K_S–K_L interference effects showing the K^0–$\bar{K}^0$ oscillations of Eq. (5). These oscillations become less pronounced with time as a consequence of the increased "path predictability".

In a second step one proceeds to entangle the object kaon with a meter kaon in the state (11). If the right going meter kaon is free to propagate in space, its lifetime "mark" is operative and precludes the observation of any strangeness oscillation of the object kaon. The following joint probabilities are then measured:

$$\begin{aligned} P[K^0(\tau_l), K_S(\tau_r^0)] &= P[\bar{K}^0(\tau_l), K_S(\tau_r^0)] = \frac{1}{2(1 + e^{\Delta\Gamma(\tau_l - \tau_r^0)})}, \\ P[K^0(\tau_l), K_L(\tau_r^0)] &= P[\bar{K}^0(\tau_l), K_L(\tau_r^0)] = \frac{e^{\Delta\Gamma(\tau_l - \tau_r^0)}}{2(1 + e^{\Delta\Gamma(\tau_l - \tau_r^0)})}. \end{aligned} \tag{16}$$

The possibility to obtain full "which width" information for the object kaon can be precluded by the third step of the quantum eraser. To erase the lifetime "mark" of the meter, it has to be measured in the $\{K^0, \bar{K}^0\}$ basis; strangeness oscillations and their complementary anti–oscillations then appear when sorting the jointly detected events according to the following probabilities:

$$
\begin{aligned}
P[K^0(\tau_l), K^0(\tau_r^0)] &= P[\bar{K}^0(\tau_l), \bar{K}^0(\tau_r^0)] = \frac{1}{4}\{1 - \mathscr{V}(\tau_l)\cos[\Delta m(\tau_l - \tau_r^0)]\}, \\
P[K^0(\tau_l), \bar{K}^0(\tau_r^0)] &= P[\bar{K}^0(\tau_l), K^0(\tau_r^0)] = \frac{1}{4}\{1 + \mathscr{V}(\tau_l)\cos[\Delta m(\tau_l - \tau_r^0)]\},
\end{aligned} \tag{17}
$$

where $\mathscr{V}(\tau_l) = 1/\cosh(\Delta\Gamma(\tau_l - \tau_r^0)/2)$. Note that this third step restores the same K_S–K_L interference effects [see Eqs. (5) and (6)] of the first step, where single kaon states [see Eq. (3)] are used.

Note also that the joint probabilities of Eqs. (17) as well as $\mathscr{K}_{\mathscr{M}}$ and $\mathscr{V}_{\mathscr{M}}$ are all of them even functions of $\tau_l - \tau_r^0$. The erasure of the lifetime "mark" on the meter kaon can thus be delayed to times after the object kaon has been detected ($\tau_r^0 > \tau_l$). This "delayed choice" mode captures the essential feature of the quantum eraser [1], which is a *proper sorting* of the jointly detected events. The neutral kaon system thus reveals to be suitable for an optimal demonstration [1, 14] of quantum erasure: (1) the "which width" information is carried by a system (the meter kaon) distinct and spatially separated from the interfering system (the object kaon); (2) as a consequence, the erasure operation can be performed in the "delayed choice" mode ($\tau_l < \tau_r^0$); (3) single–particle states (as opposed to coherent states) are detected on each side; (4) in spite of the need to entangle the object kaon with another system (the meter kaon), quantum erasure allows one to restore the same K_S–K_L interference phenomenon exhibited by a single kaon state produced as K^0 or $\bar{K}^0$.

Before concluding it is important to note that, for neutral kaons, passive measurement procedures of lifetime and strangeness (exploiting only the dynamics of kaon decays) are available in addition to the previously adopted active measurements. As demonstrated in Ref. [15], this property admits the existence of different versions of kaonic quantum erasers, some of which have no analog in any other two–level quantum system considered up to date.

An experimental test of the marking and erasure operations is advisable. The CPLEAR collaboration [11] has already done part of the work required: the two experimental points (for $|\tau_l - \tau_r^0| = 0$ and $1.2\,\tau_S$) collected by this experiment using active measurements are compatible with the joint strangeness oscillations predicted by Eqs. (17). New observations (also including passive measurement procedures) confirming with better precision these oscillations for a larger range of $\tau_l - \tau_r^0$ values, as well as the non–oscillating behaviour when "which width" information is in principle available [see Eq. (16)], are needed for a full demonstration of quantitative duality and the quantum eraser with neutral kaons. An experimental implementation of such tests is expected for the near future at the Daphne ϕ–factory [12], thanks to the next upgrade of the machine.

Work partly supported by INFN and FLAVIAnet Network.

REFERENCES

1. M. O. Scully and K. Drühl, *Phys. Rev.* **A 25**, 2208 (1982). M. O. Scully, B.-G. Englert and H. Walther, *Nature* (London) **351**, 111 (1991); ibid. *Am. J. Phys.* **67**, 325 (1999).
2. D. Greenberger and A. Yasin, *Phys. Lett.* **A 128**, 391 (1988).
3. B.-G. Englert, *Phys. Rev. Lett.* **77**, 2154 (1996); B.-G. Englert and J. A. Bergou, *Opt. Commun.* **179**, 337 (2000).
4. G. Björk and A. Karlsson, *Phys. Rev.* **A 58**, 3477 (1998).
5. S. Dürr and G. Rempe, *Opt. Commun.* **179**, 323 (2000).
6. T. J. Herzog, P. G. Kwiat, H. Weinfurter and A. Zeilinger, *Phys. Rev. Lett.* **75**, 3034 (1995); T. Tsegaye, G. Björk, M. Atatüre,A. V. Sergienko, B. E. A. Saleh, M. C. Teich, *Phys. Rev.* **A 62**, 032106 (2000); Y. H. Kim, R. Yu, S. P. Kulik, Y. Shih and M. O. Scully, *Phys. Rev. Lett.* **84**, 1 (2000); S. P. Walborn, M. O. Terra Cunha, S. Pádua and C. H. Monken, *Phys. Rev.* **A 65**, 033818 (2002); A. Trifonov, G. Björk, J. Söderholm and T. Tsegaye, *Eur. Phys. J.* **D 18**, 251 (2002); H. Kim, J. Ko and T. Kim, *Phys. Rev.* **A 67**, 054102 (2003).
7. A. Bramon and G. Garbarino, *Phys. Rev. Lett.* **88**, 040403 (2002); ibid. **89**, 160401 (2002).
8. A. Bramon, G. Garbarino and B. C. Hiesmayr, *Phys. Rev. Lett.* **92**, 020405 (2004).
9. A. Bramon, G. Garbarino and B. C. Hiesmayr, *Eur. Phys. J.* **C 32**, 377 (2004).
10. A. Bramon, G. Garbarino and B. C. Hiesmayr, *Phys Rev.* **A 69**, 022112 (2004).
11. For a comprehensive review of CPLEAR results, see A. Angelopoulos et al., *Phys. Rept.* **374**, 165 (2003).
12. A. Di Domenico (KLOE Collaboration), talk presented at the *KLOE–2 General Meeting*, Laboratori Nazionali di Frascati, July 2, 2007.
13. "*The Second DaΦne Physics Handbook*", edited by L. Maiani, G. Pancheri and N. Paver (INFN, Laboratori Nazionali di Frascati, 1995).
14. P. G. Kwiat, A. M. Steinberg and R. Y. Chiao, *Phys. Rev.* **A 49**, 61 (1994).
15. A. Bramon, G. Garbarino and B. C. Hiesmayr, *Phys. Rev.* **A 69**, 062111 (2004).

The ESR Model: Reinterpreting Quantum Probabilities Within a Realistic and Local Framework

Claudio Garola

Dipartimento di Fisica dell'Università del Salento and INFN, Sezione di Lecce
Via Arnesano, 73100, Lecce, Italy
E–mail: garola@le.infn.it

Abstract. Basing on known theoretical results (Bell's theorem, Bell–Kochen–Specker's theorem, etc.) most physicists believe that standard quantum mechanics (SQM) and local realism cannot hold together. Moreover, they also uphold that the experimental tests performed up to now confirm SQM and refute local realism. Some scholars, however, criticize this conclusion, yet maintaining that SQM conflicts with local realism. At variance with both viewpoints, an *extended semantic realism* (ESR) model has been worked out in Lecce which shows that, if quantum probabilities are interpreted as conditional rather than absolute, a picture of the microworld can be constructed that assumes local realism and embodies the mathematical formalism of SQM. We provide here a new synthetic and improved description of the ESR model, comment on its achievements and limits, and briefly compare it with other approaches.

Keywords: quantum mechanics; Bell inequalities; local realism.
PACS: 03.65.-w; 03.65.Ud

1. ORTHODOX AND ALTERNATIVE VIEWPOINTS ABOUT LOCAL REALISM

The term *local realism* has been traditionally used to denote the joint assumptions of *realism*,

- R: the values of all observables of a physical system in a given state are predetermined for any measurement context,

and *locality*,

- LOC: if measurements are made at places remote from one another on parts of a physical system which no longer interact, the specific features of one of the measurements do not influence the results obtained with the others.

Assumption R expresses a weak form of realism, since it does not require that the values of the observables be predetermined independently of the measurement context. Assumption LOC then limits the possible dependence on the measurement context by introducing the notion of part of a physical system and assuming that measurements on different parts of the same system must not influence each other if they are carried out far away [1, 2, 3, 4].

Assumptions R and LOC hold in particular within *local deterministic* (*hidden variables*) theories. There exist, however, *factorizable stochastic* [3, 5] models that replace R and LOC with

- R_S: the probabilities of the values of all observables are predetermined,

and

- LOC_S: if measurements are made at places remote from one another on parts of a physical system which no longer interact, the specific features of one of the measurements do not influence the probabilities of the results obtained with the others,

respectively, then adding a factorizability assumption F on probabilities. Moreover, assumptions LOC and LOC $_S$ have sometimes been replaced by the weaker and more precise assumption [6, 7]

- LOC': no influence can be transmitted with a speed greater than that of light.

The term *local realism* has then been used also to denote such sets of assumptions. However intended, local realism implies Bell's inequalities (BI), which contradict the predictions of standard quantum mechanics (SQM). It is then well known that, according to most quantum physicists, the experimental tests of BI show that these inequalities are

CP962, *Quantum Theory, Reconsideration of Foundations – 4*,
edited by G. Adenier, A. Yu. Khrennikov, P. Lahti, V. I. Man'ko, and Th. M. Nieuwenhuizen

violated while the predictions of SQM are confirmed, which implies that local realism must be rejected [8]. But this conclusion is not unanimously accepted (an updated bibliography on this topic can be found in a recent review by Khrennikov [9]). In particular, some scholars (see, e.g. [5, 10, 11, 12, 13]) tried to devise models that allow one to explain the obtained data without rejecting R and LOC. These models accept that a conflict exists between SQM and local realism, but show that one cannot decide which of them is correct on the basis of the experiments performed up to now.

Following a different approach, the research group of Lecce has inquired into the impossibility of reconciling SQM with R and LOC. It has then been shown [14, 15, 16, 17] that the proofs of the conflict (in particular, Bell's theorem [1, 2]) require accepting an epistemological principle (MCP) that does not fit in well with the operational philosophy of SQM. If MCP is replaced by a weaker principle (MGP) the proofs cannot be given. On this basis a new interpretation of the mathematical formalism of SQM has been provided which assumes R and LOC and formally preserves all quantum laws, yet limiting the validity of the empirical laws deduced from theoretical laws to *epistemically accessible* physical situations (i.e., physical situations in which the empirical laws can be checked, at least in principle), consistently with MGP (*Semantic Realism*, or *SR*, interpretation). But the SR interpretation is mainly based on epistemological argume nts, which may seem problematical to many physicists. The foregoing research group has therefore worked out an *extended SR* (ESR) set–theoretical model [18, 19, 20, 21, 22] which also assumes R and LOC and preserves the mathematical formalism of SQM, yet introduces an interpretation of quantum probabilities as conditional rather than absolute probabilities and avoids mentioning MCP and MGP (but it can be proved that the substitution of the former with the latter occurs in the ESR model, as a consequence of the new interpretation of quantum probabilities). The ESR model provides a more intuitive physical framework and we devote the present paper to it. Therefore we provide a new synthetic and improved description of it in Sec. 2, comment on its achievements and limits in Sec. 3, and finally compare it with other approaches in Sec. 4.

2. THE ESR MODEL

The basic notions of the ESR model can be divided into three groups.

(i) Standard primitive and derived notions: *physical system, preparing device, state, physical object*. In particular, a state of a physical system Ω is defined as a class of physically equivalent preparing devices [23, 24]. A physical object is defined as an individual sample x of Ω, obtained by activating a preparing device π (we say that "x is in the state S" if the device π belongs to the equivalence class S).

(ii) New theoretical entities: *microscopic properties*. We assume that every physical system Ω is characterized by a set $\mathscr{E}$ of microscopic properties. For every physical object x, the set $\mathscr{E}$ is partitioned in two classes, the class of properties that are possessed by x and the class of properties that are not possessed by x, independently of any measurement procedure. We briefly say that microscopic properties are *objective*.

(iii) New observative entities: (extended) *observables*. We assume that every physical system Ω is associated with a set of (extended) observables that also characterize it. Every observable A_0 is obtained in the ESR model by considering an observable A of SQM with spectrum Ξ and adding a further outcome a_0 (*no–registration outcome* of A_0) that does not belong to Ξ, so that the spectrum Ξ_0 of A_0 is given by $\Xi_0 = \Xi \cup \{a_0\}$.

The introduction of (extended) observables allows us to define the set $\mathscr{F}_0$ of all *macroscopic properties* of Ω,

$$\mathscr{F}_0 = \{(A_0, \Delta),\ A_0 \text{ observable, } \Delta \text{ Borel set on the real line } \Re\}, \tag{1}$$

and the set $\mathscr{F} \subset \mathscr{F}_0$ of all macroscopic properties associated with observables of SQM,

$$\mathscr{F} = \{(A_0, \Delta),\ A_0 \text{ observable, } a_0 \notin \Delta\}. \tag{2}$$

We now establish a link between microscopic properties of $\mathscr{E}$ and macroscopic properties of $\mathscr{F}$ by means of the following assumption.

Ax. 1. A bijective mapping $\varphi : \mathscr{E} \longrightarrow \mathscr{F} \subset \mathscr{F}_0$ exists.

By using the notions above a perfectly efficient (briefly, *idealized*) measurement can be described as follows.

Whenever a physical object x is prepared in a given state S and the observable A_0 is measured on it, the set of microscopic properties possessed by x produces a probability (which is either 0 or 1 if the model is *deterministic*) that the apparatus does not react, so that the outcome a_0 may be obtained. In this case, x is not detected and we cannot get explicit information about the microscopic properties possessed by x. If, on the contrary, the apparatus reacts, it yields an outcome $a \neq a_0$ (more generally, it informs us that the result lies in the Borel set Δ, with $a_0 \notin \Delta$) if and only if x possesses the microscopic property $\varphi^{-1}((A_0,\{a\}))$ (more generally, $\varphi^{-1}((A_0,\Delta))$; for the sake of brevity, we say that x possesses the macroscopic property (A_0,Δ) in this case). Hence, the result of the measurement provides information on, and is determined by, microscopic properties.

In order to place properly quantum probability in our picture, let us suppose that a finite set $\mathscr{S}$ of physical objects in the state S is prepared. $\mathscr{S}$ can be partitioned into subsets $\mathscr{S}^1,\mathscr{S}^2,\ldots,\mathscr{S}^n$ such that in each subset all objects possess the same microscopic properties. We briefly say that the objects in $\mathscr{S}^i$ ($i=1,2,\ldots,n$) are in some microstate S^i. This suggests associating S with a (not necessarily finite) family of microstates $S^1,S^2,\ldots$, each of which is characterized by the set of all microscopic properties that are possessed by every physical object in that microstate.

Let us now consider a physical object x in the microstate S^i, and suppose that a measurement of a macroscopic property $F=(A_0,\Delta)$, $a_0 \notin \Delta$, is performed on x. Then, our description of the measurement process implies that, whenever x is detected, it turns out to possess F if and only if it possesses the microscopic property $f=\varphi^{-1}(F)$. We are thus led to introduce the following probabilities.

$\mathscr{P}_S^{i,d}(F)$: the probability that x is detected when F is measured on it.

$\mathscr{P}_S^{i}(F)$: the conditional probability that x turns out to possess F when it is detected (which is 0 or 1 since x either possesses $\varphi^{-1}(F)$ or not).

$\mathscr{P}_S^{i,t}(F)$: the joint probability that x is detected and turns out to possess F.

Hence, we get

$$\mathscr{P}_S^{i,t}(F)=\mathscr{P}_S^{i,d}(F)\mathscr{P}_S^{i}(F). \tag{3}$$

Eq. (3) is purely theoretical, since one can never know if a physical object is in the microstate S^i. Therefore, let us consider a physical object in the state S and introduce a further conditional probability, as follows.

$\mathscr{P}(S^i|S)$: the conditional probability that x (which is in the macroscopic state S) is in the microstate S^i.

The joint probability that x is in the state S^i, is detected and turns out to possess F is thus given by $\mathscr{P}(S^i|S)\mathscr{P}_S^{i,t}(F)$, hence the joint probability $\mathscr{P}_S^t(F)$ that x is detected and turns out to possess F is given by

$$\mathscr{P}_S^t(F)=\sum_i \mathscr{P}(S^i|S)\mathscr{P}_S^{i,t}(F). \tag{4}$$

Moreover, the probability $\mathscr{P}_S^d(F)$ that x is detected when F is measured is given by

$$\mathscr{P}_S^d(F)=\sum_i \mathscr{P}(S^i|S)\mathscr{P}_S^{i,d}(F). \tag{5}$$

Let us put now

$$\mathscr{P}_S(F)=\frac{\sum_i \mathscr{P}(S^i|S)\mathscr{P}_S^{i,t}(F)}{\sum_i \mathscr{P}(S^i|S)\mathscr{P}_S^{i,d}(F)}. \tag{6}$$

Then, we get

$$\mathscr{P}_S^t(F)=\mathscr{P}_S^d(F)\mathscr{P}_S(F). \tag{7}$$

Eq. (7) is the fundamental equation of the ESR model. Let us therefore comment briefly on it. Let us consider firstly the *detection probability* $\mathscr{P}_S^d(F)$. We have seen that, since we consider idealized measurements, the occurrence of the outcome a_0 is attributed only to the set of microscopic properties possessed by x, which determines the probability $\mathscr{P}_S^{i,d}(F)$. Hence, $\mathscr{P}_S^{i,d}(F)$ neither depends on features of the measuring apparatus (flaws, termal noise, etc.) nor is influenced by the environment. Furthermore, the conditional probability $\mathscr{P}(S^i|S)$ depends only on S. Therefore, Eq. (5) implies that $\mathscr{P}_S^d(F)$ depends only on the microscopic properties of the physical objects in the state S.

Secondly, let us consider the term $\mathscr{P}_S(F)$. Because of Eqs. (3) and (6) we get $0 \leq \mathscr{P}_S(F) \leq 1$. Moreover, the interpretations of $\mathscr{P}_S^t(F)$ and $\mathscr{P}_S^d(F)$ in Eq. (7) show that $\mathscr{P}_S(F)$ can be interpreted as the conditional probability that a physical object x in the state S turns out to possess the macroscopic property F when it is detected. This interpretation of the term $\mathscr{P}_S(F)$ in Eq. (7) provides a basis for the introduction of the main assumption of the ESR model.

Ax. 2. The probability $\mathscr{P}_S(F)$ can be evaluated by using the same rules that yield the probability $p_S(F)$ of F in the state S within SQM.

It is apparent that Ax. 2 implies a new interpretation of the probabilities provided by standard quantum rules, which are now regarded as conditional instead of absolute. From an intuitive viewpoint this can be illustrated as follows. Let us interpret probabilities as the large number limits of frequencies within suitable ensembles of physical objects, and let us consider ensembles of physical objects in the state S. Then, $\mathscr{P}_S(F)$ is the limit of the ratio between the number of objects in a given ensemble that are detected and possess the property F, and the number of objects that are detected. The probability $p_S(F)$, instead, is the limit of the ratio between the number of objects in a given ensemble that possess the property F and the number of objects in the ensemble.

The difference between the new and the orthodox interpretation is conceptually relevant and implies that the predictions of the ESR model are different from the predictions of SQM if $\mathscr{P}_S^d(F) < 1$, even if it may be hard to establish experimentally which predictions are correct (the lack of efficiency of any actual device hides the term $\mathscr{P}_S^d(F)$). Notwithstanding this, all standard quantum rules for evaluating probabilities are preserved, hence in particular, states and macroscopic properties in $\mathscr{F}$ are still represented by density operators and orthogonal projection operators, respectively. But no mathematical representation is provided for microscopic properties, microstates, (extended) observables and macroscopic properties of the form (A_0, Δ), with $a_0 \in \Delta$, and no quantitative theory for $\mathscr{P}_S^d(F)$ is given.

3. LOCAL REALISM WITHIN THE ESR MODEL

It follows from Sec. 2 that microstates can be seen as possible values of an hidden variable which determines the probability of the no–registration outcome. The ESR model can then be classified as a hidden variables theory, which reminds local deterministic theories whenever it is deterministic, factorizable stochastic models whenever it is not deterministic (see Sec. 1), though its reinterpretation of quantum probabilities makes it different from all previous proposals. The new theory is still largely incomplete, as the remarks at the end of Sec. 2 show. Yet, we can already say that it recovers the mathematical formalism of SQM. Moreover, this recovery is done within a local and realistic framework, which avoids a number of old and more recent paradoxes of SQM. Indeed, let us consider $\mathscr{P}_S^{i,t}(F)$. By taking into account Eq. (3), our remarks on $\mathscr{P}_S^{i,d}(F)$ and the definition of $\mathscr{P}_S^i(F)$, we conclude that $\mathscr{P}_S^{i,t}(F)$ is completely determined by the value S^i of the hidden variable, hence it is independent of the measurement context. By considering an observable A_0, a value a of A_0 and the property $F = (A_0, \{a\})$, it follows at once that R_S and LOC_S hold within the ESR model. If the ESR model is deterministic the result itself of any measurement is predetermined, hence all macroscopic properties either are possessed or not, independently of the measurement context (that is, they are *objective*) and R and LOC follow at once. In the general case, objectivity of properties cannot be proved but it can be assumed if $\mathscr{P}_S^{i,d}(F)$ is endowed with an *ignorance interpretation* (it is sufficient to suppose that further microscopic parameters exist which do not correspond to macroscopic properties and determine, together with the microscopic properties, the values of all observables) . If this assumption is accepted, not only R_S and LOC_S but also R and LOC hold within the ESR model.

It is important to observe that assuming macroscopic objectivity imposes theoretical limits if one wants to complete the ESR model. Indeed, a minimal completion should at least introduce a mathematical representation of macroscopic observables and provide rules for evaluating $\mathscr{P}_S^d(F)$. If one puts $\mathscr{P}_S^d(F) = 1$ for any state S and property F, the ESR model reduces to SQM, and the conflict between local realism and quantum mechanics is not resolved. If instead $\mathscr{P}_S^d(F) < 1$ for some state S and property F, one can show that MCP does not hold, hence the standard proofs of the foregoing conflict cannot be carried out [18]. This result has supported our development of the ESR model. Notwithstanding this, a conflict could still occur, and we cannot indicate the general conditions that $\mathscr{P}_S^d(F)$ must fulfill in order to avoid it at the present stage of our research. However, we can pursue a more modest aim, that is, we can consider standard cases in which the conflict between local realism and SQM is exhibited and show that in each of them reasonable values of the detection probabilities can be assigned in such a way that no conflict occurs within the ESR model. Equivalently, we can say that accepting the ESR model implies predictions (in particular, on the detection probabilities) that are different from those of SQM, and look for predictions of this kind (we note that, in principle, our results can be experimentally checked, which makes the ESR model falsifiable). An example of such predictions is provided in the paper by S. Sozzo in this book, where the author shows that the standard Bell–Clauser–Horne–Shimony and Holt (BCHSH) inequalities must be replaced by *modified BCHSH inequalities* within the ESR model. This implies that one expects that the former inequalities be violated, as it actually occurs in the experiments, though the model be local and realistic. We note, however, that the ESR model applies also to the situations considered by the proofs of nonlocality of SQM provided by Greenberger *et al.* [25] or Mermin [26], even if these situations have not yet been studied at this stage of our research.

4. COMPARING THE ESR MODEL WITH OTHER APPROACHES

There are features of the ESR model which closely resemble the features of some earlier models that try to vindicate local realism by questioning the interpretation of the experimental results obtained up to now. In particular, let us consider the following assumptions.

(i) The 0 outcome is a possible result of an ideal measuring process.

(ii) The efficiencies of the detectors used in the experiments depend on the hidden variable.

Assumption (i) appears in [10], while both (i) and (ii) are explicitly stated and used in [5] and [11, 12, 13]. It is then evident that (i) anticipates our introduction of the no–registration outcome in the case of idealized measurements, while (ii) resembles our statement that the probability of detecting a physical object x in a microstate S^i depends on the microscopic properties of x, hence on the value S^i of the hidden variable (see Sec. 3).

Notwithstanding their similarities, however, the ESR model and the models mentioned above are deeply different. In order to avoid misunderstandings, we would like to close this paper by discussing this difference in more details.

Let us preliminarily note that most arguments against the interpretation of experimental data (ED) as a decisive refutation of local realism follow a common logical scheme. Indeed, one firstly observes that the inequalities that are actually tested are not standard Bell's inequalities (BI) but, rather, somewhat different inequalities (BI*) that are obtained by adding additional assumptions (AA) to R and LOC (in particular, the *renormalization of experimental probabilities*, see, e.g. [6, 7, 13]). Thus, whenever the ED violate BI*, one can write, by using standard logical symbols, the following sequence of implications: $R \wedge LOC \wedge AA \Longrightarrow BI^*$; $ED \Longrightarrow \neg BI^*$; $ED \Longrightarrow \neg(R \wedge LOC \wedge AA)$; $ED \Longrightarrow \neg(R \wedge LOC) \vee (\neg AA)$. The last implication shows that the ED do not necessarily imply ¬(R ∧ LOC). This conclusion challenges the common belief that the ED confirm SQM and refute local realism. However, it does not imply that the ED are consistent with R and LOC in all performed experiments, nor explains why the ED match the predictions of SQM. Therefore, the foregoing models complete the reasoning by denying the AA, introducing some specific hypotheses (SH) and showing that $(R \wedge LOC) \wedge (\neg AA) \wedge SH \Longrightarrow ED$.

The arguments leading instead to the unified perspective expressed by the ESR model follow a different logical scheme. Indeed, let us firstly note that the standard proofs that SQM conflicts with local realism can be schematized as follows (see also [19]): $(R \wedge LOC) \Longrightarrow BI$; $SQM \Longrightarrow \neg BI$; $SQM \Longrightarrow \neg(R \wedge LOC)$. Then, let us remind that SQM interprets quantum probabilities as absolute (Sec. 2), that is, as referring to the set of all physical objects that are prepared. Let us denote this interpretation by AP, so that $SQM \Longleftrightarrow (QM' \wedge AP)$ (where QM′ simply denotes SQM without AP). Hence the sequence above can be restated by substituting SQM with QM′ ∧ AP in it. The last implication that is thus obtained is equivalent to $(R \wedge LOC) \Longrightarrow (\neg QM' \vee \neg AP)$, which shows that R ∧ LOC needs not contradict QM′, since it could simply contradict AP. Thus , the ESR model can be contrived, which substitutes AP with a new interpretation of quantum probabilities as conditional instead of absolute, that is, as referring to the subset of physical objects that are actually detected.

The difference between the models mentioned at the beginning of this section and the ESR model is now evident. The former do not inquire into the conflict of SQM with local realism, but stress that the ED actually violate BI*, not BI, which is not sufficient for claiming that SQM holds and local realism is disproved. The latter entails instead that the ED, taking into account only those physical objects that are detected, must match the same predictions made by SQM with reference to the set of all physical objects that are prepared, hence must violate BI to the same extent predicted by SQM. Yet, this violation is compatible with local realism (which would instead be challenged if the modified BCHSH inequalities mentioned in Sec. 3 were violated).

REFERENCES

1. J. S. Bell, Physics **1**, 195 (1964).
2. J. F. Clauser, M. A. Horne, A. Shimony, and R. A. Holt, Phys. Rev. Lett. **23**, 880 (1969).
3. J. F. Clauser and M. A. Horne, Phys. Rev. D **10**, 526 (1974).
4. A. Einstein, B. Podolsky, and N. Rosen, Phys. Rev. **47**, 777 (1935).
5. A. Fine, Found. Phys. **19**, 453 (1989).
6. E. Santos, Found. Phys. **34**, 1643 (2004).
7. E. Santos, Stud. Hist. Phil. Mod. Phys. **36**, 544 (2005).
8. M. Genovese, Phys. Repts. **413**, 319 (2005).
9. A. Khrennikov, arXiv:0709.3909v1 [quant–ph] (2007).
10. L. De Caro and A. Garuccio, Phys. Rev. A **54**, 174 (1996).
11. L. E. Szabo, Found. Phys. **30**, 1981 (2000).

12. L. E. Szabo and A. Fine, Phys. Lett. A **295**, 229 (2002).
13. N. Gisin and B. Gisin, Phys. Lett. A **260**, 323 (1999).
14. C. Garola and L. Solombrino, Found. Phys. **26**, 1121 (1996).
15. C. Garola and L. Solombrino, Found. Phys. **26**, 1329 (1996).
16. C. Garola, Found. Phys. **30**, 1539 (2000).
17. C. Garola, Stud. Hist. Phil. Mod. Phys. **33**, 101 (2002).
18. C. Garola, Found. Phys. **32**, 1597 (2002).
19. C. Garola and J. Pykacz, Found. Phys. **34**, 449 (2004).
20. C. Garola, Found. Phys. Lett. **16**, 605 (2003).
21. C. Garola, Int. J. Theor. Phys. **44**, 807 (2005).
22. C. Garola, J. Pykacz, and S. Sozzo, Found. Phys. **36**, 862 (2006).
23. G. Ludwig, *Foundations of Quantum Mechanics I* (Springer, Berlin 1983).
24. E. G. Beltrametti and G. Cassinelli, *The Logic of Quantum Mechanics* (Addison–Wesley, Reading, MA, 1981).
25. D. M. Greenberger, M. A. Horne, A. Shimony, and A. Zeilinger, Am. J. Phys. **58**, 1131 (1990).
26. N. D. Mermin, Rev. Mod. Phys. **65**, 803 (1993).

The Average-Value Correspondence Principle

Philip Goyal

Perimeter Institute for Theoretical Physics 31 Caroline Street North Waterloo, Ontario N2L 2Y5, Canada

Abstract. In previous work [1], we have presented an attempt to derive the finite-dimensional abstract quantum formalism from a set of physically comprehensible assumptions. In this paper, we continue the derivation of the quantum formalism by formulating a correspondence principle, the *Average-Value Correspondence Principle*, that allows relations between measurement outcomes which are known to hold in a classical model of a system to be systematically taken over into the quantum model of the system, and by using this principle to derive many of the correspondence rules (such as operator rules, commutation relations, and Dirac's Poisson bracket rule) that are needed to apply the abstract quantum formalism to model particular physical systems.

Keywords: Quantum-classical correspondence principles, Origin of Quantum Formalism, Correspondence Rules
PACS: 03.65.-w, 03.65.Ta

INTRODUCTION

In order to obtain a quantum mechanical model for a particular physical system such as a particle moving in space, it is necessary to supplement the abstract quantum formalism (due to von Neumann and Dirac) with various rules, to which we shall refer collectively as *correspondence* rules [1], which explicitly determine the form of the operators that represent particular measurements performed on the system, or that represent particular symmetry transformations (such as displacement or rotation) of the frame of reference in which the system is being observed.

The physical origin of many of these correspondence rules is obscure. For example, the commutation relationship $[\mathsf{x}, \mathsf{p}_x] = i\hbar$ is typically obtained from Schroedinger's equation or from Dirac's Poisson bracket rule, both of whose derivations involve abstract assumptions whose physical origin is obscure.

Operator rules have been discussed in a few publications, for example in [3, 4, 5], but a derivation of the operator rules on the basis of a physical principle, taking the abstract formalism as a given, has not been successfully completed. The most recent systematic attempt to derive the commonly employed measurement and measurement–transformation commutation relations on the basis of a physical principle is found in [6, 7, 8]. However, this approach implicitly makes use of operator rules, and relies upon auxiliary assumptions (such as the assumption that certain measurement operators are unchanged by the action of particular symmetry operators) whose physical origin is unclear. In this paper, we show that, starting from the abstract quantum formalism, it is possible to derive the above-mentioned correspondence rules in a straightforward manner from a correspondence principle.

THE AVERAGE-VALUE CORRESPONDENCE PRINCIPLE

The *Average-Value Correspondence Principle* (AVCP), asserts that, in a classical experiment, if a measurement of observable A on a physical system can be implemented by an arrangement where a number of other measurements of observables $A', A'', \ldots$ are performed on the same system (perhaps performed at different times), so that the outcome of the measurement of A can be calculated as a function of the outcomes of the measurements of $A', A'', \ldots$, then we can construct a corresponding quantum experiment, where quantum measurements $\mathbf{A}, \mathbf{A}', \mathbf{A}'', \ldots$, corresponding to classical measurements of $A, A', A'', \ldots$, are performed on the same physical system, such that the same functional relation holds *on average* between the outcomes of the quantum measurements, where the average is taken over infinitely many trials of the quantum experiment.

[1] A systematic classification of the correspondence rules is given in [2]. Readers are referred to [2] for a more thorough discussion of the subject of the present paper.

CP962, *Quantum Theory, Reconsideration of Foundations – 4*,
edited by G. Adenier, A. Yu. Khrennikov, P. Lahti, V. I. Man'ko, and Th. M. Nieuwenhuizen

The precise form of the corresponding quantum experiment is arrived at by carefully considering a sequence of particular examples (see Ref. [2] for details). The key difficulty one has to face is what the quantum experiment must be in the case where the operators involved are non-commuting. For example, suppose that, in the classical framework, a measurement of C at time t can be implemented by performing measurements of A and B at time t. Suppose further that, in the quantum framework, it is known that operators A and B represent measurements of A and B respectively, and that $[\mathsf{A},\mathsf{B}] \neq 0$. How should one deal with this situation? The AVCP insists that the quantum measurements are performed on separate copies of the system. On the contrary, if the operators commute, and if the quantum measurements are not performed on sub-systems of a composite system, then the measurements are to be performed on the same copy of the system. This makes some intuitive sense: when the operators commute, we are in an essentially classical situation, and so there is no harm in having the measurements performed on the same copy of the system; when the operators do not commute, we are in a distinctively non-classical situation, in which case, to avoid the complication of one measurement's results being affected by the other's and to avoid the issue of measurement ordering, one insists that the measurements are performed on separate copies of the system.

Statement of the Principle

We shall now state the general principle. An illustrative example is given in Fig. 1.

Consider a classical idealized experiment in which a system (possibly a composite system) is prepared in some state at time t_0, and is allowed to evolve in a given background. Suppose that a measurement of $A^{(m)}$ ($m \geq 2$), performed on the system at time t_2 with outcome $a^{(m)}$, can be implemented by an arrangement where measurements of $A^{(1)}, A^{(2)}, \ldots, A^{(m-1)}$ are performed upon one copy of the system at time t_1, and their respective outcomes, denoted $a^{(1)}, a^{(2)}, \ldots, a^{(m-1)}$, are then used to compute the result $f(a^{(1)}, a^{(2)}, \ldots, a^{(m-1)})$, where f is an analytic function, so that the relation

$$a^{(m)} = f(a^{(1)}, a^{(2)}, \ldots, a^{(m-1)}) \qquad (*)$$

holds for all initial (classical) states of the system.

Consider the case where the quantum measurements $\mathbf{A}^{(1)}, \mathbf{A}^{(2)}, \ldots, \mathbf{A}^{(m)}$, with operators $\mathsf{A}^{(1)}, \mathsf{A}^{(2)}, \ldots, \mathsf{A}^{(m)}$, represent the measurements of $A^{(1)}, A^{(2)}, \ldots, A^{(m)}$, respectively. Then, consider the following idealized quantum experimental arrangement consisting of several set-ups, each consisting of identical sources and backgrounds, where, in each set-up, a copy of the system is prepared in the same initial state, v_0, at time t_0.

In one set-up, only measurement $\mathbf{A}^{(m)}$ is performed (at time t_2) and, for any i, j with $i \neq j$ and $i, j \leq m-1$, the measurements $\mathbf{A}^{(i)}, \mathbf{A}^{(j)}$ are performed (at time t_1) in:

1. the *same* set-up if both $[\mathsf{A}^{(i)}, \mathsf{A}^{(j)}] = 0$ and, if the system is composite, the measurements are performed on the same sub-system;
2. *different* set-ups if $[\mathsf{A}^{(i)}, \mathsf{A}^{(j)}] \neq 0$.

Let the outcomes of the measurements $\mathbf{A}^{(1)}, \ldots, \mathbf{A}^{(m)}$ in any given run of the experimental arrangement be denoted $a^{(1)}, \ldots, a^{(m)}$, respectively. The function $f(a^{(1)}, a^{(2)}, \ldots, a^{(m-1)})$ is defined as *simple* provided that its polynomial expansion contains no terms involving a product of eigenvalues belonging to measurements whose operators do not commute. If f is simple, then $(*)$ holds on average, the average being taken over an infinitely large number of runs of the experiment.

The above principle can be generalized in a number of ways, for example to the case where the measurements of $A^{(1)}, \ldots, A^{(m-1)}$ are not performed at the same time. However, these generalizations are unnecessary for the derivations of the usual correspondence rules of quantum theory, and are therefore not discussed here.

GENERALIZED OPERATOR RULES

We will now apply the AVCP to derive operator relations which hold when the function f takes various useful forms. In each instance of f, we shall first derive a generalized operator rule which relates the expected values of the relevant operators at *different times*. Then, taking the special case when the expected values are computed at the same time, we obtain the corresponding operator rule which relates the operators themselves.

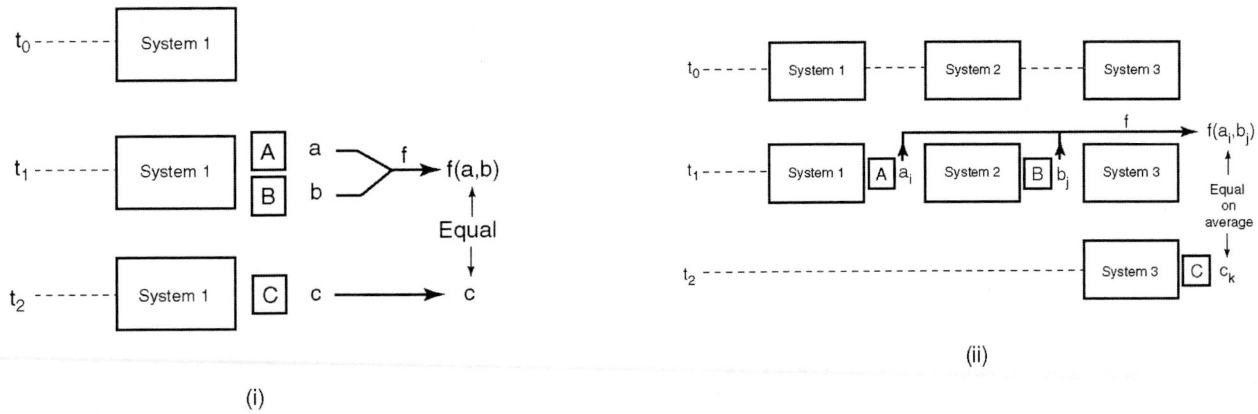

FIGURE 1. An example of the application of the AVCP. (i) A classical experiment showing the measurements of A, B and C performed at times t_A, t_B, and t_C, respectively, with outcomes denoted as a, b, and c, respectively. Here, $t_A = t_B = t_1$ and $t_C = t_2$. Suppose that one finds that the relation $c = f(a,b)$ holds for all initial states of the system. (ii) The corresponding quantum experiment. Three copies of the system are prepared in the same initial state, v_0, at time t_0, and are placed in identical backgrounds. In this example, it is assumed that the operators A and B do not commute. Hence, by the AVCP, measurements $\mathbf{A}$ and $\mathbf{B}$ are performed on different copies of the system. Measurement $\mathbf{C}$ is performed on a separate copy of the system. In any given run of the experiment, the probabilities that measurements $\mathbf{A}$, $\mathbf{B}$ and $\mathbf{C}$ yield outcome values a_i, b_j and c_k $(i,j,k = 1,\ldots,N)$, are P_i, P'_j and P''_k, respectively. The AVCP then asserts that, provided the polynomial expansion of $f(a,b)$ contains no product terms involving a and b, the relation $\overline{c} = \overline{f(a,b)}$ holds for all initial states, v_0, of the system, where the average is taken over an infinite number of runs of the experiment; that is, $\sum_k c_k P''_k = \sum_{ij} f(a_i, b_j) P_i P'_j$ for all v_0.

We shall consider a classical experiment where a system is subject to measurements of A and B at time t_1, and to a measurement of C at time t_2. We shall suppose that a measurement of C, with outcome c, can be implemented by an arrangement in which the measurements of A and B are performed, with respective outcomes a and b, and the function $f(a,b)$ then computed, so that the relation $c = f(a,b)$ holds for all initial states of the system.

In a quantum model of the appropriate experimental arrangement, let the operators that represent these measurements be denoted A, B, and C, respectively. To simplify the presentation, we shall only consider the case where these operators have finite dimension, N; the results obtained below can be readily shown to apply in the infinite dimensional case. Let the elements of orthonormal sets of eigenvectors of A, B and C be denoted $\mathsf{v}_i, \mathsf{v}'_j$, and v''_k, respectively $(i,j,k = 1,2,\ldots,N)$, let the corresponding eigenvalues be denoted a_i, b_j and c_k, and let the probabilities of the ith, jth and kth outcomes of measurements $\mathbf{A}$, $\mathbf{B}$ and $\mathbf{C}$ in any given experimental arrangement be denoted by P_i, P'_j and P''_k, respectively.

Case 1. f is a function of a only.

In this case, according to the AVCP, the quantum experiment simply consists of two set-ups, involving two copies of the system, where $\mathbf{A}$ is performed on one copy at time t_1 and $\mathbf{C}$ on the other copy at time t_2. Consequently, we obtain the relation

$$\langle \mathsf{C} \rangle_{t_2} = \langle f(\mathsf{A}) \rangle_{t_1}, \tag{1}$$

which holds for all v_0. We can summarize the above result in the form of the *generalized function rule*:

$$c(t_2) = f(a(t_1)) \mapsto \langle \mathsf{C} \rangle_{t_2} = \langle f(\mathsf{A}) \rangle_{t_1} \quad \forall \mathsf{v}_0, \tag{2}$$

where, for clarity, the times at which the outcomes are obtained has been explicitly indicated. In the special case where $t = t_1 = t_2$, we obtain the usual operator rule, the *function rule*:

$$c = f(a) \mapsto \mathsf{C} = f(\mathsf{A}). \tag{3}$$

Case 2. $f(a,b) = f_1(a) + f_2(b)$

In this case, by straightforward application of the AVCP, one finds that, whether or not the operators A, B commute, and whether or not they are performed on the same sub-system in the case of a composite system, one obtains the *generalized sum rule*:

$$c(t_2) = f_1\left(a(t_1)\right) + f_2\left(b(t_1)\right) \mapsto \langle \mathsf{C} \rangle_{t_2} = \left\langle f_1(\mathsf{A}) \right\rangle_{t_1} + \left\langle f_2(\mathsf{B}) \right\rangle_{t_1} \qquad \forall \mathbf{v}_0. \tag{4}$$

In the special case where $t = t_1 = t_2$, we obtain the *sum rule*:

$$c = f_1(a) + f_2(b) \mapsto \mathsf{C} = f_1(\mathsf{A}) + f_2(\mathsf{B}). \tag{5}$$

Case 3. $f(a,b) = f_1(a)f_2(b)$

In this case, application of the AVCP yields the *generalized product rule*:

$$c(t_2) = f_1\big(a(t_1)\big) f_2\big(b(t_1)\big) \mapsto \langle \mathsf{C} \rangle_{t_2} = \left\langle f_1(\mathsf{A}) f_2(\mathsf{B}) \right\rangle_{t_1} \quad \forall \mathbf{v}_0 \qquad \text{if } [\mathsf{A}, \mathsf{B}] = 0. \tag{6}$$

In the special case where $t_1 = t_2$, we obtain the *product rule*:

$$c = f_1(a)f_2(b) \mapsto \mathsf{C} = f_1(\mathsf{A})f_2(\mathsf{B}) \quad \text{if } [\mathsf{A}, \mathsf{B}] = 0. \tag{7}$$

Note that there is no rule in the case where A and B do not commute.

APPLICATION OF THE AVCP

Remarkably, one finds that the AVCP enables correspondence rules of each of the four types described in [2], namely operator rules, measurement commutation relations, measurement–transformation commutation relation, and transformation operators, to be obtained in a uniform manner. Consequently, one can see that the difference between these types of rules depends simply upon whether the classical relations that one is taking over into the quantum framework are relations between measurements performed at the same time (leading to the operator rules), at different times (leading to measurement commutation relations), or in different frames of reference (leading to measurement–transformation commutation relations and to the explicit forms of transformation operators). In short, from the perspective provided by the derivation, the commutation relation $[\mathsf{x}, \mathsf{p}_x] = i\hbar$ is no more elusive in its origin than fact that the classical relation $H = p_x^2/2m$ is taken over into the operator relation $\mathsf{H} = \mathsf{p}_x^2/2m$.

In this section, for reasons of space, we will pick just one simple illustrative application of the AVCP, namely the derivation of the measurement commutation relation $[\mathsf{x}, \mathsf{p}_x] = i\hbar$. The reader is referred to [2] for further applications, which include the derivation of the angular momentum commutation relations, namely $[\mathsf{L}_x, \mathsf{L}_y] = i\hbar \mathsf{L}_z$ (and cyclic permutations thereof), and Dirac's Poisson-Bracket Rule.

The position–momentum commutation relationships

Consider a particle of mass m moving in the $+x$-direction, where measurements of the x-component of position, the x-component of momentum, and the energy, are represented by the operators $\mathsf{x}, \mathsf{p}_x, \mathsf{H}$, respectively.

First, to determine the relationship of H to the operators x and p_x, we make use of the fact that the relation $H = p_x^2/2m + V(x)$ holds for all classical states (x, p_x) of the system, so that, from the function rule, it follows that $\mathsf{H} = \mathsf{p}_x^2/2m + V(\mathsf{x})$.

Next, to obtain a relation between x and p_x, we make use of the fact that, in the quantum model of the system, the expected value of x at time $t+\delta t$ can be calculated in two separate ways. First, from the definition of $\langle x\rangle_t$, the relation

$$\begin{aligned}\langle \mathsf{x}\rangle_{t+\delta t} &= \left\langle \mathsf{U}_t^\dagger(\delta t)\,\mathsf{x}\,\mathsf{U}_t(\delta t)\right\rangle_t \\ &= \left\langle \left(1+\frac{i\mathsf{H}\delta t}{\alpha}\right)\mathsf{x}\left(1-\frac{i\mathsf{H}\delta t}{\alpha}\right)\right\rangle_t + O(\delta t^2) \\ &= \langle \mathsf{x}\rangle_t + \frac{i}{\alpha}\delta t\langle \mathsf{Hx}-\mathsf{xH}\rangle_t + O(\delta t^2) \\ &= \langle \mathsf{x}\rangle_t - \frac{i}{2m\alpha}\delta t\left\langle [\mathsf{x},\mathsf{p}_x^2]\right\rangle_t + O(\delta t^2)\end{aligned} \tag{8}$$

holds for all states, v, of the system. Second, using the generalized function rule, it follows from the classical relation $x(t+\delta t) = x(t) + p_x(t)\delta t/m + O(\delta t^2)$ that the relation

$$\langle \mathsf{x}\rangle_{t+\delta t} = \langle \mathsf{x}\rangle_t + \frac{\delta t}{m}\langle \mathsf{p}_x\rangle_t + O(\delta t^2) \tag{9}$$

holds for all v.

Equating the above two expressions for $\langle x\rangle_{t+\delta t}$, we obtain

$$\mathsf{v}^\dagger\left(\mathsf{p}_x[\mathsf{x},\mathsf{p}_x] + [\mathsf{x},\mathsf{p}_x]\mathsf{p}_x\right)\mathsf{v} = \mathsf{v}^\dagger 2i\alpha\mathsf{p}_x\mathsf{v} \tag{10}$$

for all v, which implies that

$$\mathsf{p}_x[\mathsf{x},\mathsf{p}_x] + [\mathsf{x},\mathsf{p}_x]\mathsf{p}_x = 2i\alpha\mathsf{p}_x. \tag{11}$$

Representing the operators p_x and $\mathsf{A} \equiv [\mathsf{x},\mathsf{p}_x]$ as limits of infinite-dimensional matrices, and diagonalizing p_x by a unitary transformation U, one obtains

$$\mathsf{p}_x'\mathsf{A}' + \mathsf{A}'\mathsf{p}_x' = 2i\alpha\mathsf{p}_x', \tag{12}$$

where $\mathsf{p}_x' = \mathsf{U}^{-1}\mathsf{p}_x\mathsf{U}$ and $\mathsf{A}' = \mathsf{U}^{-1}\mathsf{A}\mathsf{U}$. Writing $\mathsf{p}_x' = p_i'\delta_{ij}$, where i and j range from $-\infty$ to $+\infty$, and multiplying out Eq. (12), we obtain

$$(p_i' + p_j')A_{ij}' = 2i\alpha p_i'\delta_{ij}. \tag{13}$$

Now let the eigenvalues, p_i', be chosen such that they have equal spacing δ, so that $p_{i+1}' - p_i' = \delta$, and set $p_0 = \varepsilon$ with $0 < \varepsilon < \delta$, which ensures that $(p_i' + p_j') \neq 0$ for any i, j. In this case, it follows that $\mathsf{A} = i\alpha\mathsf{I}$, which, in the limit as δ tends to zero, yields the commutation relationship

$$[\mathsf{x},\mathsf{p}_x] = i\alpha. \tag{14}$$

Although a particular system has been used to obtain this commutation relation, one can argue that it is generally valid [2]. We note that, if one instead considers a massless particle moving at the speed of light in the positive x direction, with classical Hamiltonian given by $H = cp_x$, and so that $x(t+\delta t) = x(t) + c\delta t$, then the above computation immediately yields Eq. (14).

REFERENCES

1. P. Goyal. An information-theoretic approach to quantum theory, I: The abstract quantum formalism. quant-ph/0702124, 2007.
2. P. Goyal. An information-theoretic approach to quantum theory, II: The correspondence rules of quantum theory. quant-ph/0702149, 2007.
3. P.A.M. Dirac. *Principles of Quantum Mechanics.* Oxford Science Publications, fourth edition, 1999.
4. J. von Neumann. *Mathematical Foundations of Quantum Mechanics.* Princeton University Press, 1955.
5. H. J. Groenewold. On the principles of elementary quantum mechanics. *Physica*, 12:405–60, 1946.
6. T. F. Jordan. Why $-i\nabla$ is the momentum. *Am. J. Phys.*, 43(12), 1975.
7. T. F. Jordan. *Linear Operators for Quantum Mechanics.* John Wiley and Sons, Inc., 1969.
8. L. E. Ballentine. *Quantum Mechanics : A Modern Development.* World Scientific, 1999.

Physical Concepts and Mathematical Symbols

Hans Herlof Grelland

University of Agder, Service Box 509, N-4898 Grimstad, Norway

Abstract. According to traditional empiricist philosophy of science, concepts and meaning grow out of sense experience, and the mathematical structure of a physical theory is nothing but a formalisation of a given meaning-content. This view seems to work well in classical mechanics. But it breaks down in quantum physics, where we have a self-supported mathematical structure which resists any conceptual or pictorial interpretation in the traditional sense. Thus, traditional empiricism is flawed. Quantum physics teaches us that mathematics is a language in itself which extends beyond ordinary language. To understand the meaning of this extended language, we have to explore how new concepts and intuitions grow out of mathematics, not the other way around. The symbolic structure is prior to its meaning. This point of view is called linguistic empiricism, to stress that the connection with experience is still crucial. As cases, I compare the concept of stiffness in classical mechanics and the concept of electron density in quantum mechanics. The last case demonstrates that the wave function has a richer interpretation than the probabilistic one concerning measurement of position.

Keywords: Interpretation of Quantum Mechanics, X-ray scattering, Observables in Quantum Mechanics
PACS: 01.70.+w

The traditional philosophy of physics has been empiricism, implying that physics is based on sense experience. The modifications of this simple assumption implied by the introduction of complex measuring instruments does not really change the basic empiricist attitude, that reliable knowledge is based on experience.

However, physicists traditionally has assumed more than this. Although physics always has been a mathematical science, one has made a distinction between the "physical" meaning-content of the theory and the mathematical structure in which it is expressed. This has been demonstrated by calling the mathematical structure the mathematical "formalism" of the theory. This formalism has been considered as a useful tool for doing calculations, but the physical content is not in its formalism, it is in the mind of the scientist (under the denotation of "physical intuition") and is expressed in the ordinary-language text between the mathematical equations in the textbooks.

This way of thinking put great stress on the imagery of the scientist. An experienced physicist is supposed to have a highly developed ability to imagine physical situations, an ability called "physical intuition". In doing mathematical derivations or calculations, he always try to associate this formalism with physical images. If he is unable to do this, he is assumed not really to understand the theory.

Where do the images come from? Of course, from the sense experiences of ordinary life, perhaps extended by simple, visual, tactile of kinaesthetic experiments. Thus, the development of an physical intuition depends on that the physical phenomena studied are somehow similar to such direct sensational experiences. The problem is: can this separation of intuition and formalism be sustained in theories of phenomena that are far removed from the everyday world?

The problem is deeper than it seems at first sight, if we take into account the fact that the interplay between experience and language is not adequately understood in traditional empiricism, where one assumes that the experience in itself is language-independent, and that language is used only to express or describe a given meaning-content. Today one is aware that language affects how the experienced world appears, and this is all the more pronounced in physics, which is based on a very elaborate conceptual structure. This calls for what one could call a linguistic empiricism (not to be confused with any form of constructivism). But a further step should be taken: to accept the mathematical structure as a language on its own footing. The physics community has forgotten the historical experience by Newton, where the development of (the mathematical technique of) derivation and the

CP962, *Quantum Theory, Reconsideration of Foundations – 4*,
edited by G. Adenier, A. Yu. Khrennikov, P. Lahti, V. I. Man'ko, and Th. M. Nieuwenhuizen

concept of velocity at a point was not only mutually dependent, they were one and the same thing. Without the concept of derivation, there is no concept of pointwise velocity (only the older and simpler concept of velocity in an interval). However, this role of mathematics is even more important in the quantum regime, where intuition no longer seems to work, but where the mathematics works just as good as in classical physics.

Let me explain by making the bold, although scientifically sound and reasonable, hypothesis that the mathematical structure of quantum mechanics gives an adequate description of quantum reality in a similar way that the mathematical structure of fluid dynamics gives an adequate description of liquids. It is then an important task to try to figure out what we can say about this assumed quantum reality. There are two main questions to be considered: to what extent can this mathematics be looked upon as a language in the proper sense, and what is its relation to experiments, i.e. empirical reality.

First question first. Let us start with some examples from classical physics.

We consider the question of images and language. In classical mechanics, including fluid mechanics, what can be described by the mathematics can also be imagined. I can imagine a liquid like water, for instance, with all its sensual qualities, which for physical description are limited to its appearance and its tactile qualities (one cane imagine feeling its warmth or coolness, or the feeling of the stream). These images are possible because we as human individuals have had sensible experiences with liquids in the past.

Let us then move to the electromagnetic field. It is similar to a liquid in the sense that it is described mathematically as a field, by physical quantities that are functions of space coordinates. However, we have here a part of reality which is less related to everyday experiences. We are to a greater extent dependent on indirect observation. Now, we have some direct experiences with the electromagnetic field. We can feel the attraction of an electrically hair, or the forces of one magnet on another, which we can hold in our hand. Here we use our kinaesthetic sense, by which we feel forces. Thus, we can imagine a force. So, an electromagnetic force field Fmn can be directly related to experience. However, such is not the case with the electromagnetic potential Am. Some physicist therefore consider the potential to be, in this sense, "unphysical", it is only a "mathematical" auxiliary quantity.

Nevertheless, there are obviously physical quantities in classical physics that we cannot imagine – like energy or the entropy. In such cases we go a step further and introduce measurements, which reaches beyond the immediate, sensible experience, and hence also beyond imagination. A measurements is made by an additional physical system (measurement apparatus) which are coupled to the system which we are interested in, and which somehow produces numbers indicating the value of the quantity in question. And we go further than that, we use indirect observations. We observe something from which we can calculate the desired quantities. For instance, the change in temperature of a contained gas with a constant volume can be measured by measuring its pressure.

When we turn to quantum mechanics, the situation seems to be fundamentally different. Basically this theory contains quantities recognizable from classical mechanics, like position, velocity, etc. However, due to the new phenomenon of non-compatibility, the quantities cannot be combined into an imaginable picture. One cannot have a certain position in combination with a precisely determined velocity. Instead new kind of mathematical objects appear, like the probability amplitudes. These are so named because if one measures the corresponding quantity, the square of the amplitude represents a probability distribution of outcomes of the measurement. Studies in quantum logic (i.e. the lattice of projection operators in the operator algebra) show that the states of the system are determined by the set of eigenvalues of the projection operators corresponding to the propositions compatible with the actual state of the system. These propositions have the same meaning as in classical mechanics. This shows that the amplitudes are not necessary for the formal completion of the mathematical formalism. On the other hand, the proposition lattices which are relevant can be proven (by construction) to be representable by projection operators on a Hilbert space. By making some reasonable assumptions defining a non-Boolean probability measure, one may construct the amplitudes from the Hilbert spaces vectors. The problem left is to make sense of the non-Boolean logic appearing, in particular of the phenomenon of non-compatibility.

What I want to point out is that there is more to the theory than the propositions of quantum logic. It is not true that only values of quantities corresponding to self-adjoint operators are measurable (or that all of them are!). There are many mathematical objects of this structure that can be determined experimentally, and thus which can be regarded as "observable", although they are not observables in the traditional terminology of quantum mechanics. I would like to mention two examples.

When a solid is studied by X-ray diffraction, what is "observed" (in the sense that it can be determined by the observation) is the electron density (or, rather its Fourier transform, which can be back-transformed). This density (the reduced, diagonal density matrix in space representation) is a functional of the square of the wave function, but the observation gives the density as a whole, not only a statistics of single values. Thus, the density is as good an observable as anything deserving this name.

A more interesting case is the observation of the electron structure in a molecule, using another molecule as a probe. In the cases where the Woodward-Hoffman rules come into application, there is a interplay between orbitals which may or may not have the same symmetry. Opposite symmetry leads to a high reaction barrier, and consequently to no reaction. But the symmetry in this case is not a property of the "physical" and "statistical" (and real in the mathematical sense) square of the wave function. It is a "non-physical" property of a purely mathematical object. But, since it is observable by using the reacting (or non-reacting) molecule as a probe, we are forced to conclude that even this amplitude is an "observable" in any reasonable sense of the word.

The conclusion is that one cannot select some mathematical object to represent "physical" (as opposed to "mathematical" or "formal") elements of the theory. The mathematical structure as a whole (and not only the lattice of projectors) makes the physical theory, and all objects contained in it can possibly be measured (while some "observables" in the traditional sense can not).

Thus the whole mathematical formalism of quantum mechanics should be considered as an extended linguistic system of expressions, where every element of the system has physical meaning. The new mathematical language of quantum mechanics is the only language in which we can express, precisely and adequately, the truths of the strange world of the quantum, which, as we learn from the phenomenon of de-coherence, is actually the physical world we live in.

Quantum Theory and Observer's Mathematics

Dmitriy Khots* and Boris Khots†

*3710 S. 202nd Ave, Omaha, NE 68130, USA
†Compressor Controls Corp, 4725 121st Street, Des Moines, Iowa 50323, USA

Abstract. When we consider and analyze physical events with the purpose of creating corresponding mathematical models we often assume that the mathematical apparatus used in modeling, at least the simplest mathematical apparatus, is infallible. In particular, this relates to the use of infinitely small and infinitely large numbers in arithmetic and the use of Newton's definition of a limit in analysis. We believe that is where the main problem lies in contemporary study of nature (physics, genetics). We have introduced a new concept of "Observer's Mathematics - Mathematics of Relativity" (see references). Observer's mathematics creates new arithmetic, algebra, geometry, topology, analysis and logic which do not contain the concept of continuum, but locally coincide with the standard fields. The concept of a quantum arises automatically from the human thought process, rather than from physics. From this point of view classical mechanics is brought closer to quantum mechanics.

Keywords: Observer, arithmetic, derivative, speed, probability, classical mechanics, quantum mechanics
PACS: 02.30.-f 03.65.Fd

INTRODUCTION

Mathematics of Relativity - Observer's Mathematics was introduced by authors in [1], and then was published in Khots et al [2] and Khots et al [3]. This work was created as an attempt to do away with the concept of infinity. The main concepts of Observer's Mathematics are as follows. Let W be the set of all real numbers, i.e. a set of all infinite decimal fractions. Let W_n be the set of all finite decimal fractions of length $2n$ (with n - positive integer). Visually W_n can be described as $W_n = \{\underbrace{\star\cdots\star}_{n}.\underbrace{\star\cdots\star}_{n}\}$. We consider a finite well-ordered (by n) system of observers, where each observer "lives" in W_n for some n, but "thinks" that he "lives" in W. We call this observer "naive" and denote by W_n-observer. We further assume that a W_n-observer, preserving his "naivety", sees that a W_k-observer actually sees W_k and deals with W_k, if $n > k$.

We now consider the arithmetic in W_n from the point of view of a W_m-observer with $m > n$.

For $c = c_0.c_1...c_n$, $d = d_0.d_1...d_n \in W_n$ addition and subtraction ($\pm_n$) is defined as follows.

$$c \pm_n d = \begin{cases} c \pm d, \text{ if } c \pm d \in W_n \\ \text{not defined, if } c \pm d \notin W_n \end{cases}$$

(The $\pm$ sign is a standard arithmetic sign). We will write

$$((\dots(c_1 +_n c_2)\dots) +_n c_N) = \sum_{i=1}^{N}{}^{n} c_i$$

for $c_1, \dots, c_N$ iff the contents of any parenthesis are in W_n.

Next, multiplication is defined for $c = c_0.c_1...c_n$, $d = d_0.d_1...d_n \in W_n$.

$$c \times_n d = \sum_{k=0}^{n}{}^{n} \sum_{m=0}^{n-k}{}^{n} 0.\underbrace{0...0}_{k-1}c_k \cdot 0.\underbrace{0...0}_{m-1}d_m$$

where $c, d \geq 0$, $c_0 \cdot d_0 \in W_n$, $0.\underbrace{0...0}_{k-1}c_k \cdot 0.\underbrace{0...0}_{m-1}d_m$ is the standard product, and $k = m = 0$ means that $0.\underbrace{0...0}_{k-1}c_k = c_0$ and $0.\underbrace{0...0}_{m-1}d_m = d_0$. If either $c < 0$ or $d < 0$, then we compute $|c| \times_n |d|$ and define $c \times_n d = \pm |c| \times_n |d|$, where the sign $\pm$

CP962, *Quantum Theory, Reconsideration of Foundations – 4*,
edited by G. Adenier, A. Yu. Khrennikov, P. Lahti, V. I. Man'ko, and Th. M. Nieuwenhuizen

is defined as usual. Note, if the content of at least one parentheses (in previous formula) is not in W_n, then $c \times_n d$ is not defined.

The arithmetic that we introduced above coincides with the standard arithmetic until it meets the "allowed" boundaries of space. It turns out that the "boundary conditions" could cardinally affect the computational process in general. This paper focuses on the possible application of Observer's Mathematics to Quantum Theory.

DERIVATIVES

From the point of view of W_n-observer (we will call such observers "naive", since they "think" that they "live" in W and deal with W) a real function y of a real variable x, $y = y(x)$, is called differentiable at $x = x_0$ if there is a derivative

$$y'(x_0) = \lim_{x \to x_0, x \neq x_0} \frac{y(x) - y(x_0)}{x - x_0}$$

What does the above statement mean from point of view of W_m-observer with $m > n$? It means that

$$|(y(x) -_n y(x_0)) -_n (y'(x_0) \times_n (x -_n x_0))| \leq 0.\underbrace{0\ldots01}_{n}$$

whenever

$$|y(x) -_n y(x_0)| = 0.\underbrace{0\ldots0}_{l}y_l y_{l+1} \ldots y_n$$

and

$$|(x -_n x_0)| = 0.\underbrace{0\ldots0}_{k}x_k x_{k+1} \ldots x_n$$

for $1 \leq k, l \leq n$, and x_k - non-zero digit.

We now state the main theorems.

Theorem 1 *From the point of view of a W_m-observer a derivative calculated by a W_n-observer ($m > n$) is not defined uniquely.*

Proof. Put $y'(x_0) = \pm a_0.a_1 \ldots a_p a_{p+1} \ldots a_n$ with $a_0.a_1 \ldots a_p a_{p+1} \ldots a_n \geq 0$ and $p \leq n$. Then $0.\underbrace{0\ldots0}_{l}y_l y_{l+1} \ldots y_n = a_0.a_1 \ldots a_p a_{p+1} \ldots a_n \times_n 0.\underbrace{0\ldots0}_{k}x_k x_{k+1} \ldots x_n = a_0.a_1 \ldots a_p b_{p+1} \ldots b_n \times_n 0.\underbrace{0\ldots0}_{k}x_k x_{k+1} \ldots x_n$ for any digits $b_{p+1}, \ldots, b_n$ and $p = n - k$. Hence $y'(x_0) \in V = \{\pm a_0.a_1 \ldots a_p a_{p+1} \ldots a_n | a_{p+1}, \ldots, a_n \in \{0, 1, \ldots, 9\}\}$ and $|V| = 10^k$. QED.

Theorem 2 *From the point of view of a W_m-observer with $m > n$, $|y'(x_0)| \leq C_n^{l,k}$, where $C_n^{l,k} \in W_n$ is a constant defined only by n, l, k and not dependent on $y(x)$.*

Proof. We have $\pm 0.\underbrace{0\ldots0}_{l}y_l y_{l+1} \ldots y_n = (\pm a_0.a_1 \ldots a_n) \times_n (\pm 0.\underbrace{0\ldots0}_{k}x_k x_{k+1} \ldots x_n)$ with x_k - non-zero digit and $a_0.a_1 \ldots a_p a_{p+1} \ldots a_n \geq 0$. Now, if $l > k$ then $a_0 = 0$; if $l = k$ then $a_0 \leq 9$ and if $l < k$ then $a_0 < 9 \times 10^{k-1}$. Hence

$$C_n^{l,k} = \begin{cases} 1, \text{ if } l > k \\ 10, \text{ if } l = k \\ 9 \times 10^{k-1}, \text{ if } l < k \end{cases}$$

QED.

Theorem 3 *From the point of view of a W_m-observer, when a W_n-observer (with $m > n \geq 3$) calculates the second derivative:*

$$y''(x_0) = \lim_{x_1 \to x_0, x_1 \neq x_0, x_2 \to x_0, x_2 \neq x_0, x_3 \to x_1, x_3 \neq x_1} \frac{\frac{y(x_3) - y(x_1)}{(x_3 - x_1)} - \frac{y(x_2) - y(x_0)}{x_2 - x_0}}{x_1 - x_0}$$

we get the following unequality:

$$(|x_2 -_n x_0| \times_n |x_3 -_n x_1|) \times_n |x_1 -_n x_0| \geq 0.\underbrace{0\ldots01}_{n}$$

provided that $y''(x_0) \neq 0$.

Proof. For the W_m-observer existence of $y''(x_0)$ means that $|((y(x_3) -_n y(x_1)) \times_n (x_2 -_n x_0) -_n ((y(x_2) -_n y(x_0)) \times_n (x_2 -_n x_0))) -_n y''(x_0) \times_n ((|x_2 -_n x_0| \times_n |x_3 -_n x_1|) \times_n |x_1 -_n x_0|)| \leq 0.\underbrace{0\ldots01}_{n}$, whenever

$$|(x_2 -_n x_0)| \leq 0.\overbrace{\underbrace{0\ldots0}_{k}p*\ldots*}^{n}$$

and

$$|(x_3 -_n x_1)| \leq 0.\overbrace{\underbrace{0\ldots0}_{l}q*\ldots*}^{n}$$

and

$$|(x_1 -_n x_0)| \leq 0.\overbrace{\underbrace{0\ldots0}_{s}r*\ldots*}^{n}$$

where p, q, r are non-zero digits, asterisks are any digits and $3 \leq p+q+r \leq n$. Then given $y''(x_0) \neq 0$ we have $(|x_2 -_n x_0| \times_n |x_3 -_n x_1|) \times_n |x_1 -_n x_0| \geq 0.\underbrace{0\ldots01}_{n}$. QED.

PHYSICAL INTERPRETATION

The following hypotheses illustrate possible physical interpretation of previous theorems.

Hypotheses 1 Theorem 1 could offer an explanation of why physical speed (or acceleration) is not uniquely defined and, from the point of view of a measurement system (observer), it is possible to consider speed (or acceleration) as a random variable with distribution dependend on the measurement system. Let v be the speed with $v = v_0.v_1\ldots v_{n-k} + \xi_m^{n,k}$ where $\xi_m^{n,k} \in \{0.\underbrace{0\ldots0}_{n-k}v_{n-k+1}\ldots v_n\}$ - random variable, $m > n$, and the distribution function is $F_m^{n,k}(x) = P(\xi_m^{n,k} < x)$.

Hypotheses 2 Theorem 2 could offer an explanation of why the speed of any physical body cannot exceed some constant, (the speed of light, for example). Independence of this constant on explicit expression of space-time function could offer an explanation of why the speed of light does not depend on an inertial coordinate system.

Hypotheses 3 Theorem 3 could offer an explantion of the various uncertainty principles, when a product of a finite number of physical variables has to be not less than a certain constant. This can be seen not just from consideration of second derivatives, but of any derivative.

Hypotheses 4 Theorems 1, 2, and 3 combined may provide an insight into the connection between classical and quantum mechanics.

ACKNOWLEDGMENTS

The authors thank Andrei Khrennikov for his invitation to participate in QTRF-4, Vaxjo 2007.

REFERENCES

1. Khots, B., Khots, D., Mathematics of Relativity, *Web Book*, www.mathrelativity.com, 2004.
2. Khots, B., Khots, D., An Introduction to Mathematics of Relativity, *Lecture Notes in Theoretical and Mathematical Physics*, Ed. A.V. Aminova, v. 7, pp 269-306, Kazan State University, Russia, 2006.
3. Khots, B., Khots, D., Observer's Mathematics - Mathematics of Relativity, *Applied Mathematics and Computations*, v. 187, issue 1, pp 228-238, 2007.

Moment Operators of Observables in the Balanced Homodyne Detection Scheme

Jukka Kiukas

Department of Physics, University of Turku, FIN-20014 Turku, Finland
email: jukka.kiukas@utu.fi

Abstract. We review the results of article [1], where we gave a mathematically rigorous proof that the rotated quadrature observables $\frac{1}{\sqrt{2}}(\overline{e^{-i\theta}a+e^{i\theta}a^*})$ of a single mode electromagnetic field can be determined in the so called "high amplitude limit" of the balanced homodyne detection. In the proof, we used the moment operators of observables actually measured by the detector; the high amplitude limit was described by certain limits of a sequence $(E^n)_{n\in\mathbb{N}}$ of such observables.

Keywords: balanced homodyne detection, high amplitude limit, operator integral, moment problem, positive operator measure
PACS: 03.65.-w, 03.65.Ca, 03.65.Db, 42.50.Ar, 42.50.Xa

INTRODUCTION

The balanced homodyne detection, introduced in [2, 3] is an important technique in quantum optics, because it is assumed to provide a way to measure the rotated quadrature amplitudes $\frac{1}{\sqrt{2}}(e^{-i\theta}a+e^{i\theta}a^*)$ of a single mode electromagnetic field. A perhaps most notable application of this method is the quantum state estimation, i.e. quantum tomography (See [4] for a collection of articles concerning this topic.) The significance of the homodyne measurement in that context is due to the fact that the combined measurement statistics of all the rotated quadratures determine the state of the system.

The measurement scheme is the following: A signal field is mixed with an auxiliary field by means of a 50-50 beam splitter (possibly followed by a phase shifter), and the difference of photon numbers on the output ports of the splitter is detected. The auxiliary field is taken to be an oscillator in a coherent state, operating with the same frequency as the signal field. Assuming that the beam splitter is lossless and causes no phase shift between the input modes, the process is described by a unitary operator U (see [5, 6]), transforming the field annihilation operators a (signal) and b (auxiliary) into $\tilde{a}=U^*aU=\frac{1}{\sqrt{2}}(a-b)$, and $\tilde{b}=U^*bU=\frac{1}{\sqrt{2}}(a+b)$. With respect to the initial state of the two-mode field (Heisenberg picture), the photon number difference observable is then $\tilde{b}^*\tilde{b}-\tilde{a}^*\tilde{a}=ab^*+a^*b$, provided that the photon detectors are ideal. (Of course, these formal operator relations are made precise by restricting the operators to a dense subspace.)

When the strength of the auxiliary field is high, the above mentioned photon difference statistics are considered to resemble those of a rotated field quadrature, the rotation angle being identified as the fixed phase difference between the input signal and auxiliary fields. This is usually justified by the following heuristic explanation: the strong auxiliary field can be treated "classically", by replacing b with the complex field amplitude $\beta=re^{i\theta}$ in the operator ab^*+a^*b, with θ identified as the phase relative to the input signal field. Consequently, by suitably scaling the resulting operator with a factor proportional to the intensity of the auxiliary field, one recovers the rotated quadrature $\frac{1}{\sqrt{2}}(e^{-i\theta}a+e^{i\theta}a^*)$.

A more theoretical explanation takes into account the quantum nature of the auxiliary field. In addition to the original paper [3], this has been done in [7–10]. These papers calculate the characteristic functions of the probability measures associated with the photon difference statistics, and then take the high amplitude limit at this level. Vogel [10] applies Levy-Cramer theorem to prove that the associated probability measures converge weakly. However, in the calculations leading to the above mentioned characteristic functions, some of the problematic matematical questions (like formal operator series expansions and term-by-term integration of a series) were not addressed, making it difficult to verify whether the treatment could be made rigorous.

In the following, we review the results of [1] to demonstrate that the balance homodyne detection can be understood rigorously in terms of the moment operators of the signal field observables arising from the photon difference statistics. The moment operator approach have been used in the context of homodyne detection also in [7, 11, 12], where the

CP962, *Quantum Theory, Reconsideration of Foundations – 4*,
edited by G. Adenier, A. Yu. Khrennikov, P. Lahti, V. I. Man'ko, and Th. M. Nieuwenhuizen

authors consider the effects of imperfectness of the photon detectors and the preamplification by using the concepts of [13]. Unfortunately, they dismiss the limit problem by applying the heuristic classical approximation of the auxiliary field to the characteristic functions, so that their results do not contribute to our problem.

A FEW GENERAL NOTATIONS AND DEFINITIONS

Let $\mathscr{H}$ be a complex Hilbert space and let $L(\mathscr{H})$ stand for the set of bounded operators on $\mathscr{H}$. For any (not necessarily bounded) operator A in $\mathscr{H}$, we denote by $D(A)$ the domain of definition of A. The symbol $\mathscr{B}(\mathbb{R})$ stands for the Borel σ-algebra of $\mathbb{R}$. For any two vectors $\varphi, \psi \in \mathscr{H}$, we let $|\varphi\rangle\langle\psi|$ denote the operator $\eta \mapsto \langle\psi|\eta\rangle\varphi$.

A weakly (or, equivalently, strongly) σ-additive set function $E : \mathscr{B}(\mathbb{R}) \to L(\mathscr{H})$ with $E(X) \geq 0$ for all $X \in \mathscr{B}(\mathbb{R})$, and $E(\mathbb{R}) = I$, is a *semispectral measure* (also called a normalized POM). For each $\psi, \varphi \in \mathscr{H}$, the scalar valued set function $\mathscr{B}(\mathbb{R}) \ni X \mapsto \langle\psi|E(X)\varphi\rangle \in \mathbb{C}$ is then a complex measure, and we will denote it by $E_{\psi,\varphi}$. For each positive operator T of trace one, we let $E_T : \mathscr{B}(\mathbb{R}) \to [0,1]$ be defined by $E_T(X) = \mathrm{Tr}[TE(X)]$. Then E_T is a probability measure. For a unit vector $\varphi \in \mathscr{H}$, we then have $E_{|\varphi\rangle\langle\varphi|} = E_{\varphi,\varphi}$, and we write simply E_φ to denote this probability measure.

A semispectral measure $E : \mathscr{B}(\mathbb{R}) \to L(\mathscr{H})$ is a *spectral measure*, if $E(X)$ is a projection for all $X \in \mathscr{B}(\mathbb{R})$. If A is a selfadjoint operator, we let $P^A : \mathscr{B}(\mathbb{R}) \to L(\mathscr{H})$ be the unique spectral measure of A, given by the spectral theorem.

We define the moment operators of a semispectral measure by using the operator integral introduced in [14]. If $E : \mathscr{B}(\mathbb{R}) \to L(\mathscr{H})$ is a semispectral measure, and $k \in \mathbb{N}$, there exists a unique (not necessarily densely defined) linear operator $E[k]$ on $\mathscr{H}$, such that

$$\langle\psi|E[k]\varphi\rangle = \int x^k \, dE_{\psi,\varphi}(x), \quad \psi \in \mathscr{H}, \varphi \in D(E[k]),$$

where the domain $D(E[k])$ is the set of those vectors $\varphi \in \mathscr{H}$ for which the function x^k is integrable with respect to the complex measure $E_{\psi,\varphi}$ for all $\psi \in \mathscr{H}$. The operator $E[k]$ is called the kth *moment operator* of E. It is always symmetric, but it need not be selfadjoint, or even densely defined.

We will mostly use the restriction of the operator $E[k]$ to the *square-integrability domain*

$$\{\varphi \in \mathscr{H} \mid \int x^2 \, dE_\varphi(x) < \infty\},$$

which is (in general, a proper) subspace of $D(E[k])$ [14]. We will use the symbol $\tilde{E}[k]$ to denote this restriction.

When the Hilbert space $\mathscr{H}$ is associated with a quantum system, any state of the system is represented by a positive operator T of trace one, and an observable of the system is represented by a semispectral measure $E : \mathscr{B}(\mathbb{R}) \to L(\mathscr{H})$. The associated probability measure E_T then describes the measurement statistics of E in the state T, and the moments of the measurement statistics are given by $\mathrm{Tr}[TE[k]]$, provided T is such that this trace is finite.

In the special case where E is a spectral measure, all the moment operators $E[k]$ are selfadjoint and densely defined, with $D(E[k]) = \tilde{D}(x^k, E)$ and $E[k] = E[1]^k$ for each $k \in \mathbb{N}$. Indeed, E is then the spectral measure of the selfadjoint operator $E[1]$.

The following definition is needed.

Definition 1 Let $E : \mathscr{B}(\mathbb{R}) \to L(\mathscr{H})$ be a semispectral measure.

(a) Let $\mathscr{D} \subset \mathscr{H}$ be a subspace, such that $\mathscr{D} \subset D(E[k])$ for all $k \in \mathbb{N}$. Then E is *$\mathscr{D}$-determinate*, if the positive measure E_φ is determinate[1] for each $\varphi \in \mathscr{D}$.

(b) The semispectral measure E is *determinate*, if $E' = E$ whenever $E' : \mathscr{B}(\mathbb{R}) \to L(\mathscr{H})$ is a semispectral measure with $E'[k] = E[k]$ for all $k \in \mathbb{N}$.

It is easy to see that a $\mathscr{D}$-determinate semispectral measure is *determinate*, provided that $\mathscr{D}$ is dense (see [1]).

The concepts of the above definition are essential in the moment approach, since only a determinate observable can be completely described in terms of its moment operators.

[1] Recall that a posive measure $\mu : \mathscr{B}(\mathbb{R}) \to [0,\infty)$ having finite moments of all orders is *determinate*, if $\mu = \nu$ whenever $\nu : \mathscr{B}(\mathbb{R}) \to [0,\infty)$ is a positive measure with $\int x^k \, d\nu(x) = \int x^k \, d\mu(x)$ for all $k \in \mathbb{N}$.

BALANCED HOMODYNE DETECTION

We let $\mathscr{H}$ and $\mathscr{H}_{aux}$ be separable complex Hilbert spaces corresponding to the signal and auxiliary field modes, respectively. We use the following usual notations. Fix orthonormal bases of the form $\{|n\rangle \mid n \in \mathbb{N}\}$ for both $\mathscr{H}$ and $\mathscr{H}_{aux}$, representing the photon number states. We denote

$$|nm\rangle = |n,m\rangle := |n\rangle \otimes |m\rangle \in \mathscr{H} \otimes \mathscr{H}_{aux}, \quad n,m \in \mathbb{N}.$$

Let a, a^* and b, b^* be the creation and annihilation operators for the aforementioned bases of $\mathscr{H}$ and $\mathscr{H}_{aux}$, respectively, and let $N = a^*a$ and $N_{aux} = b^*b$, be the photon number operators for the two modes. The operators $a, a^*, b, b^*, N, N_{aux}$ are considered as being defined on their natural domains, e.g.

$$\begin{aligned} D(a) = D(a^*) &= \{\varphi \in \mathscr{H} \mid \sum_{n\in\mathbb{N}} n|\langle n|\varphi\rangle|^2 < \infty\}; \\ D(N) = D(a^*a) &= \{\varphi \in \mathscr{H} \mid \sum_{n\in\mathbb{N}} n^2|\langle n|\varphi\rangle|^2 < \infty\}. \end{aligned}$$

For any $z \in \mathbb{C}$ the *coherent state* $|z\rangle \in \mathscr{H}$ is defined by

$$|z\rangle = e^{-\frac{1}{2}|z|^2} \sum_{n=0}^{\infty} \frac{z^n}{\sqrt{n!}} |n\rangle,$$

and we use the same symbols for the coherent states in $\mathscr{H}_{aux}$. Also, we use the shorthand $|z,w\rangle = |z\rangle \otimes |w\rangle \in \mathscr{H} \otimes \mathscr{H}_{aux}$. The subspace $D_{coh} := \text{lin}\{|z\rangle \mid z \in \mathbb{C}\}$ is dense in $\mathscr{H}$, and so is the corresponding subspace D^{aux}_{coh} in $\mathscr{H}_{aux}$. (Here the symbol "lin" denotes the (algebraic) linear span of the set in question.) The (algebraic) tensor product $D^2_{coh} := D_{coh} \otimes D^{aux}_{coh}$ can be identified with $\text{lin}\{|\beta, z\rangle \mid \beta, z \in \mathbb{C}\}$, which is dense in $\mathscr{H} \otimes \mathscr{H}_{aux}$.

Denote by Q and P the signal field quadrature operators $\frac{1}{\sqrt{2}}\overline{(a^* + a)}$ and $\frac{i}{\sqrt{2}}\overline{(a^* - a)}$, respectively.

For the signal field, define the *rotated quadrature operators* Q_θ, with $\theta \in [0, 2\pi)$, via

$$Q_\theta = e^{i\theta N} Q e^{-i\theta N} = \frac{1}{\sqrt{2}}\overline{(e^{-i\theta} a + e^{i\theta} a^*)} \tag{1}$$

In particular, each Q_θ is selfadjoint on its domain $D(Q_\theta) = e^{i\theta N} D(Q) \supset D(a) = D(a^*)$, and the restriction $Q_\theta|_{D(a)} = \frac{1}{\sqrt{2}}(e^{-i\theta} a + e^{i\theta} a^*)$ is essentially selfadjoint. The ordinary quadratures are given by $Q_0 = Q$ and $Q_{\frac{\pi}{2}} = P$. We state the following lemma (Lemma 2 of [1].)

Lemma 1 *For all $\theta \in [0, 2\pi)$, the spectral measure P^{Q_θ} is D_{coh}-determinate, and thereby also determinate.*

The 50-50-beam splitter is described by the unitary operator $U \in L(\mathscr{H} \otimes \mathscr{H}_{aux})$, defined by its acting in the coordinate representation (see e.g. [6]):

$$L^2(\mathbb{R}^2) \ni \Psi \mapsto \big((x_1, x_2) \mapsto \Psi(\tfrac{1}{\sqrt{2}}(x_1 + x_2), \tfrac{1}{\sqrt{2}}(-x_1 + x_2))\big) \in L^2(\mathbb{R}^2), \tag{2}$$

We have assumed that the beam splitter does not generate any phase shift[2]. Under this transform, the coherent states change according to

$$|\alpha\rangle \otimes |\beta\rangle \mapsto |\tfrac{1}{\sqrt{2}}(\alpha - \beta)\rangle \otimes |\tfrac{1}{\sqrt{2}}(\alpha + \beta)\rangle. \tag{3}$$

This implies the equally well-known transformation rules for annihilation operators:

$$\begin{aligned} U^*(a \otimes I)|_{D^2_{coh}} U &= \frac{1}{\sqrt{2}}(a - b)|_{D^2_{coh}}; \\ U^*(I \otimes b)|_{D^2_{coh}} U &= \frac{1}{\sqrt{2}}(a + b)|_{D^2_{coh}}. \end{aligned} \tag{4}$$

[2] This is not a restriction, since any shift can be easily accomplished by changing the phase of the auxiliary coherent state.

Let T be the input state of the signal field, and let $|z\rangle$ be the input coherent state of the auxiliary field. The photon difference operator is $N_- := \overline{I \otimes N_{aux} - N \otimes I}$ (where I is the identity operator), so our detection observable is the spectral measure

$$\mathscr{B}(\mathbb{R}) \ni P^{(\sqrt{2}|z|)^{-1}N_-}(X) \in L(\mathscr{H} \otimes \mathscr{H}_{aux})$$

of the scaled operator $(\sqrt{2}|z|)^{-1}N_-$, the scale $|z|$ being the amplitude of the auxiliary oscillator. (The factor $\sqrt{2}$ appears for convenience.)

Since the state emerging from the beam splitter is $U(T \otimes |z\rangle\langle z|)U^*$, the detection statistics are given by the probability measures

$$X \mapsto \mathrm{Tr}[U(T \otimes |z\rangle\langle z|)U^* P^{(\sqrt{2}|z|)^{-1}N_-}(X)],$$

For fixed coherent state $|z\rangle$, these probability measures define an observable $E^z : \mathscr{B}(\mathbb{R}) \to L(\mathscr{H})$ on the signal field according to

$$\mathrm{Tr}[TE^z(X)] := \mathrm{Tr}[T \otimes |z\rangle\langle z| P^{|z|^{-1}A}(X)], \quad X \in \mathscr{B}(\mathbb{R}), \tag{5}$$

where

$$A := \frac{1}{\sqrt{2}}\overline{(a \otimes b^* + a^* \otimes b)} = \frac{1}{\sqrt{2}} U^* N_- U.$$

Let $V_z : \mathscr{H} \to \mathscr{H} \otimes \mathscr{H}_{aux}$ be the linear isometry $\varphi \mapsto \varphi \otimes |z\rangle$. Then we have simply

$$E^z(X) = V_z^* P^{|z|^{-1}A}(X) V_z, \quad X \in \mathscr{B}(\mathbb{R}). \tag{6}$$

Now the idea in the homodyne detection is that one takes the "high amplitude limit" $|z| \to \infty$ on the observable E^z, with a fixed phase θ, i.e. $z = |z|e^{i\theta}$. We do this in the level of the moment operators $E^z[k]$ of E^z, and then apply the general "asymptotic measurement" scheme presented in the preceding section.

The following result (see Proposition 9 of [1]) justifies the consideration of moment operators $E^z[k]$ in place of the observable E^z. It should be compared to Lemma 1.

Proposition 1 *For each $z \in \mathbb{C}$, the observable E^z is D_{coh}-determinate, and thereby also determinate.*

We were only able to determine the restrictions $\tilde{E}^z[k]$ of moment operators $E^z[k]$ to the square integrability domains. It turns out, however, that this is sufficient. The result is presented in the following proposition, which is a combination of Proposition 6 and 8 of [1].

Proposition 2 *Let $z = re^{i\theta} \in \mathbb{C}$, $r \geq 1$, and $k \in \mathbb{N}$. Then*

$$\tilde{E}^z[k] = |z|^{-1} V_z^* A^k V_z \supset Q_\theta^k|_{D(a^k)} + \frac{1}{r^2} C_k(r,\theta),$$

with $D(a^k) \subset D(\tilde{E}^z[k])$, where

$$C_k(r,\theta) = \sum_{\substack{n,m \in \mathbb{N}, \\ n+m \leq k}} c_{nm}^k(r,\theta)(a^*)^n a^m,$$

and each function $c_{nm}^k : [1,\infty) \times [0,2\pi) \to \mathbb{C}$ is bounded.

The first two moment operators assume the explicit forms $\tilde{E}^z[1] \supset Q_\theta|_{D(a)}$, $\tilde{E}^z[2] \supset Q_\theta^2|_{D(a^2)} + \frac{1}{2}r^{-2}N$. (See proposition 7 of [1].) This shows, in particular, that the *intrinsic noise operator* $N(E^z) := \tilde{E}^z[2] - \tilde{E}^z[1]^2$ coincides with the selfadjoint operator $\frac{1}{2}|z|^{-2}N$ (Corollary 1 of [1]).

Proposition 2 immediately implies that $\lim_{r\to\infty} \tilde{E}^z[k]\varphi = Q_\theta^k \varphi$, if $\varphi \in D(a^k)$. It turns out that this information, together with Lemma 1, is enough to provide a satisfactory conclusion to the high amplitude limit problem, namely the following result. The reader should consult [1] for the details.

Proposition 3 *Let (r_n) be any sequence of positive numbers tending to infinity, let $\theta \in [0,2\pi)$, and set $z_n = r_n e^{i\theta}$, $n \in \mathbb{N}$.*

(a) The spectral measure P^{Q_θ} is the only observable $E : \mathscr{B}(\mathbb{R}) \to L(\mathscr{H})$ with the property that $D_{coh} \subset D(\tilde{E}[k])$ for all $k \in \mathbb{N}$, and

$$\lim_{n\to\infty} \langle \varphi | E^{z_n}[k] \varphi \rangle = \langle \varphi | E[k] \varphi \rangle, \quad \varphi \in D_{coh}.$$

(b) The sequence $(E^{z_n})_{n\in\mathbb{N}}$ *converges to* P^{Q_θ} weakly in the sense of probabilities, *that is,*

$$\lim_{n\to\infty} E^{z_n}(X) = P^{Q_\theta}(X)$$

in the weak operator topology, whenever $X \in \mathscr{B}(\mathbb{R})$ *is such that* $P^{Q_\theta}(\partial X) = O$, *where* ∂X *denotes the boundary*[3] *of the set* X.

Remark 1 (a) Since the spectral measure P^{Q_θ} has the same null sets as the Lebesgue measure, result (b) implies that

$$\lim_{n\to\infty} E^{z_n}(X) = P^{Q_\theta}(X)$$

in the weak operator topology, whenever $X \in \mathscr{B}(\mathbb{R})$ is such that ∂X has zero Lebesgue measure. One should note that in view of concrete measurements, the restriction concerning ∂X is of no consequence, as all intervals $X = (x_1, x_2)$, for example, satisfy it. In a mathematical point of view, however, it is worth mentioning that the above limit relation does indeed *not* hold for all Borel sets X. (See Remark 8 of [1].)

(b) The physical relevance of the above proposition should perhaps be stated explicitly: First of all, part (a) implies that the limits of the moments of the actual measurement statistics associated with the observables E^{z_n} determine uniquely the limiting observable (the quadrature). Secondly, part (b) implies that the actual measurement outcome probabilities $E_T^{z_n}(X)$ converge to the probabilities $P_T^{Q_\theta}(X)$ corresponding to the ideal measurement of the quadrature.

REFERENCES

1. J. Kiukas, P. Lahti, On the moment limit of quantum observables, with an application to the balanced
2. H. P. Yuen, J. H. Shapiro, Optical Communication with two-photon coherent states - part III: quantum measurements realizable with photoemissive detectors, *IEEE Trans. Inf. Theory* **26** 78-92 (1980).
3. H. P. Yuen, V. W. S. Chan, Noise in homodyne and heterodyne detection, *Opt. Lett.* **8** 177-179 (1983).
4. M. G. A. Paris, J. Řeháček (Eds.), *Quantum State Estimation*, Lect. Notes Phys. **649**, Springer-Verlag, Berlin, 2004.
5. S. Prasad, M. O. Scully, W. Martienssen, A quantum description of the beam splitter, *Opt. Commun.* **62** 139-145 (1987).
6. U. Leonhardt, H. Paul, Phase measurement and Q function, *Phys. Rev. A* **47** R2460-R2463 (1993).
7. K. Banaszek, K. Wódkiewicz, Operational theory of homodyne detection, *Phys. Rev. A* **55** 3117-3123 (1997).
8. S. L. Braunstein, Homodyne statistics, *Phys. Rev. A.* **42** 474-481 (1990).
9. G. M. D'Ariano, $SU(1,1)$ symmetry of the balanced homodyne and statistics of the output, *Nuovo Cimento* **107B** 643-651 (1992).
10. E. Vogel, *Operationale Untersuchung von quantenoptischen Meßprozessen* (doctoral dissertation produced at the University of Cologne), Shaker Verlag, Aachen, 1996.
11. M. A. Marchiolli, S. S. Mitzrahi, V. V. Dodonov, Operational approach for reconstruction of quantum distributions in a preamplified homodyne-detection scheme, *Phys. Rev. A* **56** 4278-4286 (1997).
12. M. A. Marchiolli, S. S. Mizrahi, V. V. Dodonov, Signal-to-noise ratio of preamplified homodyne detection in quantum tomography, *Phys. Rev. A* **57** 3885-3897 (1997).
13. B.-G. Englert, K. Wódkiewicz, Intrinsic and operational observables in quantum mechanics, *Phys. Rev. A* **51**, R2661 (1995).
14. P. Lahti, M. Maczyński, K. Ylinen, The moment operators of phase space observables and their number margins, *Rep. Math. Phys.* **41** (1998) 319-331.

[3] The *boundary* of a set $X \subset \mathbb{R}$ is the intersection of the closures of X and its complement.

Probability and Quantum Paradigms: the Interplay

A. F. Kracklauer

Bauhaus Universität, PF 2040; 99401 Weimar, Germany

Abstract. Since the introduction of Born's interpretation of quantum wave functions as yielding the probability density of presence, Quantum Theory and Probability have lived in a troubled symbiosis. Problems arise with this interpretation because quantum probabilities exhibit features alien to usual probabilities, namely non Boolean structure and non positive-definite phase space probability densities. This has inspired research into both elaborate formulations of Probability Theory and alternate interpretations for wave functions. Herein the latter tactic is taken and a suggested variant interpretation of wave functions based on photo detection physics proposed, and some empirical consequences are considered. Although incomplete in a few details, this variant is appealing in its reliance on well tested concepts and technology.

Keywords: quantum probability, Born interpretation, wave functions, pilot waves, de Broglie waves, photo detection
PACS: 03.65Ta, 03.65Ud, 03.65Mn, 03.75.-b

QUANTUM PROBABILITY: PEDIGREE AND PATHOLOGIES

Following Born's interpretation of the square of the modulus of quantum wave functions as a probability, the two great logical structures of Quantum Mechanics and Probability Theory have been in a troubled symbiosis. The source of dissidence is that the probabilities arising from this interpretation, although seeming quite reasonably behaved for the most part, still suffer features unknown in ordinary probability theory. Such features include non Boolean algebraic properties (i.e., the underlying wave functions for independent probabilities add, not the resulting probabilities themselves) and some are frequently not positive-definite.

In reaction to this situation, there seems to be two polar reactions as efforts to understand these features. One is, so to speak, to fix the quantum or physics interpretation and vary the probability theory. The other is the complementary tact, that is, to fix the employed probability theory and modify Born's physical interpretation. Over the decades since the discovery of Quantum Theory, many researchers have taken one or the tactic, or even a combination, in efforts to demystify the interpretation of Quantum Mechanics.

Perhaps the most systematic and far reaching study in the spirit of varying or modifying probability as applied to Quantum Mechanics was made by Khrennikov as published in his book: "The Interpretations of Probability."[1] In that work he argues that physicists implicitly, sometimes even quite explicitly, have based their considerations on the basis of Kolmogorov's formulation of probability in terms of positive measures, which has led to even greater abstraction in reasoning than that already customary in treatises on quantum foundations issues. In addition, Khrennikov has considered p-adic analysis to analyze rigorously the complications inherent in using what ideally should be infinite sets as probability spaces for situations where all applications in fact are limited to finite sets. Such efforts must be applauded; rigorous mathematics can only purge obscured errors in reasoning divining Nature's mysteries.

This writer, on the other hand, has taken, for reasons of convenience, not wisdom, the complementary tact of accepting elementary probability theory and sought to vary the Born interpretation, so as to seek resolutions of the many infamous conundrums.[2] This has been quite fruitful seemingly; there appear to be plausible alternate paradigms for the phenomena covered by quantum theory for which direct and simple probability theory is adequate. The main issue then becomes: is there empirical evidence supporting these new (occasionally old actually) alternate paradigms?

The task shall be broken in two. On the one hand, entities which classically are considered particles shall be considered as one phenotype, then on the other, those which are classically thought of as waves will be analyzed separately. The goal then becomes showing, that the particle-type evidences wave-like behavior probabilistically, and the waves likewise evidence particle-like behavior probabilistically.

CP962, *Quantum Theory, Reconsideration of Foundations – 4*,
edited by G. Adenier, A. Yu. Khrennikov, P. Lahti, V. I. Man'ko, and Th. M. Nieuwenhuizen

THE PARTICULATE CHARACTER OF WAVE FUNCTIONS

Without going into excessive details, let us recall simply that by-and-large the issue of a particulate nature of something described classically by a wave pertains to the electromagnetic interaction, i.e., to photons. The central empirical fact that underlies and supports the 'photon' paradigm is that electromagnetic radiation at very low intensity seems to be absorbed at specific point-like locations. This fact is interpreted to mean that the energy in radiation itself is contained in similarly small packets and that each detection event results from the absorption of one such packet, i.e., one 'photon.'

While this imagery is plausible, it is not "necessary" in the mathematical sense; alternate notions are equally plausible.[3] For instance, let us take explicit account of the fact that electromagnetic detection within the spectral range of primary concern is accomplished using "photo-detectors." Such detectors are devices that convert incoming electromagnetic radiation into a "photo-current" by lifting electrons from a detector's valence band to its conduction band, which, in turn, are converted in electronic circuitry into countable finite current pulses easily manipulated and, in particular, counted. Now, a photo-current is nothing else but a stream of electrons in wires. Such a current can be reduced in intensity only until there is only one remaining electron—a continuously diminishing or infinitesimal current is not possible. This reality can be the fundamental reason for the acceptance enjoyed by the photon paradigm. In other words, it is entirely possible that a fully continuous radiation wave has elicited a discrete photo current without in any way implying that the stimulating radiation itself was discrete or packaged as 'photons.'

Further, another reality is that everywhere and always there is noise. It cannot be excluded, in particular in that it might be considered as the contribution of the outside world, which cannot be totally screened off. This in turn, with respect to electromagnetic detection in photo detections, is taken into account in the basic theory of photo detection with the prescription that the intensity of an elicited photo current is in probabilistic proportion to the total electromagnetic wave intensity, i.e., the square of the electric wave amplitude, implicitly, thereby, including noise. This effect is absolutely 'local.' That is, the relevant intensity is just that exactly at the point in space and instant in time where and when the single photo electron is elevated into the conduction band of the detector after which it then flows into the counting circuitry.

Since the wave function of a photon is associated directly with electromagnetic fields, this means that the non Boolean aspect of the presumed logic of quantum mechanics is simply a reflection of the fact that photo *detection* is proportional to the local field intensity. The interference terms resulting from squaring wave functions do so as an electromagnetic phenomena having no implications for probability theory, which is brought into the story by photo detection theory, but, only *at the point* of photoelectron generation *after* electromagnetic interference has been taken into account.

Below this logic shall be extended to wave functions for particles.

POSSIBLE EMPIRICAL CONSEQUENCES

Accepting the paradigm of electromagnetic interaction as comprised of essentially waves propagated as envisioned in non quantum electrodynamics, which then appear as if comprised of parceled 'photons' as an artifact of photo detection, implies that all the properties of classical electromagnetic radiation should remain evident. In particular, this includes all phenomena resulting from the transverse nature of electromagnetic radiation, that is, mode polarization.

First, it should be noticed that the structure of polarization effects is governed by the group $SU(2)$. This group is isomorphic to $SO(3)$, the group encoding the structure of rotation on a sphere. What this implies is, insofar as the non commutivity of $SO(3)$ is geometric, and not a consequence of any fundamental hypothetical input for Quantum Mechanics (i.e.: that *Hamiltonian conjugate variables* are non commutative according to Heisenberg's uncertainty relationship $\hat{X}\hat{P} - \hat{P}\hat{X} = i\hbar$), that the non commutative structure of $SU(2)$ likewise is attributable only to geometry. Therefore, all resultant effects cannot be attributed to quantum structure.[4]

Additionally, by rejecting the photon paradigm, there are direct consequences for what should be expected in terms of correlated photons. Specifically, a parametric down conversion (PDC) crystal is thought to produce anticorrelated pairs of photons. But, if the photon paradigm is inadequate in fact, then the signals produced in a PDC crystal will have to be modeled by another paradigm, and the most likely would seem to be the classical conception of electromagnetic waves. Each "photon" equivalent signal would be a short pulse and the correlation could be expressed as two pulse polarization combinations, one with horizontal-left and vertical-right, the other with the opposite combination.

Such a combination of correlated signals admits at least two types of examinations: one with respect to the timing of the generation of the final photoelectrons, and another with respect to the correlations of the polarization directions.

If now the photon paradigm is faithful to the reality of the generation of the signals, then the individual photons constituting the pair must be generated together in the source crystal and within a time interval restricted by the Heisenberg uncertainty relationship. The random event would be the generation of the pair at the source, and might be best characterized as occurring at an instant within the interval specified by the uncertainty relationship.

In contrast, if the more correct paradigm is that described above in which it is envisioned that essentially classical pulses lift electrons from the valence to the conduction band in the detector, then the randomness is vested in two events, one for each electron. In this case the specific instant of photoelectron generation is determined both by the pulse signal from the source crystal and by a noise contribution at the detector. Insofar as neither the pulse length or the noise is constrained by Heisenberg uncertainty, the timing discrepancy between the instants at which the electrons are elevated can exceed the Heisenberg limit. To this writer's knowledge, a laboratory study of this issue has not been done.

A similar study of the nature of the polarization states of the signals generated in a PDC crystal can be made to address empirically another feature of quantum states.[5] It is this: for reasons of convenience in spectroscopy it is taken that the 'state' of the signal as specified by quantum mechanics is in the form of singlet state which is a linear combination of the two possible outcome states. This combination is rotationally invariant, as can be verified by calculation, and it is considered from this fact that this item as a unit is such that in fact it has no polarization orientation at all, until one or the other of the two signals comprising the pair encounters a measuring device. By authority then of what is called the "projection hypothesis," it is taken that the process of measurement then "projects" this neutral state onto one or the other of the possible constituents. Now, if this imagery is taken seriously, then rotational invariance means that it should make no difference at all which direction the polarizer has that the first of the pair encounters. If this polarizer 'projects' or collapses the wave function of the photon, it would be natural to presume that the polarization direction taken by the photon will be that of the polarizer; in which case, the polarization direction of the other photon in the pair must be deterministically orthogonal. In this way the correlation function for the pair will be 'rotationally invariant,' in conformity with the presumed character of the input signal.

On the other hand, if the classical electromagnetic pulse with photoelectron elicitation paradigm reigns, then the signals emitted by the PDC crystal are one or the other of the possible base signals, not some mystical superposition of mutually exclusive options. In this case, measuring the polarization of this output of the crystal on the axis of the crystal will show deterministic anticorrelation, whereas measuring it off axis will yield only statistical anticorrelation. That is, there will be individual pair events that are not anticorrelated; and, the final result is that the correlation function will not be rotationally invariant. Elsewhere this writer has published a calculation showing that this effect can be used to explain the fact that the correlation signal from PDC crystals in fact exhibits a variation in visibility.[6]

Ideally the visibility will go to zero at an azimuth of $\pi/4$ radians off the crystal axis. Actually, the only reported variation known to this writer is just a few percent.[7] While the mere presence of any variation is inexplicable in quantum terms, this low value fails to preclude all doubt that the photon paradigm is invalid. To fully reconcile this matter, however, would require detailed analysis of exactly which events are counted in the data stream analyzed for the reported figure. It could be that the coincidence circuit is so tuned that the offending events are rejected illegitimately, and so largely not included in the data analysis.

NEGATIVE PROBABILITY DENSITIES

Wigner famously long ago discovered a mapping from wave functions to phase space expressions strongly resembling density functions.[8] He did not claim, however, that these expressions could actually be considered such phase space density functions, because, for some solutions of the Schrödinger Equation, his mapping yields non positive-definite expressions. Such cannot ordinarily be interpreted as a probability. But, two possibilities open up. One, a non elementary formulation of probability in which such negative expressions arise can be introduced; or, two, a restriction can be put on wave functions to the effect, that only those wave functions—as solutions to Schrödinger's Equation—are ontologically meaningful that do yield positive-definite phase space Wigner functions.

There are several arguments supporting this latter stipulation. One of the most persuasive is, that even in non quantum statistical mechanics, the eigen states of the dynamic equation of a Markoff process do not yield positive Wigner densities.[9] Thus, the same circumstance occurring in Quantum Mechanics cannot be ascribed to peculiarities of that theory. In both theories, the remedy is the same; namely accepting as ontological solutions only those that satisfy physically meaningful boundary or initial conditions. Fortunately, it has been found that there do exist solution of Schrödinger's Equation yielding acceptable ontological solutions, namely thermal states, i.e., superpositions of eigenstates weighted with Boltzmann factors. Additionally, for the harmonic oscillator, the coherent states also meet this stipulation, in addition to being the only states that also oscillate back and forth with time as does a real oscillator. (Harmonic oscillator eigenstates oscillate up and down as a unit and otherwise do not do what a physical oscillator does.)

It is not known to this writer if similar variants of coherent (like) states exist for the Kepler problem; i.e., states that rotate around as do electrons around the nucleus. Evidently this is still a matter for research. Without such states, the contention here, that only a very limited selection of states can constitute ontological states, remains an unproven proposal, rather than a solution to this issue. Nevertheless, the historical record so far provides good reason to be optimistic.

SUMMARY: OCCAM'S PREFERENCE

Herein a paradigm for the application of probability theory to quantum wave functions has been proposed. It accounts for the non Boolean aspect of quantum probabilities by proposing that probability enters locally at a specific point, so that the fact that quantum probability functions do not add, rather that their underlying wave functions do, is a manifestation of the fact that an underlying wave function is analogous to an electric field for which the square is proportional to the probability of eliciting a photoelectron. The non Boolean character then is a consequence of the fact that electric fields add linearly but not their square or intensities. That photoelectric currents are proportional to the square of electric fields is the basic feature of the standard theory of photo-detection. In sum, thus far this paradigm is straightforward.

The extention to wave functions for particles not normally detected with photodetectors depends on the writer's proposal for understanding "de Broglie waves," which in this conception are taken to be electromagnetic pilot waves derived from interaction with a stochastic background identical to what is otherwise known as the quantized ground state of the free electromagnetic field.[10] With this model of de Broglie waves, the explanation for the point detection of point particles is a direct consequence of the notion that particles are indeed localized entities being guided by pilot waves, and that, at detection, the particle is seen by a particle detector, but not the pilot wave. The probability of entrapment at a particular position is given by the local intensity of the wave function at that position, which explains why it is that non Boolean aspect of wave functions poses no restrictions for the application of simple probability theory. The spacial distribution of these pilot waves can be revealed only by repeating the experiment often to ultimately expose the intensity pattern of the pilot wave in which the particle is, as a statistical matter, most strongly trapped to be located where the wave is most intense, etc.

Further, we propose that the non positive-definite character of Wigner functions can be remedied by limiting the set of states to be considered as ontological to those which in fact do yield non negative Wigner functions. All the other states are then reduced to the status of mathematical intermediary results without uninterpreted physical meaning. Sophisticated probability theory may not be necessary.

REFERENCES

1. Khrennikov, A. "Interpretations of Probability," (VSP, Utrecht, 1999).
2. Kracklauer, A. F., "Second Entanglement and (re)Born Wave Functions in Stochastic Electrodynamics," in (AIP Conf. Proc. **810**, Melville, 2006) pp. 343-352.[1]
3. Kracklauer, A. F., "Oh Photon, Photon, whither art thou gone?," , *Proc. of SPIE* **5866** 112-118 (2005).
4. Kracklauer, A. F., "EPR-B correlations: quantum mechanics, or just geometry?" *J. Opt. B: Quantum Semiclass. Opt.* **6** S544-S584 (2004).
5. Kracklauer, A. F., "Nonlocality, Bell's Ansatz and Probability," *Optics and Spectroscopy*, **103** (3) 451-460 (2007).
6. Kracklauer, A. F., "What do correlations tell us about photons?" , *Proc. of SPIE* **6664** OH1-5.
7. Weihs, G., "A test of Bell's Inequality with Spacelike Separation," *Foundations of Probability and Physics-4, AIP Conf. Proceedings* **889** 250-262 (2007).
8. Wigner, E. , "On the Quantum Correction for Thermodynamic Equilibrium," *Phys. Rev.* **40** 749-759 (1932).
9. Moyal, J. E., "Quantum Mechanics as a Statistical Theory," *Proc. Camb. Phil. Soc.* **45** 99-124 (1949).
10. Kracklauer, A. F., "Pilot wave steerage, a mechanism and test," *Found. Phys. Lett.* **12** (2) 441-453 (1999).

[1] Preprints of the author's papers can be downloaded at: www.nonloco-physics.000freehosting.com

EPR Paradox , Locality and Completeness of Quantum Theory

M.Kupczynski

Department of Mathematics and Statistics , Ottawa University,585 King Edward,Ottawa ON K1N 6N5 ,Canada
E-mail: mkupczyn@uottawa

Abstract. The quantum theory (QT) and new stochastic approaches have no deterministic prediction for a single measurement or for a single time -series of events observed for a trapped ion , electron or any other individual physical system. The predictions of QT being of probabilistic character apply to the statistical distribution of the results obtained in various experiments. The Copenhagen interpretation (CI) of QT acknowledged the abstract and statistical character of the predictions of QT but at the same time claimed that a state vector Ψ provided complete description of each individual physical system. The assigning the state vector to an individual physical system together with a postulate of its instantaneous reduction in the measurements was shown by Einstein Podolski and Rosen to lead to so called EPR paradox for the experiments with the entangled pairs of the particles. EPR concluded that a state vector could not provide a complete description of the individual systems and the question arose whether the probabilistic predictions of QT could be derived from some more fundamental spatio-temporal deterministic description of the invisible sub-phenomena by introduction of supplementary parameters. The experimental violation of the Bell inequalities in the spin polarization correlation experiments (SPCE) which were the implementations of Bohm and EPR gedanken experiments eliminated so called local and realistic models of the sub-phenomena.Quite often the violation of these inequalities has been incorrectly interpreted as a proof of the completeness of QT or as the violation of the locality and causality in the micro-world. The local and realistic models overlooked the fact that an experimental outcome is only the information about a particular system-system or system -instrument interaction. Namely the quantum phenomena are described in terms of the probabilities. It is well known that the probability distribution is not an attribute of a dice but it is a characteristic of a whole random experiment : " rolling a dice". In this paper we will underline: that the quantum probabilities do not describe the lack of knowledge about the physical systems but the lack of knowledge about the experiments and therefore they are "contextual". and that the statistical long range correlations between two random variables X and Y are not a proof of any causal relation between these variables. Moreover any probabilistic model used to described a random experiment is consistent only with a specific protocol telling how the random experiment has to be performed. The probabilistic model used to prove Bell inequalities implied a protocol completely inappropriate and impossible to implement for SPCE . We conclude that the meaningful question about the predictable completeness of QT is still open and it should be answered by a careful and unconventional analysis of the experimental data. The correct understanding of statistical and contextual character of QT is essential for the research in the domain of quantum information and quantum computing.

Keywords: Bell inequalities, entanglement , EPR paradox, quantum measurement, foundations of quantum theory, contextual observables, quantum information, quantum computer
PACS: 03.65. Ta, 03.67.Lx , 03.67. Dd, 03.67.Hk

INTRODUCTION

It has been shown many years ago [1-6] that the main prerequisites used to prove the Bell inequalities (BI) [7,8], or Clauser,Horn ,Shimony,Holt (CHSH) inequalities [9] such as the use of a common probability space, joint probabilility distributions for non-commuting observables etc. were inappropriate for the description of the spin polarization correlation experiments (SPCE) and the violation of BI in these experiments [10,11] and gave neither information about the non locality of QT nor about its completeness.

In spite of this, testing of BI-CHSH continued and several loopholes were indicated which could explain the apparent,but not existing violation of BI, by the imperfection of the experimental set ups [12]. Precise new experiments[13,14] permitted to close several loopholes, confirmed QT predictions for the correlation functions and even allowed to detect strange anomalies in various detection rates reported recently by Adenier and Khrennikov [15]. Instead of trying to search for new loopholes it would be more productive, in our opinion, to use the existing data in order to search for anomalies similar to those found in [15] or to study the fine structure of the data by using various purity tests proposed by us many years ago [4,16,17] . Only the discovery of some fine structure ,not accounted for by QT , would provide a decisive proof that the statistical description provided by QT of these data is incomplete ending

CP962, *Quantum Theory, Reconsideration of Foundations – 4*,
edited by G. Adenier, A. Yu. Khrennikov, P. Lahti, V. I. Man'ko, and Th. M. Nieuwenhuizen

the debate started by EPR [18] over 75 years ago.

The limitations and inapplicability of BI , CHSH and GHZ inequalities [19] to SPCE have been demonstrated now in numerous publications e.g. [20-27,46-50]. Several local models [2,3,28,29,41] were able to reproduce the QT predictions. The BI are also violated in macroscopic experiments discussed by Aerts[30,31] and in computer experiments of Accardi and Regoli [21]. The strong arguments in favor of statistical and contextual character of spin projections were given by Allahverdyan, Balian and Newenhuizen[32]. Instead of rejoicing that there was no contradiction between the QT and locality [25] many members of physics community continue to marvel at the picture of two perfect random dices giving completely correlated outcomes etc.

One may wonder why it is so ? One reason could be that they do not understand the implications of the use of a common probability space and joint probability distributions for the protocol of a random experiment . Perhaps the authors criticizing BI were using too technical or/and too condensed language e.g.:" The main hypothesis needed to prove Bell-type inequalities is the assumption that the probabilities estimated in various SPCE can be calculated from one sample space (probability space) by conditionalization." [5] .

The other reason could be that after Harry Potter and other science fiction stories everything seems to be possible and magic explanations for the correlations are much more attractive than those based on a common cause or a common sense.

Similar reasons may explain perhaps the fact known for years and explained recently in detail by Ballentine[34] that :"Once acquired, the habit of considering an individual particle to have its own wave function is hard to break.... though it has been demonstrated strictly incorrect,".

Fortunately we are at the fourth Vaxjö conference and in my opinion slowly a consensus is starting to build up around statistical contextual interpretation of QT(SCI)[20,27,33-35}].According to this interpretation the information gathered in the measurements for all different experimental contexts provides the only reliable contextual information about a "state" of the identically prepared physical systems whose properties are measured..

It is not an easy process because during the conference several discussions showed that people are attributing different meaning to the words such as : probability, contextual, observables, measurement, photon and local realism and they have their own mental images of invisible sub-phenomena underlying every physical observable phenomenon or experiment.

In this paper we will give additional arguments in favour of SCI. The paper is organized as follows. In the section 2 we define the meaning which we attach to the terms and notions such as: phenomenona, sub-phenomena, physical reality, filters, contextual , observables , probabilities etc. In the section 3 we recall shortly the EPR paradox and the explanation given by SCI. In the section 4 we explain why according to SCI strict anti-correlations of individual outcomes are not implied by the singlet state. In the section 5 we criticize in historical perspective various proofs of BI-CHCH including the most recent one given by Larsson and Gill [36] and we explain why the prerequisites used in these proofs are not valid in SPCE. In the section 6 we discuss how the predictable completeness of QT could be tested.

IMPORTANT NOTIONS

The main assumption in physics is that there exists a material world (physical reality) in which we can observe various phenomena and with which we may interact by performing various repeatable experiments. Another assumption is the existence of physical laws which are responsible for the richness of phenomena and for the regularities observed in our experiments. We have no doubt that the Moon exists when we do not look at it but of course when we look at it we can perceive it in different colors, in different phases etc.We do not have intuitive images of electrons and photons but we have the abstract mathematical model given by QT able to describe quantitatively various phenomena they are participating in. The importance of the interplay of ontic and epistemic realities in QT was recently discussed in some detail by Atmaspacher and Primas [37] and Emch [38].

The mathematical models are always of limited validity and apply to particular phenomena. To describe the motion of the Moon around the Earth we use a model of a material point following well defined trajectory obtained by solving the Newton's differential equations of motion. To explain the phenomenon of its phases the Moon is modeled as a sphere . To describe the details of the formation of a crater on the Moon, when a meteorite hits it, we need much more detailed and complicated mathematical model. This is why we will probably discover one day that at very short distances the extendedness of the hadrons plays more important role[39] than in the standard model we are using today.

Therefore the statement that a given theory for example QT provides the most complete description of the individual physical systems lacks humility since as Bohr said [40]: "The main point to realize is that the knowledge presents itself within a conceptual framework adapted to account for previous experience and that any such frame may prove too narrow to comprehend new experiences."

Phenomena and Sub-Phenomena

Phenomena produce the observable effects. The sub-phenomena are invisible. For example we can see a track left by a single charged high energy elementary particle on the picture from the bubble chamber. To describe its trajectory we can use a model of a point-like particle moving according to the laws of the classical physics but of course there is an underlying microscopic invisible sub- phenomenon leading to the track production and if we wanted to explain it in detail we should go beyond the classical electrodynamics. If this high energy elementary particle collides with the proton from a hydrogen atom in a bubble chamber then we see many out- going tracks from the collision point what is a new phenomenon : a creation of several new particles during the collision. To describe this phenomenon we have to go beyond the quantum electrodynamics. To study this phenomenon in detail we have to prepare in an accelerator a collimated beam of "identical elementary particles" and to observe several collision events in the bubble chamber. The only reproducible regularities, besides the conservation laws ,we assumed to be valid, are of statistical nature and can be predicted by some abstract mathematical model providing the probabilities for different possible outcomes of the collision. The intuitive description of invisible sub-phenomena taking place during the collision is not provided.

The phenomena described by QT are all of this nature. Any experimental set-up can be divided into three parts: a source preparing the ensemble of "identical" physical systems , an interaction/filtering part and the detectors/counters part which produces time- series of observable events : clicks on various detectors, dots on a screen, tracks in a bubble chamber etc. For the extensive discussion of the general experimental set-ups see [42]. The QT gives the probabilistic predictions for the distribution of the outcomes obtained in various phenomena without providing intuitive models of the invisible sub-phenomena. One encounters the paradoxes only if the incorrect models are used to describe these sub-phenomena . For example a light source it is not a gun and photons are not metal bullets etc.

Attributes and Contextual Properties

An attributive property is a constant property of an object which can be observed and measured at any time and which is not modified by the measurement e.g.; inertial mass, electric charge etc. A contextual property is a property revealed only in specific experiments and characterizes the interaction of the object with the measuring apparatus. Let us mention Accardi's example of a chameleon which is green on a leaf and brown on a bark of the tree [22,25].

In classical physics we assume that the measurements of various attributive properties possessed by an individual physical system are compatible what means that they can be measured simultaneously or in any sequence giving always the same results. In quantum physics the contextual properties are known after the measurement such as a click on a detector placed behind a polarization filter. The measurements of the incompatible properties cannot be performed simultaneously and the measurement of one of them destroys the information about the other one.Various sequences of these measurements lead to various probability distributions of the outcomes [42]. The measurements of attributive properties are called by Accardi [25] passive dynamical systems and those of contextual properties active dynamical systems.

In QT contextual properties of individual systems are of statistical character because they may be only deduced from the properties of the pure ensembles they are members of [41] "...a value of a physical observable ,here a spin projection, associated with a pure quantum ensemble and in this way with an individual physical system , being its member, is not an attribute of the system revealed be a measuring apparatus; it turns out to be a characteristic of this ensemble created by its interaction with the measuring device. In other words the QM is a contextual theory in which the values of the observables assigned to a physical system have only meaning in a context of a particular physical experiment."

Another argument in favor of SCI of QT comes from probability theory. The probabilities are only the "properties" of random experiments [5]: " talking about the probabilities we should always indicate the random experiment needed to estimate their values". QT provides the algorithms to find various probabilities therefore it is a contextual theory. The contextuality in this sense is fully objective property of Nature. Even if nobody observes the collisions of high energy protons they are described by the same probability distributions of the possible outcomes no matter where it happens.

The probabilities found in QT do not describe one particular random experiment but a whole class of equivalent random experiments which are assumed to be repeatable as many times as needed .

Probabilities, Correlations and Causality

There are a lot of subtle points related to the problems how to define the probability what is the randomness etc. These important topics were discussed recently in detail by Khrennikov in his stimulating book [20].

We illustrate here these problems by two simple examples, the more detailed discussion may be found in [20,22,42]

1) Let us consider a random experiment which can give only two outcomes: 1 or -1. . We repeat this experiment 2n times and we obtain a time series of the results:1,-1,1,-1,...,1,-1. By increasing the value of n the relative frequency of getting 1 can approach 1/2 as close as possible suggesting that if the repetitions of this experiment are independent then the probability of getting 1 in each experiment is equal to 1/2. Of course it is incorrect because if we analyze the time series of outcomes in detail we see that we have a succession of the couples of two deterministic experiments or one deterministic experiment keeping memory of the previous result which is called a periodic two dimensional Markov Chain.

2) Let us toss now a fair coin assigning 1 for "head" and -1 for " tail". We get again a time series: 1,1,-1,-1,-1,1,-1..with relative frequencies which tend to 1/2. There is no apparent structure in this series so the hypothesis of the independent and identical repetitions of the same experiment seems to be satisfied and we say that a complete description of this experiment is provided by a single number :a probability of getting 1 which is equal to 1/2. Of course we believe that if we knew all parameters describing the tossing experiment we could predict each individual result using the laws of the classical mechanics. The randomness is only due to the lack of control of these parameters.

The statisticians and probabilists invented many tests in order to test the randomness and independence in such time series but conclusions from any statistical study is valid only on a given level of confidence. Moreover the infinite series formed by the consecutive decimals of the number π passed all the tests of the randomness in spite of the fact that any decimal is strictly determined. Without any completely conclusive measure of randomness we have to limit ourselves for any practical applications to the generators of the pseudo-random numbers which passed the known tests of the randomness.

One of the postulates of Copenhagen interpretation was that in a measurement process the measured value of a physical observable is chosen among all possible values of this observable with a given probability and in a completely random way. It was also believed that this indeterministic behavior of quantum ensembles could not be explained by the lack of control of some hidden variables describing the deterministic interaction of the individual members with the measuring devices.This intrinsically indeterministic behavior of individual quantum "particles" was believed to provide a new standard of the randomness which could serve to produce the unbreakable keys in quantum cryptography.

The fact that the strong correlations created by the source in SPCE survive the filtration and measurement processes is a strong argument against purely random behavior of individual systems during these processes.The individual systems have to carry a memory of their preparation coded in some parameters and the measuring device described by its own uncontrollable microscopic parameters has to react in some deterministic way to produce the observable outcome without destroying completely the memory how the systems were prepared. A strong correlation between two random variables has nothing to do with a causal relation between these variables. For example the average price of oil in a given year is correlated with the average salary of Anglican priests in the same year due to the common cause which is inflation. The existence of strong correlations between non-interacting physical systems which interacted in the past was analyzed for the first time by EPR[18] and led them to conclude that the description of these phenomena provided by QT is incomplete. In the next section we discuss shortly their paper

EPR-PARADOX

EPR consider two systems I and II which are permitted to interact from t=0 to t=T and which evolve freely and independently afterwards. The state of I or II for $t \geq T$ can be found only by the reduction of wave packet. Let a_1, a_2, a_3.. be the eigenvalues of some physical observable A to be measured on the system I and u_1,u_2,u_3.. a complete set of corresponding orthogonal eigenfunctions . At the moment of measurement $T_1 \geq T$ of the observable A on the system I the wave function $\Psi(x_1,x_2)$ of the system I+II is given by

$$\Psi(x_1,x_2) = \sum_n \psi_n(x_2) u_n(x_1) \tag{1}$$

If the measurement of A gives a_k then the wave function is reduced to $c_k \psi_k(x_2)\ u_k(x_1)$ where $\psi_k(x_2)$ is, up to normalization constant, the wave function of the system II immediately after the measurement of A on the system I has been completed and the result a_k known.

If instead of A we decided at $t=T_1$ to measure on I another non-commuting observable B with the eigenvalues b_1,b_2,b_3 and a complete set of orthogonal eigenfunctions v_1,v_2,v_3... then instead of the formula (1) we would have

$$\Psi(x_1,x_2) = \sum_s \varphi_s(x_2) v_s(x_1) \tag{2}$$

If b_r was obtained then the wave of the system II immediately after the measurement of B on the system I would have been , up to normalization constant, $\varphi_s(x_2)$.

In their paper EPR conclude ." Thus it is possible to assign two different wave functions (in our example $\psi_k(x_2)$ *and* $\varphi_s(x_2)$) to the same reality (the second system after the interaction with the first" In each case the functions are assigned with certainty and without disturbing the system II. If one assumes that the wave functions are in one to one correspondence with the states of individual physical systems one obtains a contradiction called EPR Paradox.

Of course according to SCI there is no paradox because the wave functions $\psi_k(x_2)$ *and* $\varphi_s(x_2)$ describe only different sub-ensembles of the ensemble of the particles II. The eigenvalue expansions (1) and (2) being mathematical identity describe different incompatible experiments. They imply the existence of the long-range correlations between the measurements performed on the non-interacting separated physical systems . The state $\Psi(x_1,x_2)$ is not factorized and is an example of the entangled state so popular nowadays.The discussion following the EPR paper was recently reviewed in detail in [27]. Some arguments of EPR were rejected but nobody was able to prove that QT provides the complete description of individual physical systems.

SINGLET STATE

The most studied example of the entangled state is a singlet spin state. Spin version of the EPR experiment was proposed by Bohm[43]. Using SCI we analyze here only the predictions of QT for SPCE. The discussion of EPR-B paradox may be found in[27].

The singlet spin state vector for the a system of two particles has the form:

$$\Psi_0 = (|+\rangle \otimes |-\rangle - |-\rangle \otimes |+\rangle)\sqrt{1/2} \tag{3}$$

where the single particle vectors $|+\rangle$ and $|-\rangle$ denote "spin up" and "spin down " with respect to some common coordinate system.

If we "measure" the spin of the particle #1 along the unit direction vector **a** and the spin of the particle #2 along the unit direction vector **b** , the results will be correlated and for the singlet state the correlations are described by the correlation function:

$$E(\mathbf{a},\mathbf{b}) = \langle \Psi_0 | \sigma_{\mathbf{a}} \otimes \sigma_{\mathbf{b}} | \Psi_0 \rangle = -cos\, \theta_{\mathbf{ab}} \tag{4}$$

where $\sigma_{\mathbf{a}} = \sigma \bullet \mathbf{a}$ and $\sigma_{\mathbf{b}} = \sigma \bullet \mathbf{b}$ denote the components of the Pauli spin operator in the directions of the unit vector **a** and **b** .respectively and $\theta_{\mathbf{ab}}$ is the angle between the directions **a** and **b**. Since $E(\mathbf{a},\mathbf{b}) = -1$ for $\theta_{\mathbf{ab}} = 0$ it was concluded that the results of the spin projection measurements for each individual couple of particles are strictly anti-correlated. It was pointed out in [41] that this conclusion does not hold since according to SCI the state vector Ψ_0 allows only to find statistical distribution of outcomes without giving prediction for any individual outcome. The reason is that the sharp directions and angles do not exist in Nature. The fuzzy measurements in QT have been studied for years and many important results have been obtained [44,45].

In this case each spin polarization correlation experiment (A,B) is defined by two macroscopic orientation vectors **A** and **B** being some average orientation vectors of the analyzers [41,26,27]. Therefore the analyzer A is defined by a probability distribution $d\rho_A(\mathbf{a})$, where **a** are the microscopic direction vectors, $\mathbf{a} \in O_A = \left\{\mathbf{a} \in S^{(2)}; |1 - \mathbf{a}\cdot\mathbf{A}| \leq \varepsilon_A\right\}$. Similarly a polarizer B is defined by $d\rho_B(\mathbf{b})$. Therefore even if the detectors and filters were perfect , no detection loophole, the idealized QT prediction for the correlation function $E(\mathbf{A},\mathbf{B})$ would be given not by the formula (4) but by a smeared formula:

$$E(\mathbf{A},\mathbf{B}) = \int\limits_{O_A}\int\limits_{O_B} -cos\ \theta_{\mathbf{ab}}\ d\rho_A(\mathbf{a})d\rho_B(\mathbf{b}) \tag{5}$$

The quantitative effect of the smearing of $cos\ \theta_{\mathbf{ab}}$ in the formula (5) can be very small but $E(\mathbf{A},\mathbf{A}) \neq -1$ eliminating strict anti-correlations between measured polarization projections on the individual level. Of course in SPCE the formulas (4) and (5) have to include additional factors to account for the efficiencies of the detectors, various transmission coefficients etc. The formula (5) and similar formulas for the joint probabilities of detection [41,26,27,32] confirm only the fundamental contextuality of QT due to which a spin projection on a given axis is not a predetermined attribute of an individual physical system recorded by a measuring device but it is produced in the interaction of the system with this device. As we already told the existence of the long-range correlations between far away measurements suggests the existence of supplementary parameters keeping the memory of the preparation stage at the source and describing the invisible sub-phenomena during the measurement process .

BELL INEQUALITIES

Let us describe a typical SPCE in the language of observed phenomena.

A pulse from a laser hitting a non-linear crystal produces two correlated physical fields propagating with constant velocities towards the far away detectors. Each of these fields has a property that it produces clicks when hitting the photon- detector. We place two polarization analyzers A and B in front of the detectors on both sides and after interaction of the fields with the analyzers we obtain two correlated time- series of clicks on the detectors. Each analyzer is characterized by its macroscopic direction vectors, which may be changed at any time. By changing the direction vectors we have various coincidence experiments labeled by (A, B) where **A** and **B** are the macroscopic direction vectors for the analyzers A and B respectively.

In QT the crystal is described as a source of couples of photons in a spin singlet state and the ensemble of these couples is described approximately by the state vector (3). It is well known that the photons are neither localizable nor visible and they cannot be visualized as point-like particles following some classical trajectories. Nevertheless the mental picture of correlated photon pairs travelling across the experimental set-up and carrying their own unknown spins (intrinsic magnetic moments) whose projections on each direction are predetermined by a source and recognized by the polarization analyzers is commonly used in the discussions of SPCE. Knowing that such description of the sub-phenomena is inaccurate Bell tried to formulate the most abstract probabilistic local hidden variable model in order to explain the long range correlations in a singlet state predicted by QT. As we told above in the experiment (A,B) two time -series of outcomes are produced with each outcome being 1 or -1. The main assumptions in so called local realistic hidden variable model (LRHV) proposed by Bell [7] are:

1.Individual outcomes are produced locally by corresponding analyzers A and B.

2.There are some uncontrollable hidden variables $\lambda \in \Lambda$ determining the value of individual outcomes. In the experiment (A,B) they are obtained as values of some the bi-valued functions on Λ such that $A(\lambda,\mathbf{a}) = \pm 1$ and $B(\lambda,\mathbf{b}) = \pm 1$ respectively where **a** and **b** denote settings of the analyzers (we keep here for purpose the original Bell notation instead of **A** and **B** used above).

3.The space Λ , called a probability space, and the probability distribution $\rho(\lambda)$ of the hidden variables λ do not depend on **a** and **b.**

Bell wrongly believed that the assumption 3 is another consequence of the locality because the probability distribution of the hidden parameters should be prepared at the source far away from the detectors. Using the assumption (3) the expected value of the correlation function E(a,b) is given by:

$$E(\mathbf{a},\mathbf{b}) = \int\limits_{\Lambda} A(\lambda,\mathbf{a})B(\lambda,\mathbf{b})\rho(\lambda)d\lambda \tag{6}$$

Using the formula (6) for any couple of directions of analyzers BI and CHSH inequalities below can be proven.

$$|E(\mathbf{a},\mathbf{b}) - E(\mathbf{a},\mathbf{b}')| + |E(\mathbf{a}',\mathbf{b}') + E(\mathbf{a}',\mathbf{b})| \leq 2 \tag{7}$$

In 1976 we met John Bell in Geneva, and we left him few handwritten pages with our comments concerning the generality of his proof. In particular we pointed out that if the hidden variables λ describing each pair of "particles" were couples of bi-valued strictly correlated spin functions S_1 and S_2 on a sphere such that measured outcomes for each pair were the values of $S_1(\mathbf{a})$ and $S_2(\mathbf{b})$ where **a** and **b** were unit direction vectors of analyzers in a given experiment then one could not use the integration over the set of all of these functions as it is done in the formula (6). Moreover we indicated that in this case one should try to prove BI by using the estimates of the correlation functions (empirical averages) obtained by averaging the sums of the products $S_1(\mathbf{a})S_2(\mathbf{b})$ over all pairs in long runs of the corresponding experiments. If $E(\mathbf{a},\mathbf{b})$ is replaced by its estimate the proof of (7) may never be rigorous because the error bars have to be included and one has also assume that the sets of spin functions describing the couples in the runs from different experiments are exactly the same what is highly improbable due to the richness of the uncountable set of spin functions on a sphere. These ideas in their final more mature form were only published in a series of papers[4,5,16,41] in 1987.

In the meantime Pitovsky [2,3] constructed the spin functions on the sphere for which the integral (6) could not be defined and proposed a local hidden variable model based on these functions able to reproduce the predictions of QT and violating BI. Aerts[6,30] inspired by Accardi's paper[1] showed that the BI can also be violated in macroscopic experiments. De Baere[46,47] pointed out that BI might be violated due to the non-reproducibility of a set of hidden variables.

The Pitovsky model was complicated and contained some inexact statements. The model was reformulated and rendered fully contextual in [41]. The spin functions describing a pair could be strictly correlated but the individual outcomes of the measurements were not predetermined thus there was no strict anti-correlations.

In the second paper [5] we discussed why BI cannot be proven rigorously using the empirical averages and we showed that the use of the unique probability space implies the experimental protocol which is incompatible with SPCE. In conclusion we wrote:"The various SPCE cannot be replaced by one random experiment of the type discussed above and in our opinion this is the reason why the Bell inequalities do not hold. The various probabilities appearing in their proofs are counter-factual and have nothing to do with the measured ones"

Let us clarify here about which one random experiment we were talking .The k-variate random variable $X=(X_1,...,X_k)$ and joint probability distribution on a unique probability space Λ were invented in order to describe the following random experiment : we take a large random sample of members from some population we measure the values of X_i $i=1,..,k$ on each member in a sample obtaining an individual outcome for the experiment as a set of k numbers $(x_1,..,x_k)$. The empirical joint distribution for the frequencies of these outcomes gives the information about the joint probability distribution characterizing the whole population. From this joint probability distribution one may obtain by conditionalization the marginal probability distributions for any single random variable X_i or for any group of them. This is exactly the protocol of the random experiment implied by the formula (6): pick up a pair described by λ, measure the spin projections for this pair in all directions etc. what is of course impossible . We thought that the arguments presented against BI were convincing enough to stop further speculations concerning their violation but we were wrong. Probably some papers were simply unknown or not understood. With the arrival of the Quantum Information and the interest for the possible practical uses of the EPR type experiments it was necessary to provide an up- to- date refutation of BI and of non-locality what was done e.g. by Accardi et al. [22,25], Khrennikov[20], Hess and Phillip[24], Kracklauer[28] and by myself.[22,26,27]. We already explained why the formula (6) does not apply to SPCE .

The correct formula which might be used to describe locally the sub-phenomena in any particular experiment SPCE for the couple of analyzers (A,B) is:

$$E(\mathbf{A},\mathbf{B}) = \int_{\Lambda_{AB}} A(\lambda_1,\mathbf{a})B(\lambda_2,\mathbf{b})\rho(\lambda_1,\lambda_2)d\rho_A(\mathbf{a})d\rho_B(\mathbf{b})d\lambda \tag{8}$$

where $\Lambda_{AB} = \Lambda_1 \times \Lambda_2 \times \Lambda_A \times \Lambda_B$ with $(\lambda_1,\lambda_2,\mathbf{a},\mathbf{b}) \in \Lambda_{AB}$ and $d\lambda$ is a shorthand notation for the measure on $\Lambda_1 \times \Lambda_2$ for which the integral makes sense.

One of the most recent reformulations of CHSH theorem was given by Larsson and Gill[36]. Instead of Λ_{AB} they use a unique probability space Λ as in formula (6) saying :"...the particles travelling from the source carry some information about what the result would be of each possible measurement at the detectors. This information is denoted λ above,and can consist of anything , from a simple "absolute" polarization to some complicated recipe of what each result measurement will be , for each setting of the detector parameter. What it is exactly is not important the very existence of such information will be referred to as " Realism", the λ, in a sense is " the element of reality" that determines the measurement result.". Without writing explicitly the formula (6) they conclude that under these prerequisites the CHSH theorem cannot be violated and therefore the Local Realistic hidden variable models are impossible. Next authors give an interesting discussion of how various experimental factors such as visibility , efficiency etc. may prevent violation or prevent the application of the theorem getting a formula (9) which we rewrite here in a simplified form:

$$|E(\mathbf{A},\mathbf{B}|\Lambda(A,B)) - E(\mathbf{a},\mathbf{b}'|\Lambda(A,B'))| + |E(\mathbf{a}',\mathbf{b}'|\Lambda(A',B')) + E(\mathbf{a}',\mathbf{b}|\ \Lambda(A',B)| \leq 4-2\delta \quad (9)$$

where $\Lambda(C,D) \subset \Lambda$ denotes is a subset of Λ describing the experiment (C,D) and δ is some minium probability of the overlap of the subsets used in the formula (9) In spite of the fact that the authors did not list very strong assumption concerning the existence of the joint distributions on the unique probability space Λ in the proof of (7) and (8) we completely agree with the authors that if the "Realism " is understood as a strict predetermination of the experimental outcomes at the source the Local Realistic hidden variable models are impossible. All local models able to reproduce the QT predictions [21,41,25,28,29] and to violate the equalities (7) are contextual as well as contextual are all SPCE. The correct formula is (8) and there is no overlap between Λ_{AB} and Λ_{CD} if $C \neq A$ or $B \neq D$.Therefore BI or CHSH cannot be proven . The only formula which can be proven is the formula (8) with $\delta = 0$ which of course does not violate QT predictions. Using the language of Larsson and Gill the information λ it is not the information about the individual results of future measurements.The hidden information carried by the particles about the preparation at the source in the moment of the measurements is stored in (λ_1, λ_2). The information which predetermines the outcome of the measurement is stored in $(\lambda_1, \mathbf{a})$ where **a** describes a microscopic state of analyzer A in the moment of measurement . The information what the outcome would be is not produced at the source and read with mistakes by the analyzer but it is produced in the interaction with the analyzer and known only after the measurement is completed..

The hidden variable model of the underlying sub-phenomena given by (8) is intuitive, local and contextual . According to Accardi's terminology [25] such probabilistic model describes an adaptive dynamical system. Another example of a successful hidden variable model of SPCE is a model proposed by Matzov[29].

We see that the testing of BI-CHSH cannot help us to check the completeness of QT. We have to use instead the purity tests proposed by us many years ago [16,17].

PURITY TESTS AND PREDICTABLE COMPLETENESS

Let us consider an experiment in which we have a stable source producing a beam of " identical invisible particles" whose intensity is measured by the clicks on some detector. When we pass this beam by some quantum filter F we obtain a beam of different properties and of reduced intensity. The detailed discussion why the quantum filters are not the selectors of preexisting properties is given in [42]. If by repeating our experiment several times we discover that the relative frequencies converge to some number p(F) we may interpret it as a probability that an individual particle from the beam will pass the filter F. According to SCI the claim that QT gives the complete description of the individual system being a member of some pure quantum ensemble may be only understood in the sense that the probabilistic predictions of QT provide complete description of the ensembles of the outcomes of all possible measurements performed on this pure quantum ensemble.

A standard interpretation of QT did recognize the importance of a pure quantum state and defined it as a state of physical system which passed by a maximal filter or on which a complete set of commuting observables was measured .The immediate question is what to do if we do not have a maximal filter or how can we know that a filter used is a maximal one? We found this definition highly unsatisfactory and we analyzed in 1973 various general experimental set-ups containing the sources of some hypothetical particle beams, detectors(counters), filters,transmitters and instruments[42].This analysis led us to the various conclusions which are pertinent to the topic of this paper:

1) Properties of the beams depend on the properties of the devices and vice-versa and are defined only in terms of the observed interactions between them. For example a beam b is characterized by the statistical distribution of outcomes obtained by passing several replicas of this beam by all available devices d_i. A device d is defined by the statistical distribution of the results it produces for all available beams b_i. All observables are contextual and physical phenomena observed depend on the richness of the beams and of the devices.

2) In different runs of the experiments we observe the beams b_k each characterized by its empirical probability distribution. Only if an ensemble ß of all these beams is a pure ensemble of pure beams we can associate estimated probability distributions of the results with the beams b∈ß and eventually with the individual particles who are forming these beams.

3) A pure ensemble ß of pure beams b is characterized by such probability distributions s(r) which remain approximately unchanged:

(i) for the new ensembles $ß_i$ obtained from the ensemble ß by the application of the i-th intensity reduction procedure on each beam $b \in ß$

(ii) for all rich sub-ensembles of ß chosen in a random way

In order to test the validity of the Optical Theorem we decided to test whether the initial two-hadron states prepared for the high energy collision are mixed with respect to the impact parameter. Therefore we reviewed in a series of papers several non-parametric statistical purity tests which could be used [17] and together with Gajewski [51] we performed the purity tests for $\pi^- d$ charge multiplicity distributions using the raw data from the Cambridge-Cracow-Warsaw collaboration in which the deuterium filled bubble chamber was exposed to a π^- beam of momentum 21GeV/c. We wanted to find significant differences between the data obtained in the different accelerator runs. If the initial state is pure, the different channels should be randomly distributed in time , with some fixed probabilities of the appearence. If one concentrates on the appearance of two groups of the channels one can obtain the time ordered sequences of 0 and 1 such as : 1000110000... The randomness of these sequences can be tested in different ways . In one of the tests the hypothesis that the distribution was random could be rejected on the significance level:as low as of 0,0014. Since we considered our paper mainly as the illustration of various testing methods we did not insist on the importance of this result hoping that it will be confirmed by others.

In 1984 we noticed that the purity tests could be used also to test completeness of QT because :" The main feature of any theory with supplementary parameters is that the quantum pure ensembles become mixed ensembles of the individual systems characterized by the different values of these new parameters. There is a principal difference between a pure statistical ensemble and a mixed one. The pure ensemble is homogeneous , a mixed one should reveal a fine structure."[16]. If the source is producing a pure beam of particles all runs of the experiment should be highly compatible. If the source is producing the mixed ensemble the mixture could vary slightly from one run to another run or we could hope to change its composition by using some non invasive intensity reduction procedures. The purity test may be defined more rigorously as follows.

Let O be a stable source of particles and γ a measuring device of some physical observable γX. A set $S=\{x_{k;}\ k=1,...,m\}$, where x_k denote the measured values of γX, is a sample drawn from some statistical population of the random variable X associated with the observable γX. If b_i is a beam of m_i particles produced by the source O in the time interval $[t_i , t_i + \Delta t]$ we obtain a sample S_i when γX is measured the beam b_i. By using j-th beam intensity reduction procedure applied to the beam b_i we obtain a family of new beam $b_i(j)$, j=1,,,,n . Measuring γX on the beams $b_i(j)$ we obtain n new samples $S_i(j)$. We state that the beams produced by the source O are pure only if we cannot reject the hypothesis H_0 :

H_0: All the samples S_i and $S_i(j)$ for different values of t_i and Δt are drawn from the same unknown statistical population.

To test H_0 one has to use the statistical non-parametric compatibility tests such as : Wilcoxon-Mann-Whittney test, normal scores test, rank or run tests [17]. These tests can be used to analyze any already existing experimental data in particular the data from SPCE .Of course the rejection of H_0 proves only the impurity of the ensembles which were incorrectly described in QT as pure ensembles but it could be also an indication that QT is not predictably complete. It is not easy to show that QT is not predictably complete because the mathematical language it uses is very rich and flexible [27,42] allowing a good fit to the experimental data.

To prove the predictable incompleteness of QT we need something more. As we told above the results of any experiment may be represented as a time series of various possible outcomes. If there are k different outcomes possible QT describes this time series as a sample drawn at random from some particular multinomial probability distribution . The outcomes should appear therefore randomly in time with given probabilities. If one could detect some temporal fine structure in this time series or to find a stochastic model able to explain it then it would mean that QT does not provide a complete description of the experimental data obtained in this experiment. Several methods are used to study and to compare empirical time-series: frequency or harmonic analysis,periodograms etc.[52,58].

Due to the limited efficiencies of detectors and other imperfections of the experimental set-ups one has always a dilemma whether and how one should correct the data in order to obtain a "fair sample" of outcomes to be compared with the theoretical model of the underlying phenomenona. Adenier and Khrennikov analyzed recently the data from SPCE by Greg Weihs et al.[13] and found several interesting anomalies which survived various data correction attempts and which were not accounted for by the current description provided by QT. To elucidate these anomalies one would like to have more information about the calibration tests performed by the group before the experiment such as numbers of counts on the detectors: without spin analyzers on one of the sides and on both sides,with coincidence circuitry on and off, with the singlet source replaced by another source producing two beams having known spin polarization etc.

CONCLUSIONS

In spite of the experimental imperfections we do not believe that the violation of BI in SPCE is due to the unfair sampling. The main reason, is that the probabilistic models used to prove BI are not valid for SPCE .

We hope that the arguments presented in this paper will cut short all speculations about the non- locality of QT observed in SPCE and will promote the statistical contextual interpretation (SCI) of QT .

The SCI is free of paradoxes and shares Einstein's conviction that the probabilistic description of the phenomena is due to the lack of knowledge and control of the underlying sub-phenomena . At the same time SCI agrees with Bohr that the measured value of the physical observable is not predetermined and that it has only meaning in a specific experimental context which must be always included in any model aiming to describe the sub-phenomena. SCI agrees also with Bohr that QT gives only the probabilistic predictions for the phenomena but SCI does not say that more detailed description of the underlying sub-phenomena is impossible.

According to SCI the mysterious long range correlations in SPCE are due to the memory of the preparation at the source preserved in the sub-phenomena. Several local descriptions of sub-phenomena based on this idea e.g.[2,3,41,28,29] were able to reproduce the prediction of QT. Similar correlations exist in the macroscopic world. For example the violent earthquake in the middle of the ocean causes strong correlations between random variables such as :the force and the height of the Tsunami waves, force of the winds, number of victims etc. on far away shores [22]. Also in statistical physics there exist the long range correlations between coordinates and coarse -grained velocities of the Brownian particles which interacted in the past what was proven by Allahverdyan, Khrennikov and Nieuwenhuizen[53]. In view of this any speculation how the measurement performed on one photon influences the behavior of the other photon from the EPR pair is completely unfounded.

A general belief was that the language of Hilbert spaces and probability amplitudes used by QT could not be deduced from the classical theory of probability .This belief was shown to be incorrect by Khrennikov[54] who developed a probability calculus of conditional probabilities depending explicitly on experimental contexts and was able to reconstruct as a special case the probabilistic formalism of nonrelativistic quantum mechanics. In another paper [55}he showed that non relativistic quantum mechanical probabilities can be obtained form the averages of some classical random fields.

One cannot obtain a convincing proof of incompleteness of QT by constructing ad hoc local hidden variable models able to reproduce some predictions of QT . The convincing proof of incompleteness would require a construction of a general model providing a consistent description of all phenomena described by QT and able to give more detailed predictions of these phenomena than those given by QT what seems to be a formidable task.

Testing the predictable completeness of QT seems easier and more promissing. The outcomes of the physical experiments can be represented by some numerical time series. QT takes for granted the randomness of these time series and gives only the probabilities of the appearance of various outcomes.Any significant deviation from the randomness or the discovery of some reproducible fine structure in these series, not explained by QT, would prove that the QT is not predictablly complete and a search for a more complete descriptionof the phenomenon is justified.

For example to describe effectively the behavior of cold trapped ions the continuous quantum evolution had to be supplemented by some quantum jumps obeying some stochastic Lévy process what was demonstrated by Claude Cohen-Tannoudji and collaborators [39]. Similarly it seems plausible some new description may be needed in order to describe in detail the data from beautiful experiments with ultraslow propagation of coherent light pulses in Bose-Einstein condensates reported recently in [57]

. The discovery of new fine structures in the data would be important by itself but also it would give additional clear argument against treating the quantum state vectors as attributes of individual physical systems which can be manipulated instantaneously Such interpretation and instantaneous manipulations of qubits are often used in the domain of quantum computing [59].

As Bohr already pointed out [40] it is very dangerous to talk about the invisible sub-phenomena but sometimes it is inevitable. Therefore if we do it we should try to use the most precise language to avoid paradoxes and confusion.

ACKNOWLEDGMENTS

The author would like to thank Andrei Khrennikov for the warm hospitality extended to him during this enjoyable and interesting conference.

REFERENCES

1. L.Accardi, *"Topics in Quantum Probability"* ,*Phys.Rep.***77** , 169 (1981)
2. I. Pitovsky, *"Resolution of the Einstein-Podolsky-Rosen and Bell Paradoxes."Phys.Rev.Lett.***49**, 1216(1982).
3. I. Pitovsky," *Deterministic Model of Spin Statistics"* , *Phys.Rev.* **D 27**, 2316(1983)
4. M.Kupczynski,"New Test of Completeness of Quantum Mechanics", ICTP preprint IC/84/242, (1984).
5. M.Kupczynski, *"Bertrand's Paradox and Bell's Inequalities",Phys.Lett.***A 121**, 205(1987).
6. D.Aerts, *"A Possible Explanation for the Probabilities of Quantum Mechanics",J.Math.Phys.* **27**,202(1986)
7. J.S.Bell, *"On The Einstein-Podolsky-Rosen Paradox.", Physics* **1**,195(1965).
8. J.S.Bell, *Speakable and Unspeakable in Quantum Mechanics*, Cambridge, Cambridge UP, 2004
9. J.F.Clauser, M.A.Horne,A.Shiminy and R.A. Holt, *Phys.Rev.Lett.***23,** 880(1969).
10. A.Aspect, P.Grangier and G.Roger, *Phys.Rev.Lett.***47**, 460(1981); **49**, 91(1981):
11. A.Aspect, J. Dalibard and G.Roger, *Phys. Rev.Lett.***49**, 1804(1982)
12. J.F.Clauser and A.Shimony, *"Bell's Theorem: Experimental Tests and Implications",Rep.Prog.Phys* **41** ,1883(1978)
13. G. Weihs,G., T.Jennewein,.,C. Simon, C.,H.Weinfurter and A.Zeilinger., *Phys.Rev.Lett.* **81**, 5039 (1998)
14. M.A.Rowe,D.Kielpinski,V.Meyer,C.A.Sackett,W.M.Itano,C.Monroe,and D.J.Wineland, *Nature*,**409**,791(2003)
15. G.Adenier and A.Khrennikov,*"Is the Fair Sampling Assumption Supported by EPR Experiments?", J.Phys.B:At.Mol.Opt.Phys.***40**,131(2007)
16. M.Kupczynski,*"On Some New tests of Completeness of Quantum Mechanics", Phys.Lett.* **A 116**, 417 (1986)
17. M.Kupczynski,*"On Some Important Statistical Tests", Riv.Nuovo Cimento 7* ,215(1977).
18. A. Einstein, B. Podolsky and N. Rosen,*"Can Quantum-Mechanical Description of Physical Reality be Considered Complete?" Phys.Rev.***47**,777(1935).
19. D.M.Greenberger,M.A.Horne and A.Zeilinger, in *Bell's Theorem, Quantum Theory,and Conceptions of the Universe,* ed.M. Kalafatos, Dordrecht ,Kluewer,1995,pp 73-76.
20. A.Khrennikov, *Interpretation of Probability*, Utrecht, VSP, 1999.
21. L.Accardi and M. Regoli,*"Locality and Bell's Inequality"*,in: QP-XIII, Foundations of Probability and Physics, ed. A.Khrennikov, Singapore,World Scientific ,2002, pp. 1–28
22. M.Kupczynski,*"On the Completeness of Quantum Mechanics"* , arXiv:quant-ph/028061 , 2002
23. A..Khrennikov and I.V.Volovich, *"Quantum Non-locality, EPR Model and Bell's Theorem"*,in:3rd International Sakharov Conference on Physics. Proceedings, ed.. A. Semikhatov et al., Singapore, World Scientific, 2003, ,pp.260-267.
24. K.Hess and W.Philipp,*"Critique of Proofs,with and without Inequalities"*,in :Foundations of Probability and Physics-3,AIP Conference Proceedings, Volume 750, ed. Andrei Khrennikov,Melville,New York,AIP,2005, pp.516-523
25. L.Accardi,*"Some Loopholes to Save Quantum Nonlocality"*,in :Foundations of Probability and Physics-3,AIPConference Proceedings, Volume 750, ed. Andrei Khrennikov,Melville,New York,AIP,2005, pp.1-19.
26. M.Kupczynski, *"Entanglement and Bell Inequalities",Journal of Russian Laser Research* **26**, 514 (2005)
27. M .Kupczynski,*"Seventy Years of the EPR Paradox"*,in :Albert Einstein Century Conference,AIP Conference Proceedings,Volume 861, ed.Jean-Michel Alimi and André Füzfa,Melville,New York,AIP,2006, pp.516-523
28. A.F.Kracklauer,*"Bell's Inequalities and EPR-B Experiments: are They Disjoint?"*,in :Foundations of Probability and Physics-3,AIP Conference Proceedings, Volume 750, ed. Andrei Khrennikov,Melville,New York,AIP,2005, pp.219-227.
29. A.Matzkin,*"Local Hidden-Variables Can Account for EPR Quantum Correlations"*, arXiv:0706.2964v1 [quant-ph] , 2007
30. D.Aerts,*"Example of a Macroscopical Situation that Violates Bell Inequalities",Lettere al Nuovo Cimento* **34**,107(1982)
31. D.Aerts,S.Aerts,J.Broekaert and L.Gabora,*"The violation of Bell Inequalities in the Macroworld",Foundations of Physics* **30**,1387 (2000)
32. A.E.Allahverdyan,R.Balian and T.M.Nieuwenhuizen ,*" The Quantum Measurement Process in an Exactly Solvable Model"*, in :Foundations of Probability and Physics-3,AIPConference Proceedings, Volume 750, ed. Andrei Khrennikov,Melville,New York,AIP,2005, pp.26–34.
33. L.E.Ballentine,*"The Statistical Interpretation of Quantum Mechanics",Rev.Mod.Phys.* **42**, 358(1970)
34. L.E.Ballentine, *Quantum Mechanics: A Modern Development*, Singapore, World Scientific, 1998.
35. A.S.Holevo, *Statistical Structure of Quantum Theory*, Berlin,Springer-Verlag,2001
36. J.-Å.Larsson and R.D.Gill, *"Bell's Inequality and the Coincidence-Time Loophole"*, in :Foundations of Probability and Physics-3,AIPConference Proceedings, Volume 750, ed. Andrei Khrennikov,Melville,New York,AIP,2005, pp.228-235.
37. H.Atmanspacher and and H.Primas, *"Epistemic and Onctic Quantum Realities"*, in :Foundations of Probability and Physics-3,AIPConference Proceedings, Volume 750, ed. Andrei Khrennikov,Melville,New York,AIP,2005, pp.49–62.
38. G.G.Emch, *"Not what Models are,but what Models do"*, in :Foundations of Probability and Physics-3,AIPConference Proceedings,Volume 750, ed. Andrei Khrennikov,Melville,New York,AIP,2005, pp.76-86.
39. M.Kupczynski,*"Scattering of Extended Hadrons and Time-Reversal Invariance"*,Hadronic Journal **9**,215(1986)
40. N.Bohr, *Essays 1958-1962 on Atomic Physics and Human Knowledge*, New York Wiley, 1963.
41. M.Kupczynski,*"Pitovsky Model and Complementarity",Phys.Lett.***A 121**, 51(1987).
42. M.Kupczynski,*" Is Hilbert Space Language too Rich"* , *Int.J.Theor.Phys.***79**, 319(1973), reprinted in: Physical Theory as Logico-Operational Structure, ed. C.A.Hooker, Dordrecht, Reidel, 1978,p.89
43. D.Bohm, *"A Suggested Interpretation of the Quantum Theory in Tems of 'Hidden' Variables",Phys.Rev.***85**,166 (1952)
44. P.Busch,M.Grabowski,P.Lahti, *Operational Quantum Physics*,Berlin,Springer Verlag, 1995.

45. S.Gudder, *"Fuzzy Quantum Probability Theory"*, in :Foundations of Probability and Physics-3,AIPConference Proceedings,Volume 750, ed. Andrei Khrennikov,Melville,New York,AIP,2005, pp.1-19.
46. W.de Baere, *Lett.Nuovo Cimento*, **39**, 234(1984);**40,** 488(1984)
47. W.M.de Muynck, W.de Baere and H. Martens,*"Interpretations of Quantum Mechanics,Joint Measurement of Incompatible Observables and Counterfactual Definiteness",Foundations of Physics* **24**,1589(2001)
48. M.Czahor,*"On Some Class of Random Variables Leading to Violations of Bell Inequality", Phys.Lett.* **A129**,6(1988)
49. N.D.Mermin,*"Hidden Variables and the two theorems of John Bell,"Rev.Mod.Phys.* **65**,803 (1993)
50. I. Pitovsky,*" George Booles's 'Conditions of Possible Experience' and the Quantum Puzzle", Brit.J.Phil.Sci.* **45**, 95(1994)
51. J.Gajewski and M.Kupczynski,*"Purity Tests for π^-d Charge Multiplicity Distributions", Lett.Nuovo Cimento* **26**,81(1979).
52. D.Brillinger, *Time Series Data Analysis and Theory*,New York, Holt, Reinhart and Winston,Inc.,1975.
53. A.E.Allahverdyan,A.Khrennikov and T.M.Nieuwenhuizen,*"Brownian Entanglement",Phys.Rev.* **A72**,032102(2005)
54. A.Khrennikov, *"Reconstruction of Quantum Theory on the Basis of the Formula of Total Probability"*, in :Foundations of Probability and Physics-3,AIP Conference Proceedings,Volume 750, ed. Andrei Khrennikov,Melville,New York,AIP,2005, pp.187-218
55. A.Khrennikov, *"Quantum Mechanics from Time Scaling and Random Fluctuations at the "Quick Time Scale",Il Nuovo Cimento*,**121B**,1005(2006)
56. F.Bardou,,J-P. Bouchaud,A.Aspect and C.Cohen-Tannoudji,*Lévy Statistics and Laser Cooling*, Cambridge, Cambridge UP,2002.
57. S.Naomi , S. Ginsberg, S. R. Garner and L.Vestergaard Hau ,*" Coherent Control of Optical Information with Matter Wave Dynamics"* **445**, 623(2007)
58. Nina Golyandina, V. Nekrutkin and A.Zhigliavskii ,*Time-Series Analysis*,London,CRC Press,2001.
59. M. Kupczynski, *"Contextual Observables and Quantum Information"*, arXiv:quant-ph/0408002(2004).

Understanding Quanta Beyond Quantum Mechanics

Ghenadie N. Mardari

Rutgers University, 89 George St., New Brunswick, NJ 08901, USA

Abstract. Quantum Mechanics must satisfy two conditions before it can be validated as complete. Firstly, there should be some compelling reason to believe that no microscopic phenomena are possible beyond the scope of the theory. Secondly, there should be a similar reason to believe that no macroscopic observations depend on the existence of additional pre-quantum properties. Yet, the Universe appears to have some verifiable properties, which prevent the fulfillment of the second condition.

Keywords: Quantum mechanics, interpretation, foundations, completeness, pre-quantum.
PACS: 01.70.+w, 03.65.Ta, 03.65.-w

INTRODUCTION

Is quantum mechanics a complete theory? If it is, then quanta must be described as entities without well-defined properties before measurement. If it is not, then some other theory must be developed, in order to account for the pre-quantum properties of matter. Historically, this issue has been a major source of debates, and many arguments have been proposed in favor of either answer. Still, solutions are not easily found, because completeness cannot be described as an integral part of any theory. On the one hand, there are models of reality. On the other hand, there is the objective totality of material properties. Completeness is a meta-theoretical concept, which establishes the level of correspondence between theory and reality. Thus, it cannot be demonstrated mathematically. As a matter of course, completeness is a provisional label for a successful theory, granted on the basis of everything that is known at the moment. For example, the inverse square law of gravity can be treated as complete as long as it survives experimental verification at any scale of observation. Permanent validation of completeness would require some sort of proof that a slice of reality is described in full by a corresponding theory, excluding any chance for future revision. Is it possible to provide such a demonstration for quantum mechanics?

Despite numerous efforts, there is no proof that quantum mechanics is an exhaustive description of reality. Instead, there is a compelling principle that new observations beyond the scope of this theory are impossible. Accordingly, no corrections or revisions can be expected, because there can be no detectable phenomena to justify such modifications. This principle is not universally accepted, as seen throughout the discussions at our conference. However, let us assume for the purpose of the argument that it is valid. Does it follow that quantum mechanics is complete? The answer to this question depends on a methodological choice. If empirical observation is the only acceptable input for theory building, then the principle is enough for an argument in favor of completeness. If, on the other hand, rational analysis is acceptable as well, then it is insufficient. In addition to the conclusion that no microscopic observation can challenge quantum mechanics, it is also necessary to prove that no macroscopic observation can have contradictory implications.

The mere observation of macroscopic objects is not usually expected to produce reliable conclusions about fundamental elements. For one thing, too many types of initial states can produce what is observed. Why would one scenario be better then the other? However, some entities in our Universe have paradoxical qualities. When these paradoxes have unique solutions, it must be possible to get useful conclusions by solving them. Of course, quantum mechanics is already known for its paradoxes, but we need macroscopic puzzles, in order to bypass the problem of measurement. That is why I decided to look at some older philosophical problems instead. For instance, how does intelligent life evolve out of simple constituents of matter? What fundamental properties of matter are required, in order to make the emergence of consciousness possible? These are questions that seem intractable, but they entail paradoxes that make them especially valuable for the problem at hand. In what follows, I shall focus on a single aspect – the emergence of freedom – in order to show that it is indeed possible to obtain verifiable conclusions about fundamental mechanics from this approach.

CP962, *Quantum Theory, Reconsideration of Foundations – 4*,
edited by G. Adenier, A. Yu. Khrennikov, P. Lahti, V. I. Man'ko, and Th. M. Nieuwenhuizen

THE PARADOX OF FREEDOM

There are many ways to think about freedom, which is why a definition must be provided here for clarity. Our interest is in the contradiction between deterministic laws and the apparent freedom of the mind. Therefore, the most restrictive meaning of this concept needs to be used. Throughout this text, freedom should be understood as the capacity to initiate thoughts or actions in a manner that cannot be reduced to physical processes alone. In other words, freedom is understood here as "freedom of mind from matter", if not "power over matter". This definition is motivated by the idea that human reasoning must be meaningful. Yet, meaning involves an act of choice among several possible connotations for a single concept. If this choice were to be either random or determined by external processes, the mind would have to be epiphenomenal. True meaning requires a level of willful operation of the mind that is not compatible with material determinism. Accordingly, reasoned decisions cannot be linked with physical actions, unless the Universe allows for some sort of non-material causality. This conclusion cannot be stated here as a fact, because we do not know if the mind is epiphenomenal or not. It is only meant to show that one assumption ("thought is meaningful") entails another ("material determinism is insufficient"). Thus, freedom is an assumption too. Nevertheless, like any assumption, it can become a source of knowledge, if it is verifiable.

As defined above, freedom demands an exception to the laws of material determinism. But let us also consider the practical problem of freedom for a rational agent confronted with a material context. An important component of freedom is the ability to convert an intention into a well-defined action. Freedom to exercise the will implies the capacity to initiate very specific actions in view of achieving clearly formulated goals. Remarkably, such activity is impossible to guarantee, unless the material context in question is governed by unfailing causal laws. Any exception in the operation of material laws undermines the expectation that a well defined action would produce reliable outcomes. Not only the laws, the material objects should also be well-defined and with constant properties. If a material context contains bodies with arbitrarily shifting properties, their proper manipulation for the achievement of well-defined goals becomes impossible. Last but not least, the laws and the observable qualities of matter should be intelligible to the free agent, in a way that could enable appropriate predictions for well-grounded choices.

In light of this, we are faced with a remarkable paradox. On the one hand, freedom cannot be compatible with absolute material determinism. On the other hand, freedom cannot operate in a meaningful way, unless it is exercised in a material context with unfailing deterministic laws. This is not just a matter of freedom versus determinism. Freedom alone demands both an exception to determinism and its operation without exception. Is there a way out of this conundrum? To the best of my understanding, the only way to satisfy both conditions without contradiction is to allow for a Universe in which material determinism is less than absolute, but still unfailing. In other words, a material context must have pockets of indeterminism, in which no material process deviates from its laws, but still it is possible to have non-material influences.

INSUFFICIENT DETERMINISM

In complex environments, it is conceivable that several sources of energy can align such as to create a point of perfect equilibrium. If a material entity happens to be at this point, all external forces cancel out, and it is not subjected to any influence. From this position, any displacement of this entity would set it within the scope of an unbalanced force, producing a chain of events that must obey the laws of matter. Different initial displacements could translate into different outcomes. Thus, a material process would still be insufficient for an explanation of the final outcome, if it did not include this "tipping" influence. However, even the act of tipping implies a change in the momentum of the entity in equilibrium. This change would constitute a violation of material laws, if it were to be somehow "free". So, this example is not good enough as a model of insufficient determinism. The problem of reconciliation between the opposite requirements of freedom has to be solved in a manner that does not produce any contradiction. Nevertheless, this example is instructive and will be used again bellow, with a proposed solution.

Freedom cannot be added to a well-defined deterministic context (or vice-versa) without sacrifices in integrity. This may seem to suggest that the two properties are totally incompatible. However, we can step outside of the Universe, for the purpose of this analysis, and try to think of it as a virtual construct, obeying any conceivable set of rules. Is it possible to find an acceptable model in the absence of conceptual restrictions? Where would freedom fit in? If it were to be exercised over a complex entity, then the motion of the same entity would not be reducible to the motion of its constituents. A contradiction in the operation of determinism would follow. Consequently, the Universe must have a fundamental level of matter, containing indivisible discrete elements. Furthermore, freedom cannot be exercised over an entity in continuous motion, because that would entail a change of momentum in violation of deterministic displacement. Therefore, elementary indivisible entities must be "programmed" to move in discrete steps. Every step could be described as an act of choice among several possible destinations. Still, these elementary properties must lead to the observation of deterministic motion at macroscopic levels. If a fundamental entity is moving in one direction, it should not be free to change direction. Its steps should be made freely among a limited number of options in the same general direction. In other words, freedom should be limited to the way in which the particle can move forward. For example, it should be able to make every other step away from a rectilinear trajectory, choosing one out of several possible options, but also to be constrained to return without any macroscopic change of direction. (Consider the example of motion along a path with holes or obstacles. The shortest path that is available for motion must have local deflections around them. The element of choice would correspond to the decision to step around each obstacle in one direction or another). Furthermore, a fundamental entity should not be free to choose whether to make or not to make a step, or how big a step to make, because this would entail a change in the macroscopic rate of motion. The fundamental steps of motion must be constant. In effect, freedom and determinism must operate as joint criteria, resulting in a pattern of indestructible fundamental motion in discrete uniform steps.

The foregoing description is still not good enough, because it entails an ineffective mechanism for freedom. At large scales, all those elementary choices may never produce any real consequence. Moreover, if elementary entities are allowed to interact via direct collisions, the fundamental parameters of indestructible motion cannot be preserved in a self-consistent manner. Consequently, there must be a law that prevents direct contact between elementary entities, enforcing instead some mechanism for mediation. For consistency, it is necessary to postulate that the medium is continuous, but has a symmetric structure with holes, in order to explain the need for microscopic elementary choices. The elementary entities must further be encoded to maintain contact with the medium at any point in time, while the same holes enable their unimpeded motion. As a result, they behave like constant oscillators, producing waves with point-like sources (plucks) at every step on the medium. Finally, elementary motion must unfold such as to prevent any two elementary entities from coming closer than one wavelength from each other. This ensures that microscopic freedom is both protected and prevented from interfering with deterministic processes. Within such rules, elementary entities must have a constant effect on their medium, subjecting it to dynamic changes. The same entities may end up moving in complex ways, simply because the state of medium changes under their own influence.

If we now return to the example from the beginning of this section, it becomes apparent that freedom can operate without contradicting any deterministic processes. When a macroscopic entity is localized at a Lagrangian point, its elementary constituents must continue their indestructible motion. This means that they should be spinning, while making the same free choices at each step. All the external forces cancel out, which implies that "tipping" would not violate any material process, if it were to occur. Yet, this collapse of the entity in any of the possible macroscopic directions is unavoidable, because the environment is highly sensible to the slightest change in energy distribution on the medium. This means that the same elementary choices, which are inconsequential in all other contexts, translate into distinguishable outcomes via deterministic chains of events in this case. Note that the fundamental steps do not even need to be willful. They might as well be random. What matters is that deterministic processes can work without exception, but also without being sufficient for this type of outcomes. Such a model satisfies both preconditions for freedom, effectively solving the paradox described in the previous section. For example, the brain of a rationally conscious being could be a machine that constantly maintains unstable equilibriums. As such, it could serve as an effective source of free action. The mental processes would still unfold in series of materially deterministic events, but the crucial elements of choice could be triggered from these non-linear "gateways". In conclusion, it is possible to have material contexts, in which beings evolve and acquire unusual properties, such as free decision-making, without ever violating material determinism.

This demonstration implies that even intangible properties can become verifiable. Without being directly observable, freedom depends on several unique material properties that are available for detection. Thus, it is now meaningful to ask: what kind of material Universe is compatible with freedom for evolving self-consciousness beings? The answer is that, first of all, it needs to have a fundamental level of material organization, which determines the properties of all macroscopic phenomena. Second, at the fundamental level, all entities must be discrete, having constant and indestructible properties. Third, the fundamental motion must be made in constant increments at a constant rate. Fourth, fundamental interactions must be mediated. Finally, the fundamental entities and processes must be well-defined and clearly intelligible in terms of deductive reasoning. If the Universe does not have a fundamental level with such properties, then the assumption of freedom is not compatible with it. Either the Universe is real and freedom is an illusion, or the other way around. Both of these scenarios are highly counterintuitive and undesirable. Yet, the only way to find a conclusive answer is to look for empirical verification.

FUNDAMENTAL MECHANICS

What is the structural organization of matter in our Universe? Is it infinitely reducible to lower and lower levels, or does it have a fundamental level, as suggested in the previous section? This question has two aspects. One can focus on the word "infinitely", and try to establish if the Universe, as a matter of fact, has a finite or infinite number of structural levels. Alternatively, one can focus on the word "reducible", and limit the investigation to the methodological dimension. The first aspect is a good subject for interminable debates, but the second one has a straightforward answer. Reductionism implies the operation of an all-pervading governing dynamics, which can be applied consistently to each level of material organization. In plain terms, the properties of molecules can be explained by the interactions of atomic components, which are determined by the interaction of subatomic particles, and so on. The crucial point is that the lower-level processes must be sufficient to explain the supervenient dynamics in full. This implies that a top-down investigation of the properties of matter should never require the postulation of new principles of interaction for any level. In short, perfect reductionism implies infinite divisibility of matter. If it works without any break in continuity, there can be no fundamental level of matter in the Universe. Yet, the opposite is not necessarily true.

The Universe might also have one or several unusual levels, where the known principles of material organization do not apply, but this does not necessarily imply the true existence of a "bottom" level of matter. Unusual levels could also be infinite. The only question that is methodologically relevant is the ability to rely on lower levels of organization for scientific explanation. Hence, there are many types of molecules in various states at different times, but all of them can be explained by reference to a smaller set of atomic components and the interactions among them. As long as there is variation, scientists can look for theories to explain it. If their models can reproduce the details of variation, they should also predict numerous manifestations even if the elements of the theory are not directly observable. On the other hand, if there is no diversity at some level of matter, then it is no longer possible or necessary to come up with such models. If a level like this also happens to be unobservable, then theoretical work has reached a limit that cannot be overcome in any productive way. The Universe may or may not have lower levels of organization, but they become scientifically irrelevant. From a methodological point of view, this is a fundamental level of matter, where reductionism stops.

A level of matter, which is methodologically fundamental, must have a unique set of properties. It must determine the governing dynamics that applies vertically to all superior levels, but it cannot obey the same rules. If it did, it would be reducible to a lower level. Therefore, it must have a special set of rules of operation, which produce the governing dynamics for the remaining material superstructure. If it did not have any well-defined properties, it would not be able to determine anything. Presumably, its properties come from a process that is beyond the reach of any theory (for the reason explained above). If so, the fundamental level must have a set of well-defined parameters, which must be treated as theory-independent fundamental constants of Nature. To sum up, a fundamental level must have constant properties and exceptional laws of interaction, but it must determine all observables in the Universe in a consistent way.

Does our Universe have a methodologically fundamental level of matter? In order to answer this question, the following properties must be taken into account. Firstly, the gap between quantum mechanics and classical mechanics is not compatible with infinite reductionism. Quite the opposite: it indicates the possibility of a special governing dynamics, because the known macroscopic rules appear to break down. Secondly, the basic constants of Nature appear to be theory independent. There is no known mechanism to predict the exact value of the speed of light, or Planck's constant. Thirdly, the invariant speed of light is more than consistent with the assumption of indestructible fundamental motion. This is even further reinforced by the discovery of quantization, because Planck's constant is a perfect candidate for the fundamental unit of discrete action that was described earlier in this text. As a corollary, there are sufficient arguments in favor of the assumption that our Universe has a methodologically fundamental level of matter with well-defined properties. This shows that it might be possible to develop an alternative interpretation to modern quantum theory, beyond the constraints of microscopic measurements.

MEDIATION

Fundamental indestructible motion can be shown to accommodate many macroscopic properties of matter. For example, the assumption of constant motion entails energy conservation. Every elementary step must generate waves, which propagate at constant speeds without being ever reflected. A fundamental medium cannot have wave-breakers, if all entities are embedded in it. Given also the postulate that elementary entities cannot collide, all energy becomes wave energy. If the waves propagate at the same rate at which they are created, any finite volume surrounding a number of entities should have a constant amount of energy. Furthermore, the energy density of the medium can be treated in topological terms, by factoring in its effects on the motion of elementary entities. This would satisfy many known facts of the theory of relativity. The problem is that such a model cannot work without the assumption of a really existing physical medium. Wasn't this hypothesis refuted already a century ago?

The major arguments against real mediation contain methodological, as well as experimental reasons. On the one hand, the theory of relativity happens to be the best theory available for large-scale cosmic processes. It does not require a real medium. On the other hand, the Michelson-Morley experiment and its subsequent variants did not confirm the expected properties of the hypothetical ether. Nevertheless, these arguments are not threatening for the presented model. Firstly, Einstein's theory does not refute the existence of a cosmic medium. It just happens to work well without it. Other theories may still predict it (but they better accommodate the known experimental findings of relativity). Secondly, Michelson and Morley did not test the medium directly. They merely ruled out the hypothesis that the speed of light is the speed of ethereal waves. In the context of the present discussion, the speed of light corresponds to the speed of elementary particulate entities. The speed of waves corresponds to a phenomenon that has yet to be tested – the propagation of changes within static fields (electric, magnetic, or gravitational). It is the latter that need to be tested for invariance, in order to rule out a physical medium with real waves. Note that the speed of light corresponds to the rate of oscillation between magnetic and electric fields in the classical electromagnetic theory. This is not the same as the rate of propagation of changes within any of those fields.

The concept of real internal speeds for static fields may sound counterintuitive, but only from a formal point of view. For the purpose of this analysis, an elementary entity must be treated as if it were surrounded by constantly propagating waves. The amplitude of these waves diminishes with distance, depending on the number of spatial dimensions of dispersion, just like in the case of sound waves. In effect, the energy density at any point oscillates around a well-defined value. This means that any physical volume, containing sources of waves, can be mapped with scalar values for the root-mean-square magnitude of the amplitude (or any equivalent measure of potential energy) at all coordinates. From a formal point of view, there is a static scalar field. From a physical point of view, there is an underlying process of constantly running waves. Any variation in the distribution of sources must induce a variation in the properties of the field. These changes should propagate at the speed of underlying fundamental waves. Therefore, we are dealing with a verifiable hypothesis.

It seems that magnetic fields are the best candidates for initial tests, because they are easier to manipulate. A very strong magnetic pulse could be produced simply by discharging a capacitor through an electromagnet. In order to test the speed of the pulse, an experimenter would have to place two detectors in the same direction, at different distances from the source. Knowing the time of detection and the distance between detectors, it should be easy to calculate the speed of the pulse. The problem, of course, is to achieve the high accuracy that would be required for this task. Rather than using clocks to register the time of detection directly, it may be more convenient to have some electric switch connected to the detector. This way, the effect of the discharge on the detector could be accompanied by the release of a pulse of light. The pulses of light from both detectors should be aimed towards a photodetector, placed in the direction of propagation of the magnetic pulse. Hence, the magnetic pulse would reach the first detector, triggering a co-propagating pulse of light. Then it should reach the second detector, triggering the second pulse of light. If the speed of light is equal to the speed of the magnetic pulse, then the two pulses of light should get to their detector simultaneously. If the speed of the magnetic pulse is superluminal, the second pulse of light should be the first to arrive at the detector. The distance between magnetic detectors should be large enough to overcome any uncertainty in the time of emission of the pulses of light. The interval between photonic detections can be measured with a multi-channel analyzer, or with interferometric methods. Once the speed of the magnetic pulses is verified conclusively, different experiments can be devised to test it for invariance.

CONCLUSION

The concept of freedom is often treated as if were incompatible with unfailing material determinism. Yet, there is at least one model in which this was shown not to be true. The same demonstration has yielded a set of verifiable conclusions about the likely basic properties of matter in our Universe. An interesting development is that the main features of quantum mechanics support the hypothesis of a pre-quantum governing dynamics. These include the break from classical mechanics, as well as the hypothesis of quantization of energy. Even the most unusual element of the suggested model – real physical mediation – was shown to be compatible with existing knowledge. An experiment was suggested for proper verification. As a corollary, several properties of quanta can be investigated scientifically, without the constraints of microscopic measurement. This contradicts the assumption of completeness for quantum theory in its current form.

Visualization of Superposition of Macroscopically Distinct States by Joint Probability Density of Macroscopic Observables

Tomoyuki Morimae [1] and Akira Shimizu

Department of Basic Science, University of Tokyo, 3-8-1 Komaba, Tokyo 153-8902, Japan

Abstract. We consider a coarse-grained quasi joint probability density Ξ of two macroscopic observables. If a quantum many-body state has a superposition of macroscopically distinct states, the structure of the superposition can be visualized by plotting Ξ for appropriate macroscopic observables. For example, the ground state of the XY model, that of the Heisenberg antiferromagnet, and a state in Grover's quantum search algorithm, which have superpositions of macroscopically distinct states, can be visualized. Like the Wigner distribution function, Ξ can take negative values. However, if the characteristic width W of the coarse-graining is sufficiently large, Ξ becomes nonnegative. We study a sufficient magnitude of such W. We see that $W = O(\sqrt{N})$ is not sufficient, but $W = O(N)$ would be sufficient, where N is the size of the system.

Keywords: Superposition of macroscopically distinct states, quasi joint probability density, coarse-graining, magnetic models, quantum computer
PACS: 03.65.Ta, 67.40.Db, 03.67.Lx

INTRODUCTION

There are many (quasi) joint probability densities, such as the Wigner distribution function [1] and the Husimi function [2], which can visualize quantum states with *a few degrees of freedom* [3]. However, (quasi) joint probability densities of macroscopic observables, which can visualize quantum *many-body* states have not been fully developed.

Quantum many-body systems, which include quantum computers [4, 5], can have superpositions of macroscopically distinct states [6–12]. If every macroscopic superposition could be reduced to an equal-weight superposition of two macroscopically distinct states, such as the cat state $\frac{1}{\sqrt{2}}|0\cdots 0\rangle + \frac{1}{\sqrt{2}}|1\cdots 1\rangle$, visualization of macroscopic superpositions would be a trivial task. However, there are many other states in which many macroscopically distinct states are superposed with various weights [6–12]. Therefore it is important to develop good (quasi) joint probability densities which can visualize such complicated superpositions.

In this work, we consider a coarse-grained quasi joint probability density Ξ of two macroscopic observables. A superposition of macroscopically distinct states contained in a quantum many-body state can be visualized by plotting Ξ for appropriate macroscopic observables. For example, the ground state of the XY model, that of the Heisenberg antiferromagnet, and a state in Grover's quantum search algorithm [4], all of which have superpositions of macroscopically distinct states, can be visualized. Like the Wigner distribution function, Ξ can take negative values. However, if the characteristic width W of the coarse-graining is sufficiently large, Ξ becomes nonnegative. We study a sufficient magnitude of W which makes Ξ nonnegative. We see that $W = O(\sqrt{N})$ is not sufficient, but $W = O(N)$ would be sufficient, where N is the size of the system [2].

After briefly reviewing the index p of macroscopic superposition, we introduce Ξ and see some examples. We then study a sufficient magnitude of W which makes Ξ nonnegative.

---

[1] Electronic address: morimae@asone.c.u-tokyo.ac.jp
[2] $f(N) = O(N^n)$ means $\lim_{N\to\infty} f(N)/N^n = \text{constant} \neq 0$.

CP962, *Quantum Theory, Reconsideration of Foundations – 4*,
edited by G. Adenier, A. Yu. Khrennikov, P. Lahti, V. I. Man'ko, and Th. M. Nieuwenhuizen

INDEX p

A superposition of macroscopically distinct states in a pure state can be detected by the index p [6–12]. Consider an N-site lattice, where N is large but finite. We assume that the dimension of the Hilbert space on each site is independent of N. As macroscopic observables, we consider additive operators [6–12]. An additive operator $\hat{A}$ is a sum of local operators: $\hat{A} = \sum_{l=1}^{N} \hat{a}(l)$, where $\hat{a}(l)$ is a local operator on site l. We assume that $\hat{a}(l)$ is independent of N. For a given pure state $|\psi\rangle$, the index p $(1 \leq p \leq 2)$ is defined by $\max_{\hat{A}}[\langle\psi|\hat{A}^2|\psi\rangle - \langle\psi|\hat{A}|\psi\rangle^2] = O(N^p)$, where the maximum is taken over all Hermitian additive operators $\hat{A}$ [3]. If $p = 2$, $|\psi\rangle$ has a superposition of macroscopically distinct states [6–12]. We can compute p efficiently, from $e_{max} = O(N^{p-1})$, where e_{max} is the maximum eigenvalue of an $O(N) \times O(N)$ Hermitian matrix, which depends only on $|\psi\rangle$ [6–12].

COARSE-GRAINED QUASI JOINT PROBABILITY DENSITY OF MACROSCOPIC OBSERVABLES

To visualize states having $p = 2$, we introduce a quasi joint probability density of macroscopic observables. For a given pure state $|\psi\rangle$, we define the coarse-grained quasi joint probability density Ξ: $\mathbb{R} \ni X, Y \mapsto \Xi(X,Y) \in \mathbb{R}$ of two macroscopic observables $\hat{A}$ and $\hat{B}$ by

$$\Xi(X,Y) = \frac{1}{2}\langle\psi|[\hat{\mathcal{M}}(X)\hat{\mathcal{N}}(Y) + \hat{\mathcal{N}}(Y)\hat{\mathcal{M}}(X)]|\psi\rangle, \tag{1}$$

where

$$\hat{\mathcal{M}}(X) = \sum_{A \in S} w(X-A)\hat{P}_A. \tag{2}$$

Here, S is the spectrum of $\hat{A}$, $w(X-A)$ is Gaussian: $w(X-A) = \frac{1}{\sqrt{2\pi W^2}} \exp\left(-\frac{(X-A)^2}{2W^2}\right)$, and $\hat{P}_A$ is the projector onto the eigenspace of the eigenvalue A of $\hat{A}$. In other words, $\{\hat{\mathcal{M}}(X)\}_{X\in\mathbb{R}}$ is a POVM of a coarse-grained measurement of $\hat{A}$. $\hat{\mathcal{N}}(Y)$ is defined in a similar manner by using $\hat{B}$. Although Ξ satisfies some of the fundamental rules for joint probability density:

$$\int_{-\infty}^{+\infty} \Xi(X,Y)dY = \langle\psi|\hat{\mathcal{M}}(X)|\psi\rangle, \tag{3}$$

$$\int_{-\infty}^{+\infty} \Xi(X,Y)dX = \langle\psi|\hat{\mathcal{N}}(Y)|\psi\rangle, \tag{4}$$

$$\int_{-\infty}^{+\infty}\int_{-\infty}^{+\infty} \Xi(X,Y)dXdY = 1, \tag{5}$$

Ξ can take negative values. Therefore Ξ is a coarse-grained *quasi* joint probability density. If $|\psi\rangle$ has $p = 2$, the superposition of macroscopically distinct states contained in the state can be visualized by plotting $\Xi(X,Y)$ as a function of X and Y for appropriate $\hat{A}$ and $\hat{B}$. There is an efficient method of finding appropriate $\hat{A}$ and $\hat{B}$ [12].

As examples, let us consider three quantum many-body states which have $p = 2$. First, the exact ground state of the XY model: $\hat{H}_{XY} = -\sum_{\langle l,l'\rangle}[\hat{\sigma}_x(l)\hat{\sigma}_x(l') + \hat{\sigma}_y(l)\hat{\sigma}_y(l')]$ on a two-dimensional square lattice has $p = 2$ [9, 12]. Here, $\langle l,l'\rangle$ denotes the nearest neighbors. By taking $\hat{A} = \sum_{l=1}^{N}\hat{\sigma}_x(l)$ and $\hat{B} = \sum_{l=1}^{N}\hat{\sigma}_y(l)$, and plotting $\Xi(X,Y)$ as a function of X and Y, the structure of the superposition of macroscopically distinct states can be visualized: It is a superposition of all symmetry-broken states which become pure phase vacua as $N \to \infty$ [12]. For this state, Ξ takes negative values. However, if W is sufficiently large, all negative valued regions disappear, and hence Ξ becomes nonnegative [12].

Second, the exact ground state of the Heisenberg antiferromagnet:

$$\hat{H}_{HAF} = \sum_{\langle l,l'\rangle}[\hat{\sigma}_x(l)\hat{\sigma}_x(l') + \hat{\sigma}_y(l)\hat{\sigma}_y(l') + \hat{\sigma}_z(l)\hat{\sigma}_z(l')]$$

[3] same remark as in the previous footnote

on a two-dimensional square lattice also has $p=2$ [9, 12]. By taking $\hat{A}=\sum_{l=1}^{N}(-1)^l\hat{\sigma}_x(l)$ and $\hat{B}=\sum_{l=1}^{N}(-1)^l\hat{\sigma}_y(l)$, and plotting $\Xi(X,Y)$ as a function of X and Y, the superposition of macroscopically distinct symmetry-broken states is visualized [12]. Unlike in the case of the XY model, Ξ is nonnegative even when $W=0$ [12].

Finally, consider a state in Grover's quantum search algorithm [4]. Let $|G_k\rangle$ be the state which appears after k Grover iterations. If the number of the solutions is $O(N^0)$, $|G_k\rangle$'s whose k satisfies $\delta \le \frac{4k+2}{\sqrt{2^N}} \le \pi-\delta$ have $p=2$, irrespective of which numbers are the solutions [8]. Here, δ is an arbitrary small positive constant being independent of N. Among many k's which satisfy this condition, we use $k=R/2$ for even R, and $k=R/2+0.5$ for odd R, where R is the number of total Grover iterations. To be definite, we assume that the state $|1^{\otimes N}\rangle$ indexes the solution. By taking $\hat{A}=\sum_{l=1}^{N}[-\frac{1}{\sqrt{2}}\hat{\sigma}_x(l)+\frac{1}{\sqrt{2}}\hat{\sigma}_z(l)]$ and $\hat{B}=\sum_{l=1}^{N}\hat{\sigma}_y(l)$, and plotting $\Xi(X,Y)$ as a function of X and Y, the macroscopic superposition is visualized: Ξ has two peaks, which shows that two macroscopically distinct states are superposed [12]. Although Ξ takes negative values, all negative valued regions disappear, if W is sufficiently large [12].

A SUFFICIENT MAGNITUDE OF W WHICH MAKES Ξ NONNEGATIVE

Like the Wigner distribution function, Ξ can take negative values. However, if W is sufficiently large, Ξ becomes nonnegative. In this section, we study a sufficient magnitude of W which makes Ξ nonnegative. As examples, let us consider the cat state: $\frac{1}{\sqrt{2}}(|0^{\otimes N}\rangle+|1^{\otimes N}\rangle)$ and the state $|G_{R/2}\rangle$ after $R/2$ Grover iterations studied in the previous section. Here, $\hat{\sigma}_z|0\rangle=-|0\rangle$, $\hat{\sigma}_z|1\rangle=|1\rangle$, and we assume that R is even. In Figs. 1, 2, and 3, we plot

$$I(N)=\int_{-\infty}^{+\infty}\int_{-\infty}^{+\infty}\frac{\Xi(X,Y)-|\Xi(X,Y)|}{2}dXdY \tag{6}$$

as a function of N for the cat state with $W=0$, $W=\sqrt{N}$, and $W=O(N)$, respectively. $I(N)$ is the integral of Ξ over the negative valued regions. Ξ is nonnegative if and only if $I(N)=0$. We also plot $I(N)$ versus N for $|G_{R/2}\rangle$ in Figs. 4, 5, and 6 with $W=0$, $W=\sqrt{N}$, and $W=O(N)$, respectively.

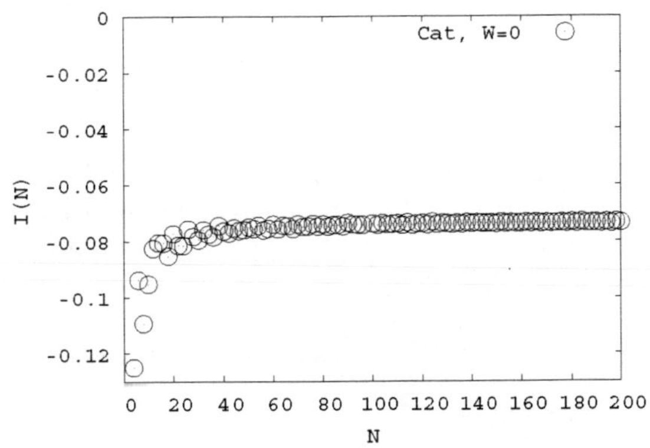

FIGURE 1. $I(N)$ versus N for the cat state. $W=0$.

From these figures, we can see that if $W=0$ and $W=\sqrt{N}$, $I(N)$ does not approach 0 as $N\to\infty$. This means that negative valued regions of Ξ do not disappear as $N\to\infty$. On the other hand, if $W=O(N)$, $I(N)$ approaches 0 as $N\to\infty$. This shows that Ξ becomes nonnegative as $N\to\infty$.

We have also confirmed that $I(N)\to 0$ as $N\to\infty$ if $W=O(N)$ for some other states. Therefore, we expect that a sufficient magnitude of W which makes Ξ nonnegative is $O(N)$. This seems reasonable, because $W=O(N)$ means that the relative error W/N of a measurement is independent of the system size N, which is a usual situation for macroscopic measurements.

On the other hand, we have seen that $W=0$ and $W=O(\sqrt{N})$ are not sufficient to make Ξ nonnegative for some states. To understand this result, let us consider two additive observables $\hat{A}=\sum_{l=1}^{N}\hat{a}(l)$ and $\hat{B}=\sum_{l=1}^{N}\hat{b}(l)$. Then, we can show

$$\left\|\left[\frac{\hat{A}}{N},\frac{\hat{B}}{N}\right]\right\|\le O\left(\frac{1}{N}\right), \tag{7}$$

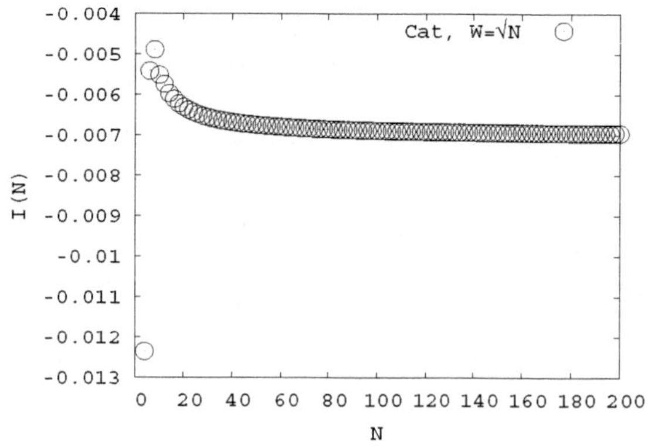

FIGURE 2. $I(N)$ versus N for the cat state. $W = \sqrt{N}$.

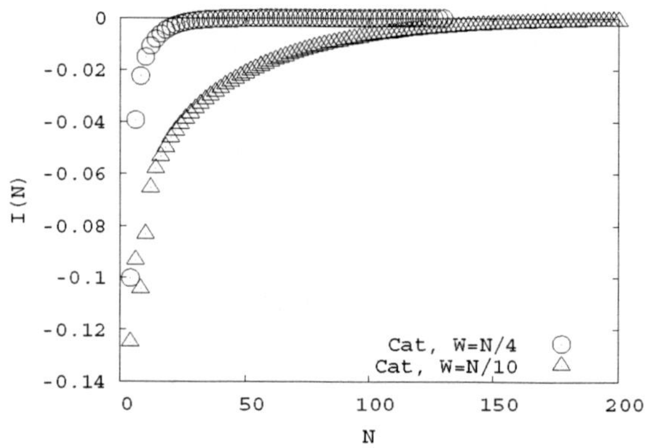

FIGURE 3. $I(N)$ versus N for the cat state. Circles represent $W = N/4$. Triangles represent $W = N/10$.

because $\hat{a}(l)$ and $\hat{b}(l')$ commute with each other: $[\hat{a}(l), \hat{b}(l')] = 0$ if $l \neq l'$ [10]. Since N is large but finite, the above commutator does not vanish in general. Therefore, noncommutativity of additive operators can be detected if resolutions of measurements are not limited. This explains why Ξ does not become nonnegative if $W = 0$ for some states. On the other hand, Eq. (7) also indicates that if a measurement is sufficiently coarse-grained, noncommutativity of additive operators could not be detected. Our result that Ξ does not become nonnegative for some states if $W = O(\sqrt{N})$ suggests that an $O(\sqrt{N})$ coarse-graining still can detect noncommutativity of macroscopic observables

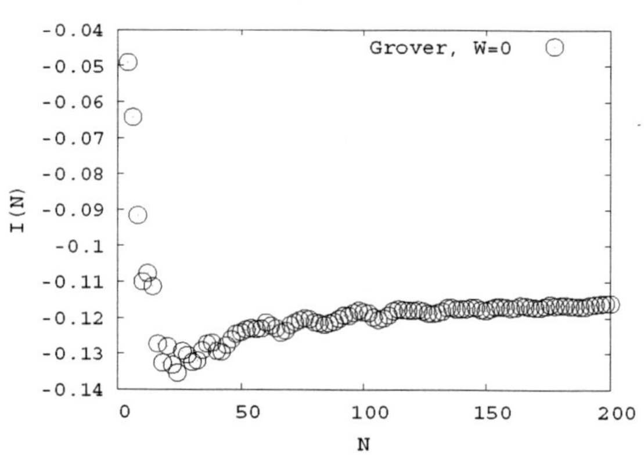

FIGURE 4. $I(N)$ versus N for the state in Grover's algorithm. $W = 0$.

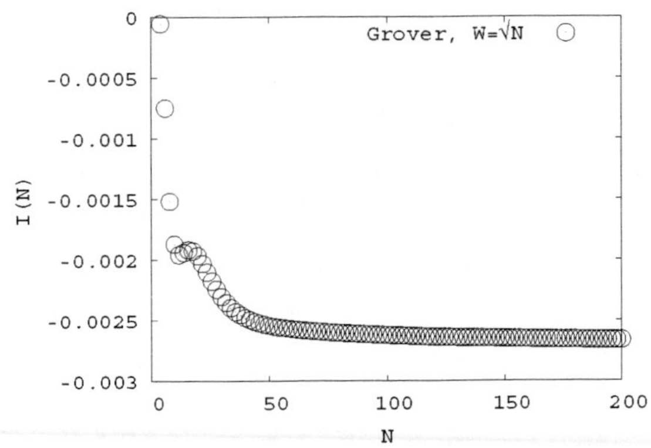

FIGURE 5. $I(N)$ versus N for the state in Grover's algorithm. $W = \sqrt{N}$.

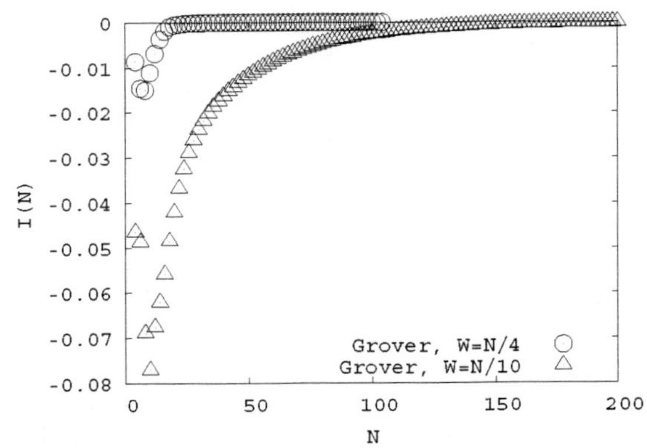

FIGURE 6. $I(N)$ versus N for the state in Grover's algorithm. Circles represent $W = N/4$. Triangles represent $W = N/10$.

for some states.

CONCLUSION

In this work, we have considered the coarse-grained quasi joint probability density Ξ of macroscopic observables, which can visualize superpositions of macroscopically distinct states in quantum many-body states. We have studied a sufficient magnitude of W which makes Ξ nonnegative. We have seen that $W = O(\sqrt{N})$ is not sufficient, but $W = O(N)$ would be sufficient.

ACKNOWLEDGMENTS

This work was partially supported by Grant-in-Aid for Scientific Research No.18-11581.

REFERENCES

1. E. Wigner, Phys. Rev. **40**, 749 (1932).
2. K. Husimi, Proc. Phys. Math. Soc. Jpn. **22**, 264 (1940).
3. L. Mandel and E. Wolf, *Optical Coherence and Quantum Optics* (Cambridge University Press, Cambridge, 1995).
4. L. K. Grover, Phys. Rev. Lett. **79**, 325 (1997).

5. M. A. Nielsen and I. L. Chuang, *Quantum computation and Quantum Information* (Cambridge University Press, Cambridge, 2000).
6. A. Shimizu and T. Miyadera, Phys. Rev. Lett. **89**, 270403 (2002).
7. A. Ukena and A. Shimizu, Phys. Rev. A **69**, 022301 (2004).
8. A. Ukena and A. Shimizu, quant-ph/0505057.
9. T. Morimae, A. Sugita, and A. Shimizu, Phys. Rev. A **71**, 032317 (2005).
10. A. Shimizu and T. Morimae, Phys. Rev. Lett. **95**, 090401 (2005).
11. A. Sugita and A. Shimizu, J. Phys. Soc. Jpn. **74**, 1883 (2005).
12. T. Morimae and A. Shimizu, Phys. Rev. A **74**, 052111 (2006).

About Essence of the Wave Function on Atomic Level and in Superconductors

A.V. Nikulov

Institute of Microelectronics Technology and High Purity Materials, Russian Academy of Sciences, 142432 Chernogolovka, Moscow District, RUSSIA

Abstract. The wave function was proposed for description of quantum phenomena on the atomic level. But now it is well known that quantum phenomena are observed not only on atomic level and the wave function is used for description of macroscopic quantum phenomena, such as superconductivity. The essence of the wave function on level elementary particles was and is the subject of heated argument among founders of quantum mechanics and other physicists. This essence seems more clear in superconductor. But impossibility of probabilistic interpretation of wave function in this case results to obvious contradiction of quantum principles with some fundamental principles of physics.

Keywords: wave function, matter waves, Aharonov-Bohm effect, force free momentum transfer, Bohm's quantum potential, superconductivity, macroscopic quantum phenomena, quantum force.
PACS: PACS 03.65.Ta; 03.75.-b; 03.65.Ud

INTRODUCTION

The wave equation proposed by Erwin Schrodinger has allowed to describe numerous atomic and other phenomena but no consensus was and is about the essence of the wave function both between founders and modem experts in quantum mechanics [1]. The assumption of Louis de Broglie on the matter waves seems to be corroborated by two-slit interference and other experiments but no universally recognized interpretation of these experiments was proposed for the present. The two-slit interference experiments manifest most obviously a duality in quantum physics but the essence of this duality is also the point at issue. Niels Bohr elevated the duality to the level of a principle he termed complementarity. This principle has satisfied some founders of quantum mechanics and most physical community. But other founders were sceptical. Einstein wrote to Schrodinger "It offers the believers a soft resting pillow from which they are not easily chased away" [2]. Schrodinger agreed, saying that "Bohr wants to 'complement away' all difficulties" [3].

The main point at issue was and is the problem of realism and possibility of a complete description of quantum phenomena. Einstein attacked the Copenhagen interpretation of quantum formalism since he kept firm belief in the aim of science as the discovery of the objective reality [1]. He believed that the reality exists and stated that any complete theory should provide *the complete description of any (individual) real situation (as it supposedly exists irrespective of any act of observation or substantiation)* [4]. According to this position, known as *realism*, measured variables exist before measurement, even if they are hidden from us, i.e. we can not measure they. Contrary to this Bohr stated that all hope of attaining a unified picture of objective reality must be abandoned. This scientific debate has already long history [1] and is still in progress up to now [5]. The works [6, 7, 8, 9, 10, 11, 12] are particular importance in history of this debate.

Einstein, Podolsky and Rosen [7] revealed first non-locality of the Copenhagen interpretation of the quantum formalism. Motivated by the EPR paper [7] Erwin Schrodinger wrote in [14]: *Maximal knowledge of a total system does not necessarily include total knowledge of all its parts, not even when these are fully separated from each other and at the moment are not influencing each other at all.* He coined the term *entanglement of our knowledge* to describe this situation [15]. Entanglement, according to Schrodinger the essence of quantum mechanics [15], can be easy describe in the limits of the quantum formalism by a wave function. But it is much more difficult to answer on the question: "What could this function describe?" For example, in the version of the EPR paradox, developed by David Bohm [16], the function

$$\Psi(r_1,r_2) = \psi_\uparrow(r_1)\psi_\downarrow(r_2) - \psi_\downarrow(r_1)\psi_\uparrow(r_2)$$

of the singlet state of two particle with spin 1/2 entangles the spin states independently of the distance $r_2 - r_1$ between the particles. This version was used by Bell [9] for the discovery a contradiction between measurement results predicted

CP962, *Quantum Theory, Reconsideration of Foundations – 4,*
edited by G. Adenier, A. Yu. Khrennikov, P. Lahti, V. I. Man'ko, and Th. M. Nieuwenhuizen

by a realistic local theory and the quantum formalism. As it is well known the experiments made with entangled photons [12, 17] testify to the quantum formalism and against a realistic local theory. Nevertheless these experimental results can not answer unambiguously on the question: *What does wave function describe?*

WHAT DOES WAVE FUNCTION DESCRIBE?

We can say, following Bohr, that the wave function can predict only a probability of experimental result and any other questions are no sense. We can also say, following Schrodinger, that the wave function describes our knowledge. In this interpretation the EPR paradox is not paradox but we must renounce on any objective reality. The Bell's experiments [12, 17] leave also a possibility of a realistic interpretation of the wave function. Bell was against locality but not against realism and his favorite example of a hidden-variables theory, Bohm theory [8], is explicitly nonlocal. Also other theories were developed, for example contextual probabilistic approach to quantum mechanics [18].

In contrast to the position of most physicists on the Bohr-Einstein debate J. Bell *felt that Einstein's intellectual superiority over Bohr, in this instance (the EPR paradox), was enormous; a vast gulf between the man who saw clearly what was needed, and the obscurantist.* Bohr stated that the atomic world differs in essence from our everyday macroscopic world and just therefore a realistic and complete description can not be valid on the atomic level. Bohr considered quantum physics as atomic physics. Therefore one ought expect that his interpretation of the wave function can come into collision with additional difficulties at description of macroscopic quantum phenomena such as superconductivity. It was noted already [19] that a direct extrapolation of quantum mechanics to the macroscopic level conflicts with some principles. In contrast to atomic level the wave function describes in superconductor no a probability but a density. Richard Feynman interpreted the wave function in superconductor as "classical" [20]. This "classical" nature of the wave function results to challenge to some fundamental principles of physics.

What does wave function describe on the atomic level?

Different interpretations of the wave function are most valid for different cases. For example the Bohr interpretation is very valid for description of electron orbits in atom.

Atom wave functions

The Schrodinger equation allowed to describe all paradoxical experimental results concerning atom orbits. It also can solved an important problem. There was a logical difficulty in the Bohr model until electron considered as a particle having a velocity since it was impossible to answer on the question: *What direction has this velocity?* The atom wave function does not say anything on a velocity or any distinguished direction. Einstein considered this as weakness of quantum formalism: *The weakness of the theory lies in the fact, that it leaves time and direction of the elementary process to chance* (the citation from the paper [21]). But thanks to this *weakness* the Bohr quantization does not violate symmetry on the atomic level. The wave functions, solutions of the Schrodinger equation, give quantum numbers, the principal one n giving energy of electron on atom orbits, orbital one l giving the orbital angular momentum and the magnetic quantum number m giving the projection of the orbital angular momentum. In fact, these quantum numbers describe all results of measurements and the wave functions give nothing in addition to these numbers having a relation to measurement results. The Bohr positivism point of view corresponds to highest degree this experimental situation. There is no sense to say not only of a velocity direction but also on a direction of the orbital angular momentum. Nevertheless the law conservation is applicable to the angular momentum. The interpretation of the magnetic quantum number m as a l projection along a specified axis is interpretation in terms of hidden variables. Thus, the Bohr interpretation keeps entirely symmetry in atom and this corresponds to the experimental corroboration of the conservation laws.

Interference experiments

The wave function $\Psi = A\exp\frac{i(pr-Et)}{\hbar}$ seems more real in the two-slit interference experiments. The probability

$$P(y) = |\Psi|^2 = |A_1\exp\frac{i(pr_1-Et)}{\hbar} + A_2\exp\frac{i(pr_2-Et)}{\hbar}|^2 = A_1^2 + A_2^2 + 2A_1A_2\cos(\frac{pwy}{d\hbar})$$

gives the interference patter predicted along y on a screen placed on a distance d from two slits separated of a distance w. In accordance with this prediction all experiments give the interference patter corresponding to the momentum p of particles. The interference patter appears just as a probability when particles pass one by one through the two-slit system [22, 23, 24]. But there is unsolved problem connected with the general unsolved problem on the collapse of the wave function at measurement. The Copenhagen interpretation refuses to answer on the question, how a particle can make its way through two slits at the same time. The Schrodinger interpretation of the wave function as a description of our knowledge is more suitable for this obvious non-locality. Some double-slit interference experiments [25] may be interpreted as a consequence of complementarity of our knowledge. But the two-slit interference problem should be considered as unsolved. The interference experiments raise also other fundamental problems. The interference of Bose-Einstein condensates [26] and atoms [27] reveals a gentle hint of challenge to the law of momentum conservation. This hint is less gentle in the Aharonov-Bohm effect [28]

The Aharonov-Bohm effects

The interference patter $|\Psi|^2 = A_1^2 + A_2^2 + 2A_1A_2\cos(\Delta\varphi)$ depends from the difference $\Delta\varphi = \varphi_1 - \varphi_2$ of the phases $\varphi_1 = \int dt\frac{E_1}{\hbar}$, $\varphi_2 = \int dt\frac{E_2}{\hbar}$ of the wave function $\Psi = A_1\exp(-i\varphi_1) + A_2\exp(-i\varphi_2)$ describing two ways motion of a particle. Since the energy of a particle with a charge q depends on electric potential V and its kinetic energy is $E_{kin} = (p-qA)^2/2m$ the phase difference, and consequently the interference patter, can be change by a potential V along one of the two ways or a magnetic flux $\Phi = \oint_l dlA$ inside the closed line l formed by two ways of the particle from source to a screen point: $|\Psi|^2 = A_1^2 + A_2^2 + 2A_1A_2\cos(\Delta\varphi_0 + q\int dtV)$ or $|\Psi|^2 = A_1^2 + A_2^2 + 2A_1A_2\cos(\Delta\varphi_0 + q\Phi/\hbar)$. Here

$$p = \hbar\nabla\varphi = mv + qA \qquad (1)$$

is the canonical momentum including the magnetic vector potential A, $\Delta\varphi_0$ is the phase difference at $V = 0$ and $\Phi = 0$.

Thus, the Aharonov-Bohm effects arise directly from the quantum formalism. But it is no mere chance that this effect was discovered by David Bohm who introduced in 1952 a quantum potential [8]. The Aharonov-Bohm experiments demonstrate that the interference pattern can be altered without any forces having acted on the particle. This experimental paradox can be easy described by the Bohm theory [8]. The interpretation of the entanglement and the wave function in this hidden-variables theory is contrary to the Schrodinger one. The hidden variables in the Bohm interpretation are simply the real configuration-space coordinates of real particles, guided in their motion by a real field, the Bohm quantum potential

$$Q_B = -\frac{\hbar^2}{4m}(\frac{\nabla^2 P}{P} - \frac{(\nabla P)^2}{P^2})$$

where $P(r) = |\Psi(r)|^2$ is a probability field. The quantum force $F_Q = -\nabla Q_B$ influences on the particle trajectories and therefore, for example, particles passing through a two-slit system give a interference patter [29].

The Bohm interpretation of quantum mechanics, favorite example of a hidden-variables theory for John Bell [13], is *hideous* nonlocal. The quantum potential $Q_B(r_1, r_2, r_3, t)$ of several particles making on a particle, for example r_1, with the force $\nabla_1 Q_B(r_1, r_2, r_3, t)$ depends on the instantaneous positions $r_2, r_3 ....$ of all the other particles of the system described by the wave function $\Psi(r_1, r_2, r_3, t)$. Bohm has resolved the EPR paradox and the the two-slit interference problem introducing a real non-locality. The quantum non-locality in the Aharonov-Bohm effect [28] is obvious [30]. But few physicist can admit that the non-locality and force-free momentum transfer could be real. Some authors [31] try to interpret the nonlocal force-free momentum transfer observed in the classical double-slit Aharonov-Bohm experiments and other experiments in the spirit of complementarity. But such interpretation is not valid for superconductors, when the wave function can not be considered in the spirit of Schrodinger as description of our knowledge.

What does wave function describe in superconductors?

All quantum phenomena observed in superconductor structures can by described by a wave function $\Psi(r) = |\Psi(r)|\exp(i\varphi(r))$ introduced by V.L. Ginzburg and L.D. Landau [32]. The phase gradient $\nabla\varphi$ is connected with the canonical momentum (1) as well as at the quantum description of elementary particles. But the $|\Psi|^2$ value is interpreted as a density of superconducting pairs but no as a probability. Such interpretation allows to explain the Meissner and quantization effects as a consequence of the demand that the complex pair wave function $\Psi(r)$ must be single-valued at any point of the superconductor

$$\oint_l dlp = \oint_l dl\hbar\nabla\varphi = m\oint_l dlv + q\oint_l dlA = m\oint_l dlv + q\Phi = \hbar 2\pi n \qquad (2)$$

The current of pairs with the charge $q = 2e$ having a density $j_s = 2e|\Psi|^2 v$ shields magnetic field in superconductor depth where the pairs velocity $v = 0$. The Meissner effect $\Phi = 0$ takes place according to (2) if the wave function has no singularity inside l and therefore the quantum number $n = 0$. When the wave function has singularity $n \neq 0$ and the flux quantization $\Phi = n\Phi_0 = n\hbar\pi/e$ takes place at a strong shielding, whereas at weak shielding the velocity circulation should have discrete permitted values

$$\oint_l dlv = \frac{\hbar 2\pi}{m}(n - \frac{\Phi}{\Phi_0}) \qquad (3)$$

Because of the velocity quantization (3) the persistent circular current [33] $I_p = sj_p = s2e|\Psi|^2 v$ should be observed in thin-walled ring or cylinder with a section s when a magnetic flux inside it is not divisible by the flux quantum $\Phi_0 = \hbar\pi/e$. The first experimental evidence of this quantum phenomenon was obtained as far back as 1962 [34].

FORCE-FREE MOMENTUM TRANSFER IN SUPERCONDUCTOR

The persistent current $I_p = s2e|\Psi|^2 v$ is much more real than a path of a single particle through a two-slit system and the wave function in superconductor can not be interpreted as description of our knowledge. Therefore the force-free momentum transfer observed is absolutely real and there is experimental evidence of a real challenge to the law of momentum conservation. For example we see in the Little-Parks experiment [34] that an equilibrium direct circular current $I_p \neq 0$ is observed at a non-zero resistance $R > 0$ and without the Faraday voltage $d\Phi/dt = 0$. The only explanation [35] of this paradoxical phenomenon is based on the change of the momentum circulation $\oint_l dlp$ of superconducting pairs on $2\pi\hbar n - 2e\Phi = 2\pi\hbar(n - \Phi/\Phi_0)$, from $2e\Phi$ to $2\pi\hbar n$, at closing of wave function in the ring with radius r, $l = 2\pi r$. This momentum change in a time unity $F_q = \hbar(n - \Phi/\Phi_0)\omega/r$ at a reiterate switching between superconducting states with different connectivity of wave function with a frequency ω, named in [35] *quantum force*, replaces the Faraday voltage and restores the force balance at an average. But the real momentum transfer $\hbar(n - \Phi/\Phi_0)/r$ because of quantization $\oint_l dlp = 2\pi n$ is force-free and nonlocal. The quantum force in superconductor [35] is more real than the Bohm quantum potential on atomic level. It has other experimental evidence [36] in additional to [34]. The problem with the force balance is also at the Meissner effect [37] and the Hall effect in superconductors [38]. Therefore a challenge to the law of momentum conservation is well deserving of attention.

FORCE-FREE MOMENTUM TRANSFER AND INTRINSIC BREACH OF SYMMETRY

The law of momentum conservation is based on symmetry demand, for example between opposite directions. Therefore it is important to note that the experimental results [39] has given evidence of intrinsic breach of right-left symmetry. The quantum oscillations in magnetic field of sign and value of the dc voltage $V_{dc}(\Phi/\Phi_0) \propto I_p(\Phi/\Phi_0) \propto \langle n \rangle - \Phi/\Phi_0$ observed on segments of asymmetric superconducting rings [39] have proved that the Bohr quantization breaks symmetry on the macroscopic level in contrast to the atomic level. This fundamental difference between atomic and macroscopic levels may be connected with a possibility to make a switch between states with different connectivity of wave function in superconducting loop whereas it is impossible in atom [40].

ACKNOWLEDGMENTS

This work has been supported by a grant "Quantum bit on base of micro- and nano-structures with metal conductivity" of the Program of ITCS department of RAS and a grant of the Program "Quantum Nanostructures" of the Presidium of Russian Academy of Sciences.

REFERENCES

1. G. Greenstein and A.G. Zajonc, *The Quantum Challenge. Modern Research on the Foundation of Quantum Mechanics*. Second Edition. Jones and Bartlett Publishers, Sudbury, Massachusetts, 2005.
2. A. Einstein, *letter to E. Schrodinger*, 31 May 1928, reprinted in *Letters on Wave Mechanics*, ed. M. Klein New York: Philosophical Library, 1967.
3. C. Moller, *Fysisk Tidsskr.* **60**, 54 (1962).
4. P.A. Schilpp, *Albert Einstein: Philosopher-Scientist* LaSalle, IL: Open Court, vol. 2, 1970, p. 666.
5. *Proceedings of International Conference "Quantum Theory: Reconsideration of Foundations"*, ed. A. Khrennikov, Vaxjo University, Sweden, 2001-2006.
6. J. von Neumann, *Mathematische Grundlagen der Quantun-mechanik* Springer-Berlin, 1932. English translation: *Mathematical Foundations of Quantum Mechanics* Princeton, NJ: Princeton University Press, 1955.
7. A. Einstein, B. Podolsky, and N. Rosen, *Phys. Rev.* **47** 777-780 (1935).
8. David Bohm, *Phys. Rev.* **85** 166-193 (1952).
9. J. S. Bell, *Physics* **1** 195-200 (1964).
10. J. S. Bell, *Rev. Mod. Phys.* **38** 447-452 (1966).
11. S. Kochen and E.P. Specker, *J. Math. Mechanics* **17**, 59-87 (1967).
12. A. Aspect et al., *Phys.Rev.Lett.* **47** 460-463 (1981); *idid* **49** 91-94 (1982); *idid* **49** 1804-1807 (1982).
13. David Mermin, *Rev. Mod. Phys.* **65** 803-815 (1993).
14. E. Schrodinger, Naturwissenschaften 23 (1935) 807; Translation published in Proc. Am. Phil. Soc. 124, p. 323 and in Quantum Theory and Measurement edited by J. A. Wheeler and W. H. Zurek (Princeton University Press, Princeton), p. 152.
15. Caslav Brukner, Marek Zukowski, Anton Zeilinger, quant-ph/0106119.
16. D. Bohm, *Quantum Theory*. New York: Prentice-Hall, 1951.
17. P. G. Kwiat et al, *Phys.Rev.Lett.* **75**, 4337-4340 (1995); G. Weihs et al., *idid* **81** 5039-5042 (1998); W. Tittel et al., *Phys. Rev. A* **57** 3229-3232 (1998).
18. Andrei Khrennikov, "Probabilistic foundation of quantum mechanics and quantum information," in *Quantum informatics* the Proceedings of SPIE, 5128, eds. Yuri Ozhigov, the International Society for Optical Engineering, Washington, USA, 2003, pp.1-10.
19. A. J. Leggett and A. Garg, *Phys. Rev. Lett.* **54**, 857-860 (1985)
20. R.P.Feynman, R.B.Leighton, M.Sands, *The Feynman Lectures on Physics*, Addison-Wesley Publishing Company, Reading, Massachusetts, 1963.
21. Anton Zeilinger, *Foundations of Physics*, **29**, 631-643 (1999).
22. A. Tonomura, J. Endo, T. Matsuda, T. Kawasaki and H. Exawa, *Amer. J. Phys.* **57**, 117-120 (1989)
23. R. Gahler and A. Zeilinger, *Amer. J. Phys.* **59**, 316-324 (1991).
24. O. Carnal and J. Mlynek, *Phys. Rev. Lett.* **66**, 2689-2692 (1991); D.W. Keith, C.R. Eckstrom, Q.A. Turchette and D.E. Pritchard, *idid* **66**, 2693-2696 (1991).
25. Anton Zeilinger, *Rev. Mod. Phys.* **71**, S288-S297 (1999).
26. M.R. Andrews, C.G. Townsend, H.-J. Miesner, D.S. Durfee, D.M. Kurn and W. Ketterle, *Science* **275**, 637-641 (1997).
27. S. Durr, T. Norm and G. Rempe, *Nature* **395**, 33-37 (1998).
28. Y. Aharonov and D. Bohm, *Phys.Rev.* **115**, 485-491 (1959).
29. C. Philippidis, C. Dewdney and B.J. Hiley, *Nuovo Cimento B* **52B**, 15-28 (1979).
30. Y. Aharonov and D. Bohm, *Phys. Rev.* **123**, 1511-1524 (1961).
31. 25. H.M. Wiseman, F.E. Harrison, M.J. Collett, S.M. Tan, D.F. Wallis and R.B. Killip, *Phys. Rev. A* **56**, 55-75 (1997); H. Wiseman and F. Harrison, *Nature* **377**, 584 (1995).
32. V.L. Ginzburg and L.D. Landau, *Zh. Eksp. Teor. Fiz.* **20** 1064-1070 (1950)
33. J. M. Blatt, *Phys. Rev. Lett.* **7** 82-83 (1961).
34. W. A. Little and R. D. Parks, *Phys. Rev. Lett.* **9** 9-13 (1962).
35. A. V. Nikulov, *Phys.Rev.B* **64** 012505 (2001).
36. A.A. Burlakov, V.L. Gurtovoi, S.V. Dubonos, A.V. Nikulov, and V.A. Tulin, *Pisma Zh.Eksp.Teor.Fiz.* **86** (2007)
37. J. E. Hirsch, *Phys.Lett. A* **315**, 474-477 (2003).
38. J.Berger, *Foundations of Physics Letters* **17** 287-293 (2004).
39. V. Dubonos et al., *Pisma Zh.Eksp.Teor.Fiz.* **77**, 439-444 (2003) (*JETP Lett.* **77**, 371-375 (2003)).
40. A.V.Nikulov, "Fundamental Differences Between Application of Basic Principles of Quantum Mechanics on Atomic and Higher Levels," in *FRONTIERS OF FUNDAMENTAL PHYSICS: Eighth International Symposium FFP8*, AIP Conference Proceedings 905 American Institute of Physics, New York, 2007, pp. 831-834.

Exact solutions of the Schrödinger equation

Börje Nilsson and Andrei Khrennikov

School of Mathematics and Systems Engineering, Sweden,
International Centre of Mathematical Modelling,
School of mathematics and Systems Engineering,
Växjö University

Abstract. Textbooks on quantum mechanics often give the impression that the Schrödinger equation can be solved exactly only for a few simple potential models. However, exact solutions are available in terms of hypergeometric functions and their confluent variants for the so-called Natanzon potentials. These potentials include the Pöschl-Teller, Manning-Rosen and Rosen-Morse potentials that are also special cases of the Eckart potential. The Natanzon potentials are reviewed and connections are made to problems in classical physics like propagation of electromagnetic waves in inhomogeneous media and of acoustic waves in a variable speed profile. The availability of exact solutions is of particular interest for the explicit construction of time evolution operators and in the solution of inverse scattering problems.

Keywords: acoustic wave propagation, analytic solution, electromagnetic wave propagation, Feynman integral, hypergeometric function, potential
PACS: 03.65.Ge, 43.20.Bi, 41.20.Jb

INTRODUCTION

The Schrödinger equation deserves and have received much interest. Typically in textbooks on quantum mechanics, the authors point out the following exactly solvable models: the harmonic oscillator, Bohrś atom, and barrier-type potentials. Then to proceed they usually use methods of perturbation theory and other advanced formalisms. Such a presentation of the material induces the impression that only a few simple cases in quantum mechanics could be handled analytically. The prime aim of the present review is to show, by recalling a number of known models, that the possibility of analytical studies is not restricted to the above mentioned simple models. Explicit expressions for the time evolution operator based on the Feynman path integral are also available for these models. A second aim for the study is to demonstrate the importance of applications to classical wave modelling, inverse scattering in particular.

In this paper the exact solutions of the Schrödinger equation are treated in one dimension for the constant energy case and the form

$$\frac{\mathrm{d}^2\psi}{\mathrm{d}x^2} + [E - V(x)]\,\psi = 0, \tag{1}$$

where units such that $\hbar^2/2m = 1$ are chosen. For quantum mechanical applications, $\psi(x)$ is the one dimensional wave function or $\psi(x) = \Psi(r)/r$, with $x = r$, where $\Psi(r)$ is the three dimensional wavefunction without angular momentum. Applications in classical physics are also possible. In the acoustic case, $\psi(x)$ is the one dimensional sound pressure $p(x)$ or $p(r)/r$ for rotationally symmetric sound pressure $p(r)$. Similarly, $\psi(x)$ is the electric field component $E_z(x)$ or magnetic field component $H_z(x)$ in electromagnetic applications.

There is an interesting class of potentials $V(x)$ for which (1) has an exact solution. These potentials are under investigation in the current paper. Since there are free constants in the analytic expressions of $V(x)$, a wide range of modelling in applications is possible.

The paper is organized in the following way. In section 2, the potentials giving solutions in terms of Bessel functions are presented. Further potentials having solutions in terms of Hypergeometric functions are described in section 3. A complete characterization of all potentials associated to solutions in Hypergeometric functions is then given in section 4. Section 5 contains references to explicit results for the time evolution operator for the above mentioned potentials, results that are based on the Feynman path integral. Finally, in section 6, applications on acoustic and electromagnetic wave propagation in inhomogeneous media are briefly considered.

CP962, *Quantum Theory, Reconsideration of Foundations – 4,*
edited by G. Adenier, A. Yu. Khrennikov, P. Lahti, V. I. Man'ko, and Th. M. Nieuwenhuizen

SOLUTIONS IN TERMS OF BESSEL FUNCTIONS

As indicated in the introduction there are potentials $V(x)$ for which the Schrödinger equation (1) has solutions in the form of special functions or other well studied mathematical functions. The general solution to (1) is

$$\psi(x) = Af(x) + Bg(x), \tag{2}$$

where f and g is a basis consisting of two linearly dependent solutions to (1); A and B are constants to be determined from boundary value conditions or other available information. For the special case $E - V(x) = \gamma^2$, a constant that can be complex, a possible choice is $f(x) = \exp(\mathrm{i}\gamma x)$ and $g(x) = \exp(-\mathrm{i}\gamma x)$.

Note that the definition of an exact solution depends on what mathematical properties or numerical routines for f and g are considered valuable. In this section results for solutions in Bessel functions are treated.

We let $\mathscr{C}_\nu(x)$ mean $f(x) = \mathrm{J}_\nu(x)$ and $g(x) = \mathrm{Y}_\nu(x)$, where J_ν and Y_ν are the Bessel functions of order ν of the first and second kind, respectively. Alternatively we may use $f(x) = \mathrm{H}_\nu^{(1)}(x)$ and $g(x) = \mathrm{H}_\nu^{(2)}(x)$, the Hankel functions of order ν of the first and second kind, respectively. Three alternatives [1] where the solution can be expressed in terms of Bessel functions are presented in (3):

$$\begin{array}{cc} E - V(x) & f(x), g(x) \\ \lambda^2 - \dfrac{\nu^2 - 1/4}{(x+x_0)^2} & \sqrt{x+x_0}\,\mathscr{C}_\nu\left(\lambda\,(x+x_0)\right) \\ \lambda^2 \mathrm{e}^{2(x+x_0)} - \nu^2 & \mathscr{C}_\nu\left(\lambda \mathrm{e}^{x+x_0}\right) \\ \lambda^2 (x+x_0)^{q-2} & \sqrt{x+x_0}\,\mathscr{C}_{1/q}\left(2\lambda\,(x+x_0)^{q/2}/q\right) \end{array} \tag{3}$$

Only those cases are considered where a constant is available to identify with the energy E. For the last result in (3), this requires $q = 3$ that corresponds to the Airy functions used in connection formulae for high energy tunneling and asymptotic solution of solutions to ordinary differential equations having a large parameter [2].

FURTHER CLASSICAL RESULTS

The beginning of the 1930s was an active period for finding exact solutions to the Schrödinger equation for various potentials. A pioneering work is that of Eckart 1930 [3] for a rather general potential. Probably independent from Eckart, special cases of this potential were considered 1932 by Rosen-Morse [4] and 1933 by Manning-Rosen [5]. One further non-trivial potential having an exact solution was presented 1933 by Pöschl-Teller [6]. In addition, variations of these potentials have been presented under individual names; one example is the Hulthén potential $V(x) = V_0[1 - \coth(x/2a)]$. The potentials of Eckart, Rosen-Morse, Manning-Rosen and Pöschl-Teller are presented in equation (4):

$$\begin{array}{cc} \textbf{Name} & V(x) \\ \text{Eckart (1930)} & \dfrac{V_1 \exp(-x/a)}{1 + b\exp(-x/a)} + \dfrac{V_2 \exp(-x/a)}{[1 + b\exp(-x/a)]^2} \\ \begin{array}{c}\text{Rosen-Morse (1932)} \\ \text{Eckart } b > 0\end{array} & \dfrac{h_0+1}{2c_1}\left(1 - \tanh\dfrac{x+x_0}{\sqrt{c_1}}\right) - \dfrac{f_0}{4c_1 \cosh^2\left((x+x_0)/\sqrt{c_1}\right)} \\ \begin{array}{c}\text{Manning-Rosen (1933)} \\ \text{Eckart } b < 0\end{array} & \dfrac{f_0+1}{2c_1}\left(1 - \coth\dfrac{x+x_0}{\sqrt{c_1}}\right) + \dfrac{h_0}{4c_1 \sinh^2\left((x+x_0)/\sqrt{c_1}\right)} \\ \text{Pöschl-Teller (1933)} & \dfrac{h_0 + 3/4}{c_1 \sinh^2\left((x+x_0)/\sqrt{c_1}\right)} - \dfrac{f_0 + 3/4}{c_1 \cosh^2\left((x+x_0)/\sqrt{c_1}\right)} \end{array} \tag{4}$$

Note that the constants in the potentials in (4) are free and may be complex that is important for applications to acoustic and electromagnetic propagation in media having losses. All potentials can have a singularity except the Rosen-Morse potential that is continuous for all possible values of the constants.

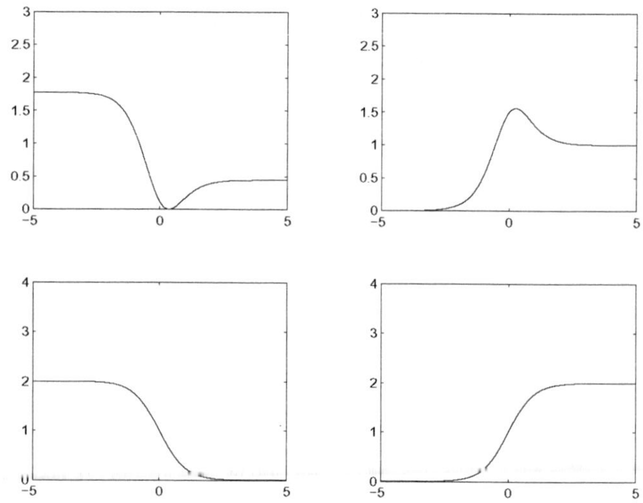

FIGURE 1. Graphs of four variants of the Rosen-Morse potential as function of the space variable x.

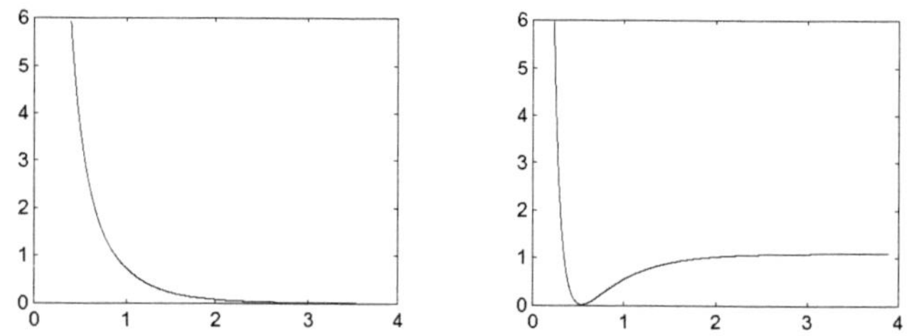

FIGURE 2. Graphs of two variants of the Manning-Rosen potential as function of the space variable x.

Described more closely in graphs are the Rosen-Morse, see Fig. 1, and the Manning-Rosen, see Fig. 2, potentials. For Rosen-Morse in the upper left picture in Fig. 1 a variant allowing free and right reflecting states as well as bound states in the potential well. Tunneling may also be present as shown in the upper right picture. However, both potential wells and potential ridges may be suppressed as shown in the two lower pictures in Fig. 1.

Whereas the Rosen-Morse potential is bounded, the Manning-Rosen potential may be unbounded as illustrated in Fig. 2. Two variants are depicted: one with a potential well and one without.

Analytical results will only be presented here for the Rosen-Morse potential. This potential is treated in a neat way by [7], see a figure there, but since we would like to keep the opportunity of having $f_0 = 0$ and $h_0 + 1 \neq 0$, which is not possible in [7], we follow instead [8]. To this end we define the following set of parameters:

$$\begin{cases} \alpha = \dfrac{1-\sqrt{1+f_0}}{2} \\ \beta = \dfrac{1+\sqrt{1+f_0}}{2} \end{cases} \quad \begin{cases} \lambda_0 = \sqrt{1+h_0-c_1 E}, \mathrm{Re}\,\lambda_0 \geq 0 \\ \lambda_1 = -\mathrm{i}\sqrt{c_1}\sqrt{E}, \ \mathrm{Im}\sqrt{E} \geq 0 \end{cases} \tag{5}$$

Then we have for the solution of the Rosen-Morse potential

$$\begin{cases} f = s^{-\lambda_0/2}(1-s)^{\lambda_1/2}F(\beta-\lambda_0, \alpha-\lambda_0, 1+\lambda_1; 1-s) \\ g = s^{-\lambda_0/2}(1-s)^{-\lambda_1/2}F(1-\beta, 1-\alpha, 1-\lambda_1; 1-s) \end{cases}, \tag{6}$$

where

$$s = \frac{1}{2}\left(1+\tanh\frac{x+x_0}{\sqrt{c_1}}\right) \tag{7}$$

and $F(a',b';c';u)$ is the hypergeometric function [1]. A number of relations exist for the hypergeometric function to get numerically efficient methods. These relations can be introduced in the numerical soft ware.

SOLUTIONS IN TERMS OF HYPERGEOMETRIC FUNCTIONS

All solutions corresponding to the potentials in (4) are given in terms of Hypergeometric functions of the form (6) with a suitable coordinate transformation like (7). Thus, it is natural to investigate the class of potentials $V(x)$ that has solutions of this type. The answer to this question is given by Natanzon [8] in the following way. The potential is expressed as

$$V(z(x)) = \frac{fz(z-1)+h_0(1-z)+h_1 z}{R(z)} - \frac{1}{2}\{z,x\}, \tag{8}$$

where $R(z)$ and $\{z,x\}$ are given by

$$R(z) = az^2+bz+c, \ \{z,x\} = \frac{z'''}{z'} - \frac{3}{2}\left(\frac{z''}{z'}\right)^2 , \tag{9}$$

and $z(x)$ is a solution to first order ordinary differential equation

$$(z')^2 = \frac{4z^2(1-z)^2}{R(z)}. \tag{10}$$

In order to avoid semi-exact solutions it is of interest to find $R(z)$ such that $z(x)$ has an explicit solution. This happens when $R(z)$ has no other zeros than 0 and 1. The classical potentials in (4) are recovered in this way.

To get Coloumb behaviour for the potential at large x confluensing of poles give a corresponding exact solution in terms of confluent hypergeometric functions [8]. Another generalization of the Natanzon theory [8] is related to non-zero angular momentum, see ,e.g., [9].

TIME EVOLUTION

The exact solutions outlined above can be used to give exact expressions for the time evolution operator. The wave function $\psi(x'',t'')$ at $x=x''$ and $t=t''$ can be expressed in terms of the wave function $\psi(x',t')$ at a previous times t' via

$$\psi(x'',t'') = \int_{\mathbb{R}} \mathrm{d}x' K(x'',x';t'',t')\psi(x',t'). \tag{11}$$

Here, K is Greenś function given as a Feynman path integral

$$K(x'',x';t'',t') = \int_{x(t')=x'}^{x(t'')=x''} Dx(t) \exp\left(\frac{\mathrm{i}}{\hbar} R[x(t)]\right), \tag{12}$$

where $Dx(t)$ is the Feynman measure [10, 11]. The action R is calculated from

$$R[x(t)] = \int_{t'}^{t''} \mathscr{L}\left(x(t),\dot{x}(t)\right) \mathrm{d}t \tag{13}$$

using the classical Lagrangian $\mathscr{L}$.

Using the Feynman path procedure, explicit expressions for Greenś function K have been constructed for the Rosen-Morse potential [12, 13, 14, 15] and more generally, the Natanzon potentials [16]. These results on time evolution operators are collected [17] in a handbook containing 945 references to the field of Feynman path integrals.

APPLICATION TO ACOUSTIC AND ELECTROMAGNETIC WAVES

Finally we will in this paper outline some applications of exact solutions to the Schrödinger equation to acoustic and electromagnetic wave propagation. The first example deals with the propagation of acoustic waves in a medium where the sound velocity varies in one direction, usually the height above ground. Whereas this is a direct scattering problem, the other example is an inverse scattering problem where the electromagnetic properties of the material is to be determined by using data from a scattering experiment. The use of exact solutions in inverse electromagnetic scattering has recently [18, 19, 20, 21] been promoted.

For both examples, the existence of exact solutions provides a deeper physical understanding, solutions for special cases in parameter selections such as high or low frequency, and increased computational speed when standard numerical routines for special functions are available. In addition, reliable results are provided to be used for the testing of numerical methods that may have other advantages. The situation for the inverse problem is further improved regarding the regularization process that is the most critical element in inverse scattering.

In the acoustic problem, with time dependence $\exp(-\mathrm{i}\omega t)$, we treat for simplicity a two-dimensional half-plane $z \geqslant 0$ in the xz plane where $z = 0$ represents the ground. Let k_0 be the speed of sound for large z, $k = \omega/c$ the wave number, $c(z)$ the speed of sound at height z and $n(z) = c_0/c(z)$ the refractive index. Then within acoustic linear theory, the sound pressure $\phi(x,z)$ can be expressed as a Fourier integral

$$\phi(x,z) = \frac{k}{2\pi}\int_{-\infty}^{\infty}\chi(z)\mathrm{e}^{\mathrm{i}kux}\mathrm{d}u, \tag{14}$$

where χ solves

$$\frac{\mathrm{d}^2\chi}{\mathrm{d}z^2} + k^2\left(n^2(z) - u^2\right)\chi = 0 \tag{15}$$

outside the source region. If $n^2(z)$ is modelled as a classical or Natanzon potential, then (15) can be solved exactly. The Fourier integral (14) is calculated numerically or analytically for large kx or kz.

In an inverse scattering problem it is $n(z)$ that is to be determined. It can be an acoustic application as above where ϕ is the sound pressure or an electromagnetic application when ϕ is one of the electric or magnetic field components. In both cases, ϕ solves (14). The possibility to construct exact solutions for the Fourier transform χ is important in the regularization process. Currently, work is in progress to determine the thickness of a bark layer using such a procedure.

REFERENCES

1. M. Abramowitz and I. Stegun (Eds), *Handbook of mathematical functions*. National Bureau of standards, Applied mathematics series 55, Washington (1964).
2. F. W. J. Olver, *Asymptotics and special functions*. Wellesley, Mass., A. K. Peters, cop. (1997)
3. C. Eckart, *Phys. Rev.*, **35**, 1303 (1930)
4. N. Rosen and P. M. Mose, *Phys. Rev.*, **42**, 210-217 (1932)
5. M. F. Manning and N. Rosen, *Phys. Rev.*, **44**, 953 (1933)
6. G. Pöschl and G. Teller, *Zeitschr. Phys.*, **83**, 143-151 (1933) (In German)
7. P. M. Morse and H. Feshbach, *Methods of theoretical physics*. McGraw-Hill, New York (1953).
8. G. A. Natanzon, *Theor. Math. Phys.*, **38**, 457-463 (1979).
9. Yu. N. Demkov and V. A. Ostrovskii, *Zh. Eksp. Teor. Fiz.*, **60**, 2011 (1971) (In Russian)
10. R. P. Feynman and A. R. Hibbs, *Quantum mechanics and path integrals*, McGraw-Hill, New York (1965)
11. S. Albeverio and R. J. Hoegh-Krohn, *Mathematical Theory of Feynman Path Integrals*, Edited by A. Dold and B. Eckmann. Lecture Notes in Mathematics **60**, Springer, Berlin, Heidelberg (1976)
12. C. Grosche, *J. Phys. A: Math. Gen.*, **22**, 5073-5087 (1989)
13. H. Kleinert, *Path Integrals in Quantum Mechanics, Statistics and Polymer Physics.* , World Scientific, Singapore (1990)
14. H. Kleinert and I. Mustapic, *J. Math. Phys.*, **33**, 643-662 (1992)
15. D.-H., *J. Phys. A: Math. Gen.*, **30**, 3201-3217 (1997)
16. C. Grosche, *J. Phys. A: Math. Gen.*, **29**, L183-L189 (1996)
17. C. Grosche and F. Steiner, *Handbook of Feynman Path Integrals*. Springer, Berlin (1988).
18. B. Nilsson, *Wave Motion,* **43**(3), 259-271 (2006)
19. B. Nilsson *2nd Conference on Mathematical Modelling of Wave Phenomena, Växjö, Sweden, AIP Conference Proceeding.* **834** 268-277 (2006)
20. T. Sjödén, S. Nordebo and B. Nilsson, *2nd Conference on Mathematical Modelling of Wave Phenomena, Växjö, AIP Conference Proceeding.* **834** 278-285 (2006)
21. S. Nordebo, M. Gustafsson and B. Nilsson, *Inverse Problems,* **23** 859-877 (2007)

Detection of Many Incompatible Quantum Properties in Double-Slit Experiment

G. Nisticò and A. Sestito

Dipartimento di Matematica, Università della Calabria, via P. Bucci 30b,87036 Rende (Italy) and Istituto Nazionale Fisica Nucleare, Italy

Abstract. In this work we theoretically show that in a double-slit experiment it is possible the simultaneous detection of properties incompatible with which-slit property and with each other, together with the measurement of the position of the final impact point. General conditions for the existence of solutions are singled out and families of solutions are provided.

Keywords: Quantum Compatibility, Quantum Detections
PACS: 03.65.Ca, 03.65.Db, 03.65.Ta

1. INTRODUCTION

Coherently with Quantum Theory [1], in a double slit experiment, it is not possible to measure both the position of the point of impact on the final screen and which slit each particle passes through, because of complementarity. Nevertheless, a lot of devices provide *indirect* knowledge of which-slit property by inferring it from a suitable property T, compatible with the final impact point (see the light electron scheme of Feymann [3], the recoiling slit of Einstein [2], etc.). This idea of *detection* has been exploited to show that Which Slit property can be detected with an incompatible one without renouncing to measure the final impact point [4, 5, 6]. In this work we find out a a larger family of these situations and show how this kind of *indirect* knowledge can be extended to a set of more properties incompatible with each other, including which-slit property.

According to [4, 5, 6] the detection at issue is possible if the Hilbert space describing the entire system can be decomposed as $\mathscr{H}_I \otimes \mathscr{H}_{II}$, where $\mathscr{H}_I$ is the Hilbert space used to represent the observable position of the particle, and $\mathscr{H}_{II}$ is the Hilbert space used to represent the observable arising from further degrees of freedom. The detection of Which Slit property E is obtained by means of an observable represented by a particular projection operator T acting on $\mathscr{H}_{II}$. The possibility of detecting a property G incompatible with E is provided by the existence of an observable represented by another projection operator Y acting on $\mathscr{H}_{II}$, but which can be measured together with T [5]. In section 2 we generalize the results obtained in [4, 5, 6] about the question of detecting three incompatible properties (which-slit property, an incompatible one and the final impact point). In section 3 we investigate the possibility of detecting four incompatible properties (which-slit property, two incompatible ones and the final impact point). As in the previous case, the existence of solutions of the problem turns out to depend on the dimension of space $\mathscr{H}_I$. Particular concrete solutions are also found; however, in these solutions the detections of the two properties incompatible with E turn out to be correlated.

2. SIMULTANEOUS DETECTION OF THREE INCOMPATIBLE PROPERTIES

We consider a physical system consisting of a localizable particle whose position observable is represented, at time t in Heisenberg picture, by an operator $Q^{(t)}$ of a suitable Hilbert space $\mathscr{H}_I$. Let us suppose that, besides the position of the centre-of-mass motion, the system possesses further degrees of freedom, described in a different Hilbert space $\mathscr{H}_{II}$. As a consequence, the Hilbert space describing the entire system can be decomposed as $\mathscr{H}_I \otimes \mathscr{H}_{II}$. The localization property "the particle passes through slit 1" (which-slit property, WS) is represented in the complete Hilbert space by the operator $E = E_I \otimes \mathbf{1}_{II}$ where E_I is a localization operator in $\mathscr{H}_I$. Given any interval Δ on the final screen we denote the projection operator representing the event "the particle hits the final screen in a point within Δ" by $F(\Delta)$. Let us suppose that the Hamiltonian operator H of the entire system is essentially independent of the degrees of freedom described by $\mathscr{H}_{II}$, so that we may assume the ideal case $H = H_I \otimes \mathbf{1}_{II}$. This implies that also $F(\Delta)$ has the form

CP962, *Quantum Theory, Reconsideration of Foundations – 4,*
edited by G. Adenier, A. Yu. Khrennikov, P. Lahti, V. I. Man'ko, and Th. M. Nieuwenhuizen

$F(\Delta) = J \otimes \mathbf{1}_{II}$ [6]. Here J is a localization operator of $\mathscr{H}_I$ but corresponding to the time of the final impact, therefore $[E, F(\Delta)] = \mathbf{0}$ [6]. Inferences about WS property E can be obtained by means of a suitable projection operator T acting on $\mathscr{H}_{II}$, whose outcomes are correlated with the outcomes of E, as expressed by the following general definition of *detector* [6]:

Definition. *A projection operator S of $\mathscr{H}$ is called a detector of property R with respect to a state ψ if*
(i) $[S, F(\Delta)] = \mathbf{0}$, *(ii)* $[S, R] = \mathbf{0}$ *and* $S\psi = R\psi$.

Condition (i) ensures that S can be measured together with $F(\Delta)$; condition (ii) allows us to infer the outcome of R (albeit not measured) from the outcome of S [7]:It implies that both conditional probabilities $p(S \mid R)$ and $p(R \mid S)$ are equal to 1.

Let $G = G_I \otimes \mathbf{1}_{II}$ be a projection operator representing a property incompatible with WS property E, i.e $[E, G] \neq 0$. The possibility of detecting G and E, together with $F(\Delta)$, is ensured if two detectors T and Y of E and G respectively exist, of the form $T = \mathbf{1}_I \otimes T_{II}$ and $Y = \mathbf{1}_I \otimes Y_{II}$, satisfying $[T, Y] = \mathbf{0}$. In such a case it is possible to infer the outcomes of E and G by means of the outcomes of T and Y [6]. Our problem can be set out as follows:
(P) Given the property $E = E_I \otimes \mathbf{1}_{II}$ we want to find a projection operator G_I of $\mathscr{H}_I$, projection operators T_{II} and Y_{II} of $\mathscr{H}_{II}$, and a state vector $\Psi \in \mathscr{H}_I \otimes \mathscr{H}_{II}$ such that the following conditions are satisfied:

(C.1) $[E, G] \neq 0$ *i.e* $[E_I, G_I] \neq 0$
(C.2) $[T, E] = 0$ *and* $T\Psi = E\Psi$
(C.3) $[Y, G] = 0$ *and* $Y\Psi = G\Psi$
(C.4) $[T, Y] = 0$
(C.5) $\psi \neq E\psi \neq 0$ *and* $\psi \neq G\psi \neq 0$
(C.6) $[Y, F(\Delta)] = 0$

Condition (C.5) is added to exclude solutions corresponding to the uninteresting case that Ψ is an eigenvector of E or G. Condition (C.1)-(C-6) can be expressed in a more useful form if the following matrix representation is adopted. Let $(e_1, e_2, e_3, \ldots; r_1, r_2, r_3, \ldots)$ be an orthonormal basis of $\mathscr{H}_I$, formed by eigenvectors of E_I, such that $E_I e_k = e_k$ and $E_I r_j = 0$ for all k and j. Therefore, every vector $\Psi \in \mathscr{H}_I \otimes \mathscr{H}_{II}$ can be uniquely decomposed as $\Psi = \sum_j \mathbf{e}_j \otimes \mathbf{x}_j + \sum_k \mathbf{r}_k \otimes \mathbf{y}_k$ where $\mathbf{x}_j, \mathbf{y}_k \in \mathscr{H}_{II}$. Condition (C.4), $[T, Y] = 0$, implies that four projection operators, A_i $(i = 1, \ldots, 4)$, of $\mathscr{H}_{II}$ exist such that $\sum_1^4 A_i = \mathbf{1}$, $T_{II} = A_1 + A_2$ and $Y_{II} = A_1 + A_3$. Thereby, we choose to represent vectors $\mathbf{x}_j, \mathbf{y}_k \in \mathscr{H}_{II}$ as $\mathbf{x}_j = (\mathbf{a}_j, \mathbf{b}_j, \mathbf{c}_j, \mathbf{d}_j)^t$ and $\mathbf{y}_k = (\alpha_k, \beta_k, \gamma_k, \delta_k)^t$ where $\mathbf{a}_j = A_1 \mathbf{x}_j$, $\mathbf{b}_j = A_2 \mathbf{x}_j$, $\mathbf{c}_j = A_3 \mathbf{x}_j$, $\mathbf{d}_j = A_4 \mathbf{x}_j$, and $\alpha_j = A_1 \mathbf{y}_j$, $\beta_j = A_2 \mathbf{y}_j$, $\gamma_j = A_3 \mathbf{y}_j$, $\delta_j = A_4 \mathbf{y}_j$. Then $\Psi \in \mathscr{H}$ shall be represented as a column vector $\Psi = (\mathbf{x}_1, \mathbf{x}_2, \ldots, \mathbf{x}_j, \ldots; \mathbf{y}_1, \mathbf{y}_2, \ldots, \mathbf{y}_k, \ldots)^t$.

If this representation is adopted, operators E, T, G and Y are represented as matrices

$$E = E_I \otimes \mathbf{1} = \begin{pmatrix} \mathbf{1}_I & \mathbf{0} & . & \mathbf{0} & \mathbf{0} & . \\ \mathbf{0} & \mathbf{1}_I & . & \mathbf{0} & \mathbf{0} & . \\ . & . & . & . & . & . \\ \mathbf{0} & \mathbf{0} & . & \mathbf{0} & \mathbf{0} & . \\ . & . & . & . & . & . \end{pmatrix} \quad \text{and} \quad T = \mathbf{1} \otimes T_{II} = \begin{pmatrix} \mathbf{T}_{II} & \mathbf{0} & . & \mathbf{0} \\ \mathbf{0} & \mathbf{T}_{II} & . & \mathbf{0} \\ . & . & . & . \\ \mathbf{0} & \mathbf{0} & . & \mathbf{T}_{II} \end{pmatrix}$$

$$G = G_I \otimes \mathbf{1} = \begin{pmatrix} . & . & . & . \\ p_{ij}\mathbf{1}_{II} & . & u_{ij}\mathbf{1}_{II} & . \\ . & . & . & . \\ v_{ij}\mathbf{1}_{II} & . & q_{ij}\mathbf{1}_{II} & . \\ . & . & . & . \end{pmatrix} \quad \text{and} \quad Y = \mathbf{1} \otimes Y_{II} = \begin{pmatrix} \mathbf{Y}_{II} & \mathbf{0} & . & \mathbf{0} \\ \mathbf{0} & \mathbf{Y}_{II} & . & \mathbf{0} \\ . & . & . & . \\ \mathbf{0} & \mathbf{0} & . & \mathbf{Y}_{II} \end{pmatrix}$$

so that condition (C.2), $T\Psi = E\Psi$, implies $\mathbf{x}_i = (\mathbf{a}_i, \mathbf{b}_i, \mathbf{0}, \mathbf{0})^t$ and $\mathbf{y_j} = \left(\mathbf{0}, \mathbf{0}, \gamma_j, \delta_j\right)^t$, and condition (C.3), $Y\Psi = G\Psi$, is equivalent to

$$(i-A)\left\{ \begin{array}{l} \sum_i p_{ji}\mathbf{a}_i = \mathbf{a}_j \\ \sum_i p_{ji}\mathbf{b}_i = \mathbf{0} \end{array} \right. \qquad (ii-B)\left\{ \begin{array}{l} \sum_l u_{jl}\gamma_l = \mathbf{0} \\ \sum_l u_{jl}\delta_l = \mathbf{0} \end{array} \right. \qquad (iii-C)\left\{ \begin{array}{l} \sum_i v_{ki}\mathbf{a}_i = \mathbf{0} \\ \sum_i v_{ki}\mathbf{b}_i = \mathbf{0} \end{array} \right. \qquad (iv-D)\left\{ \begin{array}{l} \sum_l q_{kl}\gamma_l = \gamma_k \\ \sum_l q_{kl}\delta_l = \mathbf{0}. \end{array} \right.$$

So far we have established general constraints to be satisfied by every solution of the problem. We restrict the search to the case that the two slit are symmetrical: this leads to exclude odd dimension of $\mathscr{H}_I$ and, moreover, to assume that $rank(L) = rank(I - L) = dim(\mathscr{H}_I)/2$. The existence of solutions turns out to depend on dimension of $\mathscr{H}_I$ [5]. We neglect the cases $dim(\mathscr{H})_I \leq 4$ which provide no solution or correlated ones [5, 6]. We have found the following family of solutions in the case $dim(\mathscr{H}_I) = 6$, which is dependent on seven parameters: p, q, u, μ_2, μ_3, λ_2 and λ_3.

$$G_I = \begin{pmatrix} p & -\mu_2\left(p-\frac{\mu_3^2}{1+\mu_3^2}\right) & \mu_3(1-p) & u & -\lambda_2 u & -\lambda_3 u \\ -\overline{\mu_2}(p-\frac{\mu_3^2}{1+\mu_3^2}) & |\mu|^2(p-\frac{\mu_3^2}{1+\mu_3^2}) & \mu_3\overline{\mu_2}(p-\frac{\mu_3^2}{1+\mu_3^2}) & -\overline{\mu_2}u & \lambda_2\overline{\mu_2}u & \lambda_3\overline{\mu_2}u \\ \mu_3(1-p) & \mu_3\mu_2(p-\frac{\mu_3^2}{1+\mu_3^2}) & 1-\mu_3^2(1-p) & -\overline{\mu_3}u & \lambda_2\overline{\mu_3}u & \lambda_3\overline{\mu_3}u \\ \overline{u} & -\overline{\lambda_2}\overline{u} & -\overline{\lambda_3}\overline{u} & q & -\lambda_2(q-\frac{\lambda_3^2}{1+\lambda_3^2}) & \lambda_3(1-q) \\ -\mu_2\overline{u} & \overline{\lambda_2}\mu_2\overline{u} & \overline{\lambda_3}\mu_2\overline{u} & -\overline{\lambda_2}(q-\frac{\lambda_3^2}{1+\lambda_3^2}) & |\lambda_2|^2(q-\frac{\lambda_3^2}{1+\lambda_3^2}) & \lambda_3\overline{\lambda_2}(q-\frac{\lambda_3^2}{1+\lambda_3^2}) \\ -\mu_3\overline{u} & \overline{\lambda_2}\mu_3\overline{u} & \overline{\lambda_3}\mu_3\overline{u} & \lambda_3(1-q) & \lambda_3\lambda_2(q-\frac{\lambda_3^2}{1+\lambda_3^2}) & 1-\lambda_3^2(1-q) \end{pmatrix}$$

$$\Psi = (\mathbf{x}_1, \mathbf{x}_2, \mathbf{x}_3; \mathbf{y}_1, \mathbf{y}_2, \mathbf{y}_3)^t$$

where

$$\mathbf{x}_1 = (\mu_3\mathbf{a}_3, \tfrac{\mu_2}{\mu_3^2+1}\mathbf{b}_2, \mathbf{0}, \mathbf{0})^t \qquad \mathbf{x}_2 = (\mathbf{0}, \mathbf{b}_2, \mathbf{0}, \mathbf{0})^t \qquad \mathbf{x}_3 = (\mathbf{a}_3, -\tfrac{\mu_2\mu_3}{\mu_3^2+1}\mathbf{b}_2, \mathbf{0}, \mathbf{0})^t$$
$$\mathbf{y}_1 = (\mathbf{0}, \mathbf{0}, \lambda_3\gamma_3, \tfrac{\lambda_2}{\lambda_3^2+1}\delta_2, \mathbf{0})^t \qquad \mathbf{y}_2 = (\mathbf{0}, \mathbf{0}, \mathbf{0}, \delta_2)^t \qquad \mathbf{y}_3 = (\mathbf{0}, \mathbf{0}, \gamma_3, -\tfrac{\lambda_2\lambda_3}{\lambda_3^2+1}\delta_2)^t.$$

This is a solution of problem *(P)* if $\frac{\mu_3^2}{1+\mu_3^2} < p < \frac{\mu_3^2}{1+\mu_3^2} + \frac{1}{1+|\mu_2|^2+\mu_3^2}$, $u = e^{i\theta}\sqrt{\frac{\left(p-\frac{\mu_3^2}{1+\mu_3^2}\right)-(1+|\mu_2|^2+\mu_3^2)\left(p-\frac{\mu_3^2}{1+\mu_3^2}\right)^2}{1+|\lambda_2|^2+\lambda_3^2}}$, where θ is a real number and $q = \frac{1-(1+|\mu_2|^2+\mu_3^2)\left(p-\frac{\mu_3^2}{1+\mu_3^2}\right)}{1+|\lambda_2|^2+\lambda_3^2}$. Therefore for every real number $\frac{\mu_3^2}{1+\mu_3^2} < p < \frac{\mu_3^2}{1+\mu_3^2} + \frac{1}{1+|\mu_2|^2+\mu_3^2}$, every $\mu_3, \lambda_3, \theta \in \mathbf{R}$ and every $\mu_2, \lambda_2 \in \mathbf{C}$ we have a solution of *(P)*. No constraint is imposed to $rank(G_I)$, i.e. to the trace of the projection operator G, hence these parameters are not all independent. This family is a wide extension of the family put forward in [5, 6]. where a family depending on just two parameters was found.

Until now the treatment has been carried out on a theoretical ground only. Now an ideal experiment, that realizes the detection at issue, is proposed. The experimental set-up corresponds to the particular choice of parameters $\mu_2 = \lambda_2 = \sqrt{3}$ and $\mu_3 = \lambda_3 = 1$. The system (fig. 1) consists of an electrically neutral particle of spin $\frac{3}{2}$. Let $\mathscr{H}_{II}$ be the Hilbert space describing its spin. Let us suppose that, after crossing the screen with the slits, each particle passes through a non-uniform magnetic field, with gradient $\vec{\nabla}\vec{B}$ along direction z. The beam splits into four beams

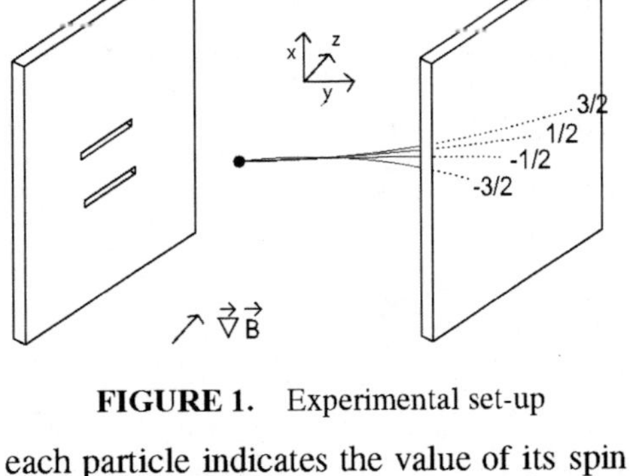

FIGURE 1. Experimental set-up

and the amount of the deflection of each particle indicates the value of its spin component. Let A be the projection operator representing the event "the spin component in the z-direction is $\frac{3}{2}$". Similarly we define operators B, C and D associated to the spin components $\frac{1}{2}$, $-\frac{1}{2}$ and $-\frac{3}{2}$, respectively. Let the state vector of the entire system be

$$\psi = \frac{1}{3}\left(1, \frac{\sqrt{3}}{2}, 0, 0, 1, 1, 0, 0, 0, -\frac{\sqrt{3}}{2}, 0, 0; 0, 0, 1, \frac{\sqrt{3}}{2}, 0, 0, 0, 1, 0, 0, 1, -\frac{\sqrt{3}}{2}\right)^t.$$

With respect to this state vector there exists a Which-Slit detector $T = \mathbf{1}_I \otimes (A+B)$. The projection operator $Y = \mathbf{1}_I \otimes (A+C)$ is a detector of property $G = G_I \otimes \mathbf{1}_{II}$, incompatible with property E, where

$$G_I = \frac{1}{30}\begin{pmatrix} 20 & -5\sqrt{3} & 10 & \sqrt{5} & -\sqrt{15} & -\sqrt{5} \\ -5\sqrt{3} & 15 & 5\sqrt{3} & -\sqrt{15} & 3\sqrt{5} & \sqrt{15} \\ 10 & 5\sqrt{3} & 20 & -\sqrt{5} & \sqrt{15} & \sqrt{5} \\ \sqrt{5} & -\sqrt{15} & -\sqrt{5} & 16 & -\sqrt{3} & 14 \\ -\sqrt{15} & 3\sqrt{5} & \sqrt{15} & -\sqrt{3} & 3 & \sqrt{3} \\ -\sqrt{5} & \sqrt{15} & \sqrt{5} & 14 & \sqrt{3} & 16 \end{pmatrix}$$

Therefore, the ideal experiment just described allows to detect both E and G, together with the measurement of the final impact point.

3. DETECTION OF FOUR INCOMPATIBLE PROPERTIES

In this section we deal with the question whether two incompatible properties, G and L, can be detected together with the further incompatible property E (Which-Slit property) and together with the measurement of the final impact point. The method used for such a research extends that presented in the previous section.

Let $G = G_I \otimes \mathbf{1}_{II}$ and $L = L_I \otimes \mathbf{1}_{II}$ be properties incompatible with WS property E. The detection of G, L and which-slit property E is possible if, with respect to the same state vector Ψ, there exists a which-slit detector $T = \mathbf{1}_I \otimes T_{II}$ of E, a detector $Y = \mathbf{1}_I \otimes Y_{II}$ of G and a detector $W = \mathbf{1}_I \otimes W_{II}$ of L. Let the detectors satisfy the condition $[T,Y] = [T,W] = [Y,W] = 0$ in such a manner that T, Y and W can be measured together; hence they provide simultaneous information about E, G and L. Formally, we have to solve the following more complicated problem:

(P') Given the property $E = E_I \otimes \mathbf{1}_{II}$ we have to find projection operators G_I and L_I of $\mathscr{H}_I$, projection operators T_{II}, Y_{II} and W_{II} of $\mathscr{H}_{II}$ and a state vector $\Psi \in \mathscr{H}_I \otimes \mathscr{H}_{II}$ such that the following conditions are satisfied:

(C.1) $[E,G] \neq 0$ *i.e* $[E_I, G_I] \neq 0$

(C.2) $[E,L] \neq 0$ *i.e* $[E_I, L_I] \neq 0$

(C.3) $[L,G] \neq 0$ *i.e* $[L_I, G_I] \neq 0$

(C.4) $[T,Y] = [T,W] = [Y,W] = 0$

(C.5) $[T,E] = 0$ *and* $T\Psi = E\Psi$

(C.6) $[Y,G] = 0$ *and* $Y\Psi = G\Psi$

(C.7) $[W,L] = 0$ *and* $W\Psi = L\Psi$

(C.8) $\psi \neq E\psi \neq 0$, $\psi \neq G\psi \neq 0$ *and* $\psi \neq L\psi \neq 0$

Following the same argument of previous section, conditions (C.6) and (C.7) are equivalent to

$$(i.A)\begin{cases} \sum_k p_{ik}\mathbf{a}_k = \mathbf{a}_i \\ \sum_k p_{ik}\mathbf{b}_k = \mathbf{b}_i \\ \sum_k p_{ik}\mathbf{c}_k = \mathbf{0} \\ \sum_k p_{ik}\mathbf{e}_k = \mathbf{0} \end{cases} \quad (ii.B)\begin{cases} \sum_k u_{ik}\delta_k = \mathbf{0} \\ \sum_k u_{ik}\zeta_k = \mathbf{0} \\ \sum_k u_{ik}\eta_k = \mathbf{0} \\ \sum_k u_{ik}\theta_k = \mathbf{0} \end{cases} \quad (iii.C)\begin{cases} \sum_k v_{ik}\mathbf{a}_k = \mathbf{0} \\ \sum_k v_{ik}\mathbf{b}_k = \mathbf{0} \\ \sum_k v_{ik}\mathbf{c}_k = \mathbf{0} \\ \sum_k v_{ik}\mathbf{e}_k = \mathbf{0} \end{cases} \quad (iv.D)\begin{cases} \sum_k q_{ik}\delta_k = \delta_i \\ \sum_k q_{ik}\zeta_k = \zeta_i \\ \sum_k q_{ik}\eta_k = \mathbf{0} \\ \sum_k q_{ik}\theta_k = \mathbf{0} \end{cases}$$

$$(i.A')\begin{cases} \sum_k m_{ik}\mathbf{a}_k = \mathbf{a}_i \\ \sum_k m_{ik}\mathbf{b}_k = \mathbf{0} \\ \sum_k m_{ik}\mathbf{c}_k = \mathbf{c}_i \\ \sum_k m_{ik}\mathbf{e}_k = \mathbf{0} \end{cases} \quad (ii.B')\begin{cases} \sum_k z_{ik}\delta_k = \mathbf{0} \\ \sum_k z_{ik}\zeta_k = \mathbf{0} \\ \sum_k z_{ik}\eta_k = \mathbf{0} \\ \sum_k z_{ik}\theta_k = \mathbf{0} \end{cases} \quad (iii.C')\begin{cases} \sum_k w_{ik}\mathbf{a}_k = \mathbf{0} \\ \sum_k w_{ik}\mathbf{b}_k = \mathbf{0} \\ \sum_k w_{ik}\mathbf{c}_k = \mathbf{0} \\ \sum_k w_{ik}\mathbf{e}_k = \mathbf{0} \end{cases} \quad (iv.D')\begin{cases} \sum_k n_{ik}\delta_k = \delta_i \\ \sum_k n_{ik}\zeta_k = \mathbf{0} \\ \sum_k n_{ik}\eta_k = \eta_j \\ \sum_k n_{ik}\theta_k = \mathbf{0}. \end{cases}$$

Now we present, by skipping the details of the derivation [8], a concrete solution, relative to the case that i, j, k take values in $\{1,2,3,4,5\}$ (i.e. $dim(\mathscr{H}_I) = 10$):

$$G_I = \frac{1}{72}\begin{pmatrix} 11 & -2 & -\frac{11}{72} & -22 & -9 & 8\sqrt{2} & -8\sqrt{2} & -8\sqrt{2} & 0 & 0 \\ -2 & 20 & 2 & -18 & -18 & -8\sqrt{2} & 8\sqrt{2} & 8\sqrt{2} & 0 & 0 \\ -11 & 2 & 11 & 9 & 9 & -8\sqrt{2} & 8\sqrt{2} & 8\sqrt{2} & 0 & 0 \\ -9 & -18 & 9 & 27 & 27 & 0 & 0 & 0 & 0 & 0 \\ -9 & -18 & 9 & 27 & 27 & 0 & 0 & 0 & 0 & 0 \\ 8\sqrt{2} & -8\sqrt{2} & -8\sqrt{2} & 0 & 0 & 19 & -10 & -19 & -9 & -9 \\ -8\sqrt{2} & 8\sqrt{2} & 8\sqrt{2} & 0 & 0 & -10 & 28 & 10 & -18 & -18 \\ -8\sqrt{2} & 8\sqrt{2} & 8\sqrt{2} & 0 & 0 & -19 & 10 & 19 & 9 & 9 \\ 0 & 0 & 0 & 0 & 0 & -9 & -18 & 9 & 27 & 27 \\ 0 & 0 & 0 & 0 & 0 & -9 & -18 & 9 & 27 & 27 \end{pmatrix},$$

$$L_I = \frac{1}{456}\begin{pmatrix} 67 & -10 & 5 & -9 & -129 & 32\sqrt{3} & -32\sqrt{3} & -32\sqrt{3} & 0 & 0 \\ -10 & 124 & -62 & -162 & -42 & -32\sqrt{3} & 32\sqrt{3} & 32\sqrt{3} & 0 & 0 \\ 5 & -62 & 139 & 153 & -87 & -32\sqrt{3} & 32\sqrt{3} & 32\sqrt{3} & 0 & 0 \\ -9 & -162 & 153 & 267 & 27 & 0 & 0 & 0 & 0 & 0 \\ -129 & -42 & -87 & 27 & 387 & 0 & 0 & 0 & 0 & 0 \\ 32\sqrt{3} & 32\sqrt{3} & -32\sqrt{3} & 0 & 0 & 171 & -114 & -99 & -9 & -129 \\ -32\sqrt{3} & 32\sqrt{3} & 32\sqrt{3} & 0 & 0 & -114 & 228 & 42 & -162 & -42 \\ -32\sqrt{3} & 32\sqrt{3} & 32\sqrt{3} & 0 & 0 & -99 & 42 & 243 & 153 & -87 \\ 0 & 0 & 0 & 0 & 0 & -9 & -162 & 153 & 267 & 27 \\ 0 & 0 & 0 & 0 & 0 & -129 & -42 & -87 & 27 & 387 \end{pmatrix},$$

$$\Psi = (\mathbf{x}_1, \mathbf{x}_2, \mathbf{x}_3, \mathbf{x}_4, \mathbf{x}_5; \mathbf{y}_1, \mathbf{y}_2, \mathbf{y}_3, \mathbf{y}_4, \mathbf{y}_5)^t,$$

where

$$\begin{aligned} \mathbf{x}_1 &= (-\tfrac{1}{3}\mathbf{a}_5, \mathbf{0}, -\tfrac{1}{3}\mathbf{c}_5, \mathbf{0}, \tfrac{1}{3}\mathbf{e}_4 + 2\mathbf{e}_5, \mathbf{0}, \mathbf{0}, \mathbf{0},)^t, & \mathbf{y}_1 &= (\mathbf{0}, \mathbf{0}, \mathbf{0}, -\tfrac{1}{3}\delta_5, \mathbf{0}, \mathbf{0}, -\tfrac{1}{3}\eta_5; \tfrac{1}{3}\theta_4 + 2\theta_5)^t, \\ \mathbf{x}_2 &= (-\tfrac{2}{3}\mathbf{a}_5, \mathbf{0}, \tfrac{1}{3}\mathbf{c}_5, \mathbf{0}, \mathbf{e}_4 + \mathbf{e}_5, \mathbf{0}, \mathbf{0}, \mathbf{0},)^t, & \mathbf{y}_2 &= (\mathbf{0}, \mathbf{0}, \mathbf{0}, -\tfrac{2}{3}\delta_5, \mathbf{0}, \mathbf{0}, \tfrac{1}{3}\eta_5, \theta_4 + \theta_5)^t, \\ \mathbf{x}_3 &= (\tfrac{1}{3}\mathbf{a}_5, \mathbf{0}, -\tfrac{2}{3}\mathbf{c}_5, \mathbf{0}, -\tfrac{2}{3}\mathbf{e}_4 + \mathbf{e}_5, \mathbf{0}, \mathbf{0}, \mathbf{0},)^t, & \mathbf{y}_3 &= (\mathbf{0}, \mathbf{0}, \mathbf{0}, \tfrac{1}{3}\delta_5, \mathbf{0}, \mathbf{0}, -\tfrac{2}{3}\eta_5, -\tfrac{2}{3}\theta_4 + \theta_5)^t, \\ \mathbf{x}_4 &= (\mathbf{a}_5, \mathbf{0}, -\tfrac{2}{3}\mathbf{c}_5, \mathbf{0}, \mathbf{e}_4, \mathbf{0}, \mathbf{0}, \mathbf{0},)^t, & \mathbf{y}_4 &= (\mathbf{0}, \mathbf{0}, \mathbf{0}, \delta_5, \mathbf{0}, \mathbf{0}, -\tfrac{2}{3}\eta_5, \theta_4)^t, \\ \mathbf{x}_5 &= (\mathbf{a}_5, \mathbf{0}, \mathbf{c}_5, \mathbf{0}, \mathbf{e}_5, \mathbf{0}, \mathbf{0}, \mathbf{0},)^t, & \mathbf{y}_5 &= (\mathbf{0}, \mathbf{0}, \mathbf{0}, \delta_5, \mathbf{0}, \mathbf{0}, \eta_5, \theta_5)^t, \end{aligned}$$

is a solution of *(P')*. However, since $(\mathbf{1} - W)Y\psi = 0$, the detections of G and L are correlated. Indeed, this equation implies that each time the outcome of G is 1, also the outcome of the detection of L must be 1 [9].

REFERENCES

1. J. von Neumann, *Mathematical Foundation of Quantum Mechanics*, Princeton University Press, 1955
2. N. Bohr, in *Albert Einstein: Philosopher-Scientist*, P.A. Schilpp, p.200, Library of Living Philosopher, Evanstone 1949.
3. R. Feynman, R. Leighton, M. Sand, *The Feynmann Lectures on Physics III*, Addison Wesley, Reading, 1965.
4. G.Nisticò, A. Sestito, Journal of Modern Optics, **51**, n. 6-7, (2004) 1063.
5. G. Nisticò, in *I Fondamenti della Meccanica Quantistica, Analisi storica e Problemi aperti*, C. Garola, A. Rossi, S. Sozzo (Eds.), World Scientific, Singapore (2006).
6. G. Nisticò, in *Foundations of Probability and Physics-4*, G. Adenier, C. Fuchs, A.Y. Khrennikov, eds, AIP Conferences Proceeding **889**, Melville 2007
7. G. Nisticò, M.C. Romania, J. Math. Phys. **35** (1994) 4534.
8. A. Sestito, arXiv:0707.3753
9. G. Nisticò, Found.Phys., **25** (1995) 1575.

Simulation of Deutsch-Jozsa Algorithm in Mathematica

Peter Nyman

International Center for Mathematical Modeling in Physics, Engineering and Cognitive science MSI, Växjö University, S-35195, Sweden

Abstract. This study examines the simulation of Deutsch-Jozsa algorithm in Mathematica. The program code implemented on a classical computer will be a straight connection between the mathematical formulation of quantum mechanics and computational methods in Mathematica. This program code will be a foundation of a universal simulation language.

Keywords: Deutsch-Jozsa algorithm, quantum computing, Mathematica
PACS: 03.67.Lx, 02.70.-c, 07.05.Tp

INTRODUCTION

A general quantum simulation language on a classical computer will open the opportunity to compare an experiential result from the developments of quantum computers with the mathematical theory. Our aim is therefore to construct a general program code where it will be easy to implement algorithms. A couple of examples of various methods for the simulation of quantum algorithms were given in [1, 2, 3]. We have in previous research [4] studied Shor's algorithm [5] and implemented this algorithm in the high-level language Mathematica. A simulation of a quantum algorithm on classical computers will give us the possibility to compare the results of quantum computers with the output of the physically more stable classical computers. In the development of quantum algorithms it will be interesting to test new algorithms on a classical computer. This article will describe the connection between future quantum computers and today's simulations of quantum computers. Thus, the computational language will include the quantum mechanical terminology such as quantum operators and quantum states. This simulation language will include the most essential operations in quantum computing. Examples of these operations will be the *Hadamard*, *Controlled not* and *Toffoli* gates. In earlier studies (see [6]) we have introduced a framework in a computational language constructed on the formulation of quantum mechanics. In this study we will go further and use this framework when we implement *Deutsch-Jozsa algorithm* as an example of a quantum algorithm. The aim is to construct a computational language and describe a straightforward connection between *Dirac's* mathematical formulation of quantum mechanics and our program code. However, the mathematically described algorithms will have a clear mathematical structure even after that we have implemented this algorithms as a program code. Moreover a program with mathematical structure will give us short program codes and this simulation of Deutsch-Jozsa algorithm will use only few lines of program code. This simulation is a link from the mathematical theory of quantum algorithms to its implementation on a quantum computer. Let us begin with a demonstration of the framework we have constructed in the high-level program language *Mathematica*, this will probably be used for future algorithms.

THE SIMULATION FRAMEWORK

This section we will introduce a framework constructed for the simulation of quantum algorithms on classical computers. We will point out that there is a symbolic similarity between our framework and the mathematical framework. This framework will be a computational dual to *Dirac's bra-ket notation*. A quantum state in n dimensions can be represented by a linear combination of n numbers of basis vectors. In the two-dimensional case a quantum state $|\phi\rangle$ is represented as a superposition of two basis vectors ,say $|0\rangle$ and $|1\rangle$, known as computational basis (computational basis, see [7, 8]). In this basis a quantum state $|\phi\rangle$ is represented as

$$|\phi\rangle = \alpha|0\rangle + \beta|1\rangle, \tag{1}$$

CP962, *Quantum Theory, Reconsideration of Foundations – 4*,
edited by G. Adenier, A. Yu. Khrennikov, P. Lahti, V. I. Man'ko, and Th. M. Nieuwenhuizen

where α and β are complex numbers such as $|\alpha|^2 + |\beta|^2 = 1$. We will introduce some new symbols for the states of the computational basis as follows: $\mathrm{e}[0] = |0\rangle$ and $\mathrm{e}[1] = |1\rangle$. This is the foundation for the structure of the program code. For more than one-qubit we will use the computational basis states $\mathrm{e}[x_1, \ldots, x_n] = |x_1 \ldots x_n\rangle$, where $x_j \in \{0,1\}$ or by using the more compact notation $\mathrm{e}[y] = |y\rangle$, where $y = x_n 2^0 + \cdots + x_1 2^{n-1}$. We will, write the state ϕ as $\mathrm{e}[\phi] = \alpha \mathrm{e}[0] + \beta \mathrm{e}[1]$, by analogy to (1). The operator A acts on the state ϕ and is often written as $A|\phi\rangle$ in the quantum mechanical literature. To match these symbols, we will use the computational symbols $A|\mathrm{e}[\phi]$ for this operation.A computational problem will be that the computer must regard expressions equal if they have identical meaning even if these notations are not identical. As an example the expression $\mathrm{e}[0, \mathrm{e}[1], 1]$ must be equal to $\mathrm{e}[0,1,1]$ in the code. We can bring in the command $\mathrm{e}[0, \mathrm{e}[1], 1] := \mathrm{e}[0,1,1]$ or the more general $\mathrm{e}[a\_\_, \mathrm{e}[b\_\_], c\_\_] := \mathrm{e}[a,b,c]$ to solve this problem. Moreover, the program code must be able to handle the linearity of the tensor product. Let $\mathrm{e}[\,.\,]$ be vectors and α a complex number. We define the tensor product as

$$\alpha(\mathrm{e}[v] \otimes \mathrm{e}[w]) = (\alpha \mathrm{e}[v]) \otimes \mathrm{e}[w] = \mathrm{e}[v] \otimes (\alpha \mathrm{e}[w]) \tag{2}$$

$$(\mathrm{e}[v_1] + \mathrm{e}[v_2]) \otimes \mathrm{e}[w] = \mathrm{e}[v_1] \otimes \mathrm{e}[w] + \mathrm{e}[v_2] \otimes \mathrm{e}[w] \tag{3}$$

$$\mathrm{e}[v] \otimes (\mathrm{e}[w_1] + \mathrm{e}[w_2]) = \mathrm{e}[v] \otimes \mathrm{e}[w_1] + \mathrm{e}[v] \otimes \mathrm{e}[w_2]. \tag{4}$$

We add two commands to the program code that will implement this definition of the tensor product. The command

```
e[a___, α_. e[x__], b___]  := α e[a, x, b]
```

will transform $\mathrm{e}[a] \otimes \alpha \mathrm{e}[x] \otimes \mathrm{e}[c]$ to $\alpha \mathrm{e}[a \otimes x \otimes b] = \alpha \mathrm{e}[a,x,b]$. This command is the computational dual to the tensor expression in Dirac's notation $|a\rangle \otimes \alpha|x\rangle \otimes |b\rangle = \alpha|axb\rangle$. The other command

```
e[a___, ξ_. (α_. e[x__] + β_. e[y__]), b___]:=ξαe[a, x, b]+ ξβe[a, y, b]
```

will transform $\mathrm{e}[a] \otimes \xi(\alpha \mathrm{e}[x] + \beta \mathrm{e}[y]) \otimes \mathrm{e}[b]$ to $\xi \alpha \mathrm{e}[a,x,b] + \xi \beta \mathrm{e}[a,y,b]$. Let U be an arbitrary one-qubit quantum gate. Then U will transform an arbitrary state $\mathrm{e}[\phi]$ which is represented in the computational basis states as $\mathrm{e}[\phi] = a\mathrm{e}[0] + b\mathrm{e}[1]$ to the state $U|\mathrm{e}[\phi] \rightarrow a(c_1 \mathrm{e}[0] + c_2 \mathrm{e}[1]) + b(c_3 \mathrm{e}[0] + c_4 \mathrm{e}[1])$, where a, b, c_i are complex numbers. We add the *Mathematica* gate U to the program code as follows $U|\mathrm{e}[0] \rightarrow c_1 \mathrm{e}[0] + c_2 \mathrm{e}[1]$ and $U|\mathrm{e}[1] \rightarrow c_3 \mathrm{e}[0] + c_4 \mathrm{e}[1]$. For example, the Hadamard gate H will be added in Mathematica as the command H:=$\{\mathrm{e}[0] \rightarrow 1/\sqrt{2}(\mathrm{e}[0] + \mathrm{e}[1]), \mathrm{e}[1] \rightarrow 1/\sqrt{2}(\mathrm{e}[0] - \mathrm{e}[1])\}$. We will define a one-qubit gate O_i as an operator which acts on the qubit in position i and leaves the other qubits unchanged. The program code must be able to operate with a gate on an arbitrary qubit. Consequently we will define an operator O_i in the Mathematica code. Defined the operator O_i as $O_i = I^{\otimes i-1} \otimes U \otimes I^{\otimes n-i}$ which acts on n-qubits, where I is the one-qubit unit operator and U is an arbitrary one-qubit operator. Then operator O_i is a function of $O_i|\mathrm{e}[v] \rightarrow \mathrm{e}[\psi]$. Similarly, we will define $O_{i,j}$ as an operator which operates as the two-qubits operator on the qubits in positions i, j and leaves the other qubits unchanged. Now we have the tools to build quantum circuits.

AN INTRODUCTION TO THE DEUTSCH-JOZSA ALGORITHM

The function $f : \{0,1\}^n \rightarrow \{0,1\}$ is constant if all output is 0 or 1 for all input. The function f is balanced if half of all outputs are 0. The Deutsch-Jozsa problem was introduced by David Deutsch and Richard Jozsa with the task to decide the function f property, which will be constant or balanced (see [9]). The most naive classical algorithm will need $2^{n-1} + 1$ outputs to decide the property of the function. Deutsch and Jozsa showed that the Deutsch-Jozsa algorithm would solve this task with only one output. In the implementation of Deutsch-Jozsa algorithm we will use to the unitary operator Uf which we defined as

$$\mathrm{Uf} : |x\rangle^{\otimes n}|y\rangle = |x\rangle^{\otimes n}|y \oplus f(x)\rangle$$

where $\oplus$ is the binary addition modulo 2. We will use the function Uf := e[i__,j_] :→ e[i, Mod[j + f[i], 2]] to apply this operator in our simulation language. Let us describe Deutsch-Jozsa problem in the following example, Bob choose a function with one of the two properties, constant or balanced, and Alice task is to decide its property. To her delight she will get Bob to calculate this function in a quantum computer. Bob prepare the register in superposition before he operates with the operator Uf that will act on all qubits at once. Before Bob measure the register and send the result to Alice he will apply the Hadamard gate to the $n-1$ first qubits. If Bob sends $|0\rangle^{\otimes n}$ to Alice then she will be sure that his function was constant, otherwise she will make a conclusion that the function was balanced.

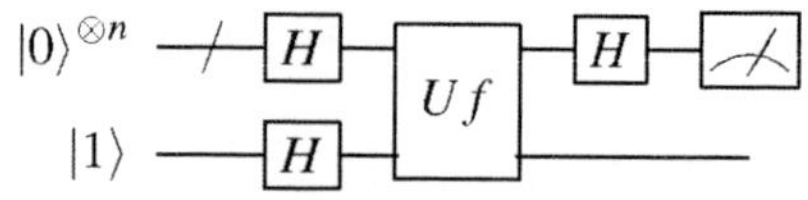

Figure 1. The Deutsch-Jozsa algorithm circuit

THE DEUTSCH-JOZSA ALGORITHM IN A SYMBOLIC LANGUAGE

Let us use our simulation language to help Alice solve her problem to decide the functions property. You may follow the algorithm in the circuit in figure 1. We start on the left with the register $\mathrm{e}[0,\cdots,0,1]$ and operate Hadamard gate $H{:=}\{\mathrm{e}[0]\rightarrow 1/\sqrt{2}(\mathrm{e}[0]+\mathrm{e}[1]),\mathrm{e}[1]\rightarrow 1/\sqrt{2}(\mathrm{e}[0]-\mathrm{e}[1])\}$ on this state which will place all qubits in superposition

$$\psi_1 = \frac{1}{\sqrt{2^{n+1}}} \sum_{x\in\{0,1\}^n} \mathrm{e}[x](\mathrm{e}[0]-\mathrm{e}[1]).$$

At this point we will use the operator Uf and the register will be in the state ψ_2

$$\psi_2 = \frac{1}{\sqrt{2^{n+1}}} \sum_{x\in\{0,1\}^n} (-1)^{f(x)}\mathrm{e}[x](\mathrm{e}[0]-\mathrm{e}[1]).$$

Before measurement the algorithm will apply the Hadamard operation on all qubits except the last one and

$$\psi_3 = \frac{1}{2^{n+1/2}} \sum_{z\in\{0,1\}^n} \sum_{x\in\{0,1\}^n} (-1)^{x\cdot z+f(x)}\mathrm{e}[z](\mathrm{e}[0]-\mathrm{e}[1]),$$

where $x\cdot z = x_1z_1+x_2z_2+\cdots+x_nz_n \mod 2$. Measurement of the first n will give the output $\mathrm{e}[0,\cdots,0]$ if and only if f is constant, otherwise it is balanced. This will follow from the fact that the probability to measure $\mathrm{e}[0,\cdots,0]$ is

$$\frac{1}{2^n}\Big|\sum_{x\in\{0,1\}^n} (-1)^{x\cdot\mathbf{0}+f(x)}\Big|^2 \tag{5}$$

and it will be one if $f(x)=0$ or $f(x)=1$ for all x where $\mathbf{0}=\{0,0,\ldots,0\}$. Moreover, it will be zero probability to measure $\mathrm{e}[0,\cdots,0]$ if f is balanced since (5) will be equal to zero.

THE DEUTSCH-JOZSA ALGORITHM IN MATHEMATICA

The Deutsch-Jozsa algorithm is implemented as follow. We will begin to define characteristic properties of quantum computers in the program code.

Listing 1: Definition of register and quantum gates in mathematica.

```
e[a___, α_. e[x__], b___] := α e[a, x, b]
e[a___, ξ_. (α_. e[x__] + β_. e[y__]), b___]:=ξαe[a, x, b]+ ξβe[a, y, b]
O_i_|v_:= Chop[Expand[v /. (e[x__] :→ ReplacePart[e[x], e[{x}[[i]]] /. O,i])]]
O_i_j_|v_:= Chop[Expand[v /. O]]
H := {e[0] :→1/√2 (e[0] + e[1]),e[1] :→1/√2 (e[0]    - e[1])}
Uf := {e[i___, j_] :→ e[i, Mod[j + f[i], 2]]}
Enlarge[ψ_,i_] := ψ /. ψ -> (ψ /.e[x_]:→ e[Sequence@@Table[0, {i - 1}], x])
```

The essential points of this framework is linearity and the tensor products. We have described before how tensor products and a register will be represented in our simulation language. The two-first lines (see listing 1) are computationally powerful and they will manage to implement some of the essential properties in simulation as linearity, superposition and tensor products. Line 3-4 in the same listing will define the operator mapping one or two qubits respectively. The Deutsch- Jozsa algorithm only need to apply the two operators H and Uf. The last line will enlarge a state of the one qubit e[1] to the n-qubit state e[0,0,0,...,1]. The operators, linearity, superposition and tensor products are now defined. The implementation of the algorithm will follow from the listing 2.

Listing 2: Deutsch-Jozsa algorithm in mathematica.

```
q = 8;
φ = Enlarge[e[1],q];
Do[φ = (H_i | φ), {i,q}];
φ = Uf_{q-1,q} | φ;
φ = φ /. {f[1, x__]:→0,f[0, x__]:→1};
Do[φ = (H_i | φ), {i,q - 1}];
φ
```

There are certain advantages to compare this listing 2 with the circuit (1). By using the command **Enlarge** on state e[1], defined in line 2, will the register will be prepared in the q-qubit state e[0,0,0,...,1]. In listing 2 on line 3-4 the Hadamard operator and the unitary function are applied on all the qubits in the register. In line 5 we will be able to choose the function property. In this example we have chosen a balanced function. The algorithm will output the result when the Hadamard operator have been applied on the $q-1$ first qubits. The output will contain zeros in $q-1$ first qubits if and only if f is constant, otherwise it is balanced.

CONCLUSION

In this study we have constructed a computational language for simulations of quantum algorithms and presented a program code for Deutsch-Jozsa algorithm. We have also demonstrated a general framework for simulation of quantum computers on classical computers. An important future challenge is to develop this computational language to include the most of the well-known quantum algorithms.

ACKNOWLEDGMENTS

I would like to thank my supervisor Prof. Andrei Khrennikov for fruitful discussions on the foundations of quantum computing. I am also grateful to Yaroslav Volovich for his involvement and ideas in this research.

REFERENCES

1. J.M. Burdis B .Juliá-Díaz and F. Tabakin. Qdensity-a mathematica quantum computer simulation. *Computer Physics Communications*, 174:914–934, 2006.
2. P. Dumai H. Touchette. Qucalc - the quantum computation package for mathematica. http://crypto.cs.mcgill.ca/QuCalc/, October 2007.
3. L. Spector. *Automatic Quantum Computer Programming: A Genetic Programming Approach*. Springer, 2004.
4. P. Nyman. Simulation of quantum algorithms with a symbolic programming language. *ArXiv e-prints*, 705, may 2007.
5. P.W. Shor. Polynomial-time algorithms for prime factorization and discrete logarithms on a quantum computer. *SIAM Journal on Computing*, 26(5):1484–1509, 1997.
6. P. Nyman. Quantum computing - declarative computational methods for simulation of quantum algorithms and quantum errors. Master's thesis, Växjö University, Sweden, 2005.
7. M.A. Nielsen and I.L. Chuang. *Quantum Computation and Quantum Information*. Cambridge University Press, October 2000.
8. M.Hirvensalo. *Quantum Computing*. Springer Series on Natural Computing. Springer, 1st edition edition, 2001.
9. D. Deutsch and R. Jozsa. Rapid solution of problems by quantum computation. *Proc Roy Soc Lond A*, 439:553–558, October 1992.

Geometrization of Classical Wave Fields

Oleg A.Olkhov

N.N.Semenov Institute of Chemical Physics, Kosygin str.,4, 119991 Moscow, Russia

Abstract. Geometrical interpretation of quantum physics is suggested. It is shown that the Dirac equation for free quantum particles can be considered as a relation describing propagation of the space topological defect with wave and corpuscular properties. This explains all "strange" properties of quantum formalism: appearance of probabilities, absence of the particles trajectories, unremovable influence of the measurement procedure, instantaneous nonlocal correlation in EPR–experiments. The same interpretation is suggested for the Maxwell equations for free electromagnetic waves. This interpretation explains independence of light velocity on the source movement.

Keywords: quantum mechanics, geometrization, topological defects, EPR-paradox, many-body problem
PACS: 03.65.Ca, 03.65.Ta

INTRODUCTION

We will show within unified approach, that the free matter field (quantum particle with spin $1/2$) and free electromagnetic waves can both be considered not as something existing into the space that plays the role of scene but as a propagating topological defects of the space itself. These defects appear to be embedded into five–dimensional space, and observable quantum object and electromagnetic wave are intersections of the defects with the three-dimensional physical space. Their wave properties are a consequence of the periodical movement of the defect in the "outer" space– just this movement attributes phase to the particle and so defines its wave properties. The classical notions of mass, energy and momentum are expressed through the parameters with dimensionality of length, and the Planck constant and the light velocity play the role of coefficients of transfer from one system of units to the other. It is shown that light velocity can be considered as a topological invariant, and this is the reason for its independence of the source motion. The uncertainty of shapes of topological manifolds is the reason of the appearance of probabilistic description. Preliminary results were published earlier in [1].

TOPOLOGICAL INTERPRETATION OF THE DIRAC EQUATION

This equation has the following symbolic form (see, e.g.,[2])

$$i\gamma^{\mu}\partial_{\mu}\psi = m\psi, \tag{1}$$

where $\partial_{\mu} = \partial/\partial x_{\mu}$, $\mu = 1,2,3,4$, $\psi(x)$ is the four-component Dirac bispinor, $x_1 = t, x_2 = x, x_3 = y, x_4 = z$, and γ^{μ} are four-row Dirac matrices. The summation in Eq.(1) goes over the repeating indices with a signature $(1,-1,-1,-1)$. Here, $\hbar = c = 1$. For definite values of 4-momentum p_{μ}, the solution to Eq.(1) has the form of the plane wave

$$\psi = u(p_{\mu})\exp(-ip_{\mu}x^{\mu}), \tag{2}$$

where $u(p_{\mu})$ is a normalized bispinor.

Different wave fields are described by different tensors realizing different representations of the Lorentz group. Dirac bispinor consists of two–component spinors that realizes two–dimensional representation of the Lorentz group, and this is the formal reason to classify the Dirac wave field of Eq.(1) as the object with "internal" angular momentum whose Z–projection equals to $\pm 1/2$ [3]. Consider, firstly, transformation properties of the Dirac bispinor from the other point of view. It is known that there is correspondence between every kind of tensors and some class of geometrical objects in the sense that these tensors define invariant properties of above objects. For example, usual vectors correspond to simplest geometrical objects—to points [4], and this is the reason why Newtonian mechanics uses vectors within its formalism. From this point of view spinors correspond to nonorientable geometrical object (see,

CP962, *Quantum Theory, Reconsideration of Foundations – 4*,
edited by G. Adenier, A. Yu. Khrennikov, P. Lahti, V. I. Man'ko, and Th. M. Nieuwenhuizen

FIGURE 1. Stochastic behavior

e.g., [5]). Therefore, we suppose that spinors are used in Eg.(1), because this equation describes some nonorientable geometrical object and "$spin = 1/2$" is a formal expression of the nonorientable property of the object.

To define properties of the proposed geometrical object more exactly we consider more precisely the symmetry properties of the solution of Eq.(1). We rewrite function (2) in the form

$$\psi = u(p_\mu)\exp(-2\pi i x^\mu \lambda_\mu^{-1}), \quad \lambda_\mu = 2\pi p_\mu^{-1}, \quad \lambda_1^{-2} - \lambda_2^{-2} - \lambda_3^{-2} - \lambda_4^{-2} = \lambda_m^{-2}, \quad \lambda_m = 2\pi m^{-1}. \tag{3}$$

Function (3) is an invariant with respect to coordinates transformations

$$x_\mu^{'} = x_\mu + n_\mu \lambda_\mu, \quad n_\mu = 0, \pm 1, \pm 2, ... \quad . \tag{4}$$

Transformations (4) can be considered as elements of the discrete group of translations operating in the 4-space where wave function (3) is defined. Then function (3) can be considered as a vector realizing this group representation.

As a bispinor, function (3) realizes representation of one more group of the symmetry transformation of 4-space. Being a four-component spinor, $\psi(x)$ is related to the matrices γ^μ by the equations (see, e.g. [3])

$$\psi^{'}(x^{'}) = \gamma^\mu \psi(x),$$

where $x \equiv (x_1, x_2, x_3, x_4)$, and $x^{'} \equiv (x_1, -x_2, -x_3, -x_4)$ for $\mu = 1, x^{'} \equiv (-x_1, x_2, -x_3, -x_4)$ for $\mu = 2$, and so on. This means that the matrices γ^μ are the matrix representation of the group of reflections along three axes perpendicular to the x_μ axis, and the Dirac bispinors realize this representation [6].

Taken together, above two groups form a group of four sliding symmetries with perpendicular axes (sliding symmetry means translations plus corresponding reflections [7]). The physical space-time does not have such symmetry. So, this group may operate only in some auxiliary space. From the other hand, it is known that discrete groups operating in some space can reflect a symmetry of geometrical objects that have nothing in common with this space. It will be the case when such space is a universal covering space of some closed topological manifold. Universal covering spaces are auxiliary spaces that are used in topology for the description of closed manifolds, because discrete groups operating in these spaces are isomorphic to fundamental groups of manifolds—groups whose elements are different classes of closed pathes on manifolds (so called π_1 group [8,9]). And we assume, that function (3) realizes a representation of the fundamental group of some closed nonorientable topological 4-manifold—a specific curved part of the space-time. Equation(1) describes this manifold, and the space-time plays also the role of an universal covering space for above manifold.

At the present time, only two-dimensional euclidean closed manifolds are classified in details, and their fundamental groups and universal covering planes are identified [8]. Therefore, we have no opportunity for rigorous consideration of specific properties of suggested psudoeuclidean 4-manifold. But qualitative properties, explaining main ideas of new interpretation, can be investigated using one of the advantages of geometrical approach—possibility of employment of low-dimensional analogies. Using these analogies we will show within elementary topology that above mentioned 4-manifold represents propagation of the space topological defect that can demonstrate specific properties of quantum particles: stochastic behavior and wave-corpuscular dualism.

STOCHASTIC BEHAVIOR

Consider the simplest example of closed manifolds—manifold homeomorphic to a circle with given perimeter length λ. The closed topological manifold is representable by any of its possible deformations without pasting and discontinuities, and we will see that just this property explains appearance of probabilities in quantum formalism. For simplicity we consider plane deformations of the circle(some of the possible deformations are shown at Fig.1).

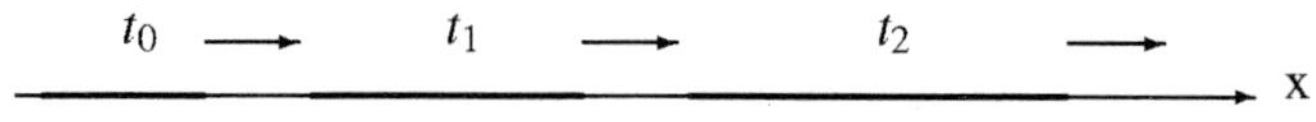

FIGURE 2. Propagation of the spreading region of the space

To use concrete simple calculations, we consider only all possible manifold's deformations that have a shape of ellipse with perimeter length λ. The equation for the ellipse on an euclidean plane has the form

$$x^2/a^2 + y^2/b^2 = 1, \tag{5}$$

where possible values of the a and b are connected with λ by the known approximate relation

$$\lambda \simeq \pi[1,5(a+b) - (ab)^{1/2}]. \tag{6}$$

This means that possible values of a is defined by the inequality $a_{min} \leq a \leq a_{max} \simeq \lambda/1,5\pi, \quad a_{min} \ll a_{max}$.

In the pseudoeuclidean "space-time," Eq.(5) for our ellipses has the form (after substitution $y = it$)

$$x^2/a^2 - t^2/b^2 = 1, \tag{7}$$

and this equation defines the dependence on time t for a position of the point x of the manifold corresponding to definite a. At $t = 0, x = \pm a$; that is, our manifold is represented by the two point sets in one-dimensional euclidean space, and the dimensions of these point sets are defined by all possible values of a. So, at $t = 0$, the manifold is represented by two regions of the one–dimensional euclidean space $a_{min} \leq |x| = a \leq a_{max}$. It can easily be shown that at $t \neq 0$ these regions increase and move along the x-axis in opposite directions (at Fig.2 this movement is shown only for positive direction).

All another possible deformations of our circle will be obviously represented by points of the same region, and every such point can be considered as a possible position of the "quantum object" described by our manifold. All manifold's deformations are realized with equal probabilities (there are no reasons for another suggestion). Therefore, all possible positions of the point–like object into the region are realized with equal probabilities. So, this example shows the possibility of the consideration of above object as a point with probability description of its positions as it suggested within standard representation of quantum particles. In fact, these points are yet a geometrical points only. Below we wille show, how this point becomes the material point. Note also that parameter λ defines the minimal size of a region$\Delta x \sim \lambda$, where the object can be localized (region at $t = 0$).

TOPOLOGICAL DEFECT. WAVE-CORPUSCULAR PROPERTIES

Above example does not explain what properties allow to differ points of the moving region from neighbour points of the euclidean space making them observable. To answer at this question we consider more complex analogy of the closed 4-manifold—two-dimensional torus. Such torus is denoted as topological production of two circles–$S^1 \times S^1$. The role of different manifold's deformations as a reason for stochastic behaviour was considered in Section 2. Therefore, now we restrict our consideration to one simplest configuration when one of S^1 is a circle in the plane XY and another is a circle in the plane ZX (we denote it as S'_1).

In pseudueuclidean space this torus looks like a hyperboloid. The hyperboloid appears if we replace the circle S'_1 by a hyperbola (as it was done in Section 2). Positions of the geometrical object described by our pseudoeucliden torus are defined by time cross-sections of the hyperboloid. These positions form an expanding circles in the two-dimensional euclidean plane (Fig.3).

Above two-dimensional pseudoeuclidean torus describes the object existing into two-dimensional space-time with one-dimensional euclidean "physical" space. This means that an observable part of the object is represented in our example by the points of intersections of above circle with $0X$ axis though, as a whole, the circle is "embedded" into two-dimensional, "external" space. This circle can be considered as a topological defect of the physical one-dimensional euclidean space. Just an affiliation of the intersection points to the topological object differs these points from neighboring points of the one-dimensional euclidean space. So, in pseudoeuclidean four-dimensional physical space-time the suggested object described by the Dirac equation looks like a topological defect of physical euclidean

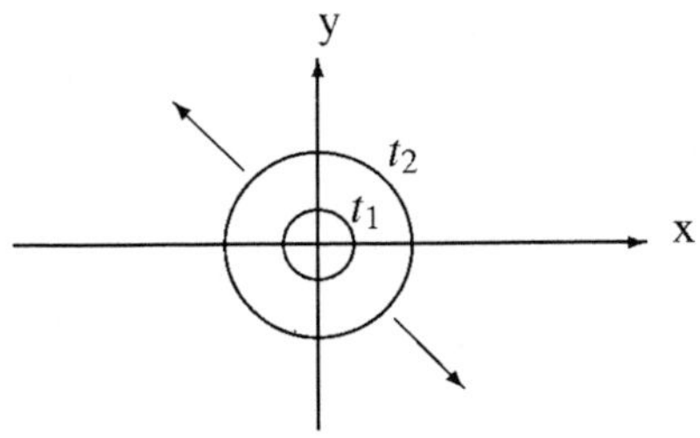

FIGURE 3. Topological defect

3-space that is embedded into 5-dimensional euclidean space, and its intersection with physical space represents an observable quantum object.

This analogy with torus does not yet demonstrate appearance of any wave-corpuscular properties of the object, represented in "physical" one-dimensional space by the moving intersection point— properties that is expressed by wave function (3). In the case of two-dimensional "space–time" this solution has the form

$$\psi = u(p)\exp(-2\pi i x^1 \lambda_1^{-1} + 2\pi i x^2 \lambda_2). \tag{8}$$

Topological defect represented by the expanding circle does not demonstrate any periodical movements when the intersection point (physical object) propagates along one-dimensional euclidean $0X$–space.

We will see below that appearance of observable wave-corpuscular properties is a consequence of nonorientable character of the topological defect. Torus is a orientable two-dimensional closed manifold and, therefore, we need to use some nonorientable low-dimensional analogy. The nonorientable Klein bottle could be such two–dimensional analogy [6,7]. In the case with torus topological defect was represented by cross-sections of pseudoeuclidean torus–plane circles. The Klein bootle is a manifold that is obtained by gluing of two Mobius strips [10]. Therefore, the Klein bootle cross-section is an edge of the Mobius strip. This edge can not be placed in the two–dimensional XY–plane without intersections, and it means that corresponding topological defect is now a closed curve embedded into three–dimensional XYZ–space.

In this case the position of the topological defect relative to its intersection with $0X$ axis (physical object) can change periodically. Parameters of this periodical movement depend on geometrical parameters λ_1, λ_2 (there are no other parameters with corresponding dimensionality). Such periodical process can be expressed by the function (8). It leads to the new interpretation of the wave function as a description of periodical movement of the topological defect relative to its projection on the physical space. Corpuscular properties of the above periodical movement appear as a result of the definition for classical notion of 4-momentum trough the wave characteristic of the topological object, namely

$$p_\mu = 2\pi/\lambda_\mu. \tag{9}$$

Substitution of these relations into (8) leads to the Dirac solution (2)

$$\psi = u(p)\exp(-ip_1x^1 + ip_2x^2). \tag{10}$$

It is important to note that within suggested geometrical interpretation the notions of the less general, macroscopic theory (4-momentums) are defined by (9) trough the notions of more general microscopic theory (wave parameters of the defect periodical movement). This looks more natural than the opposite definitions (3) within traditional interpretation.

GEOMETRIZATION OF ELECTROMAGNETIC WAVES

Let us write the Maxwell equations for electromagnetic waves in vacuum in the symbolic form analogous to the Dirac equation (1). Namely, we write these equations in the Majorana form [11]

$$i\partial \mathbf{f}^+/\partial t = (\mathbf{Sp})\mathbf{f}^+, \quad i\partial \mathbf{f}^-/\partial t = -(\mathbf{Sp})\mathbf{f}^-, \quad \mathbf{pf}^+ = 0, \quad \mathbf{pf}^- = 0,$$

$$\mathbf{f}^+ = \mathbf{E} + i\mathbf{H}, \quad \mathbf{f}^- = \mathbf{E} - i\mathbf{H}. \tag{11}$$

where Here **E** is an electric field, **H** is a magneyic field, $\mathbf{p} = -i\nabla$ and **S** is a vector-matrix

$$S_x = \begin{pmatrix} 0 & 0 & 0 \\ 0 & 0 & -i \\ 0 & i & 0 \end{pmatrix}, \quad S_y = \begin{pmatrix} 0 & 0 & i \\ 0 & 0 & 0 \\ -i & 0 & 0 \end{pmatrix}, \quad S_z = \begin{pmatrix} 0 & -i & 0 \\ i & 0 & 0 \\ 0 & 0 & 0 \end{pmatrix}. \tag{12}$$

Equations (11) may be rewritten in the form analogous to the symbolic form of the Dirac equation (1)

$$i\Gamma^{\mu}\partial_{\mu}\mathbf{f} = 0, \tag{13}$$

where bivector **f** and matrix Γ^{μ} have the form

$$\Gamma^0 = \begin{pmatrix} 0 & 1 \\ 1 & 0 \end{pmatrix}, \quad \Gamma^{1,2,3} = \begin{pmatrix} 0 & -\mathbf{S} \\ \mathbf{S} & 0 \end{pmatrix}, \quad \mathbf{f} = \begin{pmatrix} \mathbf{f}^{+} \\ \mathbf{f}^{-} \end{pmatrix}.$$

(we wrote six-row matrices trough three-row ones). Equations (13) has formally the same form as the Dirac equation (1) with $m = 0$. Only instead of the Dirac bisinor ψ we have here bivector **f**, and instead Dirac matrices γ^{μ} we have matrices Γ^{μ}. As Maxwell's Eqs.(1,) looks formally like Dirac's Eqs.(1) it seems reasonable to use for their geometrization the same arguments as we used for the Dirac equation geometrization.

For plane waves the solution of Eq.(13) has the form

$$\mathbf{f}^{+} = \mathbf{f}_k^{+} \exp i(\mathbf{kr} - \omega t), \quad \mathbf{f}^{-} = \mathbf{f}_k^{-} \exp i(\mathbf{kr} - \omega t), \quad \omega = |\mathbf{k}|. \tag{14}$$

As for topological interpretation of solution (2) of the Dirac equation we suggest that function (14) does not define any wave process into the space but describes periodical movement of the space topological defect. As function (2) solution (14) can be also considered as the realization of fundamental group of some closed topological manifold. Due to exponential factor in (14) this group contains the translation group (as for solution (2)), but now our solution are complex bivectors–not a bispinors. These vectors does not realize the representation of reflections along three different axes as it was for bispinors. These vectors $\mathbf{f}^{+}$ and $\mathbf{f}^{-}$ consist of axial and polar vectors (see (14)) and thus these vectors are transformed one into another only in result of the reflection of space axes. Therefore, solution (14) realizes representation of a sliding symmetry in 4-space only along time–axis. This distinguishes the supposed fundamental group from the fundamental group considered in previous Sections (four sliding symmetries along four perpendicular axes). Wave-corpuscular dualism of electromagnetic waves and possibility of stochastic behavior can be explained in the same manner as in Sections 2,3. But electromagnetic waves have some additional important property—their velocity does not depend on the source motion. We will show below how geometrical properties of the closed topological 4-manifold can explain this fact.

Suppose that 4-manifold corresponding to electromagnetic wave has the form $M^3(\mathbf{r}) \times M^1(\mathbf{r},t)$, that is it can be represented as a product of nonorientable three-dimensional euclidean closed manifold $M^3(\mathbf{r})$ and one-dimensional manifold $M^1(\mathbf{r},t)$ homeomorphic to a pseudoeuclidean circle. The formal reason for such representation is the distinguished role of the euclidean space within fundamental group: only into euclidean subspace translation group is combined with reflections. Consider now a low-dimensional analogy that explains an independence of light velocity on the source motion.

Instead of four-dimensional manifold $M^3(\mathbf{r}) \times M^1(\mathbf{r},t)$ we consider, as in previous Section, two-dimensional analogy—manifold $S^1 \times S_1'$, where S^1 is a one-dimensional euclidean circle and S_1' is a pseudoeuclidean circle. This manifold was considered in Section 3, and it looks like a hyperboloid. For electromagnetic waves $m = 0, E = cp$. Within our notation it leads to relation $p_1 = p_2 = p$ or $\lambda_1 = \lambda_2 = \lambda$. Therefore, there have to be only one parameter with dimensionality of length, and this will be the case if S_t^1 is a pseudoeuclidean circle of zeroth radius. Equation for such circle has the form $x^2 - t^2 = 0$, and hyperboloid is transformed into a cone. This means that the points representing in this example electromagnetic wave move with velocity equals ± 1 ($\pm c$–in chosen units system), and this velocity does not depend on coordinate frame rotations (does not depend on transfer from one moving inertial frame to another). From topological point of view this result is a consequence of the fact that the zeroth radius can be considered as topological invariant. Therefore, within geometrical approach light velocity appears to be topological invariant of the manifold representing electromagnetic wave, and this is the reason of its independence of the source motion.

Notice in conclusion that inclusion into physical world the objects from outer five-dimensional space gives principally new opportunities for understanding of this world. For example, this gives a rather simple explanation for the origin of nonlocal correlation in EPR–experiments. We see at Fig.3 that two noninteracting identical particles (intersection points) can move in opposite directions being parts of the same topological defect (the circle). In the act of measurement of one of the particles the symmetry of the whole topological defect can be changed (for example, it can become nonorientable manifold being at first orientable one). And another particle, being the part of the same defect, would "feel" at once this change without any interaction.

REFERENCES

1. O.A. Olkhov, *Journ.of Phys.:Conf.Ser.*,**67**,012037(2007); arXiv: 0706.3461
2. J.D. Bjorken, S.D. Drell, *Relativistic Quantum Mechanics and Relativistic Quantum Fields*, New York McGraw-Hill, 1964
3. A.I. Achiezer, S.V. Peletminski, *Fields and Fundamental Ineractions*, Kiev Naukova Dumka, 1986, Ch.1.
4. P. L. Rachevski, *Riemannian geometry and tensor analysis*, Ìoscow, Nauka, 1966, §55
5. V.A. Jelnorovitch, *Theory of spinors and its applications*, Ìoscow, August-Print, 2001, §1.3
6. M. Hamermesh, *Group Theory and Its Application to Physical Problems*, Argonne National Laboratory, 1964
7. H.S.M. Coxeter, *Introduction to Geometry*, N.Y.-London John Wiley and Sons, 1961
8. B.A. Dubrovin, S.P. Novikov, A.T. Fomenko, *Modern geometry*, Moscow, Nauka, 1986
9. A.S. Schvartz *Quantum Field Theory and Topology*, Moscow, Nauka, 1989
10. D. Gilbert, S. Kon-Fossen, *Nagladnaya geometriya (rus)* , Moscow, Nauka, 1981
11. À.I. Achiezer, V.B. Berestetski, *Quantum Electrodynamics*, Moscow, Nauka, 1981

Generalized Causal Quantum Theories

Claudio Parmeggiani

Mathesis - Via Lotario, 13 - 22030 Civenna (Como) - Italy
E-mail:clparm@infinito.it

Abstract. We shall show that is always possible to construct causal Quantum Theories fully equivalent (as predictive tools) to acausal, standard Quantum Theory, relativistic or not relativistic; we re-obtain, as a particular case, the usual Quantum Bohmian Theory. Then we consider the measurement process, in causal theories, and we conclude that the state of affairs is not really improved, with respect to standard theories

Keywords: Quantum Theory, Causality.
PACS: 03.65.Ta

INTRODUCTION. QUANTUM THEORIES

In acausal, standard Quantum Theories (see, for example, [1]) we try to describe the statistical behaviour of a collection $\Sigma_1, ..., \Sigma_N$ of N identical physical systems, on a fixed space-like hyperplane Π. So, if R is some, real or complex, physical quantity of the system we need a prescription to calculate, for $N \to \infty$, the average value of R (on Π): $avg(R, \Sigma_1, ..., \Sigma_N, \Pi)$.

To this purpose we associate to every bounded physical quantity a normal, bounded operator (both denoted by R); the operator acts on a separable Hilbert space $H(\Sigma)$ and we assume that, for $N \to \infty$:

$$avg(R, \Sigma_1, ..., \Sigma_N, \Pi) \to Tr(R \cdot W_\Pi), \tag{1}$$

here W_Π is a positive, trace operator on $H(\Sigma)$ and $Tr(W_\Pi) = 1$, for every Π. Then W_Π represents the collection of physical systems, on the hyperplane Π. If u is a time-like vector, orthogonal to Π, $W_{\Pi+u} = U(u) \cdot W_\Pi \cdot U^*(u)$, where U is the unitary evolution operator of the system.

A collection of systems is said homogenous if it cannot be obtained by "mixing" two sub-collections different from itself. In this case W_Π reduces to a projector on a one-dimensional subspace of $H(\Sigma)$ and, if ϕ_Π is a unit vector of this subspace, we have $Tr(R \cdot W_\Pi) = \langle \phi_\Pi | R \cdot \phi_\Pi \rangle$; obviously $\phi_{\Pi+u} = U(u) \cdot \phi_\Pi$. In the following we shall consider only homogenous collections, represented by a vector ϕ_Π, on the hyperplane Π.

When R is an orthogonal projector we prefer to speak of the "property" associate to it; and we shall "verify" that the system possess it (or not); it is also possible to build the theory starting from the properties/projectors ortho-complemented lattice ([2], [3]). In any case, if P is such a property/projector and $frq(P, \Sigma_1, ..., \Sigma_N, \Pi)$ the number of systems which possess it, divided by N (the frequency), we have, always for $N \to \infty$:

$$frq(P, \Sigma_1, ..., \Sigma_N, \Pi) \to \|P \cdot \phi_\Pi\|^2. \tag{2}$$

CP962, *Quantum Theory, Reconsideration of Foundations – 4*,
edited by G. Adenier, A. Yu. Khrennikov, P. Lahti, V. I. Man'ko, and Th. M. Nieuwenhuizen

CAUSAL QUANTUM THEORIES

Generalized Theories

In generalized, causal Quantum Theories we pretend to do more than calculate averages: we want to obtain the "exact" (but hidden) value of every quantity R, on Π, for a single physical system Σ_n, $(1 \le n \le N)$: $R_{n,\Pi}$.

To this purpose we associate to each Σ_n and Π a vector $\lambda_{n,\Pi}$ of the Hilbert space $H(\Sigma)$; the $\lambda_{n,\Pi}$ (the hidden parameters) are orthogonal and (for $N \to \infty$) : $\lambda_{1,\Pi} + ... + \lambda_{N,\Pi} \to \phi_\Pi$, $\langle \lambda_{n,\Pi} | \phi_\Pi \rangle = 1/N$; here the vector ϕ_Π always represents our homogenous collection. Then we assume that:

$$R_{n,\Pi} = \frac{\langle \lambda_{n,\Pi} | R \cdot \phi_\Pi \rangle + \langle \phi_\Pi | R \cdot \lambda_{n,\Pi} \rangle}{2 \langle \lambda_{n,\Pi} | \phi_\Pi \rangle} \tag{3}$$

and, as expected, that $\lambda_{n,\Pi+u} = U(u) \cdot \lambda_{n,\Pi}$.

Because, evidently:

$$avg(R, \Sigma_1, ..., \Sigma_N, \Pi) = \tfrac{1}{N}(R_{1,\Pi} + ... + R_{N,\Pi}) \tag{4}$$

averages (and dispersions), for all physical quantities, are now the same that in standard Quantum Theories. However the "probabilities" are simply a consequence of the fact we do not know the hidden values of our quantities. Finally we note that, if (for a fixed Π) ϕ_Π is a self-vector of R, $R \cdot \phi_\Pi = r_\Pi \cdot \phi_\Pi$, then $R_{1,\Pi} = ... = R_{N,\Pi} = r_\Pi$: in our causal Quantum Theory the fact that, for every collection of physical systems, there is always some quantity (operator) not dispersion free can be restate saying that there is always some quantity for which not all hidden values are equal.

Bohmian Theories

In causal, Bohmian Quantum Theories ([4], [5]) the Hilbert space is an $L_2(\mu, X)$, where X is a locally compact manifold (the configurations space) and μ is a Lebesgue measure on it. The family of space-like hyperplanes is parameterised by a real number, t; and the Hilbert space vectors $\lambda_{1,\Pi}, ..., \lambda_{N,\Pi}, \phi_\Pi$, are represented by functions (eventually vector valued) $t, x \mapsto \lambda_1(t,x), ..., \lambda_N(t,x), \phi(t,x)$, $x \in X$. Then we have, for a self-adjoint R:

$$R_n(t) = \Re \frac{\int_X \lambda_n^*(t,x) \cdot (R\phi)(t,x)\, d\mu(x)}{\int_X \lambda_n^*(t,x) \cdot \phi(t,x)\, d\mu(x)}; \tag{5}$$

we remember that the denominator is a constant and, in fact, $= 1/N$.

Now we assume that $\lambda_n(t,x) = g(x, \xi_n(t)) \cdot \phi(t,x)$ where the ξ_n (the hidden trajectories, one for every Σ_n) are function from the real line to X and $x, y \mapsto g(x, y)$ is a real function on X^2, significantly $\neq 0$ only when $x \approx y$. We have, then, posing $\rho_{\phi,R}(t,x) = \phi^*(t,x) \cdot (R\phi)(t,x)$ and $\rho_\phi(t,x) = \phi^*(t,x) \cdot \phi(t,x)$:

$$R_n(t) \approx \Re \frac{\rho_{\phi,R}(t, \xi_n(t))}{\rho_\phi(t, \xi_n(t))}, \tag{6}$$

and also ($U(t) = \exp(-it \cdot H)$):

$$\frac{dR_n(t)}{dt} \approx \Im \frac{\rho_{\phi,[R,H]}(t, \xi_n(t))}{\rho_\phi(t, \xi_n(t))}. \tag{7}$$

Applying the last formula to the quantities "coordinates of x" we can express the coordinates of $d\xi_n / dt$ as functions of t and of the coordinates of ξ_n; then, integrating, we can obtain the global trajectories from the initial values.

THE MEASUREMENT PROCESS

In standard Quantum Theories, acausality is a consequence of the fact that the measurement of R on ϕ (the cause) has, generally, different effects, depending on the spectre of R: in fact, after the measurement, our homogenous collection became a mixture of homogenous sub-collections, one for every self-value of R. Here we assume that the operator R have a pure discrete, simple spectrum.

In causal Quantum Theories the causality is restored assigning to every system Σ_n a different (hidden) Hilbert space vector λ_n , now also the causes are different. But, we must note that the calculated hidden values (the R_n) have, generally, nothing to do with the results of the measurement process (the self-values of R): this is the case only if the hidden parameters (the λ_n) are self-vectors of R.

Apparently, at this point, the only acceptable approach seem to admit that, along the measurement of R, the hidden parameters λ_n "jump" to some self-vectors of R, displaying the same frequencies prescribed by the standard theory: the state of affairs is, unfortunately, not really improved.

REFERENCES

1. J. von Neumann, *Mathematical foundations of Quantum Mechanics*, Princeton: Princeton University Press, 1955.
2. G. Birkhoff and J. von Neumann, "The Logic of Quantum Mechanics", *Annals of Mathematics*, **37**, 823-843 (1936).
3. G. W. Mackey, *The Mathematical Foundations of Quantum Mechanics,* Reading: The Benjamin Publishing Company, 1963.
4. D. Bohm, *Causality and Chance in Modern Physics*, London: Routledge & Kegan Paul, 1984.
5. J. S. Bell, *Speakable and Unspeakable in Quantum Mechanics*, Cambridge: Cambridge University Press, 1989.

Comments on a Discrepancy Between the Relativistic and the Quantum Concepts of Light

Claudia Pombo

Oosterpark 49-I, 1092AN Amsterdam, The Netherlands

Abstract. The realist point of view of a physical theory assumes that physical concepts must have a correspondent in the phenomenological world. We adopt a slightly modified form of realism, based on Carnap's separation of languages, in which only the observational concepts, belonging to the observational language, have a phenomenological correspondent. Other physical concepts, belonging to a theoretical language, do not correspond to entities in the physical world. This point of view is named observational realism. Based on these ideas, we review the notions of relativistic and quantum observation, independently from measurement, and show that there is a discrepancy between the concepts of wave light in relativity and in quantum mechanics.

Keywords: Observational realism, relativistic observers, light, photon
PACS: 01.70.+w, 03.30.+p, 03.65.Ta

INTRODUCTION

Observational realism is an epistemological position by itself. It is not realism in the sense of the so called scientific realism neither it is empiricism. The first does not make distinction between observational and theoretical terms and the second identify observation with measurement. Both positions, which are in the basis of the mostly known schools of thinking on physical theories, such as Copenhagen interpretation, Realist Interpretations, Many Worlds Interpretation, Statistical Interpretation, and others, have not been entirely successful. This open some space for trying alternatives. It is not our intention to discuss in deep the origins of this epistemological position in the present paper. The discussion was already initiated in a previous reference [1, 2]. But some comments on the main lines of thinking can be useful for the discussion here presented.

The difference between observation and theorization is strongly emphasized in the observational realism. The process of observation, in physics, is the collective experience of the physical world, specified by its own proper language. It is not a pure cognitive process, as a theoretical statement could be considered, neither it is based on pure experimentation as data are collected. So, it involves suitable interpretation of data, to account for the objectivity of the experience. It has influences from Analytical Psychology, where the meaning of collective experience and its relation with language is taken, and it incorporates the proposal of R. Carnap, about the separability of languages inside a theory. In this way, analysis of observation becomes analysis of language, but the semantics of the observational languages is not reducible to the quantification of the concepts and equations.

We present here two exercises of this epistemology. The first one is a review on the concept of 'relativistic phenomenon'. In this interpretation, the relativistic phenomena consist of movement of the luminous bodies, that is, bodies which carry light signals as kinematical features of them. To account for this feature, a new reference must be settled and this is also light. Therefore light has at least two different meanings in this domain. The second example discusses observational aspects in Quantum Mechanics, independently of measurement. In this other domain, light is integrant element of the quantum transitions, involving absorption and emission of energy, in which light and electromagnetic radiation, play an active role.

These theories account for different worlds of physical existence, each one has its own concepts and languages. We show that in these fields, light appears in different fashions although presenting common aspects.

THE RELATIVISTIC LIGHT

Typical relativistic situations are said to occur during decays of particles, traveling with constant velocity v, approaching the velocity of the light in vacuum c.

CP962, *Quantum Theory, Reconsideration of Foundations – 4*,
edited by G. Adenier, A. Yu. Khrennikov, P. Lahti, V. I. Man'ko, and Th. M. Nieuwenhuizen

We consider a fixed origin O, where is localized a source of light, inside a laboratory K. It is observed that half of the number of particles in a relativistic beam, traveling with constant velocity v away from the origin, decays at time t and position s from O. The relativistic feature of this phenomenon is in the fact that, with respect to a system of reference K', with respect to which the particles are at rest, the same event happens at a proper time $t' < t$ and position $s' < s$. The observation is given by

$$s = v\frac{1}{(1-\frac{v^2}{c^2})^{1/2}}t', \tag{1}$$

or could also be given by

$$s = c(t^2 - t'^2)^{\frac{1}{2}}, \tag{2}$$

where the role of light appears explicitly. We will show how these observations originate, from the integration of two points of view.

The expression 'relativistic event' implies not only that an event of the relativistic kind can be observed but that the event in question is formed by means of relativistic elements. To be relativistic is not the same as to be relative but it is to present the relativity objectively, that means, in an explicit way. In the present case, the relativity is a property of measurements, and these refers to different reference systems. Then, a relativistic event must involve quantities defined with respect to different reference systems, integrated as one phenomenon. This is the reason why the relativistic equations above involve measures from different systems of references.

Expressions such as 'observers at rest in the lab' and 'observers moving with particles' are frequently found in the literature and are intended to be psychological tools to guide the reader in the relativistic experience. The localization of the observers is intended to justify their measurements. To observe means to measure and proper times and lengths only can be measured with respect to the traveling particles. The problem with this is that the references in question are moving with respect to each other, measurements in the lab must also be done and both measurements must be compared. A person following these suggestions, is forced to 'think him/herself', as a physical body, in the moving and resting situation together, what would provoke a break in the physical unity of the person. In conclusion, a relativistic experience, reported in the usual way, becomes a life threat.

The fact is that these observations do not require physical presence of observers but thoughts in different circumstances. The constancy of the velocity of the light is a clear result of our capacity to combine observations in two different times.

Starting from the usual formula of the Doppler effect for frequencies, we can calculate the number of cycles observed at the beam, with respect to the lab. Because there is a clock inside the lab, which as a whole is a classical space, these quantities only depend on initial conditions.

$$f_b(t) = (1-\beta)f(t), \tag{3}$$

where

$$\beta = \frac{v}{c}. \tag{4}$$

Frequency is defined as the number of wave cycles during a period of time. In our example we consider constant frequencies and the number of cycles in this case is a linear numerical function of time, given by

$$f(t) = \frac{N(t)}{t}, \tag{5}$$

and

$$N(t) = n(t/sec), \tag{6}$$

for a certain fixed number n.

The increasing sequence of numbers $\{nt/s\}$ is the result of an association between two different processes: counting numbers of waves and counting cycles of a clock. During a certain period of time, equation(3) can be rewritten as

$$M(t) = (1-\beta)N(t), \tag{7}$$

with $M(t) = m(t/sec)$), for a fixed m as defined for the previous case.

Now, the origin of the difference between the classical and the relativistic views on these processes is concentrated on the interpretation given to these numerations, or continuous sequence of numbers $M(t)$. While the data are only one, the interpretations depend on the meaning given to the wave. It can be considered as a lengthy disturbance traveling in space, but it can also be considered as the transmission of a signal. And the same could be taken for the case of sound, if we had started with it, instead of a light wave. Sound waves transmit sounds from the source to a detector and the process of reception becomes the signal of the source. So, the wave can be a transmitter in both cases, what implies that the difference must be found in the concept of signal.

Indeed, there are qualitative and quantitative differences in concepts of signal. First of all, as a signal, we mean a process in time. A sound as a physical signal is a mechanical process in a device. This object is physically influenced by the wave and its resulting process is interpreted as the local information of the distance source. In this case, the quantity $M(t)$ is fixed by the common value of m.

In the case of the particles here considered, there is not a physical detector of waves with the beam. The time of the decay process is measured by the process of decay itself. At rest, the decay in question is known to happen at a certain time t', or the other way around, the specific decay means t'. Then, at the position s, as measured with respect to the lab, the quantity $M(t)$, becomes the only signal of the source. The problem now is to know how the number $M(t)$ is known from the beam, if the beam itself is just a 'code' for the cycles t'. So, the question here does not seem to be only 'what more is with the particles' but 'what more we know that is there'. Actually, this is not so difficult to answer if we know that observers are never ' *tabula rasa*, they always know something from the beginning and, it was always known, without necessity of frequent measurements, that the origin of the lab, travels away from the particles. Then, it is actually somewhere there, but in its conditions at the time t'. This is a clear example of how our observational powers substitute our bodies, with the particles.

This knowledge in question, corresponds to the observation of the source by means of the reception of light at the beam. In terms of numbers, two different quantities can represent or account for this statement about the movement of O: or just v or the quantity $N(t)$, interpreted as $N(t) = N'(t')$. It is important to notice that this quantity, which would be initial condition from both ways of observing (with respect to the lab and with respect to the beam), fix the units of time, such the second or any other of choice, as only one.

Then now the source O, with respect to the beam is represented by the quantity $M'(t')$ and if the quantity $M(t)$ is a pure number, it does not depend on the systems of reference and we have $M(t) = M'(t')$, for a certain fixed m'. For the times we can assume that there is a constant γ such that $t = \gamma t'$. And substituting these new results in the original equation(7), we find

$$M'(t') = (1-\beta)N(t). \tag{8}$$

Here, we do not have any physical explanation for the fact that the decays happen at time t, in the lab, it is just observed to be the case. Equation (8) involves two points of view and consequently it is a relativistic equation. The relative velocity between the lab and the particles is conserved and for this reason we interpret this equation as a modified Galilean conservation for relative velocities, in two different times.

After this, we need to find γ explicitly and this is done by means of the principle of conservation of information. Inside the laboratory, as a whole, there is only one clock, as mentioned before. With respect to this clock we had fixed an initial number n. With respect to the whole lab, this number must be conserved. Then, we assume that the particles approach another point on their path, inside the lab, such that

$$M''(t') = (1+\beta)M'(t') \tag{9}$$

holds. Between the two references there will always be a constant difference in time but n is conserved on the path. This gives

$$M''(t') = N(t'/\gamma) \tag{10}$$

Substituting equation (10) in equation (9) and the result in equation (8), we arrive in

$$\gamma = \frac{1}{(1-\beta^2)^{1/2}} \tag{11}$$

In a classical view of a wave, we would have Galilean corrections for the velocities of light arriving at the particles, with respect to the reference of the particles, and light leaving the beam, with respect to the laboratory's references.

But here, these classical corrections are canceled because of the difference between times. Light coming from the source (O now is departing from K') and arriving at the particles, is given by

$$c_{app.beam} = \frac{M(t)(1+\beta)(\gamma\lambda)}{t'} = c, \tag{12}$$

where the relativistic correction $\gamma\lambda$ is due to the departure of the clock (the source) at O. To calculate the other velocity we must attempt to the fact that the velocity of the light emitted by the beam, with respect to the beam, is also c. From the relation $N(t) = N'(t')$, we find that $\lambda/\lambda' = t/t'$, or $\lambda = \gamma\lambda'$. We then calculate the velocity of the light leaving the beam, with respect to the lab. This gives

$$c_{leav.beam} = \frac{N'(t)\lambda/\gamma}{t} = c, \tag{13}$$

as it was expected.

LIGHT IN THE QUANTUM DOMAIN

The lab has physical limits, which can be defined by its walls. One could conclude from this that all this relativistic argumentation would end at these limits, but it is only partially true. If $v << c$, the factor γ would not be significative. But, considering the atomic structure of the wall, the question reaches another level of inquiry. It is not possible to consider that the atoms and their constituents are at rest and this limit for γ has to be reexamined.

There are two ways to interpret the facts, considering light reaching a target K'. One is to think that the quantity c is practically infinite. As a result, we would have $t = t'$, because, with respect to the distance covered by light from both origins K and K', in a finite time, the two origins would be practically together. No new information, to add to the normal classical observation, would emerge from the presence of light. This is the classical limit.

Another very different situation is to think that v is the quantity approaching zero and c is finite. In this case, light can still be observed and, whenever this happens, a new dimension is brought with it. We can start this discussion noticing that equations (1) and (2) complement each other. Equation (1) says that moving on s there is an object in a past state and this can be interpreted as a new dimension of the object, which was absent from the classical situation at the beginning. And to have it, we did not make use of the constancy of the velocity of the light, only of the relation between times, given by equation (11). Equation (2) has another content, since it is based on the result of equation (12). It says that both times can pass while s is fixed, and its would keep the new dimension as before, without necessity of a significative v. Geometrically, we could have plenty of arguments to construct a 4-dim Minkowski structure but what actually matters in our discussion is to know that the physical space in this way constructed accounts for physical facts which would not be described otherwise. We mentioned before that we do not know the physical reasons why the decays happen at time $t > t'$, with respect to the lab. But of course this dimension is open to localize elements involved with this fact. And the fact that a moving beam is no more present should not take, from the sources, the physical properties they had from the beginning. This is how we interpret equation (2).

A relevant aspect of this new dimension, is that measurements with respect to it are performed indirectly, by means of reception of light. Another point is that there was never a dynamical argument involved in its emergence. And finally, another interesting aspect of this dimension is that it was implicitly introduced by Faraday [3], when he defined the concept of field, independently of the presence of forces or results of interaction between charged bodies. He was puzzled with the phenomenon of induction, independently from forces, and considered the possibility that there would be an entity such as a static field which would be a kinematical concept. The generalization of the concept of field, including a dynamical field, was made later with Maxwell's unification[4]. It is clear that the relativistic background permits to describe the electromagnetic field as a dynamical entity, but its main relativistic meaning, had other origins.

Details, as the ones we mention, may not be so important in general, but they are relevant for the observational realist position, because the usual local measurements do not reveal all the observational possibilities behind the experimental situations. With these considerations we can start the second exercise, about the quantum meaning of light.

The photon is said to be the particle associated to the electromagnetic field, carrying momentum and energy [5, 6]. An atomic transition is described as a process of creation or destruction of photons with certain energies and a function $\Psi(s)$ can be defined as the photon wave function. About this function, which is one of the attributes of the photon, there are divergences on meaning, considering the schools of interpretation. In mathematical terms, there are no doubts about what it is, a complex numerical function, defined on the usual space. These features already account for the main differences between the relativistic and the quantum concepts of light.

In our view, the contrast between the dynamical and the kinematical meanings are the main discrepancy between the photon and the wave light from the relativity. One is interactive and the other informative. However, the wave function is, in the end, a numerical quantity, defined as a distribution in the space. We still can introduce this object inside our spacial set up of our original discussion. A part from the influence that this function has on the dynamic transitions, there is a strong similarity between the two concepts of wave. This similarity is the numerical aspect.

First of all, numbers are observational concepts. Whatever meaning is attached to them, such as probability or others, they are still counted, what is a form of measurement too. Until the definition of the relativistic velocities of the light, in which wavelengths were introduced, no signal of wave was actually there. From equation (3) until equation (11), only numbers were observed. The possibility of non real numbers was also implicit there, since received and emitted quantities are not from the same species. And if we consider the spacial aspect of their distributions, noticing the fact that, in the relativistic case, the received and the emitted waves combine as an invariant, restoring the initial value in the absence of interaction, the similarities become more strong: $(\gamma\lambda)(\lambda/\gamma) = \lambda^2$. And another point to notice is that both kinds of wave, cannot be directly measured in which concerns their spatial existences, due to reasons involving their complex nature. It is clearly shown in the relativistic domain that complex dimension can be well observed, although not directly measured. And this is not a problem for the observational realism.

FINAL COMMENTS

We are convinced that there are strong indications about the observable condition of the wave functions. These objects can be physically real and this has consequences for our world. For instance, it can open doors for a better understanding of the limits of quantum physics. But this subject needs further considerations.

ACKNOWLEDGMENTS.

Suggestions from G. Jaeger and discussions with W. de Muynck and Th. Nieuwenhuizen are kindly acknowledged. The author greatly benefited from the hospitality in the rare interdisciplinary environment created during the present conference in Vaxjo.

REFERENCES

1. C. Pombo, Th. Nieuwenhuizen, *Foundations of Special Relativity and the Principle of Conservation of Information.* (arXiv:physics/0607199v1)
2. C. Pombo, Th. Nieuwenhuizen, *Observational Derivation of Einstein's "Law the Constancy of the Velocity of the Light in Vacuum".*(AIP Conf. Ser. 810, G. Adenier, A.YU. Khrennikov and TH. M. Nieuwenhuizen eds., American Institute of Physics, Melville, NY, 2006), pp.393.
3. M. Faraday, *Experimental Researches in Electricity* (Vol. III, Talylor and Francis, 1855).
4. P. M. Harman, *The Natural Philosophy of James Clerk Maxwell* (Cambridge University Press, 1998).
5. W. de Muynck, *Foundations of quantum mechanics, an empiricist approach.* (Kluwer Academic Publishers, Dordrecht, Boston, London, 2002).
6. E. Merzbacher, *Quantum Mechanics* (John Wiley & Sons, Inc., NY, 1970).

The Principally Weak Points of the Currently Usual Mathematical Tools and Self-inconsistence in Quantum Theory

N.Vinh Quang

Institute of Physics and Electronics, VAST, 10 Dao Tan, Hanoi, Vietnam
E-mail: nvquang@iop.vast.ac.vn

Abstract. It is explicitly shown the weak points (being not rigorous) in the reasoning of Dirac, of Landau, of Von Neumann, of Weyl, of Fadeev and of others, deriving many fundamental relations, which later became classical and are being commonly used in Quantum Theory up to now. It is because of their derivations were made when functional analysis was not yet sufficiently developed. Those weak points are of two following reasons:
1. Unbounded property of the physical operators is not yet paid by attention.
2. The " Dirac delta function" defined by Dirac is not always distinguished clearly with the Dirac distribution defined by Schwartz and with the generalized functions defined by Gel'fand.
Particularly, the formulation of the closure relation by using equality of the functionals is not rigorous. Consequently, it is necessary to improve Quantum Theory.

Keywords: Mathematical foundation of Quantum Theory.

INTRODUCTION

It is well known that two most great achievements of 20th century physics are Theory of Relativity and Quantum Theory.

For Theory of Relativity, mathematics goes before physics. Before Einstein's deriving the special theory of relativity the mathematicians Poincare, Lorentz...have established its complete mathematical tools.

For Quantum Theory, the situation was opposite: physics goes before mathematics. Before 1926, there were already the quantum mechanics papers. Only since 1926, Hilbert, in answering to Heisenberg call, started founding functional analysis -the currently usual mathematical tool for quantum theory.

When functional analysis was not yet sufficiently developed, the famous names such as Dirac, Landau (late became Nobelists in physics)... derived many fundamental relation became classical late, which are being commonly used in quantum physics. After the half-century development of the functional analysis, many weak points in the derivations became clear. They are contained in the famous books (even of the new editions)[1-7]. Hence, there are two following consequences.

1.Many readers making citation to the famous books repeat these weak points.
2. The good students lose many times to try understanding the self-inconsistence points in the famous books.

To neglect those two bad consequences, in this paper, following Ballentine [8], having a deep respect for the great scientists, we are obligatory to explicitly show those weak points.

In this paper, the phrase "the currently usual mathematical tools of Quantum Theory" means the tools presented in the most popular quantum theory books [1-9] and briefly described in the modern book of Weinberg [10] as follows.

(i) Physical states are represented by rays in Hilbert space...

(ii) Observables are represented by hermitian operators which mappings of Hilbert space into itself."

THE MEAN VALUE PROBLEM

The most important consequence of the unbounded property is the fact that the mean value of the hermitian operator at the vector of Hilbert space is not sure of being finite. The mathematicians recognized that [13].

CP962, *Quantum Theory, Reconsideration of Foundations – 4*,
edited by G. Adenier, A. Yu. Khrennikov, P. Lahti, V. I. Man'ko, and Th. M. Nieuwenhuizen

Consider the Heisenberg uncertainty relation. It can be derived by 2 methods, one of Von Neumann, other of Weyl (Fadeev followed Von Neumann). Thinking that it is proved to be valid for all physical states (normalized to 1) 1928 Ruark was the first man calling it the uncertainty principle. Then the people also do so.
Now it became clear under light of the updated functional analysis that the reasoning of Von Neumann, of Weyl, of Fadeev and of others in proving the Heisenberg uncertainty relations is not rigorous.

In the book of Von Neumann [14] the phrase"... P - ρ.1, Q $-\sigma$.1 *also have above commutation property..."* is not rigorous (ρ, σ are the mean values).

Similarly, in the book of Fadeev [5] having equality

$$\{A, B\} = \{(A - A_m), (B - B_m)\}_h$$

is not rigorous (A_m, B_m are the mean values).

In the book of Weyl [15] the phrase"...*No loss of generality is incurred by taking these mean values to be zero*" is not rigorous.

To make their reasoning rigorous it must be to add the phrase:" If the mean values are not infinite then..."

The mathematicians pointed out that the mean value at the vector of Hilbert space might be infinite. But following the super selection rule in Quantum Physics, not every vector of Hilbert space is describing a physical state [11]. Hence, it is necessary to verify either there exist the physical states, at which the mean values of the physical operators are infinite, or not?

For the first time, in [16], it is explicitly shown that there exist the physical states (normalized to 1) at which the mean values of the physical operators are infinite.

The 2nd example: At the normalized (to 1) state described by eigen-differential wave packet of electron being in constant electric field, the mean value of the position operator is infinite.

The 3rd example: At the ground state of Hydrogen atom, the mean values of the many physical operators (the momentum operator to the *n-th* power with *n* equal or greater 3) are infinite.

When the mean values are infinite even the definition of the uncertainty notion through the dispersion notion in the probability theory is irrelevant. Hence it must be to improve the concept for the uncertainty notion and, consequently, the notion of the certain value, exactly measured value.

THE DEFINITION DOMAIN PROBLEM

a) The results of the 3rd example mean that the Hydrogen atom ground state belongings to domain of $\hat{p}$, $\hat{p}^2$ but not to one of $\hat{p}^4$. Hence the definition domains of the different order powers of $\hat{p}$ may be different.

b) The space translation operator is often defined [1-7] through operators $\hat{p}$, $\hat{p}^2$ ...

$$\exp(-i\hat{p}x) \equiv \sum_{n=0}^{\infty} (-i\hat{p}x)^n \tag{1}$$

This definition is meaningful only if operators $\hat{p}$, $\hat{p}^2$ ... have a common definition domain. From a) it is not natural and hence, it must be to prove.

c) The same situation is for the time evolution operator defined [1-7,9] through $\hat{H}$, $\hat{H}^2$ ...

$$\exp\left(-i\hat{H}t\right) \equiv \sum_{n=0}^{\infty} \left(-i\hat{H}t\right)^n \tag{2}$$

Thus, using Schrödinger equation is not naturally equivalent to using the time evolution operator.

d) Having the different definition domain problem, the Schrödinger picture is not always equivalent to the Heisenberg one.

e) The 1st example shows the following. The physical operators are postulated to be hermitian only when they act on vectors in Hilbert space. Acting on got out vectors, they are not sure of being hermitian more.

f) Since the definition domain of the unbounded hermitian operator is not the whole Hilbert space, its exponential function is not sure of being unitary. Hence the basic phrase of Weinberg: (ii)"... mappings of Hilbert space into itself "is not accurate.

g) Since the different operators may have the different domains, in solving the operator equation and in making the operator transformation one must, at first, to show the definition domain of newly introduced operators as required.

THE RELATIONS CONCERNING WITH DIRAC DELTA FUNCTION

The Dirac delta function $\delta(x)$ defined by Dirac [1] (and by Landau [2]) has the following defining properties.

$$\int_{-a}^{b} \delta(x)dx = 1 \tag{3}$$

$$\int_{-a}^{b} \delta(x) f(x)dx = f(0) \tag{4}$$

where *a, b* are positive and *f(x)* is any continuous (at *x=0* in Landau definition) functions.

The equality (3) describes a point distribution at *x=0*, which is fundamentally important for physics and can be traced back to Kirchhoff, 1882 .The equality (4) describes the functional property and implies that [17] the space of testing functions (the definition domain) of the functional called" Dirac delta function" is the set of all continuous (at *x=0*) functions. The equality (3) is a particular case of (4) when $f(x) \equiv 1$.

Many relations concerning the Dirac delta function are not accurately and not sufficiently proved yet. For example, the one of the most often used relations [1,2]

$$\frac{1}{2\pi} \int_{-\infty}^{\infty} e^{ipx} \, dp = \delta(x) \tag{5}$$

is self-inconsistent. In fact, let $f(x) = exp(x^2)$ one gets a contradiction

$$\int_{-\infty}^{\infty} e^{x^2} \delta(x) \, dx = 1 \neq \frac{1}{2\pi} \int_{-\infty}^{\infty} \int_{-\infty}^{\infty} e^{x^2} e^{ipx} dxdp \tag{6}$$

Explanation. In proving Landau has used the Fourier transformation

$$f(x') = \frac{1}{2\pi} \iint f(x) e^{i(x-x')p} dxdp \tag{7}$$

(and then putted $x' = 0$) for the set of arbitrarily continuous (at *x=0*) functions. But following Kolgomorov [18] the Fourier transformation (12) is only valid for the set of absolutely integrated functions, which is much smaller than set considered by Dirac and Landau.

For the Dirac distribution, defined in the Schwartz distribution theory [19], the testing functions space is the one of the definite (vanish identically outside a bounded region) functions, which is much smaller than the space of arbitrarily continuous (at *x=0*) functions. Hence, the relations proved for the Dirac distribution of Schwartz, are not sure of being valid for the Dirac delta function.

The same situation is for the Gel'fand generalized functions theory [20,21], where the testing functions space is the Schwartz one (of functions vanishing at infinitely more rapidly than any power of 1/x).

THE CONSEQUENCE OF USING THE CLOSURE RELATION

There are three different formulations for the completeness of the orthonormal basis in Quantum Theory.

The 1st formulation is using equality of the functions or of the vectors (the Fourier expansion).

The 2nd formulation is using equality of the operators (the expansion of the identity operator as a series of the projection operators).

The 3rd formulation is using equality of the functionals, which is the closure relation being commonly used.

$$\sum_{n} \varphi_n^*(x') \varphi_n(x) = \delta(x'-x) \tag{8}$$

It is not rigorous [22,23]. Particularly, it is neglected from the new edition book of Anfken and Weber, Mathematical Methods for Physicists [22], although it was in the old edition one.

CONCLUSION

It is necessary to neglect all self-inconsistent points from Quantum Mechanics textbooks in order the students need not spend time for putting questions.

Together with using the Gel'fand triplet [8,9,11,12,21], it is necessary to improve mathematical tool for Quantum Theory more carefully, taking into account above-mentioned results. Using the tool to be improved we can wait for the new physical effects.

ACKNOWLEDGMENT

This paper is made with the contribution of Le Viet Du Khuong and N.T. Thuy Diem. This work is supported by Vietnamese Ministry of Science and Technology- the natural science basic research program 40 28 06.

REFERENCES

1. P. A. M. Dirac, *The Principles of Quantum Mechanics*, 4th ed., Oxford: Clarendon, 1974, Chap. III
2. L. D. Landau and E. M. Lifshist, *Quantum Mechanics*, 4th ed., Moscow: Science, 1989, Chap. I
3. A.C. Davydov, *Quantum Mechanics*, Moscow: Science, 1973.
4. V.V. Balashov and V.K.Dolinov, *The course of Quantum Mechanics*, Moscow: Moscow University, 1982
5. L.D.Fadeev and O.A.Iakubovskii, *The lessons on Quantum Mechanics for students-mathematicians*, Leningrad: Leningrad University, 1980, pp.36
6. A. Messiah, *Quantum Mechanics*, Vol.I, II, New York: John Wiley&Sons, 1966
7. L.T.Schiff, *Quantum Mechanics*, 3rd.ed., New York: McGraw-Hill, 1968
8. L.E.Ballentine, *Quantum Mechanics :A modern development*, Singapore: World Scientific, 1998.
9. A. Bohm, *Quantum Mechanics*, 3rd ed., New York: Springer- Verlag, 1993
10. S.Weinberg, *The Quantum Theory of Field*, vol. I, Cambridge, 1994
11. N. N. Bogoliubov, A. A. Logunov and I.T. Todorov*: Fundamentals of Axiomatic approach in Quantum Field Theory*, Moscow: Science, 1969, Ch. II.
12. J. Perina, Z. Hardil and B. Jurco: *Quantum Optics and Fundamentals of Physics*, Boston: Kluwer Academic, 1994, Ch. II.
13. *Functional analysis*, ed. by G. Krein, Moscow: Sience, 1972,p.423
14. J. von Neumann, *Mathematical Foundations of Quantum Mechanics,* Princeton: Princeton University, 1955.
15. J. Weyl, *The Theory of Groups and Quantum Mechanics*, New York: Dover, 1950.
16. N. Vinh Quang, *Do the Robertson- Schrödinger and the Heisenberg uncertainty relation imply a general physical principle*? invited talk in Feynman Festival, Maryland USA, Aug. 2002, quant-ph/0212145
17. V. S.Vladimirov, *Generalized functions in mathematical physics*, Moscow: Science, 1976
18. A. N. Kolgomorov and S.V. Fomin, *Elements of function theory and functional analysis* 2nd.ed., Moscow: Science, 1968, pp. 413.
19. L. Schwartz, *Application of distributions to the theory of elementary particles in quantum mechanics*, New York: Gordon and Breach, 1968
20. I. M. Gel'fand and G.E. Shilov, *Generalized Functions,*Vol.1, New York: Academic,1964
21. I.M.Gel'fand and N. Ya. Vilenkin, *Generalized Functions,*Vol.4, New York: Academic,1964
22. G. B. Anfken and H. J. Weber, *Mathematical Methods for Physicists*, AP, 2004.
23. N.Vinh Quang, *Laos J. Appl . Scienc*. **1,** 82-88 (2006)

Modified BCHSH Inequalities Within the ESR Model

Sandro Sozzo

Dipartimento di Fisica dell'Università del Salento and INFN, Sezione di Lecce
Via Arnesano, 73100, Lecce, Italy
E–mail: sozzo@le.infn.it

Abstract. The *extended semantic realism* (ESR) model reinterprets quantum probabilities as conditional rather than absolute probabilities, which provides a general framework that assumes local realism and incorporates the mathematical apparatus of standard quantum mechanics (SQM). We discuss expectation values, sequential measurements and correlation functions within this model. Then, we consider the standard Bell–Clauser, Horne, Shimony and Holt (BCHSH) inequalities and show that they must be replaced by *modified* BCHSH inequalities. These contain *detection probabilities* that are not predicted by SQM and do not depend on features of the measuring apparatuses or environment. In order to avoid contradiction, the detection probabilities must satisfy some no–violation conditions which can be experimentally checked, at least in principle, and make the ESR model falsifiable.

Keywords: quantum mechanics; BCHSH inequalities; ESR model; local realism.
PACS: 03.65.-w; 03.65.Ud

1. THE ESR MODEL

The *extended semantic realism* (*ESR*) model [1, 2, 3, 4] has been proposed in order to show that, contrarily to a widespread belief, the mathematical formalism of standard quantum mechanics (SQM) does not conflict with *local realism*, that is, with the joint assumptions of *realism* (R: intuitively, the values of all observables of a physical system in a given state are predetermined for any measurement context) and *locality* (LOC: intuitively, if measurements are made at places remote from one another on parts of a physical system which no longer interact, the specific features of one of the measurements do not influence the results obtained with the others [5, 6]). We refer to [7] and to the paper by C. Garola in this book [8] for a new synthetic and improved description of the ESR model and limit ourselves here to remind some of its features that are needed in the following sections.

The ESR model distinguishes from the very beginning between microscopic and macroscopic properties of a physical system and assumes that all of them are *objective* (that is, every property that can be predicated of a sample of the system either is possessed or it is not possessed by the sample, independently of the measurement context), hence R and LOC hold in it (indeed objectivity implies R and LOC, though it is not equivalent to them). The set of macroscopic properties is however broadened by adding a *no–registration outcome* to the spectrum of every standard quantum observable, interpreted as providing information about the measured system and not as expressing the inefficiency of the measuring apparatus. The model incorporates the mathematical formalism of SQM and its rules for calculating probabilities, but interpretes the latter as providing *conditional* instead of *absolute* probabilities. To be precise, all probability values predicted by SQM are recovered but they are reinterpreted as referring only to samples of the physical system which do not yield the no–registration outcome. These are not all samples, at variance with SQM, since the model assumes that some samples may yield the no–registration outcome even if the measurement is perfectly efficient. Moreover, when referring to the set of all samples the model provides predictions that are different from those of SQM and that, in principle, can be experimentally checked (though performing actual tests may be rather difficult). Hence, the ESR model constitutes the embryo of a new theory, which is still largely incomplete but already provides new results. A significant example of new prediction can be given by focusing on the Bell–Clauser, Horne, Shimony and Holt (BCHSH) inequalities. In particular, after discussing in Sec. 2 some technical notions (expectation values, sequential measurements, correlation functions) within the ESR model, we show in Sec. 3 that the model predicts that modified instead of standard BCHSH inequalities hold. The new inequalities contain *conditional expectation values* that are formally identical to standard quantum expectation values, and unknown *detection probabilities*. Hence they are never violated if suitable limits are imposed to the detection probabilities. The existence of such limits can be empirically checked, at least in principle, which makes the ESR model falsifiable.

The reinterpretation of quantum probabilities within the ESR model is the consequence of a description of the

CP962, *Quantum Theory, Reconsideration of Foundations – 4,*
edited by G. Adenier, A. Yu. Khrennikov, P. Lahti, V. I. Man'ko, and Th. M. Nieuwenhuizen

measuring process that considers microscopic properties as determining the probability of any outcome whenever a measurement is performed. We use in the following the final result of this description, which can be summed up as follows.

Let Ω be a physical system, x an individual sample of Ω (briefly called *physical object*) prepared in the state S, and A_0 the (extended) observable obtained by considering the observable A of SQM with spectrum Ξ and adding the no–registration outcome a_0. Moreover, suppose that a perfectly efficient, or *idealized*, measurement of the property $F = (A_0, \Delta)$, where Δ is a Borel set on the real line $\Re$ such that $a_0 \notin \Delta$, is performed on x (which consists in testing whether the value of A_0 lies in Δ). Then, the joint probability that x is detected and turns out to possess F is

$$\mathscr{P}_S^t(F) = \mathscr{P}_S^d(F)\mathscr{P}_S(F), \tag{1}$$

where $\mathscr{P}_S^d(F)$ is the *detection probability*, that is, the probability that x is detected when the property F is measured on it, and $\mathscr{P}_S(F)$ is the conditional probability that x turns out to possess F when it is detected. Then, the main assumption of the ESR model states that $\mathscr{P}_S(F)$ must be evaluated by using the same rules that yield the probability of F in the state S within SQM.

2. EXPECTATION VALUES AND SEQUENTIAL MEASUREMENTS

We discuss in this section some technical issues within the ESR model that are needed in order to attain our results in the following section.

Let us begin with the expectation value of an observable A_0. For the sake of simplicity, let us assume that A_0 has discrete spectrum $\Xi_0 = \{a_0\} \cup \{a_1, a_2, \ldots\}$ (the extension of our treatment to more general observables is straightforward). Measuring A_0 is equivalent to measuring the properties $F_0 = (A_0, \{a_0\})$, $F_1 = (A_0, \{a_1\})$, $F_2 = (A_0, \{a_2\})$, ... simultaneously, so that, for every F_n such that $n \in \mathbb{N}$, Eq. (1) holds with F_n in place of F. Then, we briefly write a_n instead of F_n in Eq. (1), and introduce the reasonable physical assumption that the set of observables such that the detection probability depends on the observable but not on its specific value is nonvoid (it could actually coincide with the set of all observables). Then, we convene choosing all observables within this set from now on, and write $\mathscr{P}_S^d(A_0)$ instead of $\mathscr{P}_S^d(a_n)$, so that the probability of finding the outcome a_n when measuring A_0 on x in the state S becomes

$$\mathscr{P}_S^t(a_n) = \mathscr{P}_S^d(A_0)\mathscr{P}_S(a_n). \tag{2}$$

The probability $\mathscr{P}_S^t(a_0)$ of getting the outcome a_0 is instead given by

$$\mathscr{P}_S^t(a_0) = 1 - \mathscr{P}_S^d(A_0). \tag{3}$$

Eqs. (2) and (3) can be used in order to evaluate the expectation value $\langle A_0 \rangle_S$ of A_0 in the state S. We get

$$\langle A_0 \rangle_S = a_0 \mathscr{P}_S^t(a_0) + \sum_n a_n \mathscr{P}_S^t(a_n) = a_0(1 - \mathscr{P}_S^d(A_0)) + \mathscr{P}_S^d(A_0) \sum_n a_n \mathscr{P}_S(a_n) = a_0(1 - \mathscr{P}_S^d(A_0)) + \mathscr{P}_S^d(A_0)\langle A \rangle_S. \tag{4}$$

The term $\langle A \rangle_S = \sum_n a_n \mathscr{P}_S(a_n)$ in Eq. (4) deserves special attention. Indeed, because of the main assumption of the ESR model stated in Sec. 1, it formally coincides with the standard quantum expectation value in the state S of the observable A from which A_0 is obtained. Yet, its physical interpretation is different. For, it represents the mean value of A_0 whenever only detected physical objects are considered. Hence, we call it the *conditional expectation value* of A_0 in the following.

We can now suppose, without loss of generality, that $a_0 = 0$ (hence, for every $n \in \mathbb{N}$, $a_n \neq 0$). Indeed, whenever $a_0 \neq 0$, we can substitute A_0 with $\chi(A_0)$, where χ is a Borel function on $\Re$ which is bijective on Ξ_0 and such that $\chi(a_0) = 0$, and $\chi(A_0)$ is defined as the observable obtained from $\chi(A)$ by adjoining the no–registration outcome 0 and the detection probability $\mathscr{P}_S^d(\chi(A_0)) = \mathscr{P}_S^d(A_0)$. Then we get

$$\langle A_0 \rangle_S = \mathscr{P}_S^d(A_0)\langle A \rangle_S. \tag{5}$$

Let us now consider two discrete observables A_0 and B_0 with spectra $\{a_0\} \cup \{a_1, a_2, \ldots\}$ and $\{b_0\} \cup \{b_1, b_2, \ldots\}$, respectively, and let us suppose that idealized measurements of these observables exist that satisfy the following conditions.

(i) The measurement may change the state of the physical object x on which it is performed but it does not destroy x, even if x is not detected.

(ii) If the physical object x is detected, its state after a measurement can be predicted by using the projection postulate of SQM.

We can then calculate the probabilities $\mathscr{P}^t_S(a_n,b_p)$, $\mathscr{P}^t_S(a_n,b_0)$, $\mathscr{P}^t_S(a_0,b_p)$, $\mathscr{P}^t_S(a_0,b_0)$ (with $n,p\in\mathbb{N}$) of obtaining the pairs of outcomes (a_n,b_p), (a_n,b_0), (a_0,b_p), (a_0,b_0), respectively, in a sequential measurement of A_0 and B_0 on a physical object x in the state S. To this end, let us denote by S_n the state of x after a measurement of A_0 yielding outcome a_n (which can be predicted by using the projection postulate of SQM because of (ii)), by S_0 the state of x after a measurement of A_0 yielding outcome a_0 (that cannot be predicted by using the rules of SQM) and by $\mathscr{P}_S(a_n,b_p)$ the quantum probability of obtaining the pair (a_n,b_p) when firstly measuring A_0 and then B_0. We get

$$\mathscr{P}^t_S(a_n,b_p)=\mathscr{P}^t_S(a_n)\mathscr{P}^t_{S_n}(b_p)=\mathscr{P}^d_S(A_0)\mathscr{P}_S(a_n)\mathscr{P}^d_{S_n}(B_0)\mathscr{P}_{S_n}(b_p)=\mathscr{P}^d_S(A_0)\mathscr{P}^d_{S_n}(B_0)\mathscr{P}_S(a_n,b_p), \tag{6}$$

$$\mathscr{P}^t_S(a_n,b_0)=\mathscr{P}^d_S(A_0)\mathscr{P}_S(a_n)(1-\mathscr{P}^d_{S_n}(B_0)), \tag{7}$$

$$\mathscr{P}^t_S(a_0,b_p)=(1-\mathscr{P}^d_S(A_0))\mathscr{P}^d_{S_0}(B_0)\mathscr{P}_{S_0}(b_p), \tag{8}$$

$$\mathscr{P}^t_S(a_0,b_0)=(1-\mathscr{P}^d_S(A_0))(1-\mathscr{P}^d_{S_0}(B_0)). \tag{9}$$

The probabilities $\mathscr{P}^d_S(A_0)$, $\mathscr{P}^d_{S_n}(B_0)$, $\mathscr{P}^d_{S_0}(B_0)$, $\mathscr{P}_{S_0}(b_p)$ in Eqs. (6)–(9) cannot be evaluated by using the rules of SQM. We can eliminate some of them by introducing further selection of observables and reasonable physical assumptions, as follows.

Firstly, we assume that all observables that are considered are perfect filters, in the sense that, whenever the physical object x is not detected, its state is not changed by the measurement. Hence, $S_0=S$, so that, in particular, $\mathscr{P}^d_{S_0}(B_0)=\mathscr{P}^d_S(B_0)$ and $\mathscr{P}_{S_0}(b_p)=\mathscr{P}_S(b_p)$. Since the latter probability can be evaluated by using the rules of SQM, our assumption reduces the unknown probabilities to $\mathscr{P}^d_S(A_0)$, $\mathscr{P}^d_{S_n}(B_0)$ and $\mathscr{P}^d_S(B_0)$.

Secondly, we observe that we are mainly interested in this paper to the special case of a compound physical system Ω made up by two far apart subsystems Ω_1 and Ω_2, with A_0 and B_0 observables of the component subsystems Ω_1 and Ω_2, respectively. Thus, we refer to this case from now on. Then, objectivity of properties implies that the change of the state of Ω induced by a measurement of A_0 on Ω_1 must not affect the detection probability associated with the measurement of B_0 on Ω_2.[1] We therefore assume that $\mathscr{P}^d_{S_n}(B_0)=\mathscr{P}^d_S(B_0)$.

Because of the foregoing assumptions, only two probabilities remain in Eqs. (6)–(9) that cannot be evaluated by using the rules of SQM, that is, $\mathscr{P}^d_S(A_0)$ and $\mathscr{P}^d_S(B_0)$.

To close up, let us introduce a *generalized correlation function* $P(A_0,B_0)$, defined as follows.

$$P(A_0,B_0)=\sum_{n,p}a_nb_p\mathscr{P}^t_S(a_n,b_p)+\sum_n a_nb_0\mathscr{P}^t_S(a_n,b_0)+\sum_p a_0b_p\mathscr{P}^t_S(a_0,b_p)+a_0b_0\mathscr{P}^t_S(a_0,b_0). \tag{10}$$

By using the assumptions introduced above we get from Eqs. (6)–(9)

$$\begin{aligned}P(A_0,B_0)=&\sum_{n,p}a_nb_p\mathscr{P}^d_S(A_0)\mathscr{P}^d_S(B_0)\mathscr{P}_S(a_n,b_p)+\sum_n a_nb_0\mathscr{P}^d_S(A_0)(1-\mathscr{P}^d_S(B_0))\mathscr{P}_S(a_n)\\&+\sum_p a_0b_p(1-\mathscr{P}^d_S(A_0))\mathscr{P}^d_S(B_0)\mathscr{P}_S(b_p)+a_0b_0(1-\mathscr{P}^d_S(A_0))(1-\mathscr{P}^d_S(B_0)).\end{aligned} \tag{11}$$

Eq. (11) can be simplified reasoning as in the case of Eq. (4) and choosing, without loss of generality, $a_0=0=b_0$ (hence, for every $n,p\in\mathbb{N}$, $a_n\neq 0\neq b_p$). Then, the generalized correlation function is given by

$$P(A_0,B_0)=\mathscr{P}^d_S(A_0)\mathscr{P}^d_S(B_0)\langle AB\rangle_S, \tag{12}$$

where

$$\langle AB\rangle_S=\sum_{n,p}a_nb_p\mathscr{P}_S(a_n,b_p). \tag{13}$$

[1] Note that objectivity of properties implies an epistemic (or *ignorance*) interpretation of quantum probabilities. Hence the transition from a state S to a state S_n that occurs whenever a measurement of A_0 on a physical object x yields outcome a_n, modifying our information about x, may change, for every property F, the probability that x possesses F, but might not change the (partially unknown) set of properties that are possessed by x.

Because of the main assumption of the ESR model stated in Sec. 1, $\langle AB\rangle_S$ formally coincides with the standard quantum expectation value in the state S of the product of the (compatible) observables A and B from which A_0 and B_0, respectively, are obtained. Yet, as we have already observed by referring to $\langle A\rangle_S$, its physical interpretation is different, and we call it the *conditional expectation value of the product* A_0B_0.

As usual, $P(A_0, B_0)$ may provide an index of the correlation among the outcomes that are different from a_0 and b_0.

3. THE MODIFIED BCHSH INEQUALITIES

It is well known that the BCHSH inequalities are obtained by assuming R and LOC (Sec. 1). For the sake of brevity, let us consider only the original inequality provided by Clauser, Horne, Shimony and Holt [9], that we write as follows,

$$|P(\vec{a},\vec{b}) - P(\vec{a},\vec{b}')| + |P(\vec{a}',\vec{b}) + P(\vec{a}',\vec{b}')| \leq 2 \tag{14}$$

and briefly call *standard BCHSH inequality*. The four terms on the left in inequality (14) are *correlation functions*, all of which are defined in the same way and differ only because of the choice of the parameters $\vec{a}$, $\vec{a}'$, $\vec{b}$, $\vec{b}'$. Let us therefore discuss only the first of them. This is given by

$$P(\vec{a},\vec{b}) = \int_\Lambda d\lambda \rho(\lambda) A(\lambda,\vec{a}) B(\lambda,\vec{b}), \tag{15}$$

where λ is a *hidden variable*, the value of which ranges over a domain Λ when different samples of a physical system Ω in a given state S are considered, $\rho(\lambda)$ is a probability distribution on Λ, $\vec{a}$ and $\vec{b}$ are fixed parameters, $A(\lambda,\vec{a})$ and $B(\lambda,\vec{b})$ are the values of two dichotomic observables $A(\vec{a})$ and $B(\vec{b})$, respectively, each of which can be 1 or -1. Furthermore, Ω is assumed to be a compound physical system made up by two component subsystems Ω_1 and Ω_2, and $A(\vec{a})$ and $B(\vec{b})$ are observables of Ω_1 and Ω_2, respectively.

Let us resume now the orthodox viewpoint about inequality (14). One considers all terms in the sum as expectation values of products of compatible observables. These values can be calculated within SQM. Yet, if one puts them into inequality (14), one easily sees that there are physical choices of the parameters $\vec{a}$, $\vec{a}'$, $\vec{b}$, $\vec{b}'$ and of the state S such that the inequality is violated. This leads one to the conclusion that the assumptions from which the inequality is deduced, *i.e.*, R and LOC, are not consistent with SQM.

Let us come to the viewpoint introduced by the ESR model. Here all properties are objective, hence R and LOC hold, but the no–registration outcome must be adjoined to the spectrum of every observable. We can then follow the standard procedures used for getting inequality (14), yet substituting the dichotomic observables $A(\vec{a})$, $B(\vec{b})$, $A(\vec{a}')$, $B(\vec{b}')$ by the trichotomic observables $A_0(\vec{a})$, $B_0(\vec{b})$, $A_0(\vec{a}')$, $B_0(\vec{b}')$, respectively, in each of which a no–registration outcome 0 is adjoined to the outcomes $+1$ and -1. Thus, it can be shown [7] that

$$|P(A_0(\vec{a}),B_0(\vec{b})) - P(A_0(\vec{a}),B_0(\vec{b}'))| + |P(A_0(\vec{a}'),B_0(\vec{b})) + P(A_0(\vec{a}'),B_0(\vec{b}'))| \leq 2. \tag{16}$$

By using Eq. (12) we obtain

$$\begin{aligned} &|\mathscr{P}_S^d(A_0(\vec{a}))\left(\mathscr{P}_S^d(B_0(\vec{b}))\langle A(\vec{a})B(\vec{b})\rangle_S - \mathscr{P}_S^d(B_0(\vec{b}'))\langle A(\vec{a})B(\vec{b}')\rangle_S\right)| \\ &+|\mathscr{P}_S^d(A_0(\vec{a}'))\left(\mathscr{P}_S^d(B_0(\vec{b}))\langle A(\vec{a}')B(\vec{b})\rangle_S + \mathscr{P}_S^d(B_0(\vec{b}'))\langle A(\vec{a}')B(\vec{b}')\rangle_S\right)| \leq 2. \end{aligned} \tag{17}$$

The *modified BCHSH inequality* (17) replaces the standard BCHSH inequality (14) within the ESR model. It contains explicitly four detection probabilities and four conditional expectation values. The latter can be calculated by using the rules of SQM because of the main assumption of the ESR model, and formally coincide with expectation values of SQM (Sec. 2). Nevertheless, if one puts them into inequality (17), one can neither obtain a contradiction nor an agreement, since there is as yet no theory that allows us to calculate the detection probabilities. However, should one be able to perform measurements that are close to ideality, the detection probabilities could be determined experimentally [2] and then inserted into inequality (17). Two possibilities occur.

[2] A major difficulty when performing an experiment for determining a detection probability is counting the physical objects that are actually produced, even if they are not detected by the measurement. Another diffic ulty is distinguishing the detection probability occurring because of intrinsic features of the physical object from the detection inefficiency occurring because of features of the physical apparatus. Of course, we are only pointing out some theoretical possibilities here, and do not aim to suggest how measurements can actually be done.

(i) There exist states and observables such that the conditional expectation values violate inequality (17). In this case one must reject the ESR model (hence R and LOC), or the additional assumptions introduced in Sec. 2 in order to attain Eq. (12), or both.

(ii) For every choice of states and observables the conditional expectation values fit in with inequality (17). In this case the ESR model is confirmed.

The above alternatives show that the ESR model is, in principle, falsifiable. Let us suppose that case (ii) occurs and that the ESR model is accepted. Then, no conflict emerges between R and LOC, which hold within the model, and the reinterpreted quantum probabilities, which are embodied in it. In addition, the ESR model predicts that the conditional expectation values violate the standard BCHSH inequality, since they formally coincide with standard quantum expectation values.

The implications of inequality (17) discussed above can be better understood by studying particular examples. Let us consider, for instance, a typical situation in the literature, in which Ω_1 and Ω_2 are two spin–$\frac{1}{2}$ quantum particles, S is the singlet spin state represented by the unit vector

$$|\eta\rangle = \frac{1}{\sqrt{2}}(|+,-\rangle - |-,+\rangle), \tag{18}$$

$A(\vec{a})$ (or $A(\vec{a}')$) is the observable "spin of particle Ω_1 along the direction $\vec{a}$ (or $\vec{a}'$)" represented by the self–adjoint operator $\vec{\sigma}(1)\cdot\vec{a}$ (or $\vec{\sigma}(1)\cdot\vec{a}'$), and $B(\vec{b})$ (or $B(\vec{b}')$) is the observable "spin of particle Ω_2 along the direction $\vec{b}$ (or $\vec{b}'$)" represented by the self–adjoint operator $\vec{\sigma}(2)\cdot\vec{b}$ (or $\vec{\sigma}(2)\cdot\vec{b}'$; for the sake of simplicity we have obviously omitted a factor $\frac{\hbar}{2}$ in the above representations). Then, it is well known that $\langle A(\vec{a})B(\vec{b})\rangle_S = -\vec{a}\cdot\vec{b}$, and similarly $\langle A(\vec{a})B(\vec{b}')\rangle_S = -\vec{a}\cdot\vec{b}'$, $\langle A(\vec{a}')B(\vec{b})\rangle_S = -\vec{a}'\cdot\vec{b}$, $\langle A(\vec{a}')B(\vec{b}')\rangle_S = -\vec{a}'\cdot\vec{b}'$ in SQM. In addition, the rota tional invariance of the vector $|\eta\rangle$ and the choice of the observables suggest that the four detection probabilities in inequality (17) must be identical in the situation that we are considering. Let us therefore put $\mathscr{P}_S^d(A_0(\vec{a})) = \mathscr{P}_S^d(A_0(\vec{a}')) = \mathscr{P}_S^d(B_0(\vec{b})) = \mathscr{P}_S^d(B_0(\vec{b}')) = p_\eta^d$. Then, we get from inequality (17)

$$(p_\eta^d)^2 \leq \frac{2}{|\vec{a}\cdot\vec{b} - \vec{a}\cdot\vec{b}'| + |\vec{a}'\cdot\vec{b} + \vec{a}'\cdot\vec{b}'|}\,. \tag{19}$$

Inequality (19) shows that the ESR model predicts, under suitable assumptions, an upper bound for the probability that a spin–$\frac{1}{2}$ particle be detected when the spin along an arbitrary direction is measured on it and the compound system is in the singlet spin state. Since R and LOC hold within the model, this bound obviously cannot depend on the spin observables that are measured on the compound system, hence it coincides with the minimum value of the right member in inequality (19), which is $\frac{1}{\sqrt[4]{2}} \approx 0.841$ (the maximum value of the denominator in this inequality is indeed $2\sqrt{2}$). Thus, we get a prediction that, in principle, can be confirmed or falsified by means of experiments, even if it is difficult to imagine how this can be done because of the problems pointed out in footnote 2.

Finally, let us note that our procedures in this section establish a general paradigm that can be followed for dealing with further Bell's inequalities, obtaining a number of examples of non–standard results and testable predictions.

REFERENCES

1. C. Garola, Found. Phys. Lett. **16**, 605 (2003).
2. C. Garola and J. Pykacz, Found. Phys. **34**, 449 (2004).
3. C. Garola, Int. J. Theor. Phys. **44**, 807 (2005).
4. C. Garola, J. Pykacz, and S. Sozzo, Found. Phys. **36**, 862 (2006).
5. A. Einstein, B. Podolsky, and N. Rosen, Phys. Rev. **47**, 777 (1935).
6. J. S. Bell, Physics **1**, 195 (1964).
7. C. Garola and S. Sozzo, *quant–ph/0703260* (2007).
8. C. Garola, in *Quantum Theory: Reconsideration of Foundations 4*, edited by A. Khrennikov (Växjö University Press, Växjö, 2007).
9. J. F. Clauser, M. A. Horne, A. Shimony, and R. A. Holt, Phys. Rev. Lett. **23**, 880 (1969).

A Novel Interpretation of the Klein-Gordon Equation

K. B. Wharton

Department of Physics, San Jose State University, San Jose, CA 95192-0106, USA
wharton@science.sjsu.edu

Abstract. The relativistically-correct Klein-Gordon equation requires twice the boundary conditions of the non-relativistic Schrödinger equation and does not have an accepted configuration-space interpretation. Instead of interpreting its solution as a probability wave determined by an initial boundary condition, this paper considers the possibility that the solutions are determined by both an initial and a final boundary condition. By constructing a relativistically invariant probability weight from the size of the solution space, it is shown that the usual probability distributions can nearly be recovered in the non-relativistic limit. The non-exact results imply small but potentially measurable deviations from standard quantum mechanics.

Keywords: Time Symmetry, Quantum Mechanics, Time Energy Uncertainty, General Boundary
PACS: 03.65.Pm, 03.65Ta

1. INTRODUCTION

Although quantum field theory (QFT) appears to be a successful relativistic generalization of standard quantum mechanics (QM), it is a curious fact that the higher-level axioms of QFT do not shed much light on the well-known interpretational difficulties of QM. For example, QFT does not explain why the square of a QM wavefunction should correspond to the probability of a position measurement [1]. The connection between QM and QFT is that QM's Schrödinger equation (SE) is usually considered to be the non-relativistic limit of QFT's Klein-Gordon equation (KGE). But the reason that QM does not necessarily follow from QFT is that there is no accepted configuration-space interpretation of the KGE. The largest obstacles are that 1) twice the number of boundary conditions are required (compared to the SE), and 2) there is no obvious way to construct a conserved four-vector probability current. Recently-proposed solutions to this latter obstacle [2, 3] do not address the missing boundary condition.

In this paper I will attempt to take the logically natural (but daunting) approach of reversing the historical QM→QFT inferences – in other words, constructing a new interpretation of QM based only upon the higher-level QFT. Although I cannot complete such a task in this single paper, substantial progress can be made using only a few basic principles from QFT. This procedure requires one to set aside all of the axioms of non-relativistic QM, but it is possible that such a dramatic step is required to securely place quantum mechanics on a relativistic framework.

An insight that will serve as a starting point is to take CPT symmetry seriously, not just for the underlying equations but also for the boundary conditions on those equations. For neutral spin-0 particles, CPT is effectively just time-symmetry – something that is notably lacking in non relativistic QM. Not only is the SE formally time-asymmetric, but the collapse postulate is arguably the most egregious example of time-asymmetry in all of modern physics. I have previously published an example of how non-relativistic QM might be re-interpreted in a time-symmetric manner by imposing two consecutive boundary conditions corresponding to a preparation and a measurement [4].

For the relativistic case, using this same two-boundary methodology allows one to solve the time-symmetric KGE on a classical field ϕ. The uncertainty principle prevents one from solving the KGE as an initial boundary problem, because one cannot measure both the amplitude and phase of the field at the same time (or both ϕ and $\partial\phi/\partial t$), and these two initial conditions are needed to solve any second-order-in-time differential equation. This problem is naturally solved in a time-symmetric framework by imposing two consecutive boundary conditions on two *different* space-like hypersurfaces (corresponding to two external measurements). The solution to the KGE is then well-defined over the space-time volume between the two measurements. In order to impose an extra boundary condition and still find general exact solutions, ϕ must necessarily include the so-called "negative frequency" solutions to the KGE. Because ϕ cannot be determined by the initial measurement alone, it is not possible to use the post-determined solution as a predictive tool; instead, one can weight the solution space to determine the relative likelihood of different pairs of measurements.

The central result of this paper is to demonstrate the existence of a relativistically-invariant weighting scheme that,

CP962, *Quantum Theory, Reconsideration of Foundations – 4*,
edited by G. Adenier, A. Yu. Khrennikov, P. Lahti, V. I. Man'ko, and Th. M. Nieuwenhuizen

in the non-relativistic limit, gives a probability distribution very similar (but not identical) to the standard probabilities. While the results in this paper have not yet been generalized to the point where this is a full alternative interpretation of quantum mechanics, the indications are promising that this approach might put quantum mechanics on a solid relativistic footing while also helping to explain many open interpretational questions.

2. THE TWO-BOUNDARY KLEIN GORDON EQUATION

In the absence of a potential, the Klein-Gordon equation and its solutions can be written

$$\left(\frac{\partial^2}{\partial t^2} - c^2\nabla^2 + \frac{m^2c^4}{\hbar^2}\right)\phi = 0, \tag{1}$$

$$\phi(\boldsymbol{r},t) = \int_{-\infty}^{\infty} a(\boldsymbol{k})e^{i(\boldsymbol{k}\cdot\boldsymbol{r}-\omega t)} + b(\boldsymbol{k})e^{i(\boldsymbol{k}\cdot\boldsymbol{r}+\omega t)}\boldsymbol{dk}. \tag{2}$$

Here $a(\boldsymbol{k})$ and $b(\boldsymbol{k})$ are complex amplitudes, and ω is assumed to be a positive function of $\boldsymbol{k}$; $\omega(\boldsymbol{k}) = \sqrt{c^2k^2 + m^2c^4/\hbar^2}$. Disregarding non-relativistic QM, there is no reason to eliminate the $b(\boldsymbol{k})$ terms as unphysical; indeed they are needed to have sufficient free parameters to impose two boundaries.

In order to impose boundary conditions on this wavefunction, it would be convenient to use the measurement theory from standard QM, where a measurement constrains ϕ to be an eigenfunction of some operator. One typically constructs such an operator from the position operator $\mathbb{X}$ and the momentum operator $\mathbb{P}$. In position space, one makes the assignments $\boldsymbol{X} \to \boldsymbol{x}$, $\boldsymbol{P} \to -i\hbar\nabla$, allowing one to construct a general operator $Q(\mathbb{X},\boldsymbol{P})$ that operates on the function $\phi(\boldsymbol{r})$ at some particular time.

This approach is forced to make two changes to the usual methodology. First, in a relativistically correct theory, one may only impose a boundary condition "at an instant" in some particular reference frame. So in general these boundaries may only be imposed on a space-like hypersurface $\mathbb{s}$. The second change is more substantial; one is reduced to only considering time-even operators of the less-general form $Q(\mathbb{X},\boldsymbol{P}^2)$. This is because the $\boldsymbol{P} \to -i\hbar\nabla$ operator has been constructed for the SE, not the KGE. Applying it to solutions of the form (2) leads to unphysical results, because the $b(\boldsymbol{k})$ terms physically propagate in the opposite direction as determined by the eigenvalues of $\mathbb{P}$. The $\boldsymbol{P}^2$ operator does not encounter this difficulty, so here I only consider a general time-even measurement $Q(\mathbb{X},\boldsymbol{P}^2)$.

One can now attempt to impose two consecutive boundaries on the KGE. In the simplest case, suppose the eigenfunction of the initial measurement operator on the hypersurface $t=0$ is $F(\boldsymbol{x})$, and the eigenfunction of the final measurement operator on the hypersurface $t=t_0$ is $G(\boldsymbol{x})$. Fourier-expanding the general boundary conditions, one simply gets

$$\phi(\boldsymbol{x},0) = F(\boldsymbol{r}) = \int_{-\infty}^{\infty} F(\boldsymbol{k})e^{i\boldsymbol{k}\cdot\boldsymbol{x}}\boldsymbol{dk}, \tag{3}$$

$$\phi(\boldsymbol{x},t_0) = G(\boldsymbol{r}) = \int_{-\infty}^{\infty} G(\boldsymbol{k})e^{i\boldsymbol{k}\cdot\boldsymbol{x}}\boldsymbol{dk}. \tag{4}$$

Comparing these equations to (2), the uniqueness of the Fourier transform implies

$$a(\boldsymbol{k}) + b(\boldsymbol{k}) = F(\boldsymbol{k}), \tag{5}$$

$$a(\boldsymbol{k})e^{-i\omega t_0} + b(\boldsymbol{k})e^{i\omega t_0} = G(\boldsymbol{k}). \tag{6}$$

Here are the correct number of equations to solve for the coefficients $a(\boldsymbol{k})$ and $b(\boldsymbol{k})$, but there is a problem for the particular values of $\boldsymbol{k}$ at which $\omega(\boldsymbol{k}_n)t_0 = n\pi$. At these discrete values, (5) and (6) imply that the solutions for $a(\boldsymbol{k})$ and $b(\boldsymbol{k})$ have infinite poles.

But quantum field theory is no stranger to infinite poles – the propagator for the KGE contains a term $(E^2 - p^2c^2 - m^2c^4)^{-1}$ which blows up on the mass shell. The standard prescription for solving this problem is to make the assignment $m^2 \to m^2 - i\varepsilon$. After performing the integral, one takes the limit $\varepsilon \to 0$. Because this method is known to give correct results in QFT, it seems reasonable to use precisely the same method here. Implementing this change in m^2 changes the solutions to the KGE;

$$\phi(\boldsymbol{x},t) = \int_{-\infty}^{\infty} a(\boldsymbol{k})e^{i(\boldsymbol{k}\cdot\boldsymbol{x}-\omega t)}e^{-\varepsilon t} + b(\boldsymbol{k})e^{i(\boldsymbol{k}\cdot\boldsymbol{x}+\omega t)}e^{\varepsilon t}\boldsymbol{dk}. \tag{7}$$

where, as usual, all positive constants are absorbed into ε itself. (Here I have also absorbed a $1/\omega$ into ε, which must be taken into account if one is not using a non-relativistic approximation.)

Now it is possible to solve the two boundary problem, using the new form of the solution (7). The original boundary conditions (5) and (6) now appear as

$$a(\boldsymbol{k}) + b(\boldsymbol{k}) = F(\boldsymbol{k}), \tag{8}$$

$$a(\boldsymbol{k})e^{-i\omega t_0}e^{-\varepsilon t_0} + b(\boldsymbol{k})e^{i\omega t_0}e^{\varepsilon t_0} = G(\boldsymbol{k}). \tag{9}$$

These are now exactly solvable for any two generic boundary conditions, yielding

$$a(\boldsymbol{k}) = \frac{F(\boldsymbol{k})e^{i\omega t_0}e^{\varepsilon t_0} - G(\boldsymbol{k})}{e^{i\omega t_0}e^{\varepsilon t_0} - e^{-i\omega t_0}e^{-\varepsilon t_0}}, \tag{10}$$

$$b(\boldsymbol{k}) = \frac{G(\boldsymbol{k}) - F(\boldsymbol{k})e^{-i\omega t_0}e^{-\varepsilon t_0}}{e^{i\omega t_0}e^{\varepsilon t_0} - e^{-i\omega t_0}e^{-\varepsilon t_0}}. \tag{11}$$

3. THE PROBABILITY AMPLITUDE

In relativistic quantum mechanics, one finds that the charge density $\rho(\boldsymbol{r},t)$ is a useful concept, so this is a reasonable place to begin searching for a probability weight for the entire solution ϕ. For the Klein-Gordon equation, ρ is given by

$$\rho(\boldsymbol{x},t) = \frac{i\hbar}{2mc^2}\left(\phi^*\frac{\partial\phi}{\partial t} - \phi\frac{\partial\phi^*}{\partial t}\right), \tag{12}$$

and is obviously real, although not obviously an invariant (it is normally viewed as one component of a four-current). But with a two-boundary problem one does not need a current; one needs an invariant probability weight. It is not possible to make ρ invariant without some reference unit four-vector that defines the time direction.

Fortunately, the boundary conditions themselves define two hypersurfaces, $\boldsymbol{s}_1$ and $\boldsymbol{s}_2$. These surfaces have associated inward-pointing normal four-vectors η_1 and η_2 (pointing forward in time from the initial boundary and backward in time from the final boundary). Generalizing to any closed hypersurface defined by the two boundary conditions (e.g. the two infinite planes $t = 0$ and $t = t_0$), the 3-D surface integral

$$W = \frac{\hbar}{mc}\oint_{\boldsymbol{s}_1,\boldsymbol{s}_2} Im(\phi\eta^\mu\partial_\mu\phi^*) \tag{13}$$

is a parameter that is plausibly related to probability weight, motivated both by relativistic invariance and by known results from relativistic quantum mechanics. (In this notation, η^μ is the inward-pointing four-vector unit normal to the integration surface.)

In the special case that the closed hypersurface is defined by the planes $t = 0$ and $t = t_0$, W can be evaluated using (7). Plugging in $a(\boldsymbol{k})$ and $b(\boldsymbol{k})$ using (10) and (11), to lowest surviving orders in ε becomes

$$W = \int_{-\infty}^{\infty} \frac{\omega\varepsilon t_0\left[|F(\boldsymbol{k})|^2 + |G(\boldsymbol{k})|^2\right] - 2\omega\varepsilon t_0 Re[F(\boldsymbol{k})G^*(\boldsymbol{k})]cos(\omega t_0) + \varepsilon sin(\omega t_0)(\ldots)}{\varepsilon^2 t_0^2 + sin^2(\omega t_0)} d^3k, \tag{14}$$

to within an overall constant. This does not disappear as $\varepsilon \to 0$ because one gets a periodic delta function from the identity

$$\lim_{\varepsilon\to 0}\frac{\varepsilon t_0}{\varepsilon^2 t_0^2 + sin^2(\omega t_0)} = \sum_{n=0}^{\infty}\pi\,\delta(\omega t_0 - n\pi). \tag{15}$$

The $sin(\omega t_0)$ term in the numerator of (14) was not fully shown because it is odd over the delta function, and vanishes; the $cos(\omega t_0)$ term is either 1 or -1 depending on n. Expanding d^3k in spherical $\boldsymbol{k}$-coordinates ($k = |\boldsymbol{k}|$) reduces the radial integral to a sum, leaving only a 2-D integral;

$$W = \int\sum_{n=n_0}^{\infty}\left(|F(\boldsymbol{k_n})|^2 + |G(\boldsymbol{k_n})|^2 - 2Re[F(\boldsymbol{k_n})G^*(\boldsymbol{k_n})](-1)^n\right)\frac{\omega_n^2}{k_n t_0}k_n^2 sin\theta_k d\theta_k\, d\phi_k. \tag{16}$$

Here $k_n = |\boldsymbol{k_n}|$ is the solution to $\omega_n(\boldsymbol{k_n}) = n\pi/t_0$, and n_0 is the smallest value of n for which there is a solution. At this point, W is still a function of θ, the unknown relative phase between the boundary conditions $F(\boldsymbol{r})$ and $G(\boldsymbol{r})$. This phase does not affect the first two terms, but it does affect the final term – the term that looks promisingly similar to the standard probability distribution.

I propose that an invariant probability weight can be constructed from W according to

$$P = (W_{max} - W_{min})^2. \tag{17}$$

Here W_{max} is the maximum value W can attain when varying any free parameters (i.e. θ); similarly W_{min} is the minimum value. But standard probability is given by a 3-D integral over k-space, not a 2-D integral and a discrete sum. Still, the sum in (16) approximates an integral if neither $F(\boldsymbol{k})$ nor $G(\boldsymbol{k})$ change very rapidly when the magnitude $|\boldsymbol{k}|$ changes by $k_{n+1} - k_n$; the precise conditions are derived in the appendix of Ref. [5]. Given these conditions, in the non-relativistic limit,

$$\int \sum_{n=n_0}^{\infty} Re[F(\boldsymbol{k_n})G^*(\boldsymbol{k_n})](-1)^n \frac{k_n}{t_0} sin\theta_k d\theta_k d\phi_k \approx \frac{\hbar}{2m} \int_{-\infty}^{\infty} Re[F(\boldsymbol{k})G^*(\boldsymbol{k})] cos(\omega t_0)\, d^3k. \tag{18}$$

Because both $F(\boldsymbol{r})$ and $G(\boldsymbol{r})$ are eigenfunctions of real maximal operators, they can both be constrained to be entirely real if their relative phase $exp(i\theta)$ is added explicitly. In this case, the approximation (18) simplifies the probability weight (17) into the expression

$$P = \left(\int_{-\infty}^{\infty} Re[F(\boldsymbol{k})G^*(\boldsymbol{k})e^{i\theta}] cos(\omega t_0)\, d^3k \right)^2_{max}. \tag{19}$$

This assumes that one is in the non-relativistic limit such that the ω_n^2 in (16) is roughly constant and can be pulled out of the integral.

This proposed probability weight (19) is a function of both the initial and the final boundary condition. Holding the initial boundary condition fixed (as a given), it then becomes a function of only the final boundary condition – or the probability which one calculates in ordinary quantum mechanics. Therefore, it is promising that (19) is reminiscent of the traditional measurement probability

$$P_0 = \left| \int_{-\infty}^{\infty} F(\boldsymbol{k})G^*(\boldsymbol{k}) e^{-i\omega t_0} d^3k \right|^2, \tag{20}$$

although they are not obviously equivalent. But because the overall phase freedom has been used to constrain $F(\boldsymbol{x})$ and $G(\boldsymbol{x})$ to be real, both $Im[F(k)G^*(k)] cos(\omega t_0)$ and $Im[F(k)G^*(k)] sin(\omega t_0)$ are odd functions in k, so their integrals cancel out of (19) and (20). Extracting the k-independent portion $\omega_0 = mc^2/\hbar$ from $\omega(k) = \omega_0 + \omega_1(k)$ simplifies the above expressions to

$$P = \left(\int_{-\infty}^{\infty} Re[F(\boldsymbol{k})G^*(\boldsymbol{k})] cos\theta\, [cos(\omega_0 t_0) cos(\omega_1 t_0) - sin(\omega_0 t_0) sin(\omega t_0)]\, d^3k \right)^2_{max}, \tag{21}$$

$$P_0 = \left(\int_{-\infty}^{\infty} Re[F(\boldsymbol{k})G^*(\boldsymbol{k})] cos(\omega_1 t_0)\, d^3k \right)^2 + \left(\int_{-\infty}^{\infty} Re[F(\boldsymbol{k})G^*(\boldsymbol{k})] sin(\omega_1 t_0)\, d^3k \right)^2. \tag{22}$$

While similar, these two expressions are unfortunately not equivalent if t_0 is constrained to be a precise value. But is practically impossible to constrain t_0 to an accuracy less than ω_0^{-1} ($\sim 10^{-20}$ seconds for an electron, and even less for more massive particles). No current experiments can determine the measurement time to a sub-attosecond accuracy, so an effective equivalence between P and P_0 can be shown to exist if t_0 itself is added to the list of parameters over which W should be maximized. In other words, so long as the constrained solution space of ϕ still allows t_0 to vary by at least ω_0^{-1}, a maximization of (21) over the unspecified angles $\omega_0 t_0$ and θ leads to the equivalence

$$P = \left(\int_{-\infty}^{\infty} Re[F(\boldsymbol{k})G^*(\boldsymbol{k})] cos(\omega_1 t_0)\, d^3k \right)^2 + \left(\int_{-\infty}^{\infty} Re[F(\boldsymbol{k})G^*(\boldsymbol{k})] sin(\omega_1 t_0)\, d^3k \right)^2 = P_0. \tag{23}$$

4. CONCLUSIONS

The conclusion of this preliminary research is that it is likely that one can recover standard (non-relativistic) quantum mechanics from this general approach. Furthermore, because the formalism is relativistically correct (and maintains a spacetime representation) these results should be generalizable to fields in curved spacetime.

However, the derived equivalence between the predicted probability distribution P and the traditional probability P_0 only holds in the limit of 1) non-relativistic velocities and 2) the validity of the approximation (18). This approximation in turn is equivalent to something that looks like a time-energy uncertainty principle [5], $t_0\sigma_E >> \hbar$, where σ_E is the uncertainty in either the initial or final energy measurement. As the time between measurements decreases, therefore, one should start to expect an experimental differentiation between this approach and standard quantum mechanics. This is an exciting development which will be explored further, particularly if it appears possible to design an experiment to test this small deviation.

This research also holds the promise of an appealing new interpretation of quantum mechanics itself. The key would be to forge a novel compromise between the apparently contradictory "epistemic" and "realistic" viewpoints concerning quantum mechanics. This ongoing philosophical disagreement concerns whether or not the solution to the Schrödinger equation ψ corresponds to reality, or is merely an epistemic construct concerning our knowledge of a system. The "epistemic" viewpoint, espoused by Spekkens and others [6], points out that constraining our knowledge of a system leads to behavior that is qualitatively similar to quantum mechanics, implying that the wavefunction collapse is more a matter of updating our knowledge rather than anything discontinuous and time-asymmetric happening in reality. (Intriguingly, Spekken's model starts with the assumption that one can only know half of the parameters of the quantum system – precisely equivalent to the above formalism.) But the "realist" viewpoint also has a strong case, in that that some of this mathematics must correspond to what is actually happening in reality, and until now the only reasonable mathematical object to ascribe to reality has been ψ itself.

This research program offers a middle ground; perhaps ψ is epistemic, because the Schrödinger equation is just an approximation of the underlying second-order field equations, and only uses half of the total boundary conditions on the system. Meanwhile, the solution to the full field equations can be deduced after the experiment is complete (and the final boundary becomes known); it is this field to which a "reality" can be ascribed. Without knowing the future boundary condition, the best approximation one can make concerning the field would be a guess that looks something like ψ, but after the fact, one can now reconstruct what actually happened. The advantage of finding a middle ground supported by the strongest arguments from both the "epistemic" and "realistic" philosophies hopefully outweighs the potential philosophical disadvantage of requiring ϕ to depend on future events. Indeed, if all of the "weirdness" of quantum mechanics could be explained by a wavefunction that partially depends on future events, such a dramatic confirmation of CPT symmetry would hopefully be embraced, not rejected on philosophical grounds.

Regardless of philosophy, the ultimate test for any theoretical proposal is experiment. Perhaps the strongest reason to pursue this line of research is that it actually leads to results which contradict standard quantum mechanics (albeit only in experimentally-difficult regimes). I hold out hope that this research program, by incorporating interacting fields and a relativistically correct measurement theory, will lead to practical experiments which can confirm or deny the validity of this general approach.

ACKNOWLEDGMENTS

The author is eternally grateful to J. Finkelstein for carefully and thoughtfully critiquing an earlier version of this manuscript. Further thanks go to W. Wharton, R. Schafer, and L.S. Schulman.

REFERENCES

1. H. Nikolic, "Quantum mechanics: Myths and facts", *Found. Phys. (in press)*, arXiv:quant-ph/0609163.
2. G. Horton, C. Dewdney and A. Nesteruk, "Time-like flows of energy-momentum and particle trajectories for the Klein-Gordon equation" *J. Phys. A* **33**, 7337 (2000).
3. H. Nikolić, "Bohmian particle trajectories in relativistic fermionic quantum field theory" *Found. Phys. Lett.* **18**, 549 (2005).
4. K.B. Wharton, "Time-Symmetric Quantum Mechanics", *Found. Phys.* **37**, 159 (2007).
5. K.B. Wharton, "A novel interpretation of the Klein-Gordon Equation", arXiv:0706.4075 [quant-ph].
6. R. Spekkens, "Evidence for the epistemic view of quantum states: A toy theory", *Phys. Rev. A* **75**, 032110 (2007).

Probability Paradoxes and a Strategy to Search for Solutions

Richard Woesler

Florian-Geyer-Str. 96, 12489 Berlin, Germany, woesler@uni-bremen.de, telephone +49 30 67068670

Abstract. Extremely many worlds would exist mathematically where the results of quantum processes happen to occur from now on outside earth always according to completely other probability distributions then the known ones. At first glance it seems to be extremely improbable to perceive ourselves in a world where the probability distributions will remain those according to the simple rules of quantum theory. This is called a probability paradox, and further similar paradoxes are presented. A strategy is described to search for solutions, an information theoretic approach. In that approach infinitely many worlds are considered in parallel.

Keywords: probability, quantum theory, paradoxes
PACS: 03.65.-w, 03.65.Ta

1 INTRODUCTION

The search for a theory of physics is difficult [1] and led to many approaches, e.g. string/M-theory [2] and loop quantum gravity [3], which might in future be unified. These approaches are testable [3,4]. Loop quantum gravity is a discrete theory and its discreteness was not just assumed but derived from theoretical principles [3], and further results favor a discrete theory [5]. Loop quantum gravity can be compared with a computer program (cf. appendix B) which emulates a world [3], leading also to an emulation interpretation of quantum physics, cf. [6] and appendices A and B. There are many computer games that emulate simple worlds. All possible input programs are considered, and they yield many possible output results. The output of each input program is called a world in the present text. The conceivable case is considered that our world is such theoretical computer emulation (e.g., quantum computers can be included, cf. appendix B), e.g., with infinite computational steps and infinite memory. Also conceivable is a sequence where first a finite spacetime region is computed, and then larger and larger regions with other boundary and initial conditions applying a pseudo-random generator (PRG), cf. appendix B, are computed automatically. Analysis can be done considering only worlds in which perceiving persons exist, and this is done in the present text.

2 PROBABILITY PARADOX 1: DISTURBED WORLDS

The conceivable case is considered that the input program is changed so that (e.g., in a given spacetime region) in the emulated world a small additional subprogram is conducted which changes some bits. Afterward, e.g., the main but modified program is continued. Considering a computer game, this could cause a breakdown of the game, or, e.g., a change of a color of one stored point of the world. For the given correct program an infinite number of changed programs exist where small parts are changed by such additional disturbing sub-programs. Considering the conceivable case that our world is such theoretical program, the question occurs, why do we live in a world where, as far as we know and as far as we look into our world, there are no such disturbing deviations? It seems that for each undisturbed correct program for a world there are infinitely many ones that are disturbed. Thus at first glance it seems to be extremely improbable that we perceive a completely undisturbed world (although of course with at least pseudo randomness in an emulation quantum theory [6], or true randomness when emulating many worlds, cf. appendix B and section 3.1). This will be called probability paradox 1.

CP962, *Quantum Theory, Reconsideration of Foundations – 4*,
edited by G. Adenier, A. Yu. Khrennikov, P. Lahti, V. I. Man'ko, and Th. M. Nieuwenhuizen

2.1 General Solution Strategy: Information Theoretic Approach

Computable worlds have been studied and the coding-theorem suggests that the probability might be especially large that we perceive a world which has got an especially short shortest code [7]. Note that an undisturbed program exists additionally as a part of long random input programs more often than longer disturbed programs do. Roughly speaking, disturbed programs have got a longer shortest code than the undisturbed program, because each deviation from the undisturbed program has to be encoded in an additional disturbing sub-program. Thus, the coding-theorem [7] may enable a solution to probability paradox 1 (appendices A and B).

3 PROBABILITY PARADOX 2: QUANTUM PROBABILITIES

A similar probability paradox is given by considering, e.g., Copenhagen interpretation of quantum theory. In such a world, e.g., quantum measurements yield results according to probability distributions. From mathematics point of view there are many further worlds where just other results happen to occur in quantum measurements according to the known probability distributions. However, also worlds would exist mathematically where the results of quantum measurements etc. happen to occur always with other probability distributions. There are infinitely many probability distributions that could exist in such other worlds. At first glance it seems to be extremely improbable to find ourselves in a world where always and everywhere the probability distributions are those according to the simple rules of quantum theory. This is called probability paradox 2. Answers to the questions how to emulate true randomness, a probability distribution etc. are given in appendix B.

3.1 Many World Theory Probability Paradox

Can traditional many world quantum theory claim that it shows how probabilities emerge, because Hilbert space 'volume' or norm of the set of 'maverick' worlds, where the observed probability distributions do not hold, approaches zero [8]? The probability distributions emerge within that theory (which is a very valuable result), but the probability paradox of the present text remains: when taking into account all mathematical structures that theory does not show how the probability distributions do emerge really. How they do emerge really might only be found out when taking into account at least all mathematical structures within which all logically possible probability distributions exist separately at first, and computing an overall probability distribution appropriately afterwards. Then it is conceivable that especially small probability to live within 'maverick' worlds, where the observed probability distributions do not hold (e.g., in most future spacetime regions) might emerge really when applying the coding-theorem [7], because the simple rules of known quantum theories can be encoded in especially few bits, and they may be appropriate so that perceiving persons 'emerge relatively frequently' in our world.

4 PROBABILITY PARADOXES OF A 'SECOND KIND'

Probability paradoxes of a 'second kind' arise as follows. Why does our world not start to behave completely irregular approximately here and now, because extremely many irregular worlds can be imagined which start to behave irregular about now in comparison with the maybe extremely small fraction of worlds which continue to behave so extremely regular? It is proposed that the solution to these probability paradoxes of second kind are suggested analogously by the coding-theorem [7], because such disturbing subprograms would need further bits which encode time and place when they start. The study of computable universes [7] is closely related to a 'theory of everything' [9,7,6]. Also for continuous theories like, e.g., the standard model, MSSM (which can be encoded by a finite number of bits, e.g., in a text book, e.g., when assuming that the parameters of the models are numbers (like, e.g., certain rational numbers) which can be encoded within a finite number of bits) and superstring theories [2] many similarly disturbed versions are conceivable, i.e. conceivably similar results hold for existence proofs of continuous worlds, too, where short shortest existence proofs could be used instead of short shortest codes.

5 CONCLUSION: INFORMATION THEORETIC APPROACH IS A GENERAL SOLUTION STRATEGY FOR PROBABILITY PARADOXES

Generally, probability paradoxes might be solvable by the information theoretic approach [7,6] (cf. appendix A) with appendix B when taking into account more and more mathematical structures, i.e. mathematical worlds. Taking into account those mathematical structures with relatively short shortest codes (or existence proofs) might already yield sufficiently good results due to the coding theorem [7]. The concept yields predictions falsifiable in principle with present and future experiments [6,3]. A further experiment idea is presented next. Many coherent photons can be created and be detected with high spatiotemporal resolution. If the detection pattern is not random but pseudo-random (taking detector efficiency and disturbing events into account) always according to the same pseudo-random generator, then this might be called a hint favoring this approach.

Further ideas arise from the concept. It was argued that we might perceive 3+1 spacetime dimensions, because other dimensions would be unfavorable for persons [9]. However, intelligent self-aware agents might emerge in computer programs using genetic algorithms, neural networks etc. in a few decades, and such agents might conceivably also emerge within emulated 'spacetimes' of other dimensions with extremely simplified 'laws of physics' as existing in present computer games. Speculatively, the length of a short such computer program might conceivably be a first rough guess on an upper bound for the length of the shortest code of our world (when assuming that the latter code is an especially short one as sketched above emulating only our single world (or conceivably many world theory) theoretically).

Similar to the paradoxes, which were presented above, questions about the probabilities in conceivable travel faster than c and/or in conceivable time travel [10-13] in general relativity are conceivable. An experiment proposal using the Casimir effect at a cylinder of rotating discs with holes is presented in appendix C. Questions about the probabilities might conceivably be solvable using future generalizations of the concept [7] including future theories of physics.

APPENDIX A. INFORMATION THEORETIC APPROACH TO QUANTUM PHYSICS

Special concepts regarding quantum theory are described in appendix B. Discrete theories are considered, and worlds are studied which can be computed, i.e. emulated, on a theoretical computer, e.g., using infinite computation time and infinite output memory. A conceivable final correct theory of physics of our world is not yet known. Thus, if it can be emulated, it is not known which kind of generalized Turing machine, e.g. including quantum computers (see appendix B), might be necessary to emulate it. Thus one can start analyzing some kinds of Turing machines, and extend it in future, if advances in theoretical physics show that extensions are necessary. Kolmogorov complexity theory was applied to a set of computable worlds by Schmidhuber [7]. In a first simple study, the following simplifying assumptions were made. Each world evolves on a discrete time scale, any world's state at a given time is describable by a finite number of bits, and our world would be one of these worlds. TM denotes an arbitrary universal Turing machine with unidirectional output. Input and output symbols are "0", "1" and ",". Input programs are "", "0", "1", ",", "00", "01", "0,", etc. A_k denotes TM's k^{th} program in this list, and its output is a finite or infinite string using the alphabet {"0","1",","}. This sequence of 0/1-bitstrings separated by "," symbols is called the evolution E_k of world W_k, where the state at the mth 'time' step, i.e. a 0/1-bitstring, is called W_{km}. The special compiler choice is not very important (if the compiler is short) due to the compiler theorem according to which different compilers can be transformed by changing the input program accordingly. All worlds *could* be computed by dove-tailing running A_1 for one instruction every second step, A_2 for one instruction every second step of the remaining steps etc. $P_U(s)$, (e.g., conceivably our world's history so far) which is called the a priori probability of a finite symbol string s, is defined to be probability of guessing a halting program that computes s on a universal Turing machine U. Guessing involves here that the input tape consists of no square. Initially, and when the scanning head of the program tape shifts to the right at a last square (I) append a new square, (II) with probability 1/3 fill it with "0", or "1", or ",", and execute the program. Programs are self-delimiting, i.e. there exists a halting instruction terminating the program. This yields that $P_U(s)$ is the sum of $(1/3)^{|p|}$ over all logically possible programs p where U computes s from p and halts ($|p|$ is the number of symbols in string p). Using this and all stated assumptions, the coding theorem suggests [7] that our world and we would be run by a theoretical algorithm which has got an especially short shortest code, of course except for the trivial fact that the code has to be at least about so long that it enables emergence of perceiving persons in the world. In a further study, generalizations of Turing machines have been analyzed [7] providing similar results, e.g., that any future corresponding to a history without any short description given by an appropriate generalized Turing machine type is suggested to be necessarily unlikely.

Arbitrary compilers, computing machines, interpretations of output etc. could be considered in parallel trying to compute an overall probability distribution. Lengths of a short code (or existence proof) of many world quantum theory and of a single quantum world with a short PRG could be compared. It may be remarked that our world might be infinitely large now. However, it can also *seem* to be infinitely large, and be a theoretical emulation with a space which is now *finite* and, e.g., extremely many times larger than the astronomically visible world.

APPENDIX B. HOW TO EMULATE TRUE RANDOMNESS AND A PROBABILITY DISTRIBUTION ETC.

Theoretical emulation of a single world (e.g., of many world theory) on a Turing machine (TM) using an input program of finite code length cannot yield true randomness [7,6], because an input program of *infinite* length is necessary to produce true randomness (when considering only that single world, e.g., including infinitely many quantum processes). For example, digits of π look random but can be computed by short algorithms, thus are pseudo random, not random. Such an algorithm is a pseudo-random number generator (PRG).

However, true randomness and a probability distribution can hold *within* an emulation of more than one world, i.e. when neglecting other mathematical structures. This will be shown next. The assumption is made that perceiving persons can be emulated in principle, e.g., considering an emulation of a future loop quantum gravity theory. An existence proof for a continuous world (e.g., using MSSM or in future M-theory) will also be called an emulation. Can traditional theoretical quantum computers be emulated on traditional computers and on a TM? They can be simulated, e.g., using a PRG, various appropriate discretizations, truncating conceivable infinite series to obtain appropriate finite series etc. It is not yet known, whether or not our world contains structures which correspond to true theoretical quantum computers – instead they may correspond to such pseudo-random discretizations. For simplification assumptions are made. There would be a simulation of the Schrödinger equation which would contain perceiving persons. In this simulation there would be the quantum experiment 1 which can yield 2 different results, i.e. 2 worlds. According to Schrödinger equation result 1 would occur with probability p_1, and this would constitute world 1. Result 2 would occur with probability p_2, and this would constitute world 2. Other worlds are neglected here. However, the values p_1 and p_2 are not just assumed but remain to be realised in the simulation, which would then be the desired emulation of true randomness and of a simple probability distribution. Different cases are considered. Case 1) $p_1+p_2=1$, $m\,p_1= n\,p_2$, where m and n are natural numbers. How can this case be emulated? World 1 is emulated n times and world 2 is emulated m times. Considering only this emulation, the trivial probability that a person perceives himself within world 1 would be p_1, thus, in this sense true randomness would occur. Case 2) $p_1+p_2=1$, p_1 irrational. How can this case be emulated? Better and better approximations n_k / m_k exist converging to p_1 / p_2 when k would diverge to infinity with $m_k \geq m_{k-1}$ and $n_k \geq n_{k-1}$. In step 1 world 2 is emulated m_1 times and world 1 is emulated n_1 times. In step k world 2 is emulated $m_k - m_{k-1}$ times and world 1 is emulated $n_k - n_{k-1}$ times and so on.

Remark: In case 2), if rearrangements of the simulations would be allowed, this would yield other results. However, many rearrangements would need additional bits encoding a new order, thus conceivably 'lower probability' of many rearrangements would emerge when considering the coding-theorem [7]. However, our world might conceivably even be computable in a generalized sense [7] having a short shortest input program, thus our special world might conceivably contain no true randomness (when considering only that special world).

Trivially, the above examples for emulations are not the reason for true probabilities of the worlds, because true probabilities cannot be computed when further logically possible corresponding emulations of such worlds are not taken into account [7].

APPENDIX C. CASIMIR EFFECT AT CYLINDER OF DISCS (WITH HOLES) WHICH ROTATE IN OPPOSITE DIRECTIONS

Many publications exist with arguments that transmission of classical information faster than c might eventually be achievable in future, e.g., using general relativity and quantum theory, cf. [10-12]. At a well conducting plate the Casimir effect yields a modified stress-energy tensor $T_{\mu\nu}$ with negative energy region close to the plate. Thus it was considered to be an interesting idea [12] using such a plate with a hole through which a particle should travel on a geodesic. However, approximate computations indicate that the particle might not travel appropriately faster than c [13], because on the geodesic $T_{\mu\nu}$ is modified too much by the hole.

To check whether no arrangement would yield transmission faster than c over large distances, because many plates yield larger effects, the following setup 1 is presented. Setup 1 is a long pack of n circular well conducting plates of D=0.8m diameter aligned parallel to the x-axis, where each plate has a small cylindrical hole in the middle and where neighboring plates are, e.g., fixed at their circular boundaries using further conducting material. $M=\{(x,y,z,t),\ x_1-1\text{m}<x<x_2+1\text{m},\ x_1<x_2,\ -0.1\text{m}<y<0.1\text{m},\ -10\text{km}<z<10\text{km},\ t_3-10\text{s}<t<t_2+10\text{s},\ t_3<t_1<t_2\}$ is approximately a Minkowski spacetime region. From $(x_1,0,0,t_1)$ particles are sent to the leftmost plate hole at $(x_1,-1\text{m},0,t_1+g)$, $g>0$, some might travel through all plate holes to $(x_2,-1\text{m},0,t_2-g)$, and would then be sent to $(x_2,0,0,t_2)$.

All parameters D, g, n, the width h of each hole, the distance d between each plate in x-direction and the angle b between the direction, which is perpendicular to each plate, and the direction of the x-axis, with b=arcsin(k) and

$0 \leq k < 2h/d < 1$, vary from one thought experiment to another and parameter values with maximally fast transmission could be computed in future. Additionally, maximally fast transmission can be computed including further parameters F, H, J and D_h as follows: each disc has got diameter H, thickness J and rotates with constant high frequency F around the axis which contains the centers of the discs. Each hole is approximately at non-zero distance D_h from the center of the disc in such way that a particle, which is sent through the holes, would follow a geodesic through all plate holes, so that the particle travels through regions with correspondingly modified stress-energy tensor $T_{\mu\nu}$, with μ=1,2,3,4 and ν=1,2,3,4; the computations can also be done for a setup where each disc is replaced by two such discs rotating in opposite direction at same frequency (in this case each axis of rotation is perpendicular to each plate). The holes should be at those places in the discs so that the geodesic of a particle which travels through all plate holes is in good approximation parallel to the x-axis, in order to minimize the overall travel time. The reason to analyze also a setup with fast rotating discs is, that particles which travel on a geodesic toward a plate hole are on the geodesic a bit closer to the region of large negative energy, which is close to the plate apart from the hole. The reason to analyze also neighboring discs which rotate in opposite directions is, that fast rotating masses can be assumed to yield a small additional acceleration on the particle, as it has been computed for similar cases in general relativity ([14] page 101, equation 118). A part of the small additional acceleration would be about into that direction, into which the nearby part of the plate moves currently. I.e., the geodesics for particles which shall travel through all plate holes would be deformed accordingly and this might lead to slight increase of travel time. Thus using neighboring plates rotating in opposite directions might reduce this effect and thus might conceivably shorten the overall travel time of the particle. Possible particles which shall travel through the plate holes are photons. Sending many particles simultaneously through the holes (and/or using many copies of the setup 1), particles arriving at $(x_2,0,0,t_2)$ could be exploited to transmit a classical bit. Bit value 0 is transmitted as follows: no particle is sent through the holes, and no particle is found arriving at $(x_2,0,0,t_2)$. Bit value 1 is transmitted as follows: the particles are sent through the holes, and a particle which is found arriving at $(x_2,0,0,t_2)$ indicates that the bit value is 1. Detailed computations according to general relativity and quantum field theory, and experimental tests, remain to be done in future to show whether or not the transmission in setup 1 can occur faster than c.

It may be remarked that a copy of the whole setup 1 at different y-value y=+1m moving at appropriate positive speed in x-direction can be used to send particles arriving at about $(x_2,0,0,t_2)$ fast to this copy of the setup and within that setup analogously back to x_1. Conceivable transmission faster than c, and, e.g., from $(x_1,0,0,t_1)$ to $(x_1,0,0,t_1-□)$, □>0, of that world or of other worlds of many world theory might yield questions about the probabilities (similar as those in the present main text) which might be solvable using the information theoretic approach [7], too, because it is an approach to estimate probability distributions of worlds.

REFERENCES

1. T. R. Geroch and J. B. Hartle, *Found. Phys.* **16** (6), 533W (1986).
2. R. Easther, B. R. Greene, M. G. Jackson and D. Kabat, *JCAP* **0502**, 009 (2005); R. Easther, B. R. Greene, M. G. Jackson and D. Kabat, *JCAP* **0401**, 006 (2004).
3. J. Baez, *Nature* **421**, 702 (2003); L. Smolin, http://arxiv.org/pdf/hep-th/0605052 (accessed 2007-9-30) for inclusion in "Approaches to Quantum Gravity - toward a new understanding of space, time, and matter", edited by D. Oriti, to be published by Cambridge University Press (2006).
4. J. C. Long, H. W. Chan, A. B. Churnside, E. A. Gulbis, M. C. M. Varney, J. C. Price, *Nature* **421**, 922-925 (2003).
5. J. D. Bekenstein, *Sci. Am.* **289** (2), 58-65 (2003).
6. R. Woesler, "Problems of quantum theory may be solved by an emulation theory of quantum physics" in *Foundations of Probability and Physics-3 -2004*, edited by A. Yu. Khrennikov, AIP Conference Proceedings **750**, American Institute of Physics, Melville, NY, 2005, pp. 378-381.
7. J. Schmidhuber, *LNCS* **1337**, 201-208 (1997); *Int. J. Found. Comput. Sci.* **13** (4), 587-612 (2002); *Nature* **439**, 392 (2006); *Am. Sci.* **94**, 364 (2006).
8. B. DeWitt, *Physics Today* **23**, 30-35 (1970). http://www.hedweb.com/everett/everett.htm#probabilities (accessed 2007-9-30); cf. J. B. Hartle, *American J. Phys.* **36** (8), 704-712 (1968).
9. M. Tegmark, *Ann. Phys.* **270**, 1R (1998).
10. M. S. Morris, K. S. Thorne and U. Yurtsever, *Phys. Rev. Lett.* **61**, 1446-1449 (1988).
11. J. P. S. Lemos, F. S. N. Lobo and S. Quinet de Oliveira, *Phys. Rev. D* **68**, 064004 (2003).
12. K. D. Olum, *Phys. Rev. Lett.* **81**, 3567-3570 (1998); K. D. Olum and N. Graham, *Phys. Lett. B* **554**, 175-179 (2003).
13. N. Graham and K. D. Olum, *Phys. Rev. D* **72**, 025013 (2005).
14. A. Einstein, *Grundzuege der Relativitaetstheorie*, copyright by Estate of Albert Einstein, O. Nathan, Executor, 55 East 10th Street, New York, N.Y., USA, 1956, Braunschweig, Germany: Vieweg, 1969.

Decoherent Histories Approach Might be Predictive even when the Decoherence Condition Does Not Hold

Norifumi Yamada

Department of Information Science, University of Fukui, 3-9-1 Bunkyo, Fukui, Fukui 910-8507, Japan

Abstract. The decoherent (or consistent) histories approach is applied to the arrival time problem. Since the decoherence condition does not hold for arrival time, a probability distribution of arrival times cannot be defined. Instead, a quasidistribution, which is not positive definite, can be defined in the histories approach. It is shown that well-known Kijowski's distribution, which is positive definite and often claimed to be the correct distribution of arrival times, is obtained from the quasidistribution when the momentum wave function has certain properties. This suggests that the histories approach still holds some predictive power even when the decoherence condition does not hold.

Keywords: decoherent histories approach, consistent histories approach, arrival time, passage time, tunneling time
PACS: 03.65.Xp, 03.65.Ca, 03.65.Ta

INTRODUCTION

The type of probabilities treated in the standard formulation of quantum mechanics is rather limited. The standard formulation assigns probabilities for a set of events whose members (events) are restricted to a single moment of time. For example, let $E_i(t_0)$ be the event that a particle is found in the i-th interval on the x axis at a moment of time t_0. The set $\{E_i(t_0)|i=-\infty,\cdots,\infty\}$ is an example of a set of events whose members are restricted to a single moment of time, which is now t_0. The standard formulation then assigns probabilities to this set; the probability of the occurrence of the i-th event is given by $|\psi(x,t_0)|^2$ integrated over the i-th interval, where ψ is the wave function of the particle. Clearly, the fact that all the events are restricted to a single moment of time is a re-expression of the normalization $\int dx|\psi(x,t)|^2=1$, where time t is fixed. From the viewpoint of Newtonian spacetime, the events lie on a surface of constant time and the probabilities are normalized on the surface.

In classical stochastic process, we find different type of probabilities, that is, probabilities for a set of events whose members are not restricted to a single moment of time. For example, the classical Brownian motion allows not only the probability distribution of particle's positions at a given moment of time, $P_t(x)$ with normalization $\int dx P_t(x)=1$, but also the probability distribution of particle's arrival times at a given position, $P_x(t)$ with normalization $\int dt\, P_x(t)=1$. For the case of position distribution, the events lie on a surface of constant time, $\mathscr{S}_t$, and the distribution is normalized on the surface, while for the case of arrival time distribution, the events lie on a surface of constant position, $\mathscr{S}_x$, and the distribution is normalized on it. Since the arrival time events do not lie on a surface of constant time, it is not immediately clear how one can deal with an arrival time distribution in the standard formulation of quantum mechanics. More generally, the standard formulation is not suitable, or at least not convenient, for arguing about probabilities when the events are not necessarily restricted to a single moment of time.

The decoherent (or consistent) histories approach [1–4] is a modest generalization of the standard formulation of quantum mechanics. The aim of the generalization, especially that by Gell-Mann and Hartle [3] and also by Yamada and Takagi [4], is to make the status of time variable in quantum mechanics much closer to that of the other physical variables. It is thus expected that the histories approach is suitable for studying quantum mechanical probabilities for a set of events that are not necessarily restricted to a single moment of time. For instance, the definability of probability distributions of arrival times [5] and tunneling times [6] is a natural subject to study with the histories approach. The present author applied the histories approach to the arrival time problem [7] and to the tunneling time problem [8–10], hoping to obtain some probabilistic descriptions of these times. The results were, however, not attractive for those who wish to know the probability distributions of these times. It was shown that both arrival time and tunneling time do not allow probabilistic descriptions due to the interference between events. This is not a rare exception. Although the histories approach was developed to extend the predictive power of quantum mechanics, a sufficient number of examples has not been known that demonstrate the usefulness of the approach. This has made the histories approach less appealing than it should be. The decoherence condition, the criterion that must be satisfied in order for probabilities

CP962, *Quantum Theory, Reconsideration of Foundations – 4*,
edited by G. Adenier, A. Yu. Khrennikov, P. Lahti, V. I. Man'ko, and Th. M. Nieuwenhuizen

to be definable is very restrictive; it is hardly satisfied except for trivial cases.

Despite the strict negative results regarding the definability of probability distributions of arrival times and tunneling times, there are some reasons to believe that the histories approach is still informative about these times. In Ref. [10], the present author showed that the famous four tunneling times can be derived from the decoherence functional in a unified manner (the decoherence functional [3] is an important tool introduced in the histories approach as the measure of interference between events). In Ref. [11], Anastopoulos and Savvidou showed a way of deriving well-known Kijowski's distribution [12], which is positive definite and often claimed to be the correct distribution of arrival times, from the decoherence functional by considering the effects of measuring device through time coarse grainings. These results imply that, even when the decoherence condition does not hold, the decoherence functional still contains useful information about the physical quantities under consideration.

The purpose of the present paper is to add an evidence that shows the usefulness of the histories approach, especially the histories approach for closed system[1] that uses Feynman paths as histories[2] [3, 4], even when the decoherence condition does not hold. In particular, we show that Kijowski's distribution of arrival times can be reproduced in the histories approach without the help of measuring devices. This shows that the histories approach for closed system still hold some predictive power even when the decoherence condition does not hold. We proceed as follows: We first calculate the decoherence functional for arrival time and show that the decoherence condition does not hold. We then define and calculate a quasidistribution of arrival times from the decoherence functional, where no measuring devices are brought into the discussion. We obtain an explicit expression for the quasidistribution in the form of momentum integrals. We show that the quasidistribution of arrival times is indistinguishable from Kijowski's distribution in the situation under consideration.

DECOHERENCE FUNCTIONAL FOR ARRIVAL TIME

Consider the motion of a free particle in one dimensional space. We assume that the wave packet for the particle is localized in $x < 0$ at $t = 0$ and moves rightward. For $t > 0$, the particle obeys the free Schrödinger equation. We then ask "When will the particle be found at $x = d$ $(> x_0)$? This is the simplest version of the arrival time problem and we are interested in defining a distribution of arrival times in this simplest setup. We assume that the initial packet is far away from the reference point $x = d$, so that $\psi(x,0)$ is negligible for $x \geq d$.

To apply the histories approach to the arrival time problem, we must calculate the decoherence functional $D(\tau;\tau')$. This is a measure of the quantum mechanical interference between the event that the particle arrives at $x = d$ at time τ and the event that it arrives at $x = d$ at time τ'. To calculate the decoherence functional, we follow Ref. [7] to use a method based on Feynman paths classifications. Another method based on projection operators has been developed in Ref. [11]. Now, let us consider the following quantity:

$$P(t) \equiv \int_d^{\infty} dx |\psi(x,t)|^2. \tag{1}$$

This is the probability of finding the particle in the region $x \geq d$. Under the present situation, $P(t)$ is an increasing function of t and it tends to unity as t goes to infinity, simply because the packet passes through the reference point from left to right. Now we substitute

$$\psi(x,t) = \int dx_0 K(x,t;x_0,0)\psi(x_0,0) \tag{2}$$

into the right-hand side of Eq.(1), where $K(x,t;x_0,0)$ is the free propagator, and the range of x_0 integration, which is originally $(-\infty,\infty)$, is now effectively $(-\infty,0)$ because of the localization of the initial packet in $x < 0$. We then have

$$P(t) = \int_d^{\infty} dx \left| \int_{-\infty}^{0} dx_0 K(x,t;x_0,0)\psi(x_0,0) \right|^2. \tag{3}$$

---

[1] In Ref. [11], a measuring device is brought into the discussion from outside the histories approach. In this sense, it is not a closed system that is treated there. Halliwell [13] studied the arrival time problem by applying the histories approach of closed system to the composite system that consists of a particle and the environmental degrees of freedom interacting with the particle.

[2] The histories approach may be classified into two versions from the viewpoint of the type of histories. One is the histories approach that uses products of projection operators as histories [1–3], and the other is the one that uses Feynman paths as histories [3, 4].

Here comes the crucial step for deriving the decoherence functional. Recall that Feynman's path integral (real time path integral in configuration space) gives a sum-over-paths interpretation to the propagator; $K(x,t;x_0,0)$ is the sum of $e^{iS/\hbar}$ (S being the classical action) over those paths that connect $(x_0,0)$ and (x,t). All the paths contributing to this sum cross $x=d$ at least once (note that a Feynman path can cross a spatial point more than once in general). If we consider a class of paths, $\mathscr{C}_{t'}$, whose first arrival at $x=d$ occur at time t', we can write the sum over paths as the sum over the paths in $\mathscr{C}_{t'}$ followed by an integration over t' from 0 to t, giving

$$K(x,t;x_0,0)=\int_0^t dt' K(x,t;x_0,0|t'), \tag{4}$$

where $K(x,t;x_0,0|t')$ is the result of the restricted sum over paths. Next, we decompose the paths in $\mathscr{C}_{t'}$ into two parts: (i) the part from $(x_0,0)$ to (d,t'), where the reference point $x=d$ is never reached before time t' and (ii) the part from (d,t') to (x,t), where there are no restrictions except that the paths connect the end points. This decomposition of paths gives

$$K(x,t;x_0,0|t')=K(x,t;d,t')F(d,t';x_0,0), \tag{5}$$

where $F(d,t';x_0,0)$ corresponds to the first part and the free propagator $K(x,t;d,t')$ correspond to the second part. From Eqs. (4) and (5), we have

$$K(x,t;x_0,0)=\int_0^t dt' K(x,t;d,t')F(d,t';x_0,0). \tag{6}$$

Substituting the above equation into the right-hand side of Eq. (3), and then exchanging the order of integrations, we can rewrite $P(t)$ as follows:

$$\begin{aligned} P(t)=&\int_0^t dt' \int_0^t dt'' \int_d^\infty dx\, K(x,t;d,t')K^*(x,t;d,t'') \\ &\times \int_{-\infty}^0 dx_0\, F(d,t';x_0,0)\psi(x_0,0) \int_{-\infty}^0 dx_0\, F^*(d,t'';x_0,0)\psi^*(x_0,0). \end{aligned} \tag{7}$$

The integrand of the double t integral is the decoherence functional $D(t';t'')$. To know it, therefore, we must carry out the x integral and the x_0 integrals. We can carry out the x integral by using the explicit expression for the free propagator. This gives

$$\int_d^\infty dx\, K(x,t;d,t')K^*(x,t;d,t'')=\frac{1}{2}\sqrt{\frac{m}{2\pi i\hbar(t''-t')}}, \tag{8}$$

where note that the result does not depend on t. To carry out the x_0 integrals, we need to know the function F. Since $K(x,t;d,t')$ depends on t and t' only through $t-t'$, Eq. (6) is the Volterra integral equation of the first kind for function F, which can be solved by the method of Laplace transform. We obtain the following result [8]:

$$F(d,t';x_0,0)=\sqrt{\frac{(d-x_0)^2}{2\pi i(t')^3}}\exp\left\{i\frac{(d-x_0)^2}{2t'}\right\}=\frac{\hbar}{im}\frac{\partial K(y,t';x_0,0)}{\partial y}\bigg|_{y=d}. \tag{9}$$

Before proceeding, it should be pointed out that Eq. (6) with Eq. (9) is a simple example of the path decomposition expansion (PDX) [14, 15]. The readers may raise the question whether it is possible to calculate the function F without using PDX. It is indeed possible to do so; we can calculate F just by summing complex weights over those paths that start from $(x_0,0)$ and arrive at (d,t') without reaching $x=d$ before t'. This sum over paths is not easy to carry out if the usual time slicing definition of Feynman's path integral is used. We can instead use a quantum random walk definition of path integral [16], where not only time but also space is discretized. We carry out the summation on the spacetime lattice and then take an appropriate continuum limit to obtain Eq. (9) [17]. This is the direct way of calculating the function F.

We return to the calculation of the decoherence functional. It follows from Eqs. (9) and (2) that

$$\int_{-\infty}^0 dx_0\, F(d,t';x_0,0)\psi(x_0,0)=\frac{\hbar}{im}\frac{\partial\psi(y,t')}{\partial y}\bigg|_{y=d}. \tag{10}$$

Substituting Eqs. (8) and (10) into Eq. (7), we have

$$P(t)=\int_0^t dt' \int_0^t dt'' D(t';t'') \tag{11}$$

with $D(t';t'')$ given by [7]

$$D(t';t'') = \frac{\hbar^2}{2m^2}\sqrt{\frac{m}{2\pi i\hbar(t''-t')}}\left.\frac{\partial\psi(y,t')}{\partial y}\right|_{y=d}\left.\frac{\partial\psi^*(y,t'')}{\partial y}\right|_{y=d}. \tag{12}$$

The histories approach defines probabilities for a given set of events if the decoherence condition holds. For the present problem, the decoherence condition is written as

$$\mathrm{Re}\,D(t';t'') \propto \delta(t'-t''). \tag{13}$$

The histories approach claims that, if the above condition holds, then a probability distribution of arrival times is definable and the distribution is given by the proportionality coefficient between $\mathrm{Re}\,D(t';t'')$ and $\delta(t'-t'')$, which we denote by $p(t')$. It is convenient to express in one equation both the decoherence condition and the rule of finding the distribution when the condition holds:

$$\mathrm{Re}\,D(t';t'') = p(t')\delta(t'-t''). \tag{14}$$

Equation (12) clearly shows that the decoherence condition does not hold. Therefore, a probability distribution of arrival times cannot be defined in the histories approach of closed system.

QUASIDISTRIBUTION OF ARRIVAL TIMES

In Ref. [11], Anastopoulos and Savvidou arrived at the decoherence functional, Eq. (12), in a somewhat different way, and constructed a positive definite probability distribution of arrival times from the decoherence functional. The key point in their construction is the use of a time coarse graining procedure which they attribute to a finite time resolution of a measuring device. The interference is then smeared out and a positive definite distribution appears. The obtained distribution, not exactly the same as Kijowski's distribution, is indistinguishable from Kijowski's distribution when the momentum wave function has certain properties. In this section, we define a quasidistribution of arrival times and obtain its explicit expression. A connection of between the obtained quasidistribution and Kijowski's distribution will be clarified in the next section.

If there exists a probability distribution of arrival times, which we denote by $p(t)$, it must satisfy

$$P(t) = \int_0^t dt'\, p(t'), \tag{15}$$

where $P(t)$ is defined by Eq. (1). The fact that the decoherence condition does not hold means that Eq. (11) cannot be rewritten in the form of Eq. (15) with a positive definite $p(t)$. If we do not demand that $p(t)$ in Eq. (15) be positive definite, then there are various ways of rewriting Eq. (11) into the form of Eq. (15). We call such p's the quasidistributions of arrival times. For example,

$$\int_0^t dt'' D(t';t'') \tag{16}$$

is a quasidistribution, which is immediately understood from Eq. (11). This quasidistribution is, however, complex valued. One would be more interested in real valued quasidistributions, expecting that they would be closer to experimental tests than complex valued quasidistributions would be.

A real valued quasidistribution can be constructed by integrating the decoherence functional with respect to $t''-t'$. To do this, we switch to a set of new variables

$$u \equiv t''-t', \quad s \equiv \frac{t''+t'}{2}. \tag{17}$$

Since the Jacobian associated with this change of variables is unity, Eq. (11) with $t\to\infty$ can be rewritten as

$$1 = \int_0^\infty ds \int_{-2s}^{2s} du\, D(s;u) \tag{18}$$

where the left-hand side is $P(\infty)$, and $D(s;u) \equiv D(t'(s,u);t''(s,u))$. We consider the case $t \longrightarrow \infty$ just for simplicity. Equation (18) allows us to identify $\int_{-2s}^{2s} du\, D(s;u)$ as a quasidistribution. In the rest of this paper, the symbol $p(s)$ is used to represents this quasidistribution:

$$p(s) = \int_{-2s}^{2s} du\, D(s;u). \tag{19}$$

This is not positive definite but is real valued because of the Hermiticity of the decoherence functional, $D(t',t'') = D^*(t'',t')$, or equivalently $D(s,u) = D^*(s,-u)$.

Let us calculate the right-hand side of Eq. (19) as much as possible. Equation (12) now reads

$$D(s;u) = \frac{\hbar^2}{2m^2}\sqrt{\frac{m}{2\pi i\hbar u}}\left.\frac{\partial\psi(y,s-u/2)}{\partial y}\right|_{y=d}\left.\frac{\partial\psi^*(y,s+u/2)}{\partial y}\right|_{y=d}. \tag{20}$$

We write ψ as a superposition of plane waves:

$$\psi(x,t) = \frac{1}{\sqrt{2\pi}}\int dk\,\phi(k)e^{ikx-i\omega t}, \qquad \omega = \frac{\hbar k^2}{2m}, \tag{21}$$

where $\phi(k)$ is the wave function in k space. We use Eq. (21) on the right-hand side of Eq. (20). Calculate the y derivatives, and substitute the resultant $D(s;u)$ into the right-hand side of Eq.(19). This yields the following integral:

$$\int_{-2s}^{2s} du\sqrt{\frac{m}{2\pi i\hbar u}}e^{i(\omega+\omega')u/2} = 2\mathrm{Re}\int_{0}^{2s} du\sqrt{\frac{m}{2\pi i\hbar u}}e^{i(\omega+\omega')u/2}, \tag{22}$$

where $\omega\,(=\hbar k^2/2m)$ and $\omega'\,(=\hbar k'^2/2m)$ are the angular frequencies that come from the plane wave expansion of $\psi(y,s-u/2)$ and $\psi^*(y,s+u/2)$, respectively. Here we recall the definition of Fresnel integrals:

$$C(z) = \int_0^z d\tau\cos\left(\frac{\pi}{2}\tau^2\right), \quad S(z) = \int_0^z d\tau\sin\left(\frac{\pi}{2}\tau^2\right). \tag{23}$$

It is clear that Eq. (22) can be expressed in terms of these Fresnel integrals (put $(\omega+\omega')u/2 = \pi\tau^2/2$). A straightforward calculation shows that Eq. (22) gives

$$2\sqrt{\frac{m}{\hbar(\omega+\omega')}}\left\{C\left(\sqrt{\frac{2(\omega+\omega')s}{\pi}}\right)+S\left(\sqrt{\frac{2(\omega+\omega')s}{\pi}}\right)\right\}. \tag{24}$$

Using the above result, we obtain

$$p(s) = \frac{\hbar}{\sqrt{2\pi m}}\int dk\int dk'\,\phi(k)\phi^*(k')\frac{kk'}{\sqrt{k^2+k'^2}}\left\{C\left(\sqrt{\frac{2(\omega+\omega')s}{\pi}}\right)+S\left(\sqrt{\frac{2(\omega+\omega')s}{\pi}}\right)\right\}e^{i(k-k')d-i(\omega-\omega')s}, \tag{25}$$

where kk' comes from the y derivatives. This is the explicit expression for the quasidistribution of arrival times.

RECOVERY OF KIJOWSKI'S DISTRIBUTION

Now we apply some approximations to simplify Eq. (25). Following Anastopoulos and Savvidou [11], we assume that $\phi(k)$ is sharply peaked at some k_0. Then, in the region on the kk' plane where the integrand makes a dominant contribution to the double k integral, we can approximate $(k^2+k'^2)/2 \approx \sqrt{k^2k'^2} = |k||k'|$, which leads to the following factorization [11]:

$$\frac{kk'}{\sqrt{k^2+k'^2}} \approx \frac{1}{\sqrt{2}}\frac{k}{\sqrt{|k|}}\frac{k'}{\sqrt{|k'|}}. \tag{26}$$

The second approximation is

$$C\left(\sqrt{\frac{2(\omega+\omega')s}{\pi}}\right)+S\left(\sqrt{\frac{2(\omega+\omega')s}{\pi}}\right) \approx 1, \tag{27}$$

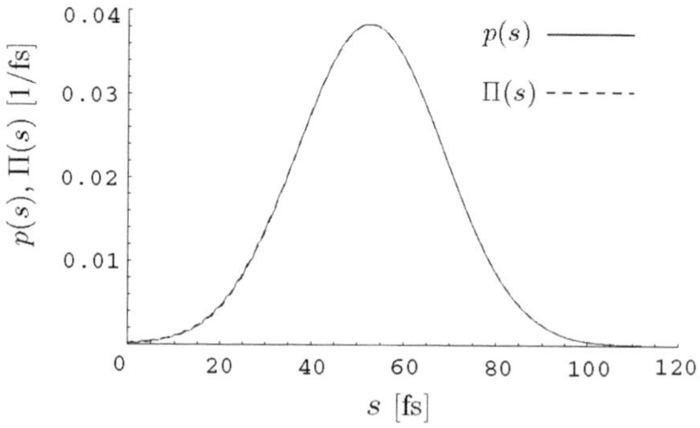

FIGURE 1. Quasidistribution $p(s)$ and Kijowski's distribution $\Pi(s)$. The two curves are indistinguishable in this scale.

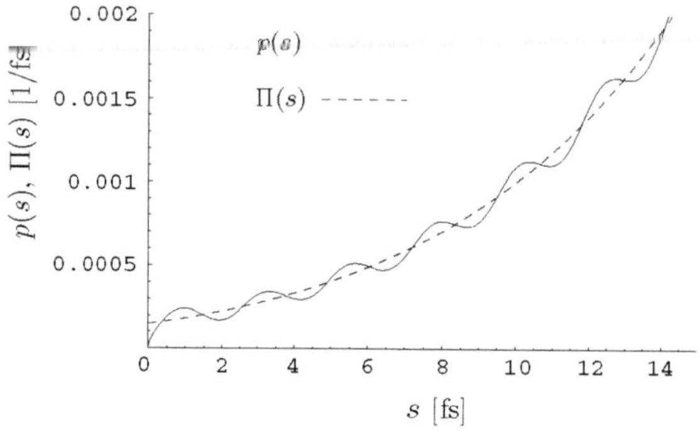

FIGURE 2. Quasidistribution $p(s)$ and Kijowski's distribution $\Pi(s)$ at small times.

which is valid for $s \gg \hbar/E_0$ ($E_0 = \hbar^2 k_0^2/2m$), because $C(\infty) = S(\infty) = 1/2$. With these approximations, the k integral and the k' integral are completely factorized in Eq. (25), giving

$$p(s) \approx \frac{\hbar}{2\pi m} \left| \int dk\, \phi(k)\, \frac{k}{\sqrt{|k|}} e^{ikd - i\omega s} \right|^2 . \tag{28}$$

Let us compare Eq. (28) with Kijowski's distribution $\Pi(s)$, which is given by

$$\Pi(s) = \frac{\hbar}{2\pi m} \left\{ \left| \int_{-\infty}^{0} dk\, \phi(k) \sqrt{-k}\, e^{ikd - i\omega s} \right|^2 + \left| \int_{0}^{\infty} dk\, \phi(k) \sqrt{k}\, e^{ikd - i\omega s} \right|^2 \right\} . \tag{29}$$

This is not the same as Eq. (28). Recall, however, that we are considering the situation where a wave packet moves rightward and the probability of finding the particle in $x > d$ is unity at sufficiently late times. Thus $\phi(k)$ is negligible for $k < 0$. In this case, Eq. (28) indistinguishable from Eq. (29). If $\phi(k)$ is sharply peaked in the positive k region, Kijowski's distribution and the quasidistribution agree very well for arrival times much longer than $\hbar/E_0$.

Figure 1 shows an example of the quasidistribution and Kijowski's distribution. Figure 2 is a magnification of Fig. 1 for small values of s. We have used the origin $x = 0$ as the reference point (d is set to be zero), and assumed a particle of mass $m = 0.088 m_0$ (m_0 being the bare electron mass) prepared in the following Gaussian state:

$$\psi(x,0) = \frac{1}{\sqrt[4]{2\pi(\delta x)^2}} \exp\left[-\frac{(x - x_0)^2}{4(\delta x)^2} + ik_0 x \right], \qquad (k_0 > 0) \tag{30}$$

where $x_0 = -100$ [nm], $\delta x = 30.0$ [nm], and $E_0 \equiv \hbar^2 k_0^2/2m = 0.9$ [eV]. In this case, $\phi(k)$ is sharply peaked in the positive k region. The quasidistribution is almost positive definite in the present example. Although the two distributions are indistinguishable in the scale of Fig. 1, we can see a clear difference between the two in Fig. 2.

Although the quasidistribution oscillates at small s, Kijowski's distribution does not show such a oscillating behavior. One can confirm that the oscillating behavior disappears if Eq. (27) is used in Eq. (25). Therefore, the Fresnel integrals in Eq. (25) are responsible for the oscillating behavior. Another difference between the two distributions is that, while $p(s)$ is vanishing at $s = 0$, $\Pi(s)$ is not vanishing there. This reflects the difference in normalization. The quasidistribution is normalized on the half line $0 < s < \infty$, while Kijowski's distribution is normalized on the full line $-\infty < s < \infty$, which has been criticized in Ref. [18].

CONCLUSION

A real valued quasidistribution of arrival times can be constructed from the decoherence functional even though the decoherence condition does not hold for arrival time. The quasidistribution agrees very well with Kijowski's distribution for arrival times much longer than $\hbar/E_0$ if $\phi(k)$ is sharply peaked in the positive k region. We have obtained these results without bringing measuring devices into discussion. The fact that the histories approach for closed system reproduces Kijowski's distribution (under certain circumstances) seems to imply that the histories approach for closed system is still predictive even when the decoherence condition does not hold. The construction of a quasidistribution might generally be an effective way of extracting useful information from the decoherence functionals. Finally, it is interesting to observe that the histories approach for closed system gives Kijowski's distribution *as the result of approximations*, which might in turn shed light on the nature and the status of Kijowski's distribution.

ACKNOWLEDGMENTS

This research was supported by the Ministry of Education, Science, Sports and Culture, Grant-in-Aid for Scientific Research (C), 17540353, 2005, 2006 and 2007.

REFERENCES

1. R. B. Griffiths, *J. Stat. Phys.* **36**, 219 (1984).
2. R. Omnès, *J. Stat. Phys.* **53**, 893 (1988);**53**, 933 (1988):**53**, 957 (1988).
3. M. Gell-Mann and J. B. Hartle, "Quantum Mechanics in the Light of Quantum Cosmology" in *Proceedings of the 3rd International Symposium on the Foundations of Quantum Mechanics in the Light of New Technology*, edited by S. Kobayashi *et al.*, Physical Society of Japan, Tokyo, 1990, pp. 321-343.
4. N. Yamada and S. Takagi, *Prog. Theor. Phys.* **85**, 985 (1991); **86**, 599 (1991).
5. For the arrival time problem in quantum mechanics, see for example, J. G. Muga and C.R. Leavens, *Phys. Rep.* **338**, 353 (2000), and references therein.
6. For the tunneling time problem, see for example, E. H. Hauge and J. A. Støvneng, *Rev. Mod. Phys.* **61**, 917 (1989); V. S. Olkhovsky and E. Recami, *Phys. Rep.* **214**, 339 (1992); R. Landauer and Th. Martin, *Rev. Mod. Phys.* **66**, 217 (1994), and references therein.
7. N. Yamada, "An Approach to the Arrival Time Problem with Gell-Mann and Hartle's Decoherence Functionals" in *Meeting Abstracts of the Physical Society of Japan*, **59**, 240 (2004) (in Japanese).
8. N. Yamada, *Phys. Rev. A* **54**, 182 (1996).
9. N. Yamada, *Phys. Rev. Lett.* **83**, 3350 (1999).
10. N. Yamada, *Phys. Rev. Lett.* **93**, 170401 (2004).
11. C. Anastopoulos and N. Savvidou, *J. Math. Phys.* **47**, 122106 (2006).
12. J. Kijowski, *Rep. Math. Phys.* **6**, 361 (1974); *Phys. Rev. A* **59**, 897 (1999).
13. J. Halliwell, *Prog. Theor. Phys.* **102**, 707 (1999).
14. A. Auerbach and S. Kivelson, *Nucl. Phys.* **B257**, 799 (1985).
15. P. van Baal, "Tunnelling and the Path Decomposition Expansion" in *Lectures on path integration: Trieste 1991*, edited by H. A. Cerdeira, S. Lundqvist, D. Mugnai, A. Ranfagni, V. Sa-yakanit, and L. S. Schulman, World Scientific, Singapore, 1993, p. 54-83.
16. N. Yamada and S. Takagi, "Non-Relativistic Sum Over Paths on Real-Spacetime Lattice" in *Quantum Control and Measurement*, edited by H. Ezawa and Y. Murayama, Proceedings of the ISQM Satellite Workshop, Japan, 1992, North-Holland, Amsterdam, 1993, pp. 87-98.
17. N. Yamada, "First Passage Time Amplitude in Nonrelativistic Quantum Mechanics" in *Path Integrals: Dubna'96*, edited by V. S. Yarunin and M. A. Smondyrev, Proceedings of the Dubna Joint Meeting of International Seminar "Path Integrals: Theory and Applications" and 5th International Conference "Path Integrals from meV to MeV," JINR E96-321, Dubna, 1996, pp. 144-148.
18. C. R. Leavens, *Phys. Lett. A* **303**, 154 (2002).

A

B

C

D

F

G

H

J

K

M

N

O

P

Q

R

S

U

V

W

Y